Organic Reactions

Organic Reactions

VOLUME 22

JOHN WILEY & SONS, INC.
NEW YORK · LONDON · SYDNEY · TORONTO

ATTENTION

SPECIAL PREFACE TO VOLUME 22

This issue of *Organic Reactions*, the twenty-second volume in the series, initiates a program of updating earlier reviews. Although many previous reviews warrant such treatment, the initial attack is directed toward those reactions in which important advances have been reported. Three of the four chapters in the present volume fall within this category.

The plan was to present only a brief report on the newer developments, together with a critical, but not complete, survey of the literature. This was done in the chapters dealing with the Clemmensen Reduction and the Reformatsky Reaction. Such a review was not possible in the area related to the Claisen Rearrangement, and the present chapter is a comprehensive coverage of Claisen and Cope Rearrangements since the reaction was reviewed in Volume 2.

Chapter 2, on the other hand, deals with a subject of recent vintage, "Substitution Reactions Using Organocopper Reagents." Future volumes will continue to present a mix of updated and new chapters.

The Editorial Board, which selected the first chapters to be updated, welcomes your suggestions about which earlier reactions warrant a current coverage. Your recommendations should list the developments that have been made in the area so that a basis for review can be evaluated. Send your suggestions directly to the Editor-in-Chief. The Editorial Board also welcomes comments on the updated reviews in this volume. Such comments may serve as a guide in the Board's future plans.

WILLIAM G. DAUBEN
EDITOR-IN-CHIEF

June 1974
Department of Chemistry,
University of California,
Berkeley, California

PREFACE TO THE SERIES

In the course of nearly every program of research in organic chemistry the investigator finds it necessary to use several of the better-known synthetic reactions. To discover the optimum conditions for the application of even the most familiar one to a compound not previously subjected to the reaction often requires an extensive search of the literature; even then a series of experiments may be necessary. When the results of the investigation are published, the synthesis, which may have required months of work, is usually described without comment. The background of knowledge and experience gained in the literature search and experimentation is thus lost to those who subsequently have occasion to apply the general method. The student of preparative organic chemistry faces similar difficulties. The textbooks and laboratory manuals furnish numerous examples of the application of various syntheses, but only rarely do they convey an accurate conception of the scope and usefulness of the processes.

For many years American organic chemists have discussed these problems. The plan of compiling critical discussions of the more important reactions thus was evolved. The volumes of *Organic Reactions* are collections of chapters each devoted to a single reaction, or a definite phase of a reaction, of wide applicability. The authors have had experience with the processes surveyed. The subjects are presented from the preparative viewpoint, and particular attention is given to limitations, interfering influences, effects of structure, and the selection of experimental techniques. Each chapter includes several detailed procedures illustrating the significant modifications of the method. Most of these procedures have been found satisfactory by the author or one of the editors, but unlike those in *Organic Syntheses* they have not been subjected to careful testing in two or more laboratories.

Each chapter contains tables that include all the examples of the reaction under consideration that the author has been able to find. It is inevitable, however, that in the search of the literature some examples will be missed, especially when the reaction is used as one step in an extended synthesis. Nevertheless, the investigator will be able to use the tables and their accompanying bibliographies in place of most or all of the literature search so often required.

vii

Because of the systematic arrangement of the material in the chapters and the entries in the tables, users of the books will be able to find information desired by reference to the table of contents of the appropriate chapter. In the interest of economy the entries in the indices have been kept to a minimum, and, in particular, the compounds listed in the tables are not repeated in the indices.

The success of this publication, which will appear periodically, depends upon the cooperation of organic chemists and their willingness to devote time and effort to the preparation of the chapters. They have manifested their interest already by the almost unanimous acceptance of invitations to contribute to the work. The editors will welcome their continued interest and their suggestions for improvements in *Organic Reactions*.

Chemists who are considering the preparation of a manuscript for submission to *Organic Reactions* are urged to write either secretary before they begin work.

CONTENTS

Organic Reactions

CHAPTER 1

THE CLAISEN AND COPE REARRANGEMENTS

Sara Jane Rhoads and N. Rebecca Raulins

University of Wyoming, Laramie, Wyoming

CONTENTS

1

ACKNOWLEDGEMENT

The assistance of Sona Janjigian of the DuPont Experimental Station in searching the literature is gratefully noted.

INTRODUCTION

Since the first observation of a thermally induced rearrangement of a vinyl allyl ether to the corresponding homoallylic carbonyl compound

(Eq. 1) by Claisen in 1912,[1] rearrangements of vinyl and aryl allylic ethers have been extensively studied and exploited for their synthetic value. The corresponding rearrangement of substituted 1,5-hexadienes (Eq. 2), first recognized by Cope[2] in 1940 as the carbon analog of the Claisen rearrangement, has enjoyed comparable attention in the ensuing three decades. Today it is recognized that such transformations fall within the general

$$\text{(Eq. 1)}$$

$$\text{(Eq. 2)}$$

category of [3,3]sigmatropic reactions[3,4] and that considerable variation may be accommodated in the basic requirement of a system of six atoms with terminal unsaturated linkages (Eq. 3).

$$\text{(Eq .3)}$$

This chapter attempts to survey the vast accumulation of Claisen and Cope rearrangements recorded since the first coverage of the Claisen rearrangement in *Organic Reactions* appeared in 1944.[1] Included are those reactions which fulfill the basic requirement of involving thermal [3,3]-sigmatropic migrations, namely, the familiar *ortho* and *para* Claisen rearrangements in aromatic systems, rearrangements in open-chain systems (aliphatic Claisen), their nitrogen and sulfur analogs (amino- and thio-Claisen), the Cope rearrangement, the oxy-Cope rearrangement, and other variants of the Cope rearrangement in which the unsaturated linkages are isocyanato and imino functions. Claisen and Cope rearrangements in which one of the unsaturated linkages of the six-atom system is acetylenic are also included. Excluded are [1,3]sigmatropic rearrangements, "photo-Claisens and Cope rearrangements," electrocyclic reactions of conjugated

---

[1] D. S. Tarbell, *Org. Reactions*, **2**, 2 (1944).

[2] A. C. Cope and E. M. Hardy, *J. Amer. Chem. Soc.*, **62**, 441 (1940).

[3] G. B. Gill, *Quart. Rev.* (London), **22**, 338 (1968).

[4] R. B. Woodward and R. Hoffmann, *The Conservation of Orbital Symmetry*, Academic Press, New York, 1970.

trienes, rearrangements which require acid catalysis, and those which clearly proceed by ionic or homolytic dissociation and recombination pathways.

The chemical literature has been searched from 1943 to January 1972 with special attention to instances of synthetic utility and novelty. A number of reviews of various aspects of the Claisen and Cope rearrangements have appeared in recent years; the interested reader is referred to them for more intensive discussions of mechanism, stereochemistry, and specific applications than can be provided in this survey.[5-19]

MECHANISM AND STEREOCHEMISTRY

Although the overall mechanistic picture of the Claisen rearrangement as a cyclic process involving simultaneous bond-making and -breaking processes accompanied by relocation of the unsaturated bonds was specifically described by Claisen as early as 1925,[20] a detailed understanding of these reactions has developed only since about 1950. Experimentally, the problems posed by the "no-mechanism" nature of the Claisen and Cope rearrangements have been attacked using labeling techniques, stereochemical probes, kinetic analyses, inter- and intra-molecular "crossing" experiments, and, in the aromatic Claisen rearrangements, by the detection and direct study of the dienone intermediates.* Theoretical interpretations

---

* Full discussions of these experiments are found in references 5–18.

[5] E. N. Marvell and W. Whalley in *Chemistry of the Hydroxyl Group*, S. Patai, Ed., Vol. 2, Interscience, New York, 1971, Chap. 13.

[6] G. G. Smith and F. W. Kelley in *Progress in Physical Organic Chemistry*, A. Streitweiser, Jr., and R. W. Taft, Eds., Vol. 8, Wiley-Interscience, New York, 1971, p. 75.

[7] H.-J. Hansen and H. Schmid, *Chimia*, **24**, 89 (1970).

[8] H.-J. Hansen and H. Schmid, *Chem. Brit.*, **5**, 111 (1969).

[9] A. Jefferson and F. Scheinmann, *Quart. Rev.*, (London), **22**, 391 (1968).

[10] B. Miller in *Mechanisms of Molecular Migrations*, B. S. Thyagarajan, Ed., Vol. I, Interscience, New York, 1968, p. 247.

[11] D. L. Dalrymple, T. L. Kruger, and W. N. White in *Chemistry of the Ether Linkage*, S. Patai, Ed., Interscience, New York, 1967, Chap. 14.

[12] B. S. Thyagarajan in *Advances in Heterocyclic Chemistry*, A. R. Katritzky and A. J. Boulton, Eds., Vol. 8, Academic Press, New York, 1967, p. 143.

[13] H. J. Shine, *Aromatic Rearrangements*, Elsevier, New York, 1967, p. 89.

[14] H. Schmid. *Österr. Chem.-Ztg.*, **65**, 109 (1964).

[15] E. Vogel, *Angew. Chem., Int. Ed. Engl.*, **2**, 1 (1963).

[16] W. von E. Doering and W. R. Roth, *Angew. Chem., Int. Ed. Engl.*, **2**, 115 (1963).

[17] S. J. Rhoads in *Molecular Rearrangements*, P. deMayo, Ed., Vol. I, Interscience, New York, 1963, Chap. 11.

[18] H. Schmid, *Gazz. Chim. Ital.*, **92**, 968 (1962).

[19a] D. J. Faulkner, *Synthesis*, **1971**, 175.

[19b] H.-J. Hansen in *Mechanisms of Molecular Migrations*, B. S. Thyagarajan, Ed., Vol. 3, Wiley-Interscience, New York, 1971, p. 177.

[20] L. Claisen and E. Tietze, *Chem. Ber.*, **58**, 275 (1925).

of the rearrangements based on a variety of molecular orbital approaches have been advanced,[21-25] and calculations of activation parameters and transition-state geometries for some examples of the Cope rearrangement have been carried out.[26-28]

The normal course of Claisen and Cope rearrangements can be illustrated with examples from the aromatic categories (Scheme 1). In an allyl aryl ether, the first cyclic rearrangement occurs with bonding of the γ-carbon atom of the allylic portion at the *ortho* carbon atom of the ring to generate an *ortho*dienone, A, in which the migrating allyl group has undergone a structural inversion. If the *ortho* substituent, R, is hydrogen, rapid enolization may occur at that stage, leading to an *ortho* allyl phenol (*ortho* rearrangement). Alternatively, realignment of the *ortho* allyl group may take place, positioning the terminal unsaturated bond opposite the *para* carbon of the ring.

A second cyclic reorganization (now a Cope rearrangement) leads to the *para* dienone, B. Once more, inversion of structure in the migrating group

SCHEME 1

[21] R. Hoffmann and R. B. Woodward, *J. Amer. Chem. Soc.*, **87**, 4389 (1965)

[22] K. Fukui and H. Fujimoto, *Tetrahedron Lett.*, **1966**, 251.

[23] K. Fukui and H. Fujimoto, *Bull. Chem. Soc. Jap.*, **40**, 2018 (1967).

[24] M. J. S. Dewar in *Aromaticity*, Chem. Society, London, 1967, p. 177.

[25] P. Beltrame, A. Gamba, and M. Simonetta, *Chem. Commun.*, **1970**, 1660.

[26] M. Simonetta, G. Favini, C. Mariani, and P. Gramaccioni, *J. Amer. Chem. Soc.*, **90**, 1280 (1968).

[27] A. Brown, M. J. S. Dewar, and W. Schoeller, *J. Amer. Chem. Soc.*, **92**, 5516 (1970).

[28] M. J. S. Dewar and D. H. Lo, *J. Amer. Chem. Soc.*, **93**, 7201 (1971).

occurs so that the original structure of the allyl side chain is restored. When the *para* substituent, R', is hydrogen, rapid enolization follows with the formation of a *para*-substituted phenol (*para* rearrangement). The *ortho* rearrangement, then, is accomplished by a Claisen rearrangement of an allyl aryl ether, whereas the *para* rearrangement is, in fact, a sequence of two rearrangements, a Claisen and a Cope. The invariable structural inversion in the *ortho* rearrangement and structural retention in the *para* rearrangement have been amply verified as have the strict intramolecularity of the rearrangements, the intervention of the dienone intermediates, and the complete reversibility of the processes when the final enolization step is prohibited.[9,17]

The aliphatic Claisen and Cope rearrangements depart from this general pattern only in that the final product is a homoallylic carbonyl compound (Eq. 1) or an isomeric 1,5 doubly unsaturated chain of six atoms (Eq. 2). Kinetic studies show the rearrangements of both aromatic[17,29-34] and aliphatic[17,35-51] systems to be unimolecular processes with activation enthalpies, entropies, and volumes in harmony with a concerted cyclic process having a highly ordered transition-state geometry.

For suprafacial-suprafacial [3,3]sigmatropic processes exemplified by the vast majority of Claisen and Cope rearrangements, two possible geometries have been considered for the cyclic transition state, the four-centered or chairlike arrangement, C, and the six-centered or boatlike arrangement,

[29] S. Marcinkiewicz, J. Green, and P. Mamalis, *Tetrahedron*, **14**, 208 (1961).

[30] K. R. Brower, *J. Amer. Chem. Soc.*, **83**, 4370 (1961).

[31] C. Walling and M. Naiman, *J. Amer. Chem. Soc.*, **84**, 2628 (1962).

[32] J. Mirek, *Zesz. Nauk Uniw. Jagiellon Pr. Chem.*, **1964**, 77 [*C.A.*, **66**, 37116f (1967)].

[33] W. N. White and E. F. Wolfarth, *J. Org. Chem.*, **35**, 3585 (1970).

[34] B. W. Bycroft and W. Landon, *Chem. Commun.*, **1970**, 168.

[35] F. W. Schuler and G. W. Murphy, *J. Amer. Chem. Soc.*, **72**, 3155 (1950).

[36] L. Stein and G. W. Murphy, *J. Amer. Chem. Soc.*, **74**, 1041 (1952).

[37] C. Walling and H. J. Schugar, *J. Amer. Chem. Soc.*, **85**, 607 (1963).

[38] G. S. Hammond and C. D. DeBoer, *J. Amer. Chem. Soc.*, **86**, 899 (1964).

[39] A. Amano and M. Uchiyama, *J. Phys. Chem.*, **69**, 1278 (1965).

[40] H. M. Frey and A. M. Lamont, *J. Chem. Soc.*, *A*, **1969**, 1592.

[41] H. M. Frey and D. H. Lister, *J. Chem. Soc.*, *A*, **1967**, 26.

[42] H. M. Frey and D. C. Montague, *Trans. Faraday Soc.*, **64**, 2369 (1968).

[43] H. M. Frey and B. M. Pope, *J. Chem. Soc.*, *B*, **1966**, 209.

[44] H. M. Frey and R. K. Solly, *Trans. Faraday Soc.*, **64**, 1858 (1968).

[45] P. S. Wharton and R. A. Kretchmer, *J. Org. Chem.*, **33**, 4258 (1968).

[46] K. Humski, T. Strelov, S. Borčić, and D. E. Sunko, *Chem. Commun.*, **1969**, 693.

[47] K. Humski, R. Malojčić, S. Borčić, and D. E. Sunko, *J. Amer. Chem. Soc.*, **92**, 6534 (1970).

[48] I. R. Bellobono, P. Beltrame, M. G. Cattania, and M. Simonetta, *Tetrahedron*, **26**, 4407 (1970).

[49] P. Leriverend and J.-M. Conia, *Bull. Soc. Chim. Fr.*, **1970**, 1040.

[50] A. Viola and J. H. MacMillan, *J. Amer. Chem. Soc.*, **92**, 2404 (1970).

[51] W. von E. Doering, V. G. Toscano, and G. H. Beasley, *Tetrahedron*, **27**, 5299 (1971).

D. It is now abundantly clear that for molecules which can readily adopt

C

D

X = O or —C—

either arrangement, the chairlike geometry, C, is strongly favored. Moreover, of two alternative chairlike arrangements, that one which minimizes 1,3-pseudo-diaxial interactions is preferred.[7–9,16,17,19,52–56*] This is clearly illustrated by the stereoselectivity shown in the aliphatic Claisen rearrangement of the isomeric crotyl propenyl ethers.[54] The rearrangement of the *trans,cis* ether, for example, proceeds through a chairlike transition

CROTYL PROPENYL ETHER → 2,3-DIMETHYLPENT-4-ENAL

	Erythro	*Threo*
trans,cis	97 ± 1%	3 ± 1%
cis,cis	2.2 ± 0.1%	97.8 ± 0.1%
trans,trans	2.2 ± 0.7%	97.8 ± 0.7%

state to produce the *erythro* isomer with a free energy of activation advantage of about 3 kcal/mole over that of the boat-like transition state (Scheme 2, p. 8).

Parallel results had been demonstrated earlier for the Cope rearrangement of *meso-* and *rac-*3,4-dimethylhexa-1,5-diene.[57] For these reactions a free energy of activation difference of about 6 kcal/mole favors the chairlike geometries and, of the two chairlike arrangements available to the

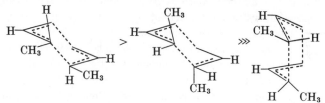

racemic isomer, the one in which the pseudo-1,3-diaxial interactions are minimized is preferred by about 2 kcal/mole. A similar ordering of the

---

* When inversion to the more favorable chairlike conformation is precluded by structural restraints, the energy requirement for the attainable chairlike geometry may lie very close to that of the boatlike arrangement. In such a case, the attendant stereoselectivity of the rearrangement may be expected to decrease. See, for example, the contrasting stereoselectivities in the *para* rearrangements of *trans* and *cis* crotyl ethers of 2,6-dimethylphenol.[106, 215]

[52] R. K. Hill and N. W. Gilman, *Chem. Commun.*, **1967**, 619.

[53] R. K. Hill and N. W. Gilman, *Tetrahedron Lett.*, **1967**, 1421.

[54] P. Vittorelli, T. Winkler, H.-J. Hansen, and H. Schmid, *Helv. Chim. Acta*, **51**, 1457 (1968).

[55] D. J. Faulkner and M. R. Petersen, *Tetrahedron Lett.*, **1969**, 3243.

[56] C. L. Perrin and D. J. Faulkner, *Tetrahedron Lett.*, **1969**, 2783.

[57] W. von E. Doering and W. R. Roth, *Tetrahedron*, **18**, 67 (1962).

Chair-like
transition state   *Erythro* (97%)

Boat-like
transition state   *Threo* (3%)

SCHEME 2

energies for possible transition state geometries has been revealed for the
amino-Claisen rearrangement.[53]

The stereoselectivity of concerted [3,3]sigmatropic processes has also
been demonstrated by asymmetric induction in optically active molecules
for the Cope rearrangement,[52,58] the aliphatic Claisen,[59] the aromatic
Claisen,[60] and the amino-Claisen rearrangements.[53] The generally very
high stereoselectivity and retention of optical purity in these reactions
recommend them for synthetic purposes.

Although the chairlike transition-state geometry is clearly preferred, it
is equally clear that the boatlike arrangement represents the only accessible
pathway for certain sterically constrained molecules and that it, too, may
be achieved without excessive expenditure of energy. The facile Cope
rearrangements of *cis* divinylcyclopropanes[15,16,61-67] and cyclobutanes,[68]

[58] H.-J. Hansen, J. Zsindely, and H. Schmid, unpublished work quoted in ref. 8.
[59] R. K. Hill and A. G. Edwards, *Tetrahedron Lett.*, **1964**, 3239.
[60] H. L. Goering and W. I. Kimoto, *J. Amer. Chem. Soc.*, **87**, 1748 (1965).
[61] W. von E. Doering and W. R. Roth, *Tetrahedron*, **19**, 715 (1963).
[62] E. Vogel, K.-H. Ott, and K. Gajek, *Ann. Chem.*, **644**, 172 (1961).
[63] C. Cupas, W. E. Watts, and P. von R. Schleyer, *Tetrahedron Lett.*, **1964**, 2503.
[64] J. M. Brown, *Chem. Commun.*, **1965**, 226.
[65] G. Ohloff and W. Pickenhagen, *Helv. Chim. Acta*, **52**, 880 (1969).
[66] K. C. Das and B. Weinstein, *Tetrahedron Lett.*, **1969**, 3459.
[67] (a) A. W. Burgstahler and C. M. Groginsky. *Trans. Kansas Acad. Sci.*, **72**, 486 (1969);
C. M. Groginsky, *Diss. Abstr. Int. B*, **31**, 6504 (1971).
[67] (b) J. M. Brown, B. T. Golding, and J. J. Stofko, Jr., *Chem. Comun.*, **1973**, 319.
[68] E. Vogel, *Ann. Chem.*, **615**, 1 (1958).

their isocyanato[69–72] and imino[73] counterparts, bicyclic derivatives in which the unsaturated linkages are incorporated in the ring system,[74–77] the retro-Claisen rearrangements of *cis* vinyl cyclopropane carboxaldehydes,[78,79] and the Claisen rearrangements of 3,4-dihydro-2H-pyranylethylenes[80] all demand a boatlike transition-state geometry.

Finally, mention should be made of the possibility of antarafacial-antarafacial [3,3]sigmatropic rearrangements. Although not normally competitive with suprafacial-suprafacial chairlike or boatlike interactions, such a twistlike process has been invoked[4] to explain the Cope rearrangement observed in the [3.2.0] bicyclic system **1** wherein steric restraints render suprafacial-suprafacial interaction impossible.[81] It is noteworthy,

however, that a corresponding antarafacial-antarafacial [3,3] rearrangement failed in the [3.3.0] system **2**,[82] and that alternative mechanisms for rearrangements of the [3.2.0] systems have been advanced.[8,82]

The effect of polarity of the solvent on the rates of some *ortho*-Claisen rearrangements has been examined in several studies.[33,83,84] With the

[69] W. von E. Doering and M. J. Goldstein, *Tetrahedron*, **5**, 53 (1959).

[70] E. Vogel, R. Erb, G. Lenz, and A. A. Bothner-By, *Ann. Chem.*, **682**, 1 (1965).

[71] I. Brown, O. E. Edwards, J. M. McIntosh, and D. Vocelle, *Can. J. Chem.*, **47**, 2751 (1969).

[72] T. Sasaki, S. Eguchi, and M. Ohno, *J. Amer. Chem. Soc.*, **92**, 3192 (1970).

[73] H. A. Staab and F. Vögtle, *Chem. Ber.*, **98**, 2701 (1965).

[74] R. B. Woodward and T. J. Katz, *Tetrahedron*, **5**, 70 (1959)

[75] P. Yates and P. Eaton, *Tetrahedron*, **12**, 13 (1961).

[76] K. N. Houk, *Tetrahedron Lett.*, **1970**, 2621.

[77] K. N. Houk and R. B. Woodward, *J. Amer. Chem. Soc.*, **92**, 4143 (1970).

[78] M. Rey and A. Dreiding, *Helv. Chim. Acta*, **48**, 1985 (1965).

[79] S. J. Rhoads and R. D. Cockroft, *J. Amer. Chem. Soc.*, **91**, 2815 (1969).

[80] G. Büchi and J. E. Powell, Jr., *J. Amer. Chem. Soc.*, **92**, 3126 (1970).

[81] T. Miyashi, M. Nitta, and T. Mukai, *J. Amer. Chem. Soc.*, **93**, 3441 (1971).

[82] J. E. Baldwin and M. S. Kaplan, *J. Amer. Chem. Soc.*, **93**, 3969 (1971).

[83] H. L. Goering and R. R. Jacobson, *J. Amer. Chem. Soc.*, **80**, 3277 (1958).

[84] W. N. White and E. F. Wolfarth, *J. Org. Chem.*, **35**, 2196 (1970).

exception of hydroxylic and phenolic media, the rate enhancement with increasing polarity is modest. To the extent that they have been investigated, Cope rearrangements of substituted 1,5-hexadienes likewise show little response to variation in polarity of the solvent.[85a,85b] Such results accord with the low polarity of the transition state in these concerted, intramolecular processes. However, the rate acceleration observed for the rearrangement of allyl *p*-tolyl ether on changing the solvent from a non-polar hydrocarbon to an aqueous-alcoholic or -phenolic system is quite appreciable, ranging from 35- to 100-fold.[33] Conceivably, this effect is the manifestation of a superimposed "acid-catalyzed" process induced by the hydrogen-bonding capacity of the hydroxylic solvent and the basic nature of the ether oxygen. The effect of such a change in solvent on the rate of the Cope rearrangement is much less impressive.[85a,*]

Transition-metal catalysis of a few Cope and Claisen rearrangements has been reported. *cis*-1,2-Divinylcyclobutane, for example, rearranges to *cis*-1,5-cyclooctadiene in quantitative yield at 24° when treated with a nickel catalyst bearing the tri-(2-biphenyl)phosphite ligand.[86] These conditions may be contrasted with the purely thermal process which requires temperatures of 80–120°. Similarly, platinum and palladium complexes have been shown to mediate the rearrangement of *cis,trans*-1,5-cyclodecadiene to *cis*-divinylcyclohexane.[87] The catalyzed rearrangements proceed at room temperature, whereas the thermal process requires a temperature of 150°. It has been suggested that bis-π-allylic complexes are involved as intermediates in these rearrangements.[86,87]

An especially instructive example of the advantages that may result from the use of transition-metal catalysis in promoting such rearrangements is provided by the behavior of 2-crotyloxypyridine.[88] The thermal

(Quant.)		
	(37%)	(30%)

i-PrOH / H₂PtCl₆ / 125   Me₂NC₆H₅ / 250°

* From a preparative point of view, a more important consequence of the nature of the solvent is its influence on the product composition. See pp. 24–27, and 48–49.

85 (a) D. C. Wigfield and S. Feiner, *Can. J. Chem.*, **48**, 855 (1970).

85 (b) D. C. Berndt, *J. Chem. Eng. Data*, **14**, 112 (1969).

86 P. Heimbach and W. Brenner, *Angew. Chem., Int. Ed. Engl.*, **6**, 800 (1967).

87 J. C. Trebellas, J. R. Oleckowskii, and H. B. Jonassen, *J. Organometal. Chem.*, **6**, 412 (1966).

88 H. F. Stewart and R. P. Seibert, *J. Org. Chem.*, **33**, 4560 (1968).

rearrangement proceeds in dimethylaniline at about 250° to give a mixture of the N- and C-allylated products, 1-α-methylallyl-2-pyridone and 3-α-methylallyl-2-pyridone.[89] In the presence of 1% chloroplatinic acid in isopropyl alcohol, the rearrangement proceeds quantitatively to give only the N-allylated product.

SCOPE AND LIMITATIONS

Recognition of the high stereoselectivity of the Claisen and Cope rearrangements coupled with the development of more versatile methods for preparing systems suitable for rearrangement has strongly stimulated synthetic applications in recent years. Advances in the use of the aliphatic rearrangements in the syntheses of natural products are especially noteworthy. Angular alkylations in polycyclic systems may be accomplished stereoselectively by rearrangement of suitably constructed allyl vinyl ethers.[90-95] The elegant syntheses of juvenile hormone[96] and of squalene[97-99] by Johnson and co-workers employ aliphatic Claisen rearrangements as key steps, whereas the synthetic approach to terpenoid materials devised by Thomas involves sequential aliphatic Claisen and Cope rearrangements.[100-102] In many of these synthetic procedures, the rearrangement-prone allyl vinyl ether system is not isolated but is prepared and rearranged *in situ*. These newer methods are described and compared in the following section.

Preparation of Starting Materials

Allyl and Propargyl Aryl Ethers. The preparation of allyl aryl ethers by modifications of the general Williamson synthesis has been described earlier and the complications of competing C-allylation pointed out.[1] Further complications may arise when the allylic halide is substituted

[89] F. J. Dinan and H. Tieckelmann, *J. Org. Chem.*, **29**, 892 (1964).

[90] A. W. Burgstahler and I. C. Nordin, *J. Amer. Chem. Soc.*, **83**, 198 (1961)

[91] M. Torigoe and J. Fishmann, *Tetrahedron Lett.*, **1963**, 1251.

[92] K. Mori and M. Matsui, *Tetrahedron Lett.*, **1965**, 2347.

[93] R. F. Church, R. E. Ireland, and J. A. Marshall, *J. Org. Chem.*, **31**, 2526 (1966).

[94] D. J. Dawson and R. E. Ireland, *Tetrahedron Lett.*, **1968**, 1899.

[95] W. G. Dauben and T. J. Dietsche, *J. Org. Chem.*, **37**, 1212 (1972).

[96] W. S. Johnson, T. J. Brocksom, P. Loew, D. H. Rich, L. Werthemann, R. A. Arnold, T.-T. Li, and D. J. Faulkner, *J. Amer. Chem. Soc.*, **92**, 4463 (1970).

[97] W. S. Johnson, L. Werthemann, W. R. Bartlett, T. J. Brocksom, T.-T. Li, D. J. Faulkner, and M. R. Petersen, *J. Amer. Chem. Soc.*, **92**, 741 (1970).

[98] L. Werthemann and W. S. Johnson, *Proc. Nat. Acad. Sci. U.S.A.*, **67**, 1465 (1970).

[99] L. Werthemann and W. S. Johnson, *Proc. Nat. Acad. Sci. U.S.A.*, **67**, 1810 (1970).

[100] A. F. Thomas, *Chem. Commun.*, **1967**, 947.

[101] A. F. Thomas, *Chem. Commun.*, **1968**, 1657.

[102] A. F. Thomas, *J. Amer. Chem. Soc.*, **91**, 3281 (1969).

so as to allow rearrangement in the side chain. This is particularly trouble-some when α-substituted allyl derivatives of hindered phenols are desired. For example, in the reaction of the sodium salt of 2-carbomethoxy-6-methyl-phenol with α-ethylallyl chloride in methanol, the reaction mixture was found to contain the four possible C- and O-allylated derivatives shown, all produced as primary products.[103] Similar results have been recorded in the preparation of α-substituted allyl ethers of other hindered phenols.[104-106]

Separation of the ethereal and phenolic fractions of such reaction mix-tures is best accomplished by Claisen's alkali.[1] Even when the phenolic products are fairly acidic, extraction with aqueous sodium hydroxide is slow and often incomplete; with weakly acidic phenolic products, such treatment is virtually ineffective.[103] Separation of the isomeric components of the ethereal fraction of such a mixture conceivably can be achieved without rearrangement by appropriately mild chromatographic methods; otherwise, one may resort to a preferential rearrangement of the more labile α-substituted allyl ether.[103]

Propargyl aryl ethers have been prepared from the appropriate pro-pargyl halide and phenol by the usual method of heating with potassium carbonate in acetone solution.[107-111] Alternatively, the propargyl aryl

[103] S. J. Rhoads, R. Raulins, and R. D. Reynolds, J. Amer. Chem. Soc., 76, 3456 (1954).

[104] S. J. Rhoads and R. L. Crecelius, J. Amer. Chem. Soc., 77, 5183 (1955).

[105] E. N. Marvell, A. V. Logan, L. Friedman, and R. W. Ledeen, J. Amer. Chem. Soc., 76, 1922 (1954).

[106] Gy. Frater, A. Habich, H.-J. Hansen, and H. Schmid, Helv. Chim. Acta, 52, 335 (1969).

[107] W. N. White and B. E. Norcross, J. Amer. Chem. Soc., 83, 1968 (1961).

[108] I. Iwai and J. Ide, Chem. Pharm. Bull. (Tokyo), 10, 926 (1962) [C.A., 59, 2759e (1963)].

[109] I. Iwai and J. Ide, Chem. Pharm. Bull. (Tokyo), 11, 1042 (1963) [C.A., 59, 13930b (1963)].

[110] J. Zsindely and H. Schmid, Helv. Chim. Acta, 51, 1510 (1968).

[111] R. D. H. Murray, M. M. Ballantyne, and K. P. Mathai, Tetrahedron, 27, 1247 (1971).

ether itself may be alkylated.[107,110] Propargyl ethers serve not only as

$$ArOCH_2C\equiv CH \xrightarrow[\text{liq. NH}_3]{\text{RX, NaNH}_2} ArOCH_2C\equiv CR$$

starting materials for Claisen rearrangements leading to chromenes via o-allenyl phenols[108–110] but also as precursors for difficultly accessible allyl ethers. Reduction of disubstituted triple bonds of propargyl ethers by the Lindlar method provides cis-disubstituted allyl ethers.[107,110] Propargyl ethers bearing α substituents smoothly reduce to the corresponding α-substituted allyl ethers in good yield. Since the formation of propargyl ethers from α-substituted propargyl halides is not attended by a significant degree of rearrangement in the propargyl unit, the complication noted above in the direct preparation of α-substituted allylic ethers may be avoided by this method. The accompanying sequence is illustrative of this procedure.[111]

$$HC\equiv CC(CH_3)_2Cl \xrightarrow[\text{Acetone}]{\substack{\text{ArOH,}\\ \text{K}_2\text{CO}_3}} \underset{(77\%)}{HC\equiv CC(CH_3)_2OAr} \xrightarrow[\text{reduction}]{\text{Catalytic}} \underset{(96\%)}{CH_2=CHC(CH_3)_2OAr}$$

Traditionally, the solvents commonly employed for the preparation of aryl ethers have been acetone, methyl ethyl ketone, and alcohols.[1] When competing C-allylation is a problem, however, the use of an aprotic solvent of high dielectric constant, such as dimethylformamide, and homogeneous reaction conditions offer a practical advantage in promoting O-allylation.[106,112] Heterogeneous reaction conditions, likely to be encountered in hydrocarbon solvents, promote C-allylation and are to be avoided.[1,113,*]

Allyl and Propargyl Vinyl Ethers. Acid-catalyzed dealcoholation of diallyl and dipropargyl acetals and ketals[114–118] and base-catalyzed dehydrohalogenation of β-haloalkylallyl (or propargyl) ethers[119–122] represent some of the older methods of preparing aliphatic ethers capable of undergoing the Claisen rearrangement. In favorable cases, the ether

---

* The general problem of reactions of ambident anions has been treated by numerous workers in recent years. For discussion and leading references see reference 112 and S. J. Rhoads and R. W. Holder, *Tetrahedron*, **25**, 5443 (1969).

[112] N. Kornblum, R. Seltzer, and P. Haberfield, *J. Amer. Chem. Soc.*, **85**, 1148 (1963).
[113] N. Kornblum and A. P. Lurie, *J. Amer. Chem. Soc.*, **81**, 2705 (1959).
[114] C. D. Hurd and M. A. Pollack, *J. Org. Chem.*, **3**, 550 (1939).
[115] K. C. Brannock, *J. Amer. Chem. Soc.*, **81**, 3379 (1959).
[116] E. R. H. Jones, J. D. Loder, and M. C. Whiting, *Proc. Chem. Soc.*, **1960**, 180.
[117] N. B. Lorette and W. L. Howard, *J. Org. Chem.*, **26**, 3112 (1961).
[118] P. Cresson, *Bull. Soc. Chim. Fr.*, **1964**, 2618.
[119] S. M. McElvain, H. I. Anthes, and S. H. Shapiro, *J. Amer. Chem. Soc.*, **64**, 2525 (1942).
[120] P. Cresson, *C.R. Acad. Sci.*, Ser. C, **261**, 1707 (1965).
[121] E. Demole and P. Enggist, *Chem. Commun.*, **1969**, 264.
[122] S. J. Rhoads and J. M. Watson, *J. Amer. Chem. Soc.*, **93**, 5813 (1971).

itself need not be isolated but may undergo rearrangement directly to produce the homoallyl or homoallenyl carbonyl compound.

Preparation of allyl and propargyl vinyl ethers has also been accomplished by addition of the corresponding alcohols to acetylenic bonds.[123–128a] In this connection, mention may be made of the reactions of ynamines with allyl,[125] propargyl,[126] and allenyl alcohols[127] in the presence of a borontrifluoride etherate catalyst. Formation and rearrangement of the adduct occur in a single operation, usually at room temperature, and the resulting amide can be isolated directly from the reaction mixture. The reactions are facilitated by the boron trifluoride catalyst but do not require it. Comparable yields of rearrangement products may be obtained by heating the components under reflux in benzene or toluene in the absence of the catalyst. The accompanying reaction with furfuryl alcohol is useful for introducing a side chain at the 3 position of a furan nucleus.[127]

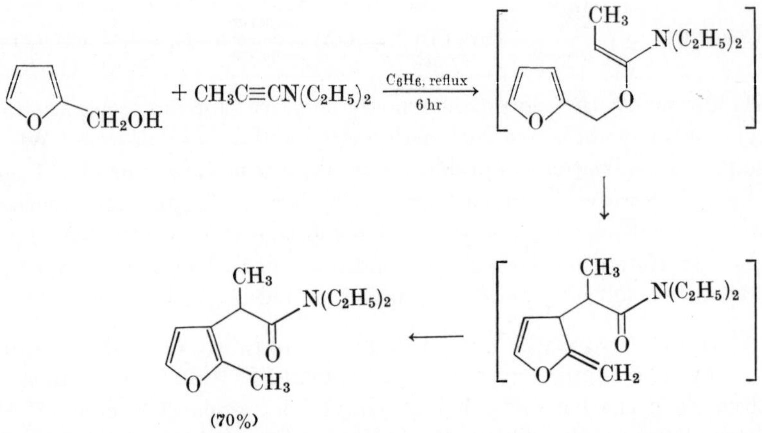

(70%)

Highly enolic compounds and phenols may add to the activated double bond of 2-methoxy-1,3-butadiene to produce allyl vinyl ethers which then suffer a Claisen rearrangement *in situ* as illustrated for the reaction of dimedone.[128b] When phenols are employed, the final product is a methoxychromane formed by ring closure of the *ortho*-substituted phenolic product.[128b]

[123] R. Paul, G. Roy, M. Fluchaire, and G. Collardeau, *Bull. Soc. Chim. Fr.*, **1950**, 121.

[124] J. W. Ralls, R. E. Lundin, and G. F. Bailey, *J. Org. Chem.*, **28**, 3521 (1963).

[125] J. Ficini and C. Barbara, *Tetrahedron Lett.*, **1966**, 6425.

[126] J. Ficini, N. Lumbroso-Bader, and J. Pouliquen, *Tetrahedron Lett.*, **1968**, 4139.

[127] J. Ficini and J. Pouliquen, *C.R. Acad. Sci., Ser. C*, **268**, 1446 (1969).

[128] (a) C. G. Krespan, *Tetrahedron*, **23**, 4243 (1967).

[128] (b) L. J. Dolby, C. A. Elliger, S. Esfandiari, and K. S. Marshall, *J. Org. Chem.*, **33**, 4508 (1968).

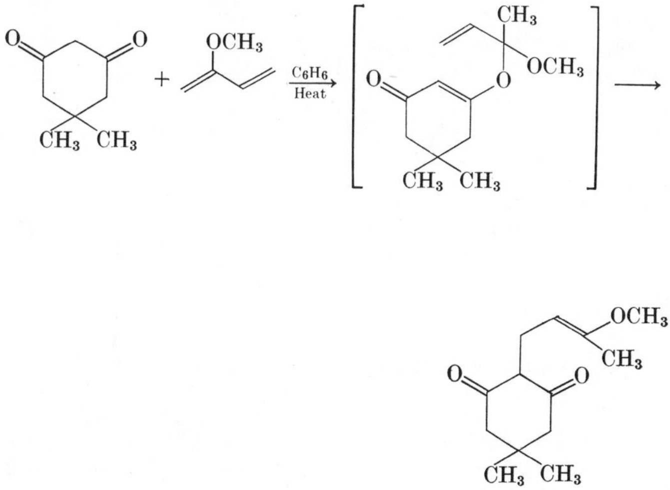

O ... O

CH_3 CH_3

$+$

OCH_3

$\xrightarrow[\text{Heat}]{C_6H_6}$

CH_3

OCH_3

O

O

CH_3 CH_3

$\longrightarrow$

OCH_3

CH_3

O ... O

CH_3 CH_3

(92%)

OCH_3

CH_3

O

H

$\longrightarrow$

OCH_3

CH_3

OH

$\longrightarrow$

CH_3 OCH_3

O

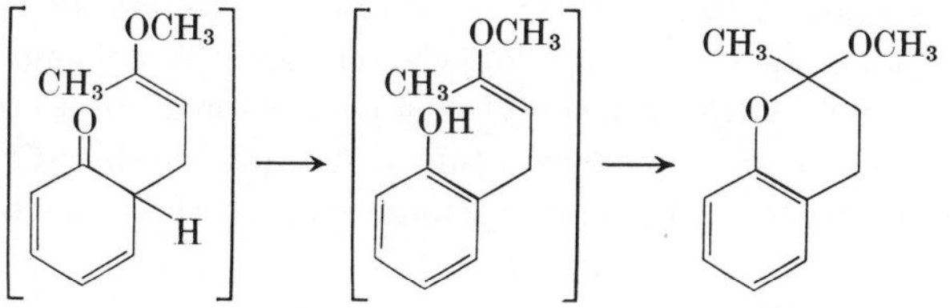

(82%)

The scope of the aliphatic Claisen rearrangement has been greatly enlarged by the development of superior methods of transvinyletherification in recent years. In principle, all of these methods involve the treatment of an allyl or propargyl alcohol with a vinyl ether derivative (or its ketal precursor) and are assumed to proceed through the stages of ketal formation and dealcoholation to the allyl or propargyl ether system (Scheme 3, p. 16). The resulting ether derivative may sometimes be isolated, but it generally undergoes rearrangement *in situ* to the rearranged product in one operation from the starting alcohol. Most transetherifications are catalyzed by acids. In the original method of Watanabe and Conlon,[129] transvinylation was accomplished with vinyl alkyl ethers ($G = R$ or H) in the presence of the Lewis acid, mercuric acetate, and the allyl vinyl ether was isolated after neutralization of the equilibrated mixture or by slow distillation from the equilibrated mixture. Modifications of this procedure

---

[129] W. H. Watanabe and L. E. Conlon, *J. Amer. Chem. Soc.*, **79**, 2828 (1957).

$$
\begin{array}{c}
\text{A—OH} \\
\text{or} \\
\text{P—OH}
\end{array}
+ \quad \underset{\text{and/or}}{\overset{\text{OR}}{\underset{}{C=C}}}\overset{\text{OR}}{\underset{\text{G}}{}} \rightleftharpoons
\left[\underset{\overset{|}{\text{H}}}{-\overset{|}{\text{C}}} - \underset{\text{O—A (OP)}}{\overset{\text{OR}}{\overset{|}{\text{C}}}} - \text{G} \right]
$$

$$
\underset{\overset{|}{\text{H}}}{-\overset{|}{\text{C}}} - \underset{\text{OR}}{\overset{\text{OR}}{\overset{|}{\text{C}}}} - \text{G}
$$

$$
\underset{\text{OA (OP)}}{\overset{\text{G}}{C=C}} \quad + \text{ ROH}
$$

Rearrangement products

A = allyl, P = propargyl, R = alkyl, G = H, R, OR, NR$_2$

SCHEME 3

use dry hydrogen chloride,[130] phosphoric acid,[131,132] p-toluenesulfonic acid,[133] or oxalic acid[55] as the catalytic agent and induce *in situ* re-arrangements.[55,100–102,130–133]

The three methods developed by Johnson and co-workers and applied with such success in the synthesis of terpenoid materials incorporate an additional function, G, in the vinylating reagent of Scheme 3. In the orthoacetate method the transvinylating agent is methyl or ethyl ortho-acetate (G = OR), and the rearrangement product, produced *in situ*, is a γ,δ-unsaturated ester. A trace of propionic acid serves as the catalyst.[97]

$$
\underset{\overset{|}{\text{H}}}{\overset{\text{R}}{\underset{\text{CH}_2}{\diagup}}}\!\!\!\overset{\text{OH}}{\underset{\text{R}'}{}} + \text{CH}_3\text{C(OC}_2\text{H}_5)_3 \xrightarrow[138°, \, 1 \text{ hr}]{\text{C}_2\text{H}_5\text{CO}_2\text{H}}
$$

$$
\left[\begin{array}{c}
\overset{\text{OC}_2\text{H}_5}{\underset{\text{O}}{\text{CH}_2}} \\
\underset{\underset{\text{R}}{}}{\text{CH}_2} \diagdown \underset{\text{R}'}{\overset{\text{H}}{}}
\end{array} \right] \longrightarrow
\begin{array}{c}
\overset{\text{OC}_2\text{H}_5}{\underset{\text{O}}{\text{CH}_2}} \\
\underset{\underset{\text{R}}{}}{\text{CH}_2} \diagdown \underset{\text{R}'}{\overset{\text{H}}{}}
\end{array}
$$

(92%)

R = CH$_3$, R' = CH$_2$CH$_2$C(CH$_3$)=CH$_2$

The olefinic ketal method employs a trace of 2,4-dinitrophenol as catalyst and permits the introduction of an unsaturated linkage α,β to the

[130] S. Julia, M. Julia, H. Linarès, and J.-C. Blondel, *Bull. Soc. Chim. Fr.*, **1962**, 1947.
[131] G. Saucy and R. Marbet, *Helv. Chim. Acta*, **50**, 2091 (1967).
[132] R. Marbet and G. Saucy, *Helv. Chim. Acta*, **50**, 2095 (1967).
[133] G. Saucy and R. Marbet, *Helv. Chim. Acta*, **50**, 1158 (1967).

carbonyl group of the homoallylic carbonyl product. The latter may be reduced to the corresponding allyl alcohol, and repetition of the sequence elongates the carbon chain by another four-carbon unit (Eq. 4). The juvenile hormone precursor ($R = C_2H_5$, $R' = CO_2CH_3$, $R'' = CH_3$) was prepared in this manner.[96]

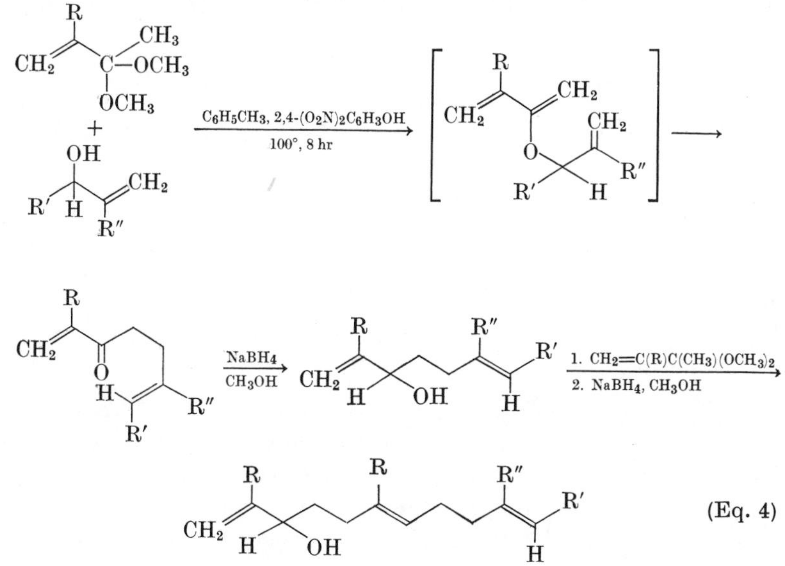

(Eq. 4)

The chloroketal method uses the dimethyl ketal of 3-chloro-3-methyl-2-butanone as the transvinylating agent to introduce a chlorodimethyl-carbinyl group which can be transformed later into a terminal isopropyl-idene unit.[98,99]

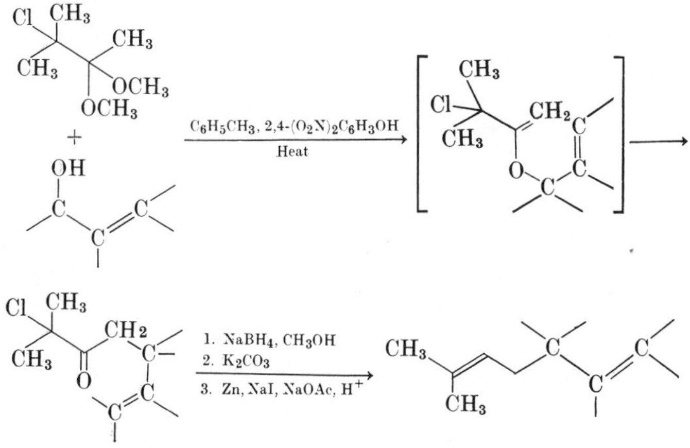

All three methods developed by the Johnson group show a higher degree of stereoselectivity than do the rearrangements of the corresponding simple allyl vinyl ethers. Application of the chloroketal method to the C_{20}-allylic diol, **3**, for example, led through a double Claisen rearrangement to a C_{30} derivative which could be converted to the triterpene, squalene, with better than 97% all-*trans* geometry.[98,99]

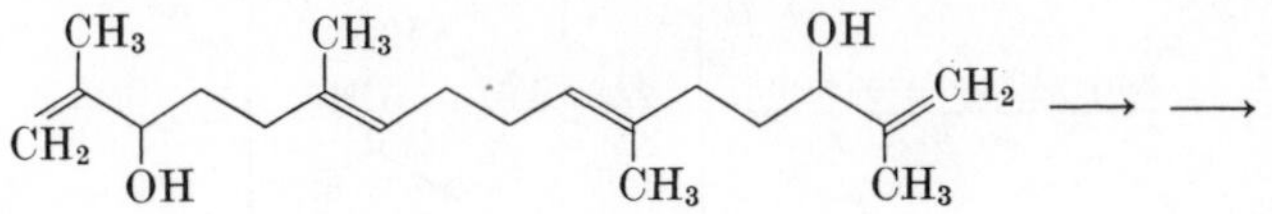

3

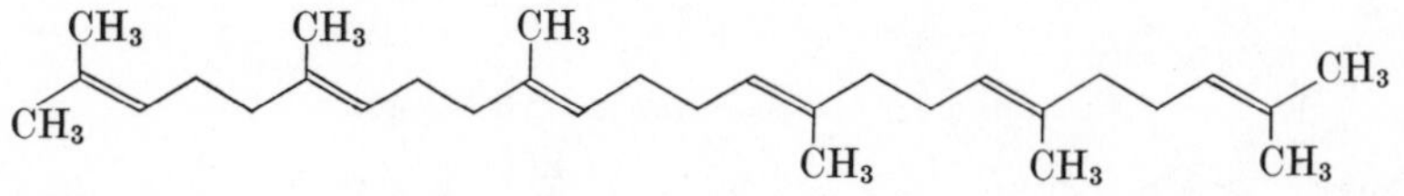

The Meerwein-Eschenmoser method for transvinyletherification employs a mixture of 1-dimethylamino-1-methoxyethene and the corresponding dimethyl acetal of N,N-dimethylacetamide [G = $N(CH_3)_2$, Scheme 3] as the vinylating agent and has the advantage that an acidic catalyst is not required.[134-136] The rearrangement product, formed *in situ* in refluxing solvent, is a γ,δ-unsaturated amide.

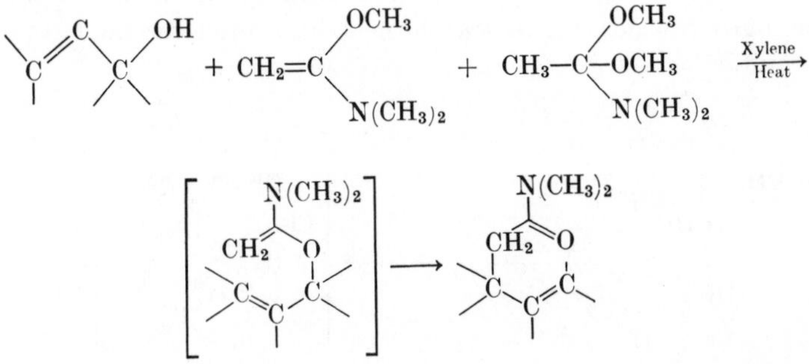

Interesting and useful variants of the *in situ* preparation and rearrangement of allylic vinyl ethers have been developed for terpene syntheses by

[134] H. Meerwein, W. Florian, N. Schön, and G. Stopp, *Ann. Chem.*, **641**, 1 (1961).

[135] D. Felix, K. Gschwend-Steen, A. E. Wick, and A. Eschenmoser, *Helv. Chim. Acta*, **52**, 1030 (1969).

[136] A. E. Wick, D. Felix, K. Steen, and A. Eschenmoser, *Helv. Chim. Acta*, **47**, 2425 (1964).

Thomas[100-102] and by Faulkner and Petersen.[55] Both methods employ alkoxy derivatives of isoprene as the vinylating agent, thereby adding a functionalized isoprene unit to the three-carbon chain of the starting allylic alcohol. The Thomas method uses an isoprene molecule bearing a 1-alkoxy substituent, whereas the Faulkner-Petersen procedure uses an isoprene unit bearing an alkoxy group at C_3.

Vinyl ether exchange of an allyl alcohol by the Thomas method leads to an allyl dienyl ether, which, after undergoing a Claisen rearrangement, possesses the 1,5-hexadiene structure necessary for a Cope rearrangement. The successive *in situ* rearrangements ultimately give rise to a dienic aldehyde in which the functionalized isoprene unit is attached to C_1 of the original allylic alcohol as shown in the accompanying sequence.

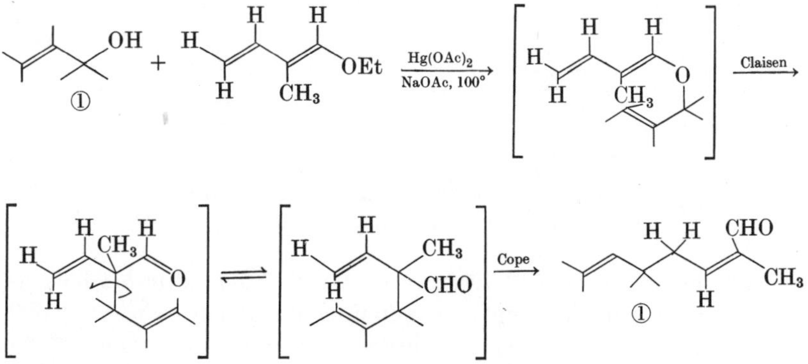

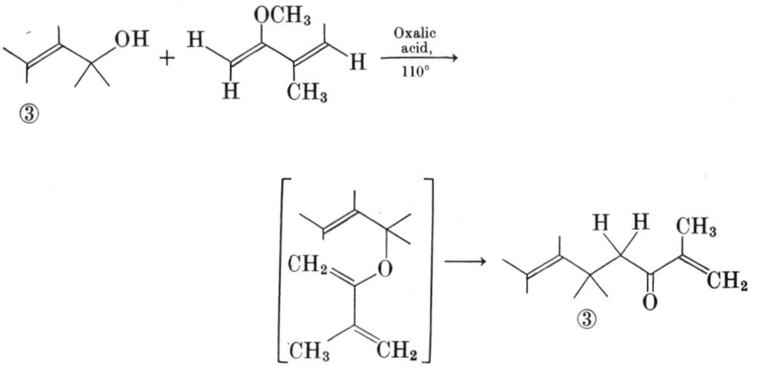

On the other hand, the Faulkner-Petersen method ("methoxyisoprene" method) leads, through a single *in situ* Claisen rearrangement, to a dienic aldehyde in which the functionalized isoprene unit is attached to C_3 of the original allylic alcohol. Both methods show the stereoselectivity characteristic of [3,3]sigmatropic processes and can be used to fashion molecules of predictable geometry. Application of the Thomas method to the triene alcohol **4** produced β-sinensal (**5**) as the sole product in 43 % yield.[102]

Derivatives of 1,5-Hexadienes, 1,5-Hexenynes, and 1,5-Hexadiynes. Unsaturated systems suitable for Cope rearrangements are available in diverse ways. The systems originally studied by Cope and coworkers were prepared by alkylation of alkylidene malonic acid derivatives with allylic halides (Scheme 4).[2,137–143] Alkylidene ketones may be substituted for the malonic acid derivative[144–150] and propargyl halides may

Y and/or Z = CO_2R, CN, C=O, A = allyl, P = propargyl

SCHEME 4

replace the allyl halides.[151] Hydrocarbons with 1,5-unsaturated linkages have been prepared by coupling the appropriate allyl or propargyl halide over magnesium[57,152–155] or in the presence of nickel carbonyl,[156] by

[137] A. C. Cope and E. M. Hancock, *J. Amer. Chem. Soc.*, **60**, 2644, 2903 (1938).

[138] A. C. Cope, K. E. Hoyle, and D. Heyl, *J. Amer. Chem. Soc.*, **63**, 1843 (1941).

[139] A. C. Cope, C. M. Hofmann, and E. M. Hardy, *J. Amer. Chem. Soc.*, **63**, 1852 (1941).

[140] A. C. Cope and L. Field, *J. Amer. Chem. Soc.*, **71**, 1589 (1949).

[141] A. C. Cope, L. Field, D. W. H. MacDowell, and M. E. Wright, *J. Amer. Chem. Soc.*, **78**, 2547 (1956).

[142] A. C. Cope, J. E. Meili, and D. W. H. MacDowell, *J. Amer. Chem. Soc.*, **78**, 2551 (1956).

[143] D. E. Whyte and A. C. Cope, *J. Amer. Chem. Soc.*, **65**, 1999 (1943).

[144] J.-M. Conia and P. LePerchec, *Tetrahedron Lett.*, **1964**, 2791.

[145] J.-M. Conia and P. LePerchec, *Tetrahedron Lett.*, **1965**, 3305.

[146] J.-M. Conia and P. LePerchec, *Bull. Soc. Chim. Fr.*, **1966**, 273.

[147] J.-M. Conia and P. LePerchec, *Bull. Soc. Chim. Fr.*, **1966**, 278.

[148] J.-M. Conia and P. LePerchec, *Bull. Soc. Chim. Fr.*, **1966**, 281.

[149] J.-M. Conia and P. LePerchec, *Bull. Soc. Chim. Fr.*, **1966**, 287.

[150] J.-M. Conia and A. Sandré-LeCraz, *Tetrahedron Lett.*, **1962**, 505.

[151] D. K. Black and S. R. Landor, *J. Chem. Soc.*, **1965**, 6784.

[152] W. D. Huntsman, J. A. DeBoer, and M. H. Woosley, *J. Amer. Chem. Soc.*, **88**, 5846 (1966).

[153] W. D. Huntsman and H. J. Wristers, *J. Amer. Chem. Soc.*, **89**, 342 (1967).

[154] H. P. Koch, *J. Chem. Soc.*, **1948**, 1111.

[155] H. Levy and A. C. Cope, *J. Amer. Chem. Soc.*, **66**, 1684 (1944).

[156] M. F. Semmelhack, *Org. Reactions*, **19**, 115 (1972).

reductive coupling of allylic alcohols in the presence of titanium tetra-chloride and methyllithium,[157] by direct allylation of the anions of allyl-benzene[155] and of allyl sulfides,[158] by allylation of phosphorus ylides,[159] and by [2,3]sigmatropic rearrangements of sulfur ylides derived from sulfonium salts carrying two allyl groups as shown in the accompanying formulation.[160,161] In reactions which lead to sulfur or phosphorus deriv-atives of 1,5-dienyl systems, a final step may be reduction to remove the heteroatom.[161]

The boronate fragmentation reaction has been used to advantage in the synthesis of cyclic and acyclic 1,5-dienes, as illustrated (p. 22) for the prep-aration of 1,5-*trans,trans*-cyclodecadienes[162] and for *trans*-1,5-octadiene.[163]

1,5-Enynes have been prepared by dehydrohalogenation of suitable vicinal dibromides. The process may be combined with alkylation of a resulting acidic alkyne to produce unsymmetrically substituted enynes in one operation.[152]

---

* THF is the abbreviation for tetrahydrofuran.

[157] K. B. Sharpless, R. P. Hanzlik, and E. E. van Tamelen, *J. Amer. Chem. Soc.,* **90**, 208 (1968).

[158] J. F. Biellmann and J. B. Ducep, *Tetrahedron Lett.,* **1969**, 3707.

[159] E. H. Axelrod, G. M. Milne, and E. E. van Tamelen, *J. Amer. Chem. Soc.,* **92**, 2139 (1970).

[160] G. M. Blackburn, W. D. Ollis, J. D. Plackett, C. Smith, and I. O. Sutherland, *Chem. Commun.,* **1968**, 186.

[161] J. E. Baldwin, P. S. Hackler, and D. P. Kelly, *Chem. Commun.,* **1968**, 537.

[162] J. A. Marshall and G. L. Bundy, *Chem. Commun.,* **1967**, 854.

[163] J. A. Marshall, *Synthesis,* **1971**, 229.

Allylated cyclohexadienones, useful for the study of concurrent Cope and retro-Claisen rearrangements, are readily accessible by the direct C-allylation of phenoxides in benzene at room temperature or lower.[164-166] The Wittig reaction, Hofmann elimination, and the Cope amine-oxide reaction also have been widely used to introduce olefinic bonds in desired positions.

The oxy-Cope rearrangement, synthetically valuable for the preparation of δ,ϵ-unsaturated carbonyl compounds and for α,δ-dicarbonyl compounds, requires hydroxy substituents at the 3 position or the 3 and 4 positions of the 1,5-hexadiene system. Such structures are commonly prepared by the action of allyl or propargyl Grignard reagents on the appropriate α,β-unsaturated carbonyl compounds.[50,167-169] Alternatively, vinyl Grignard reagents may be used with β,γ-unsaturated carbonyl[170,171] or α-dicarbonyl compounds.[172-177] These synthetic approaches are generalized in Scheme 5.

164 D. Y. Curtin and R. J. Crawford, J. Amer. Chem. Soc., 79, 3156 (1957).

165 B. Miller, J. Org. Chem., 35, 4262 (1970).

166 B. Miller, J. Amer. Chem. Soc., 87, 5115 (1965).

167 A. Viola, E. J. Iorio, K. K. Chen, G. M. Glover, U. Nayak, and P. Kocienski, J. Amer. Chem. Soc., 89, 3462 (1967).

168 A. Viola and L. A. Levasseur, J. Amer. Chem. Soc., 87, 1150 (1965).

169 A. Viola and J. H. MacMillan, J. Amer. Chem. Soc., 90, 6141 (1968).

170 J. A. Berson and M. Jones, Jr., J. Amer. Chem. Soc., 86, 5019 (1964).

171 E. N. Marvell and W. Whalley. Tetrahedron Lett., 1970, 509.

172 E. Brown and J.-M. Conia, Bull. Soc. Chim. Fr., 1970, 1050.

173 E. Brown, P. Leriverend, and J.-M. Conia, Tetrahedron Lett., 1966, 6115.

174 P. Leriverend and J.-M. Conia, Tetrahedron Lett., 1969, 2681.

175 P. Leriverend and J.-M. Conia, Bull. Soc. Chim. Fr., 1970, 1060.

176 E. N. Marvell and T. Tao, Tetrahedron Lett., 1969, 1341.

177 E. N. Marvell and W. Whalley, Tetrahedron Lett., 1969, 1337.

A = allyl, P = propargyl, V = vinyl

SCHEME 5

Bimolecular reduction of α,β-unsaturated carbonyl compounds using a zinc-copper couple[178,179] or other reducing metals[180–183] may also be used in the preparation of symmetrical 3,4-dihydroxy-1,5-hexadienes.

Aromatic Claisen Rearrangements

Allyl Ethers

Complications in rearrangements of allyl aryl ethers may arise from competitive *ortho* and *para* migrations, the occurrence of abnormal rearrangements leading to structural and geometric isomerization in the migrating group, subsequent double-bond shifts and coumaran formation, out-of-ring migrations, and, occasionally, the formation of stable dienones and the incursion of retro-Claisen rearrangements. To some extent, these processes can be controlled by a proper choice of solvent and rearrangement conditions as discussed in the following sections.

ortho-para Migrations. Detailed examinations of systems in which both *ortho* and *para* positions are open[184–187] have shown that rearrangements to these positions can be competitive and that mixed products are

[178] R. A. Braun, *J. Org. Chem.*, **28**, 1383 (1963).

[179] W. G. Young, L. Levanas, and Z. Jasaitis, *J. Amer. Chem. Soc.*, **58**, 2274 (1936).

[180] J. Chuche and J. Wiemann, *C.R. Acad. Sci., Ser. C*, **262**, 567 (1966).

[181] J. Chuche and J. Wiemann. *Bull. Soc. Chim. Fr.*, **1968**, 1491.

[182] J. Kossanyi, *Bull. Soc. Chim. Fr.*, **1965**, 714.

[183] J. Wiemann and S.-L. Thuan, *Bull. Soc. Chim. Fr.*, **1959**, 1537.

[184] J. Borgulya, H.-J. Hansen, R. Barner, and H. Schmid, *Helv. Chim. Acta*, **46**, 2444 (1963).

[185] E. N. Marvell, B. J. Burreson, and T. Crandall, *J. Org. Chem.*, **30**, 1030 (1965).

[186] E. N. Marvell, B. Richardson, R. Anderson, J. L. Stephenson, and T. Crandall, *J. Org. Chem.*, **30**, 1032 (1965).

[187] F. Scheinmann, R. Barner, and H. Schmid, *Helv. Chim. Acta*, **51**, 1603 (1968).

formed much more commonly than had been appreciated in earlier investigations.[1] The *ortho/para* ratio is conditioned by the bulk of the substituents in the migrating allyl group,[184,187] the number, size, and location of other ring substituents,[184–186,188–191] and the solvent.[184,187] That the solvent can exert a profound effect is illustrated by the results for the rearrangement of the γ-methylallyl ether of 3,5-dimethylphenol in solvents

Solvent	Product Composition (%)	
	ortho	*para*
Decalin	38	42
Diethylaniline	79	21
Dimethylformamide	91	1.5

of differing polarity.[184] These effects have been accounted for in terms of steric interactions in the first-formed *o*-dienone which, by hindering the usually rapid enolization step, allow migration to the *para* position to become competitive. Polar solvents facilitate the enolization and restore the *ortho* rearrangement to its usual prominence.[184]

Abnormal Claisen Rearrangement.[19b]   The abnormal rearrangement leading to structural[1,5,9] and geometric[192–194] isomerization in the migrating allyl group is commonly observed to accompany the *ortho* rearrangement of ethers bearing γ-alkyl substituents on the allyl group. The abnormal product, in fact, is produced in a subsequent rearrangement of the normal *o*-allyl phenol[195] and is formed through an intermediate spirocyclopropyl-cyclohexadienone resulting from hydrogen transfer from the phenolic function to the terminal carbon atom of the allylic group (Scheme 6). Reversal of this process (a 1,5-hydrogen shift), but involving a hydrogen from the γ-alkyl group, leads to the abnormal product.[192,195–198]

[188] E. D. Burling, A. Jefferson, and F. Scheinmann, *Tetrahedron*, **21**, 2653 (1965).

[189] A. Dyer, A. Jefferson, and F. Scheinmann, *J. Org. Chem.*, **33**, 1259 (1968).

[190] S. C. Sethi and B. C. Subba Rao, *Indian J. Chem.*, **2**, 323 (1964) [*C.A.*, **61**, 14492e (1964)].

[191] B. D. Tiffany, *J. Amer. Chem Soc.*, **70**, 592 (1948).

[192] A. Habich, R. Barner, R. M. Roberts, and H. Schmid, *Helv. Chim. Acta*, **45**, 1943 (1962).

[193] Gy. Frater and H. Schmid, *Helv. Chim. Acta*, **49**, 1957 (1966).

[194] E. N. Marvell and B. Schatz, *Tetrahedron Lett.*, **1967**, 67.

[195] E. N. Marvell, D. R. Anderson, and J. Ong, *J. Org. Chem.*, **27**, 1109 (1962).

[196] W. M. Lauer, G. A. Doldouras, R. E. Hileman, and R. Liepins, *J. Org. Chem.*, **26**, 4785 (1961).

[197] W. M. Lauer and T. A. Johnson, *J. Org. Chem.*, **28**, 2913 (1963).

[198] A. Habich, R. Barner, W. von Philipsborn, H. Schmid, H.-J. Hansen, and H. J. Rosenkranz, *Helv. Chim. Acta*, **48**, 1297 (1965).

In summary, in the abnormal product, the original β-carbon atom of the side chain is attached to the ring, the original α-carbon atom appears as a saturated β substituent, and the double bond has shifted to a position between the original γ-carbon atom and its hydrogen-bearing alkyl group. The interconversion of normal and abnormal products through such acyl cyclopropyl intermediates is quite common[199] and is recognized as a

SCHEME 6

special case of a general phenomenon, the "enolene rearrangement." [200] *cis,trans* Isomerization in the side chain of *o*-allyl phenols has been shown to occur through the same intermediate.[194] An analogous abnormal course has been reported in the rearrangement of the acyclic γ-ethylallyl vinyl ether.[114]

Most abnormal rearrangements are considerably slower than the formation of the normal *o*-allylic phenol, and mild reaction conditions and shorter reaction times can eliminate, or at least minimize, formation of the abnormal product.[195–197] In certain systems, however, such precautions do not suffice, and the normal product can be isolated only by intercepting it with a reactive trapping agent.[201] The solvent also, by its effect on the *ortho*/*para* ratio, can play an important role in controlling the product composition in systems in which the abnormal reaction is possible. For

[199] R. M. Roberts and R. G. Landolt, *J. Org. Chem.*, **31**, 2699 (1966).

[200] R. M. Roberts, R. G. Landolt, R. N. Greene, and E. W. Heyer, *J. Amer. Chem. Soc.*, **89**, 1404 (1967).

[201] A. Jefferson and F. Scheinmann, *J. Chem. Soc., C.*, **1969**, 243.

example, in the rearrangement of γ,γ-dimethylallyl phenyl ether, the distribution of products is strongly dependent on the solvent polarity.[187] In dimethylformamide, a medium which promotes enolization of the initially formed o-dienone and thereby opens the way for the abnormal rearrangement, the abnormal o-substituted phenol 6 accounts for 89% of the rearranged products. Under otherwise identical reaction conditions in diethylaniline, however, the major product is the p-substituted phenol (7, 72%), formed in a competitive transposition of the o-dienone to the less sterically congested p-dienone.

Double-Bond Migration, Coumaran Formation, and Cleavage. Accompanying both normal and abnormal products in the *ortho* rearrangement, isomeric propenyl phenols and coumarans resulting from ring closure are often observed as by-products arising from the initially formed o-allylic phenols.[1] The extent of these secondary reactions is strongly dependent on the experimental conditions employed and of these, once more, the nature of the solvent is especially noteworthy. In an analysis of the products of the rearrangement of β-methylallyl phenyl ether (8), it has been demonstrated that coumaran formation is promoted by phenols and primary aromatic amines, whereas isomerization of the double bond into conjugation with the aromatic ring is especially facilitated by primary aromatic amines.[202] Tertiary aromatic amines, on the other hand, minimize these secondary reactions and, of the various amines of this type which were tested, dimethylaniline proved superior. The data shown are representative of the results obtained when the rearrangements were carried to 95% completion under identical conditions of concentration at about 200°.

202 A. T. Shulgin and A. W. Baker, *J. Org. Chem.*, **28**, 2468 (1963).

8

Solvent		Product Composition (%)		
None	53	12		26
2,6-Xylenol	6	10		73
2,6-Xylidine	8	44		32
Dimethylaniline	90	2		1

Cleavage of substituted allyl ethers to phenols and dienes can be antici-pated when excessive temperatures are applied or when the ether is heavily substituted.[1] The decompositions reported for the γ,γ-dimethylallyl ethers of various hydroxycoumarins may be attributed to such causes.[203]

Effects of Nuclear and Side-Chain Substituents. The electronic nature of ring substituents has only minor effects on the ease of rearrange-ment of allyl aryl ethers,[17] and there is no discernible pattern that suggests a consistent directive influence on the part of a given substituent. Thus, when nonequivalent *ortho* positions and the *para* position are open, migration to any or all of them may be expected. Indeed, other experi-mental variables such as solvent, reaction temperature, and duration appear to exert a greater control over product composition than do the nature and location of ring substituents. Since rearrangements to the *ortho* and *para* positions are, in fact, reversible processes, one might expect that prolonged heating would ultimately lead to a thermodynamically controlled reaction mixture. An indication of such control is provided by experiments[184,194] that show that *ortho-* and *para*-substituted allyl phenols do slowly interconvert (with inversion) when held under the usual re-arrangement conditions for long periods of time ("allyl phenol rearrange-ment").[184]

Isolated instances in which specific interactions of adjacent substituents can control the direction[204,205] and even the realization[206] of rearrangement are found in aromatic systems in which strong chelation forces effectively deactivate an open *ortho* position. The acetophenone derivative **9**, for example, failed to rearrange at 190° and resinified at higher temperatures. The corresponding tosylate **10**, however, underwent smooth rearrangement at the lower temperature.[206]

[203] B. Chaudhury, S. K. Saha, and A. Chatterjee, *J. Indian Chem. Soc.*, **39**, 783 (1962) [*C.A.*, **59**, 2628b (1963)].

[204] W. Baker and O. M. Lothian, *J. Chem. Soc.*, **1935**, 628. See discussion in ref. 1, p. 14.

[205] R. Aneja, S. K. Mukerjee, and T. R. Seshadri, *Tetrahedron*, **2**, 203 (1958).

[206] R. Aneja, S. K. Mukerjee, and T. R. Seshadri, *Chem. Ber.*, **93**, 297 (1960).

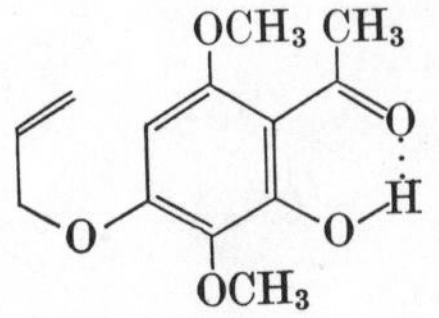

9 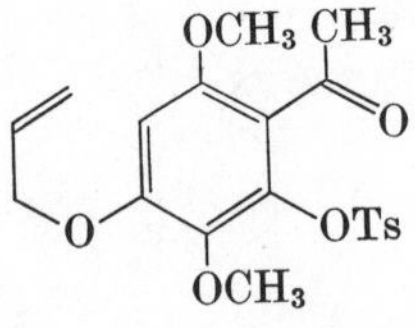   10

Little quantitative information about the effect of substitution in the allylic side chain on the ease of rearrangement seems to be available. The relative rates for the o-rearrangement of allyl, α-methylallyl, β-methylallyl, and γ-methylallyl phenyl ethers have been reported as 1.52:21.1:1.32:1.62 at 185° in diphenyl ether.[83] With the exception of α-substituents, then, the effect of alkyl substitution in the allyl group appears to be negligible. A corresponding accelerating effect of α-alkyl substituents is implicit in the success of the preferential rearrangements in mixtures of α- and γ-substituted derivatives in *para* rearrangement studies.[103,104]

Out-of-Ring Migrations. When the aromatic ring of an allyl phenyl ether carries a conjugated olefinic substituent in the *ortho* or *para* position, out-of-ring rearrangement routes become available to the intermediate dienones. First observed by Claisen and Tietze,[1] the migration of an allyl group to the β-carbon atom of an o-propenyl side chain was later shown to be an intramolecular process and to occur with overall retention of structure in the migrating group.[207,208] In analogy to the *para* rearrangement, the process is most simply viewed as a sequence of sigmatropic shifts as illustrated for the γ-methylallyl ether of 2,4-dimethyl-6-propenyl phenol.[207]

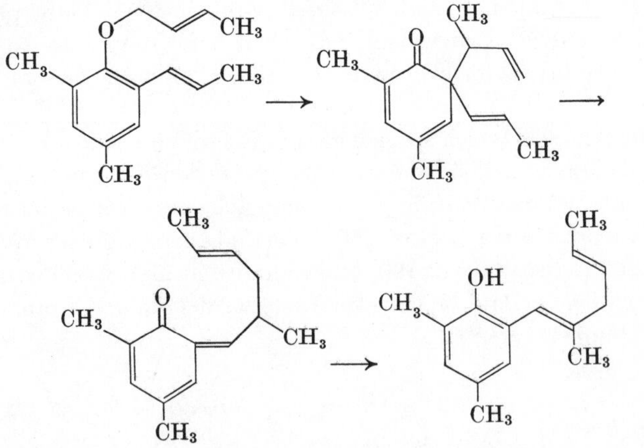

[207] W. M. Lauer and D. W. Wujciak, *J. Amer. Chem. Soc.*, **78**, 5601 (1956).
[208] K. Schmid, P. Fahrni, and H. Schmid, *Helv. Chim. Acta*, **39**, 708 (1956).

A three-stage process in which an allyl group migrates out-of-ring to a *para*-substituted propenyl side chain has been reported.[209] Presumably, it

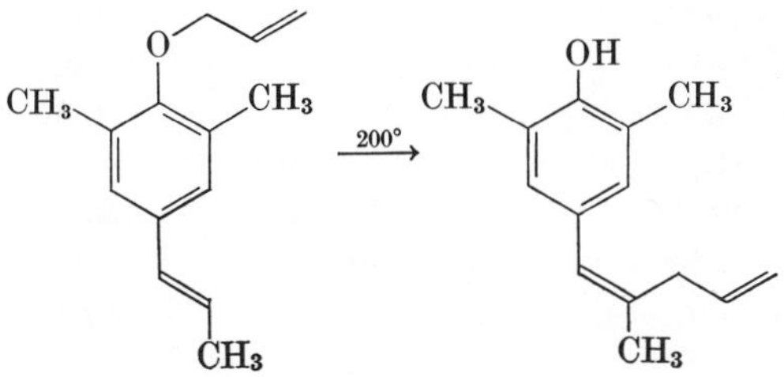

passes over the *o*-dienone and *p*-dienone intermediates to give overall inversion of structure in the migrating group. Attempts to induce rearrangements in similar systems in which the *p*-propenyl group was replaced by a phenyl ring failed; only cleavage of the ether resulted.[209]

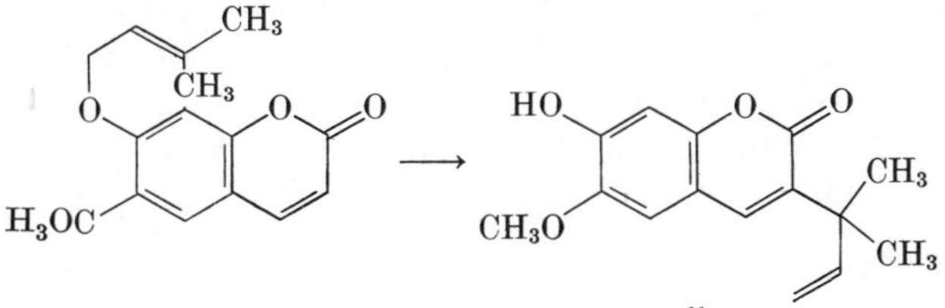

11

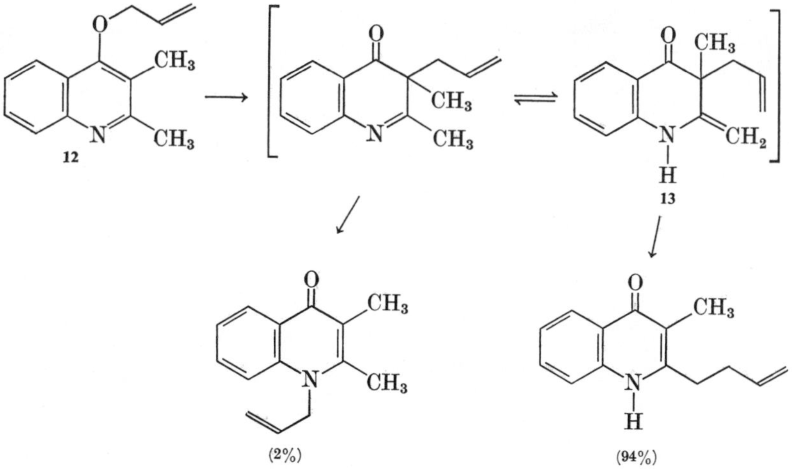

[209] A. Nickon and B. R. Aaronoff, *J. Org. Chem.*, **29**, 3014 (1964).

Another example of what appears to be a triple rearrangement is observed with 6-methoxy-7-(γ,γ-dimethylallyloxy)coumarin.[210,211] In addition to the normal *ortho* rearrangement product, its corresponding coumaran and some of the abnormal *ortho* rearranged product, the out-of-ring product (11) was formed in 14% yield. In this case, the inversion of structure in the migrating allylic group expected of three consecutive [3,3]sigmatropic rearrangements is observable.

Out-of-ring migrations actually predominate in some quinoline derivatives, exemplified by 12, in which migration to a methyl group *meta* to the ether oxygen successfully competes with the *para* migration to nitrogen.[212] The out-of-ring rearrangement clearly must occur over the enamine tautomer, 13.

Other Anomalies: The Retro-Claisen Rearrangement, Isolation of Stable Dienones, the ortho-ortho' Rearrangement. Because of the thermodynamic bias in favor of phenolic products, neither the retro-Claisen rearrangement nor the isolation of dienones is ordinarily a serious limitation in synthetic work. Such results have been observed, however, in rather unusual structures. The *o*-allylic spirocyclohexadienone, 14, quantitatively undergoes a retro-Claisen to the more stable aromatic system 15,[213] and the 2-allyloxynaphthalene derivative 16 equilibrates with the isolable 1,1-diallyl dienone.[214] Systems in which retro-Claisen

[210] M. M. Ballantyne, R. D. H. Murray, and A. B. Penrose, *Tetrahedron Lett.*, **1968**, 4155.

[211] M. M. Ballantyne, P. H. McCabe, and R. D. H. Murray, *Tetrahedron*, **27**, 871 (1971).

[212] Y. Makisumi, *J. Org. Chem.*, **30**, 1989 (1965).

[213] M. F. Ansell and V. J. Leslie, *Chem. Commun.*, **1967**, 949; *J. Chem. Soc.*, *C*, **1971**, 1423.

[214] J. Green and D. McHale, *Chem. Ind.* (London), **1964**, 1801.

rearrangements compete with Cope rearrangements to *p*-dienones have been studied.[165,215]

First detected in a radioactive tracer study of the reversibility of the Claisen rearrangement,[216] the intramolecular *ortho-ortho'* rearrangement leads to an *ortho* rearrangement product without the usual structural inversion of the allyl group. "Forbidden" as a concerted process by orbital symmetry considerations, the rearrangement has been formulated as a stepwise reaction involving the intermediacy of an internal Diels-Alder adduct, **17**.[8] It must be emphasized that the *ortho-ortho'* rearrangement should rarely be encountered in usual synthetic applications of the

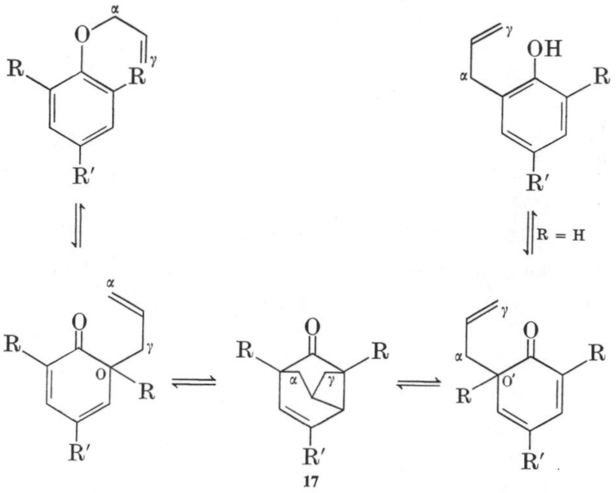

Claisen rearrangement, since ordinarily it could not compete with the more rapid steps of enolization of an *o*-dienone or of rearrangement to a *p*-dienone and subsequent enolization. Only a few *o*-allyl products with noninverted structure have been obtained under conditions that appear to rule out an alternative explanation of dissociation and recombination.[217]

Propargyl Ethers. Rearrangement of propargyl aryl ethers proceeds smoothly in boiling diethylaniline to produce Δ^3-chromenes, **19**.[108,109] That the *o*-allenyl phenol **18** is, indeed, the precursor of the chromene is supported by the isolation of internal Diels-Alder adducts, **20**, from propargyl 2,6-dimethylphenyl ethers[110] and by the fact that *o*-allenyl phenols

[215] A. Wunderli, T. Winkler, H.-J. Hansen, and H. Schmid, unpublished work quoted in ref. 7.

[216] P. Fahrni and H. Schmid, *Helv. Chim. Acta*, **42**, 1102 (1959).

[217] J. Green, S. Marcinkiewicz, and D. McHale, *J. Chem. Soc., C*, **1966**, 1422.

themselves cyclize thermally by a 1,5-hydrogen shift and electrocyclic
ring closure sequence to chromenes.[7]

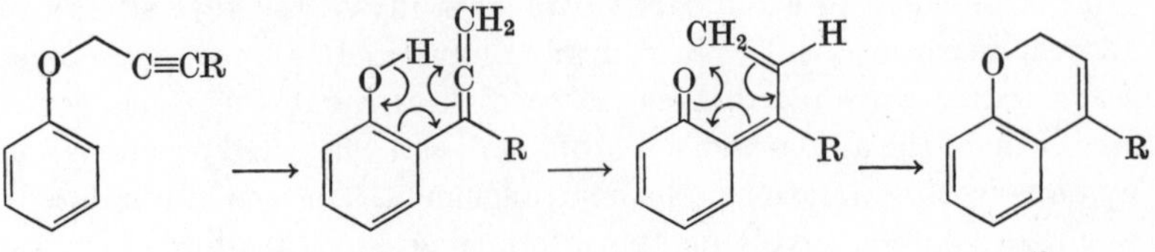

18   19

R = H, CH₃

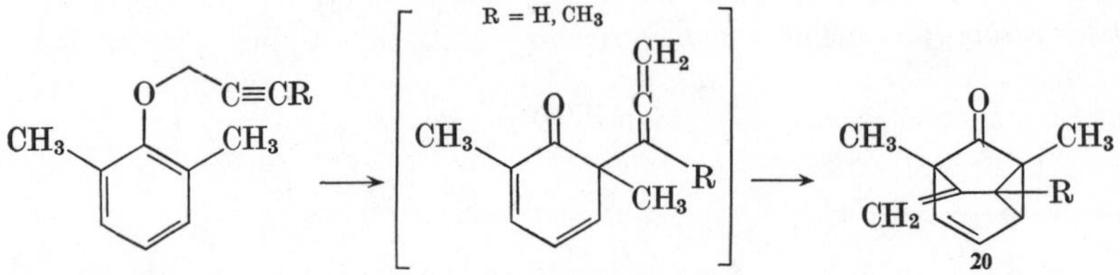

20

R = H, CH₃

An interesting application of the propargyl aryl ether rearrangement is
found in the thermal conversion of 1,4-bis(phenoxy)-2-butyne to the
benzofurobenzopyran **21**.[218] This reaction proceeds through the formation
of the Δ³-chromene followed by a second rearrangement of an allyl aryl
ether and finally coumaran ring closure of the o-allyl phenol.

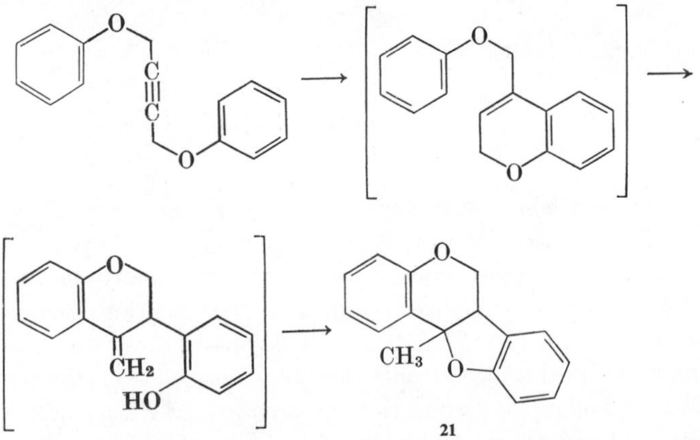

21

The only *para* migration of a propargyl side chain reported thus far is
that of butynyl 2,6-dimethylphenyl ether. In addition to the internal
Diels-Alder adduct **20** (R = CH₃), a small amount of the *p*-rearrangement
product **22** (p. 33) was found.[110]

²¹⁸ B. S. Thyagarajan, K. K. Balasubramanian, and R. B. Rao, *Tetrahedron*, **23**, 1893
(1967).

OH

CH₃ ⟍ ⟋ CH₃

C≡C–CH₃

22

Experimental Conditions. Although many Claisen rearrangements have been accomplished simply by heating the aryl ethers in the temperature range 150–200°, better yields and more consistent results are obtained by using a solvent of the appropriate boiling point.[1] It has been pointed out in the foregoing discussion that the nature of the solvent can strongly affect the product distribution by its influence on the relative rates of rearrangement to available open positions and the extent of secondary reactions; these factors should be carefully considered in choosing the reaction medium. The traditional use of tertiary aromatic amines such as dimethylaniline and diethylaniline[1] appears well justified for most reactions, although dimethylformamide offers advantages in certain systems and other aprotic, polar solvents could well prove equally useful. Reaction temperature and duration also are important and should, of course, be minimized in order to achieve maximum yields of the normal product. The use of trapping agents such as butyric anhydride to capture unusually labile *ortho*-rearrangement products has been successful.[201,219] Isolation of the rearrangement products usually offers no complications; the value of Claisen's alkali[1] for separation of weakly acidic phenolic products from neutral material has already been emphasized.

Aliphatic Claisen Rearrangements

Aliphatic Claisen rearrangements have been successful not only with open-chain systems but also with structures in which the vinyl or the allyl group is part of a ring and with systems in which the allyl portion of the ether is replaced by propargyl or allenyl groups (refs. 116, 120, 126, 127, 133, 151, 220–222). The ether oxygen itself may be part of a ring as in derivatives of partially reduced furans and pyrans.[80, 121, 122, 223] Under certain circumstances the double bond of the allyl moiety may be part of

[219] R. D. H. Murray and M. M. Ballantyne, *Tetrahedron*, **26**, 4667 (1970).
[220] S. Julia, M. Julia, and P. Graffin, *Bull. Soc. Chim. Fr.*, **1964**, 3218.
[221] R. Gardi, R. Vitali, and P. P. Castelli, *Tetrahedron Lett.*, **1966**, 3203.
[222] R. Vitali and R. Gardi, *Gazz. Chim. Ital.*, **96**, 1125 (1966) [*C.A.*, **66**, 28976j (1967)].
[223] S. J. Rhoads and C. F. Brandenburg, *J. Amer. Chem. Soc.*, **93**, 5805 (1971).

an aromatic system; benzyl and furfuryl vinyl ethers have been successfully rearranged (refs. 101, 119, 127, 136, 224, 225).

Acyclic Allyl Vinyl Ethers. Relatively few limitations on the rearrangement of open-chain systems have been reported. Generally, the rearrangements proceed smoothly at moderate temperatures, most conveniently by *in situ* methods, to give good yields of the homoallylic carbonyl compounds.

A competitive elimination in the rearrangement of the intermediate allyl vinyl ether derivatives prepared by the *in situ* method of Ficini and Barbara has been observed in structures of type **23**.[125] The elimination can be minimized by omitting the boron trifluoride catalyst and conducting the reaction at a higher temperature.

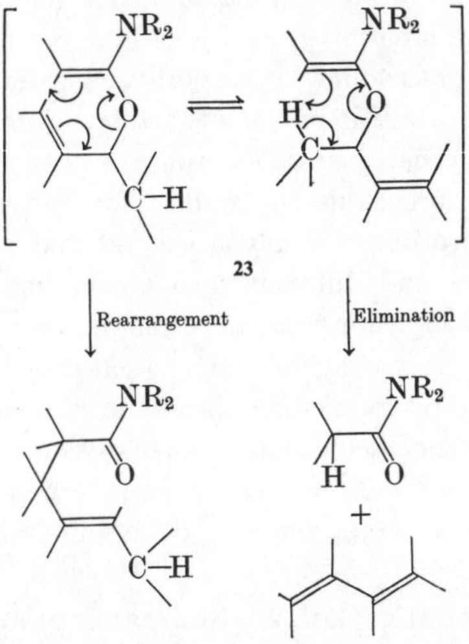

The general method of Thomas, involving sequential Claisen and Cope rearrangements, also is subject to a limitation when, in place of the dienic ether derived from isoprene, the transetherifying agent bears a hydrogen atom on C-2 of the 1-alkoxy-1,3-diene system.[102] In such reactions the initial Claisen rearrangement product (a β,γ-unsaturated aldehyde) may preferentially isomerize to an α,β-unsaturated system incapable of undergoing the Cope rearrangement. For example, when nerol was treated with 1-ethoxy-1,3-butadiene under the conditions of the Thomas *in situ* method, the Claisen rearrangement product **24** partitioned to **25** and **26** in a 4:1 ratio.

[224] W. J. Le Noble, P. J. Crean, and B. Gabrielsen, *J. Amer. Chem. Soc.*, **86**, 1649 (1964).
[225] A. F. Thomas and M. Ozainne, *J. Chem. Soc*, *C*, **1970**, 220.

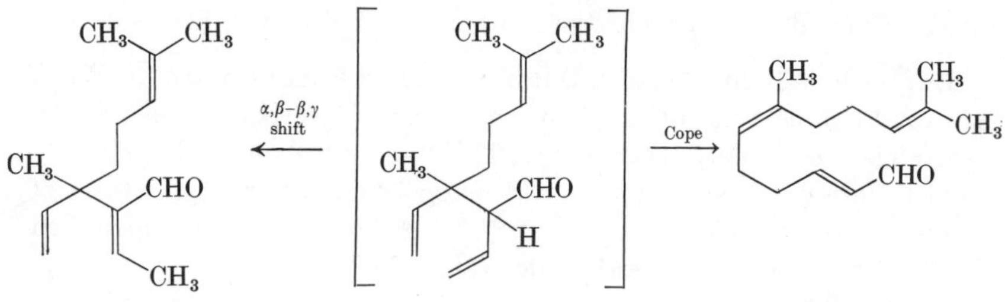

Modified products may be anticipated when the initially formed rearrangement product contains other reactive functional groups. Thus application of the Meerwein-Eschenmoser *in situ* method to the diol **27** produced only a γ-lactone.[226]

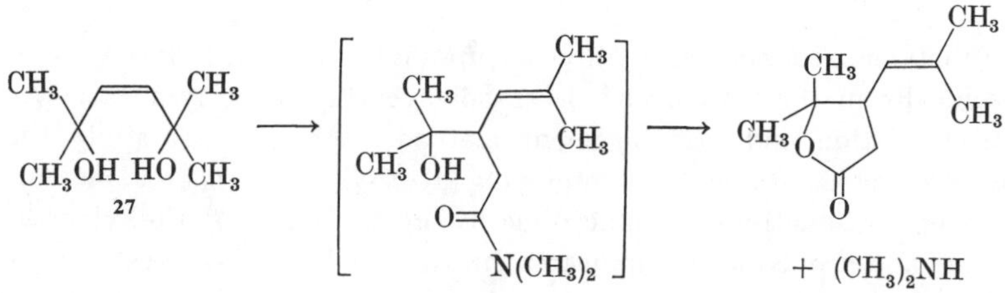

Vinyl divinylmethyl ethers **(28)** and vinyl ethynylvinylmethyl ethers **(29)** in which alternative rearrangement pathways are open have been examined in some detail.[227–232] In general, when two allyl systems are

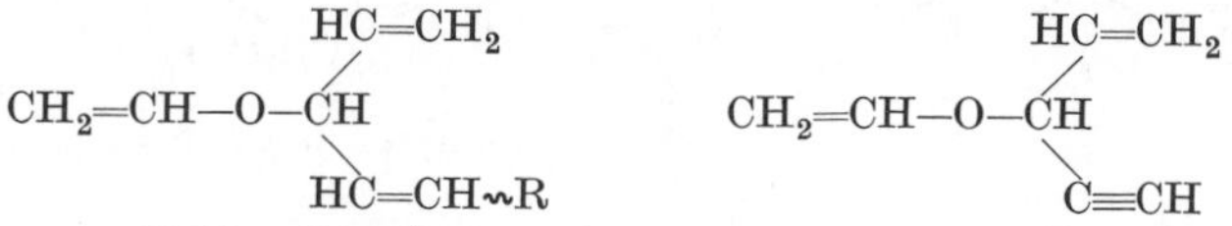

available as in **28**, the major rearrangement course involves the less heavily substituted allyl system; moreover, the selectivity is greater when the substituent, R, is *cis* than when it is *trans*. In systems such as **29**, which pit an allyl against a propargyl group, the rearrangement occurs exclusively through the allyl system to produce the skeletal structure $C\equiv C-C=C-C-C-C=O$. It is noteworthy that rearrangements in systems

[226] W. Sucrow and W. Richter, *Tetrahedron Lett.*, **1970**, 3675.
[227] P. Cresson and L. Lacour, *C.R. Acad. Sci.*, *Ser. C*, **262**, 1157 (1966).
[228] P. Cresson and M. Atlani, *C.R. Acad. Sci.*, *Ser. C*, **262**, 1433 (1966).
[229] P. Cresson and S. Bancel, *C.R. Acad. Sci.*, *Ser. C*, **266**, 409 (1968).
[230] S. Bancel and P. Cresson, *C.R. Acad. Sci.*, *Ser. C*, **268**, 1535 (1969).
[231] S. Bancel and P. Cresson, *C.R. Acad. Sci.*, *Ser. C*, **268**, 1808 (1969).
[232] S. Bancel and P. Cresson, *C.R. Acad. Sci.*, *Ser. C*, **270**, 2161 (1970).

such as **28** and **29** occur with ease in boiling ethyl vinyl ether ($\sim$35°).[233]

Ethers in Which the Allyl Double Bond Is Part of a Ring. When the allyl double bond of an aliphatic system is incorporated in a cycle, rearrangement usually occurs easily. This type of system has been widely used to introduce side chains in a stereospecific manner in structures of interest in the natural products area.[90–94, 234–239] An example is furnished by the formation of the decalone derivative **30**.[234]

30 (75%)

Functionalized angular methyl groups have been introduced stereo-specifically in the octalin and other polycyclic systems by a variety of transvinylation and rearrangement methods.[90–94] A recent study that compares the *in situ* methods employing (a) ethyl vinyl ether and mercuric acetate, (b) dimethylacetamide dimethyl acetal, and (c) triethyl ortho-acetate and propionic acid indicates that all may be used successfully with 3-octalol derivatives to place angular functions at the 5 position.[95]

(a) R = H, (b) R = N(CH$_3$)$_2$, (c) R = OC$_2$H$_5$

Application of these same *in situ* conditions to the hydrindenyl ring system, however, led mainly to elimination with the formation of mixtures

R = *t*-C$_4$H$_9$

[233] S. F. Reed, Jr., *J. Org. Chem.*, **30**, 1663 (1965).
[234] R. F. Church, R. E. Ireland, and J. A. Marshall, *J. Org. Chem.*, **27**, 1118 (1962).
[235] R. E. Ireland and J. A. Marshall, *J. Org. Chem.*, **27**, 1620 (1962).
[236] G. Büchi and J. D. White, *J. Amer. Chem. Soc.*, **86**, 2884 (1964).
[237] H. Muxfeldt, R. S. Schneider, and J. B. Mooberry, *J. Amer. Chem. Soc.*, **88**, 3670 (1966).
[238] F. E. Ziegler and G. B. Bennett, *Tetrahedron Lett.*, **1970**, 2545.
[239] F. E. Ziegler and J. G. Sweeny, *Tetrahedron Lett.*, **1969**, 1097.

of dienes. For this ring system, better results were obtained when the simple vinyl ether was isolated, purified, and then thermolyzed in decalin at 160° or higher.[95]

Only limited success has been realized in attempts to rearrange simple vinyl benzyl ethers. Vinyl benzyl ether itself fails to rearrange,[240] but derivatives in which the aromatic ring bears one or two *meta*-methoxy groups do undergo rearrangement to give the expected products.[224] An

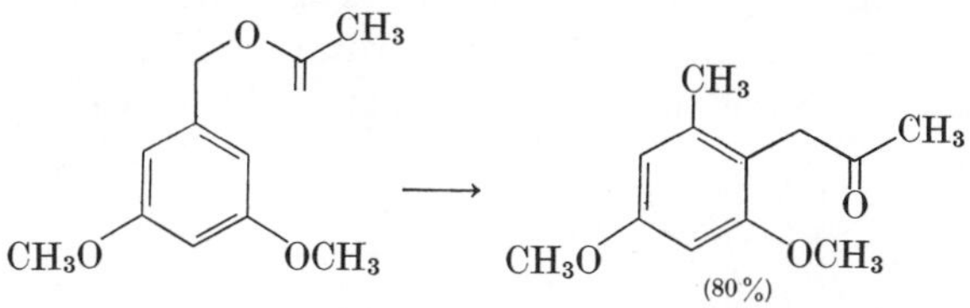

attempt to extend the rearrangement to the corresponding phenyl benzyl ether **31** failed.[241] In 1942, McElvain and co-workers recorded an

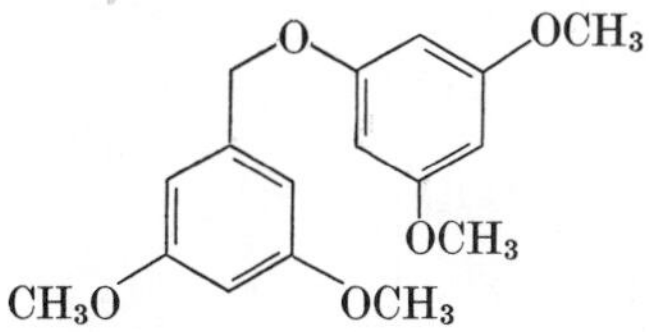

31

in situ rearrangement of an unsubstituted benzyl vinyl ether derivative when the dibenzyl bromoacetal **32** was subjected to dehydrohalogenation.[119] Reminiscent of this early success is the recent report of the use of

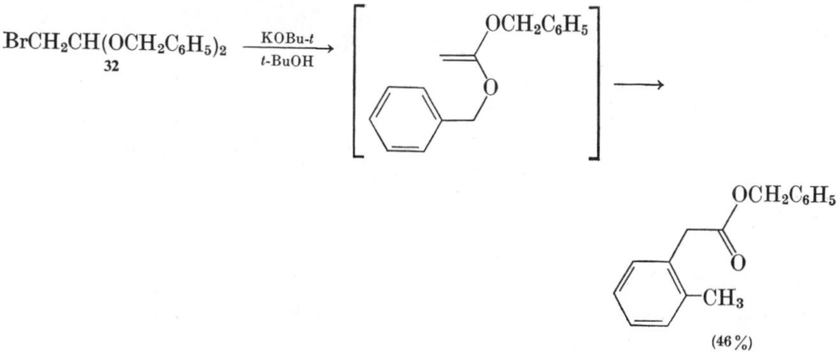

the Meerwein-Eschenmoser method which smoothly brings about the rearrangement of derivatives of benzene, naphthalene, and furan nuclei.[135]

[240] A. W. Burgstahler, L. K. Gibbons, and I. C. Nordin, *J. Chem. Soc.*, **1963**, 4986.
[241] W. J. Le Noble and B. Gabrielsen, *Chem. Ind.* (*London*), **1969**, 378.

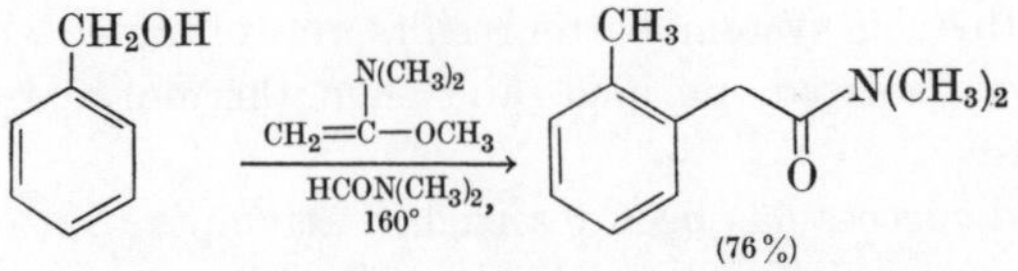

(76%)

Application of the Thomas method to 2-furylmethanol produces the 2-substituted product **33** expected of consecutive Claisen-Cope rearrangements, accompanied by a smaller amount of the 3-substituted derivative **34** which results from a competing aromatization of the first rearrangement product.[225] Similar treatment of 3-furylmethanol leads to the corre-

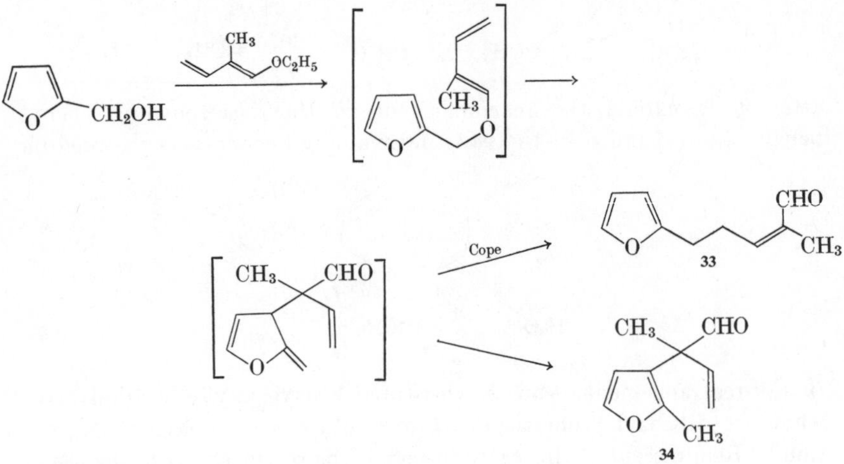

sponding ether **35** (*cis-trans* mixture) which can be induced to rearrange only at elevated temperatures. Application of the same method to 2-thienylmethanol gives only traces of the rearrangement product, **36**.[225]

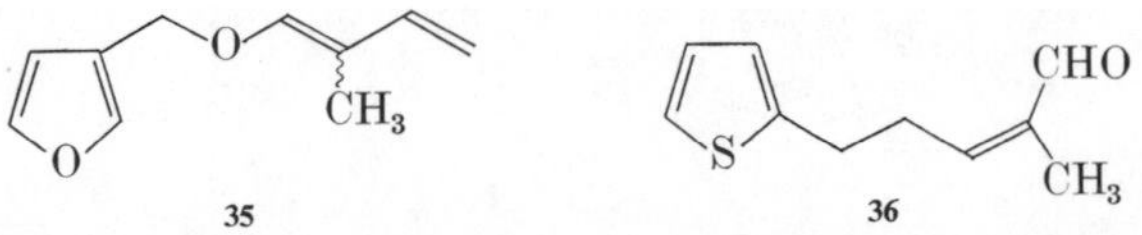

Ethers in Which the Vinyl Double Bond Is Part of a Ring.

Noteworthy in this category are tropolone derivatives. Rearrangement of the allyl ether of 3,5,7-trimethyltropolone gives rise to the normal product **37** in good yield.[242] The latter, on prolonged heating, forms internal Diels-Alder adducts which are analogous to the proposed intermediates in the *ortho-ortho'* rearrangement in the aromatic series. Rearrangement of the

[242] M. M. Al Holly and J. D. Hobson, *Tetrahedron Lett.*, **1970**, 3423.

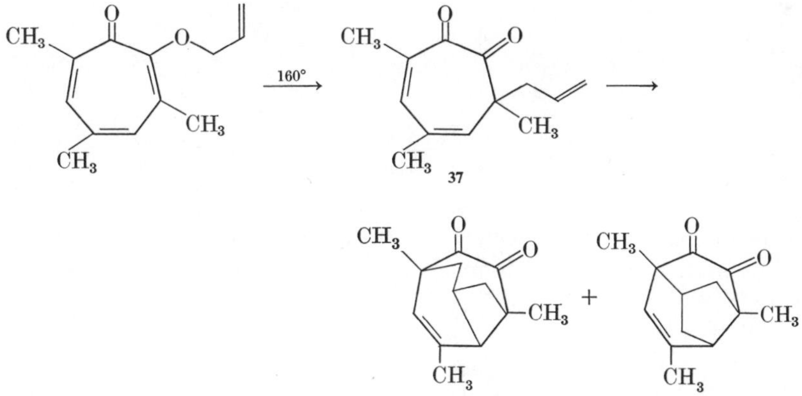

tropolone derivative **38** was reported to produce the *"para"* rearrangement product in poor yield.[243]

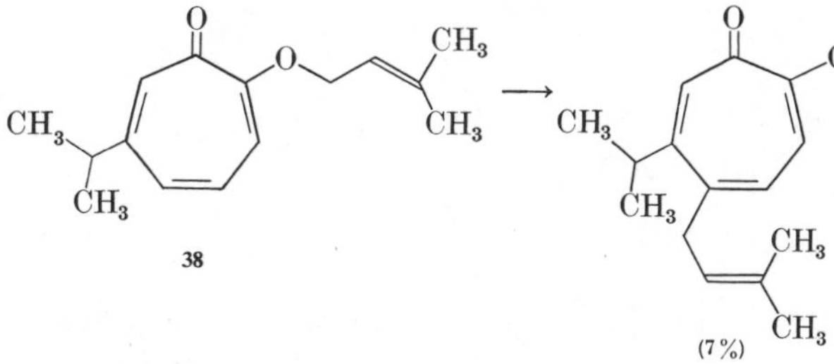

Allyloxytropilidene derivatives also rearrange by Claisen processes after a series of preliminary 1,5-hydrogen shifts to produce the allyl vinyl ether system.[244] 7-Allyloxytropilidene undergoes a remarkable sequence of thermal reactions at 200° to give rise, ultimately, to the tricyclic systems **39** and **40** in 83% yield.[245] Their formation may be explained by a series

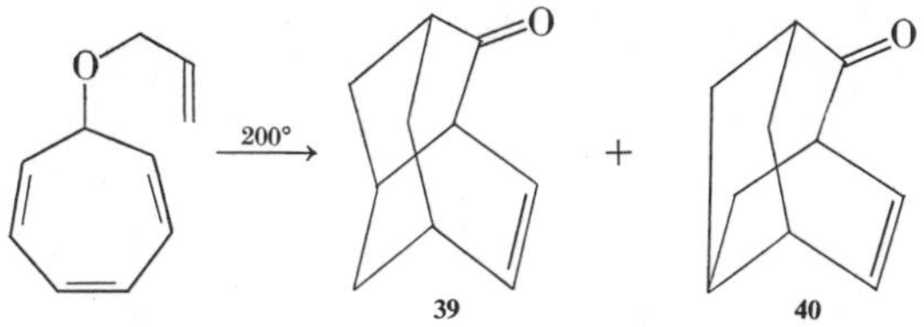

of 1,5-hydrogen shifts followed by a Claisen rearrangement and internal Diels-Alder reactions.

[243] Y. Kitahara and M. Funamizu, *Bull. Chem. Soc. Jap.*, **31**, 782 (1958).
[244] E. Weth and A. S. Dreiding, *Proc. Chem. Soc.*, **1964**, 59.
[245] C. A. Cupas, W. Schumann, and W. E. Heyd, *J. Amer. Chem. Soc.*, **92**, 3237 (1970).

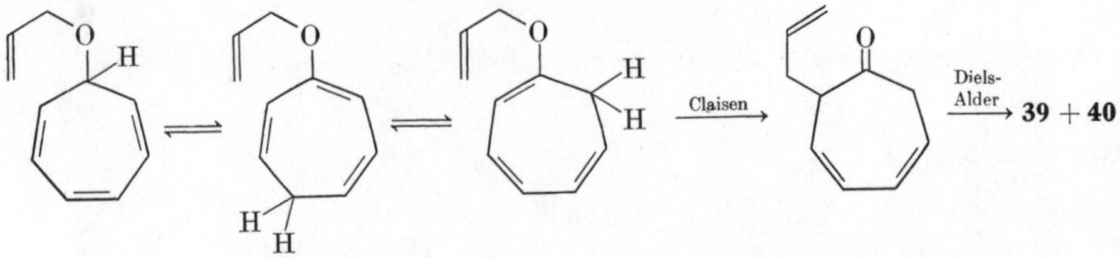

An interesting rearrangement involving imine-enamine tautomers in a cyclic system has been observed in O-allyl and O-cinnamylhexanolactims.[246]

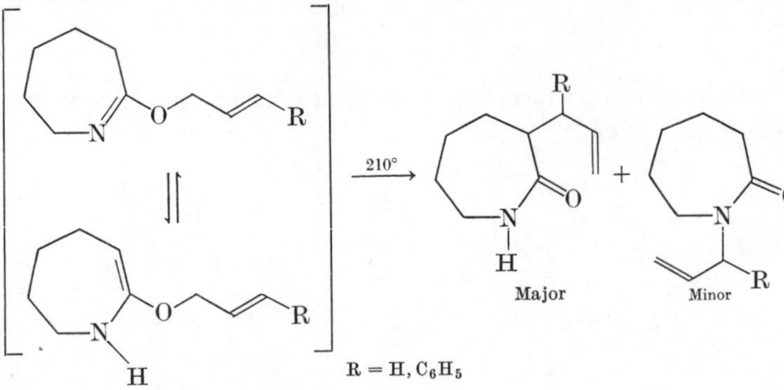

Synthetically useful rearrangements of systems in which both the vinyl group and the ether oxygen are incorporated in a ring have been developed recently. Dihydropyran derivatives of the general structure **41** rearrange to acyl cyclohexenes when pyrolyzed in sealed tubes at 200–250° or in a flow system at 400–450°.[80] This method makes accessible certain

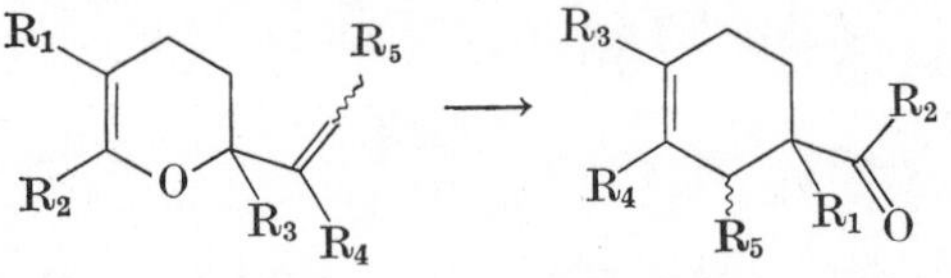

41

cyclohexene derivatives that are not available by the usual Diels-Alder route from dienes and unsaturated carbonyl compounds. An *in situ* generation and rearrangement of tetrahydrofuran derivatives like **43** has been utilized to prepare 4-cycloheptenone derivatives.[121, 122] The accessibility of the 2-bromomethyl-5-vinyltetrahydrofuran precursors, **42**, makes this an attractive route to a variety of substituted 4-cycloheptenones.

246 D. St. C. Black and A. M. Wade, *Chem. Commun.*, **1970**, 871.

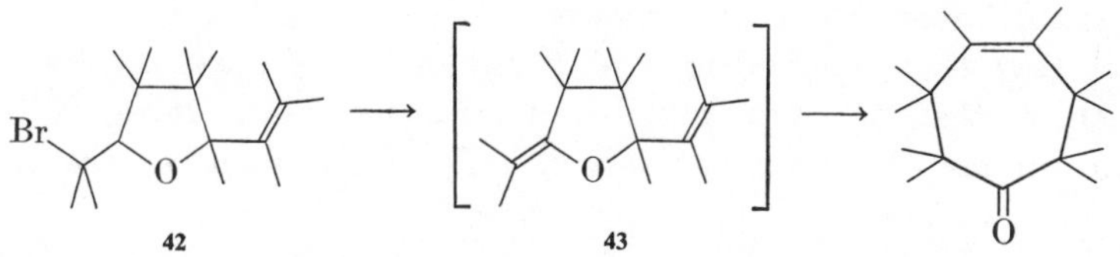

42   43

Propargyl and Allenyl Vinyl Ethers. Propargyl vinyl ethers rearrange readily to form homoallenyl carbonyl compounds (refs. 116, 120, 126, 128a, 133, 151, 220–222, 247, 248). Since the latter are easily isomerized by base, the sequence of reactions can be employed for the preparation of conjugated dienic carbonyl systems. The synthesis of ψ-ionone **(44)** is a good example.[133] This general scheme has also been used

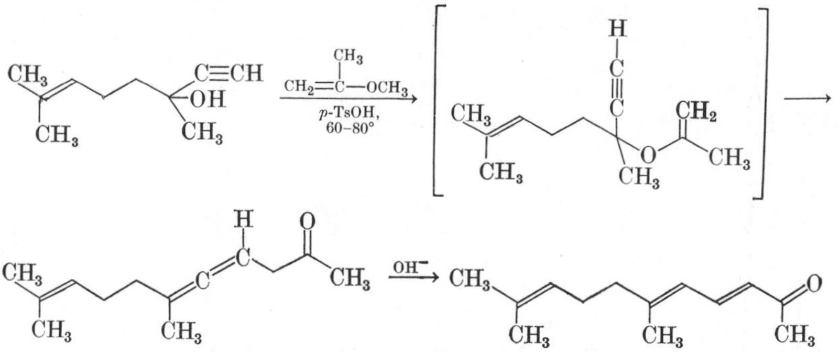

to prepare conjugated dienic amides from the corresponding allenyl amides.[126]

The rearrangement of allenyl vinyl ether derivatives has been reported by Ficini and Pouliquen.[127] The boron trifluoride-catalyzed addition of (1-methylallenyl)carbinol to ynamines results in the isolation of moderate yields of the rearrangement products **45**.

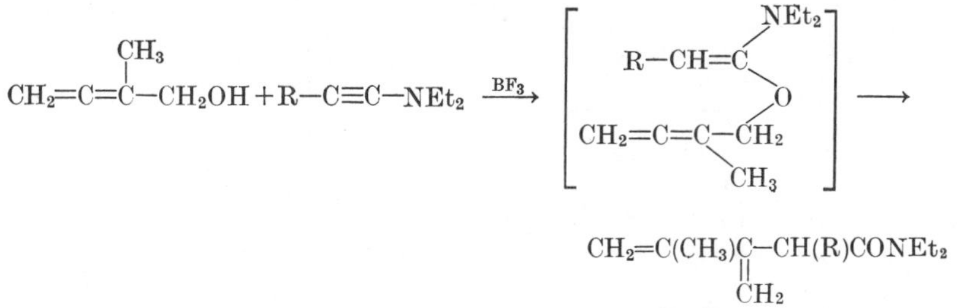

$$CH_2=C(CH_3)\underset{\underset{CH_2}{\|}}{C}-CH(R)CONEt_2$$

45 (R = CH_3, C_6H_5)

[247] J. Corbier, P. Cresson, and P. Jelenc, *C.R. Acad. Sci., Ser. C*, **270**, 1890 (1970).
[248] J. K. Crandall and G. L. Tindell, *Chem. Commun.*, **1970**, 1411.

Retro-Claisen Rearrangements. A few retro-Claisen rearrangements have been reported in aliphatic systems. Excellent thermal conversions of the Diels-Alder adducts of various fulvenes, **46**, to dihydropyran systems, **47**, have been realized.[249] The vinylcyclopropanecarboxal-

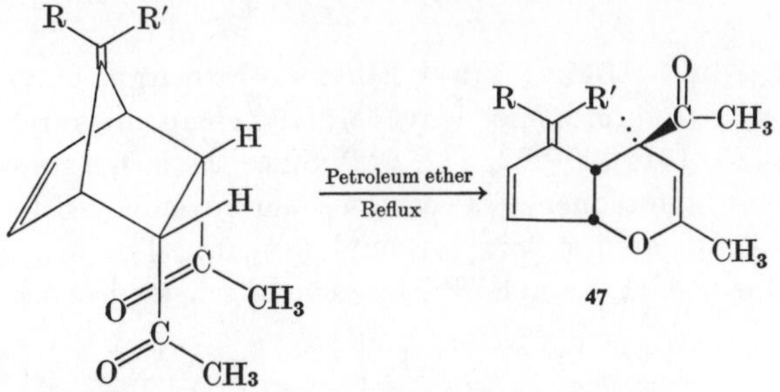

dehyde systems **48**[78] and **49**[79] have been shown to exist in equilibrium with their retro-Claisen isomers at room temperature.

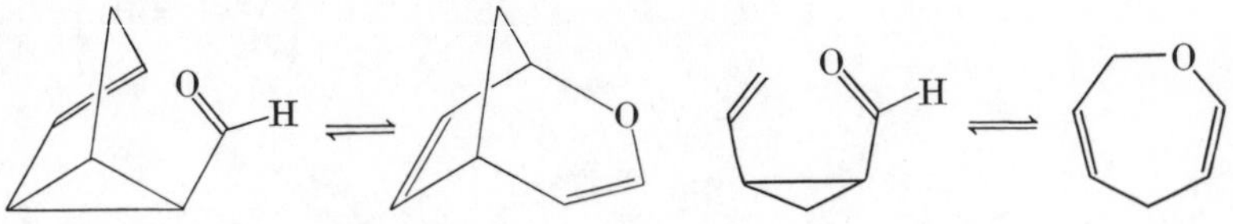

Amino-Claisen Rearrangements

In general, amino-Claisen rearrangements are considerably less facile than those of their oxygen counterparts. In aromatic systems, only two purely thermal rearrangements have been recorded, that of N-allyl-1-naphthylamine[29] and that of the phenylaziridine **50**.[250] Simple N-allyl-

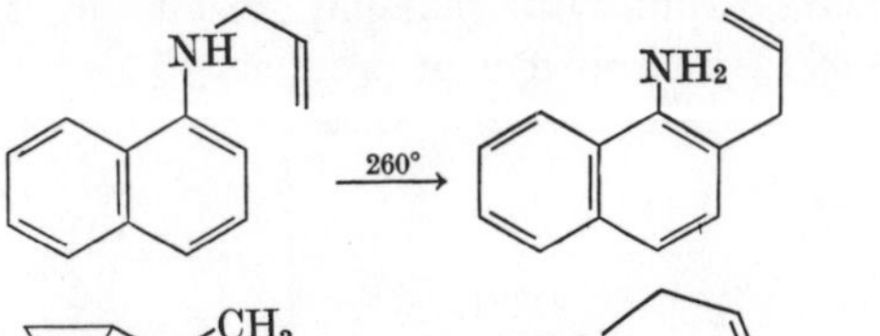

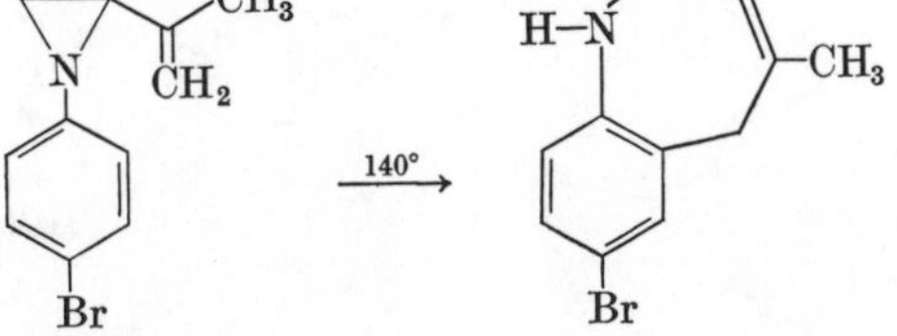

[249] M. T. Hughes and R. O. Williams, *Chem. Commun.*, **1968**, 587.
[250] P. Scheiner, *J. Org. Chem.*, **32**, 2628 (1967).

aniline derivatives are recovered unchanged after thermal treatment,[251] or they undergo fragmentation at more elevated temperatures.[1] The success of the rearrangements mentioned has been attributed to the lowering of the activation energy for the concerted rearrangement process below that of the cleavage reaction by the enhanced olefinic character of the 1,2 bond in the naphthalene derivative and by the relief of ring strain in the aziridine.

More successful have been rearrangements in aliphatic systems. Various N-allylic enamine derivatives have been rearranged thermally, and the reactions show the characteristics of concerted [3,3]sigmatropic processes.[53, 125, 247] The quantitative conversion of the enamine **51** to the homoallylic imine proceeds with the high stereoselectivity associated with the preferred chairlike transition-state geometry.[53] A related rearrangement is that of the allyl amine **52** to the amidine.[125] The N-propargyl

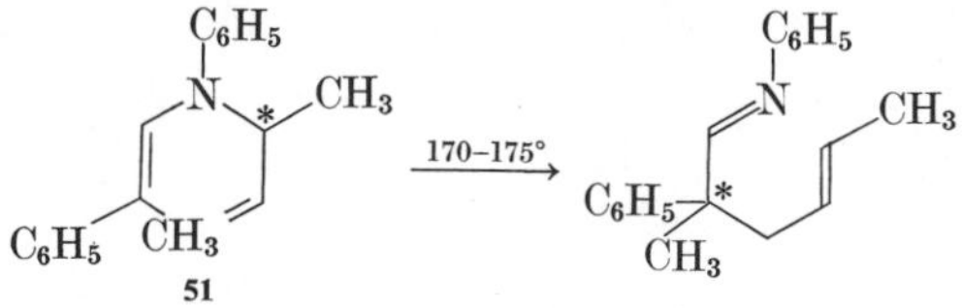

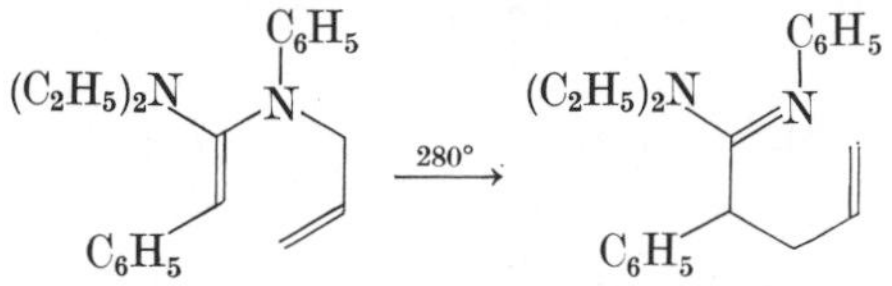

enamine **53** rearranges quantitatively to the homoallenyl imine, but it is noteworthy that the reaction gave only poor conversion or failed completely with substituted propargyl derivatives.[252]

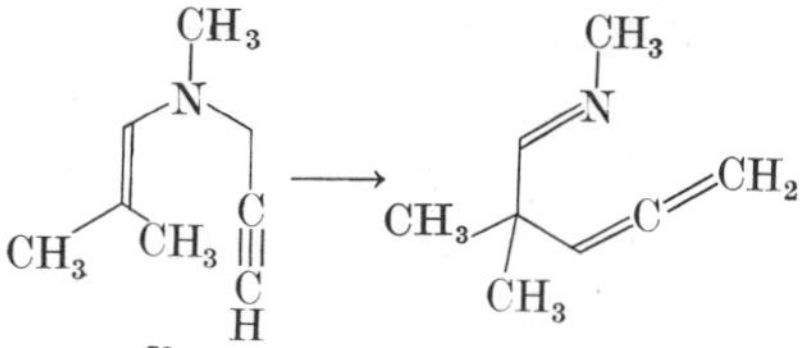

The temperatures required for amino-Claisen rearrangements usually are about 100–150° higher than those which suffice for rearrangement of the corresponding oxygen compounds. The lower activation energy

[251] M. Elliot and N. F. Janes, *J. Chem. Soc.*, **1967**, 1780.
[252] J. Corbier and P. Cresson, *C.R. Acad. Sci., Ser. C*, **270**, 2077 (1970).

associated with the allyl (or propargyl) vinyl ether rearrangement is demonstrated by the systems **54** and **55** in which alternative rearrangement routes over nitrogen and oxygen are available. In each case, the only product observed was that formed by rearrangement through the ether system.[247]

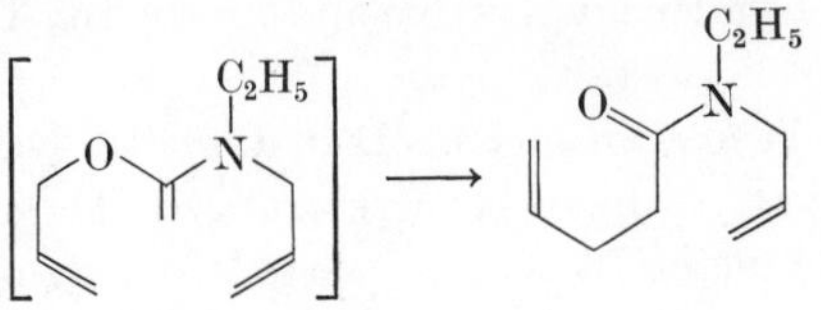

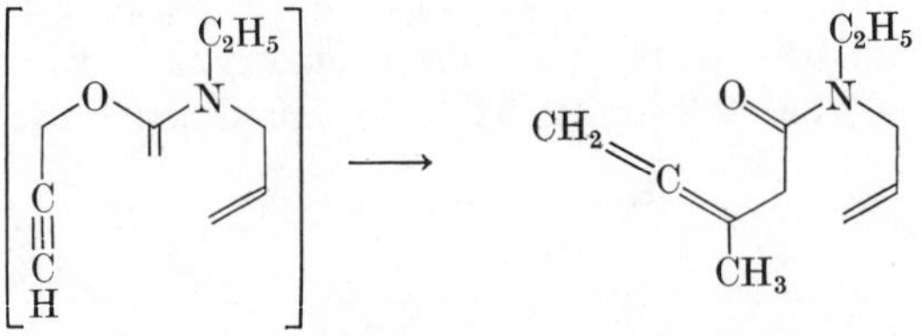

Amino-Claisen rearrangements have been observed in a few N-heterocyclic systems.[253-256] Pyrazolinone derivatives of the general structure **56**

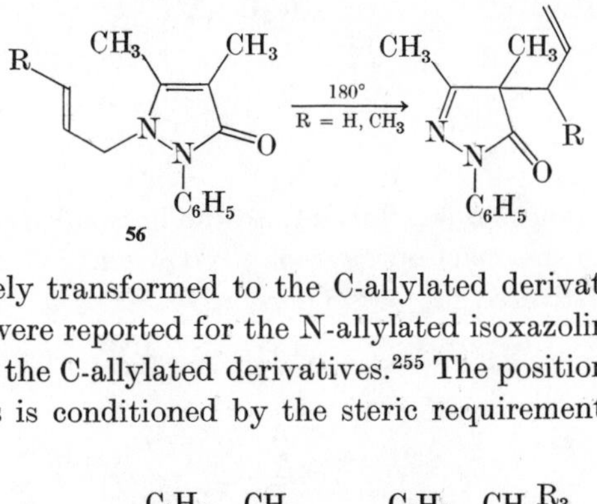

are quantitatively transformed to the C-allylated derivatives at 180°.[254] Similar results were reported for the N-allylated isoxazolinones **57**, which equilibrate with the C-allylated derivatives.[255] The position of equilibrium in such systems is conditioned by the steric requirements of the allylic group.

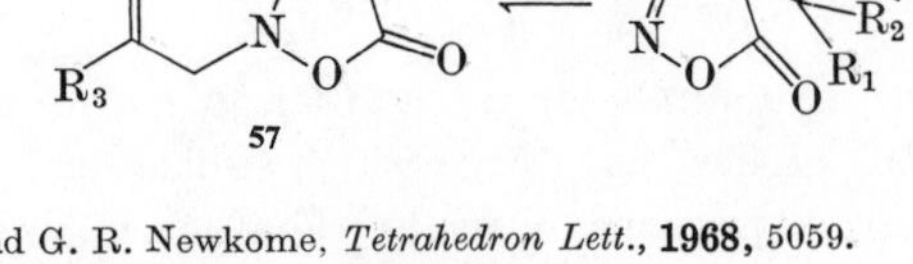

[253] R. K. Hill and G. R. Newkome, *Tetrahedron Lett.*, **1968,** 5059.

[254] Y. Makisumi, *Tetrahedron Lett.*, **1966,** 6413.

[255] Y. Makisumi and T. Sasatani, *Tetrahedron Lett.*. **1969,** 543.

[256] B. A. Otter, A. Taube, and J. J. Fox, *J. Org. Chem.*, **36,** 1251 (1971).

A useful application of the aliphatic amino-Claisen rearrangement to the preparation of derivatives of quinoline and indolenine has been described.[253] The anhydrobases, **58** and **59**, which become readily available by the action of dilute alkali on the corresponding quaternary salts, rearrange to cyclic imines in moderate to excellent yield. The reaction sequence furnishes a method of alkylating an active α-carbon atom without recourse to unusually strong base.

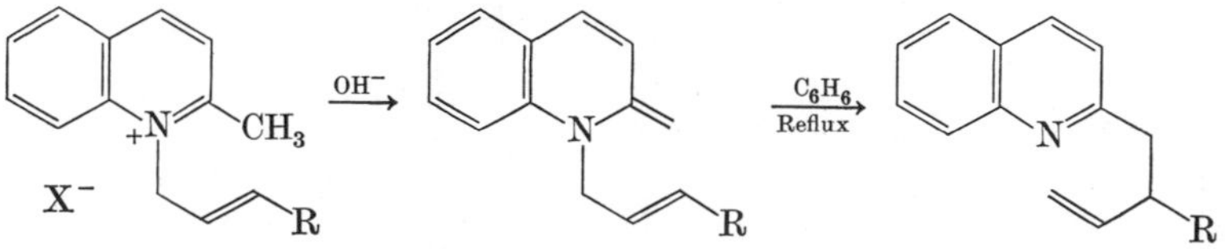

58 (R = H, CH$_3$, C$_6$H$_5$)

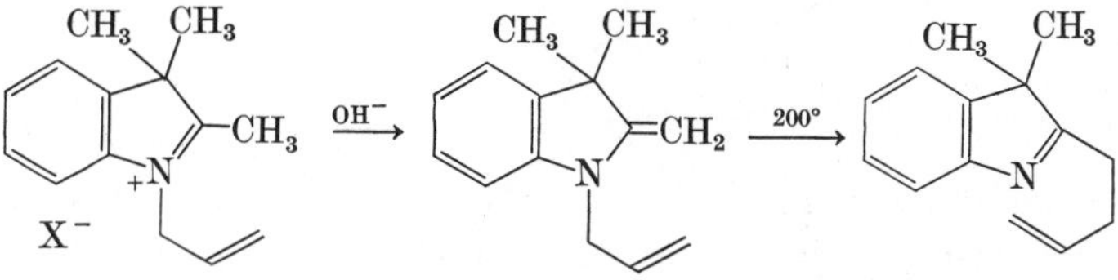

59

Only limited success has been realized in the *in situ* rearrangements of quaternary salts prepared by treatment of tertiary enamines with allyl or propargyl halides.[252, 257-261] The *in situ* rearrangement of the quaternary salt **60**, formed from the corresponding enamine and crotyl bromide in refluxing acetonitrile, yielded the aldehyde **61** after hydrolysis.[257] However, the aldehydic hydrolysis product expected from the rearrangement of the N-allyl quaternary ion is often accompanied by an isomeric aldehyde which arises from a competing C-allylation of the original enamine

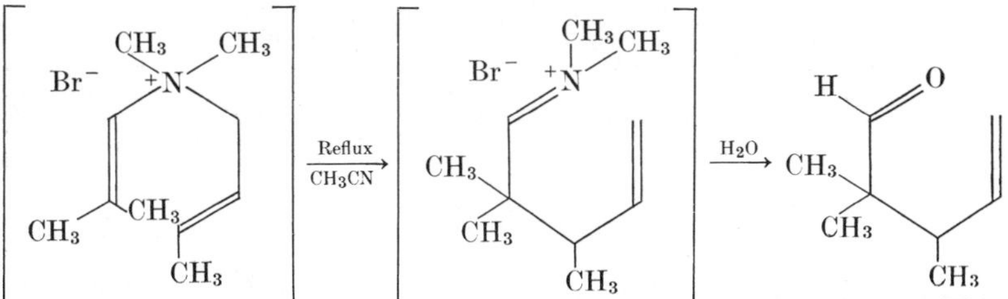

[257] K. C. Brannock an d R. D. Burpitt, *J. Org. Chem.*, **26**, 3576 (1961).

[258] P. Cresson and J. Corbier, *C.R. Acad. Sci.*, Ser. C, **268**, 1614 (1969).

[259] G. Opitz, *Ann. Chem.*, **650**, 122 (1961).

[260] G. Opitz and H. Mildenberger, *Ann. Chem.*, **649**, 26 (1961).

[261] G. Opitz, H. Hellmann, H. Mildenberger, and H. Suhr, *Ann. Chem.*, **649**, 36 (1961).

(Scheme 7).[252, 258–261] This alternative alkylation route limits the synthetic usefulness of the rearrangement. Generally, the yields and purities of carbonyl compounds obtained by the aliphatic amino-Claisen rearrangement in acyclic enamines and ammonium derivative are inferior to those

SCHEME 7

derived from the corresponding rearrangements of allyl and propargyl vinyl ethers. From a preparative point of view, then, the oxygen systems are usually preferred.

Thio-Claisen Rearrangements

The thermal behavior of allyl phenyl sulfides, like that of N-allyl-anilines, departs from the usual pattern observed in the oxygen system. The sulfur analogs also exhibit high thermal stability but undergo cleavage when heated neat at about 300°.[262, 263] In solution, preferably in quinoline or diethylaniline, rearrangement does occur.[262, 264, 265] Allyl phenyl sulfide is slowly transformed in boiling quinoline to a mixture of equal parts of 2-methylthiacoumaran and thiachroman.

Despite earlier conflicting data, it now seems established that *ortho*-allyl thiophenol (62) is indeed the initial product of this rearrangement but

[262] H. Kwart and M. H. Cohen, *J. Org. Chem.*, **32**, 3135 (1967).
[263] H. Kwart and E. R. Evans, *J. Org. Chem.*, **31**, 413 (1966).
[264] H. Kwart and M. H. Cohen, *Chem. Commun.*, **1968**, 319.
[265] H. Kwart and C. M. Hackett, *J. Amer. Chem. Soc.*, **84**, 1754 (1962).

that under the rearrangement conditions it cyclizes rapidly to the observed products.[263] Trapping experiments permitted the isolation of the

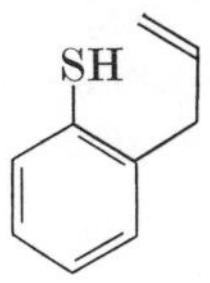

62 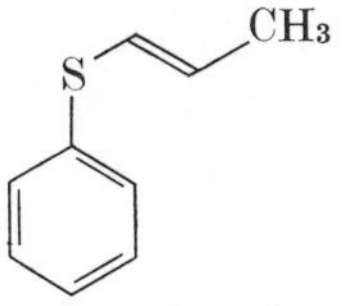   63

thiophenol as the corresponding methyl sulfide; moreover, it has been shown that the thiophenol **62** (prepared independently) produces in refluxing quinoline the same products in the same ratio as does allyl phenyl sulfide.[266] The formation of the cyclized products has been rationalized as the result of competitive ionic and radical additions of the thiol function to the allylic double bond in **62**.[267] The possibility that the thiacoumaran derivative arises from the isomeric propenyl phenyl sulfide (**63**) may be discounted since the sulfide is recovered unchanged under the reaction conditions.[265] There is disagreement concerning the interconvertibility of 2-methylthiacoumaran and thiachroman.[263, 268]

Allyl phenyl selenide has been reported to yield 2-methylselenacoumaran in refluxing quinoline.[269]

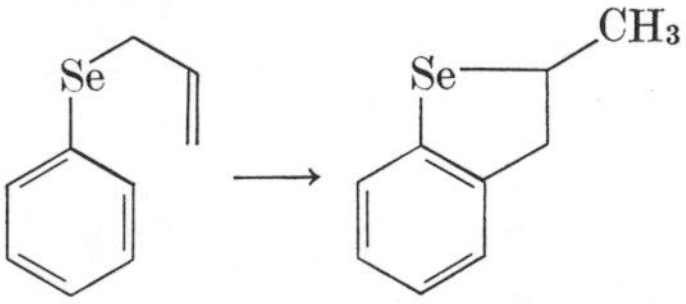

There has been one report on the thermal behavior of propargyl phenyl sulfides.[270] When held at 250° for 30 minutes in quinoline, propargyl phenyl sulfide gives rise to the cyclized derivatives shown in the accompanying reaction. Complications arising from propargyl-allenyl isomerization in the

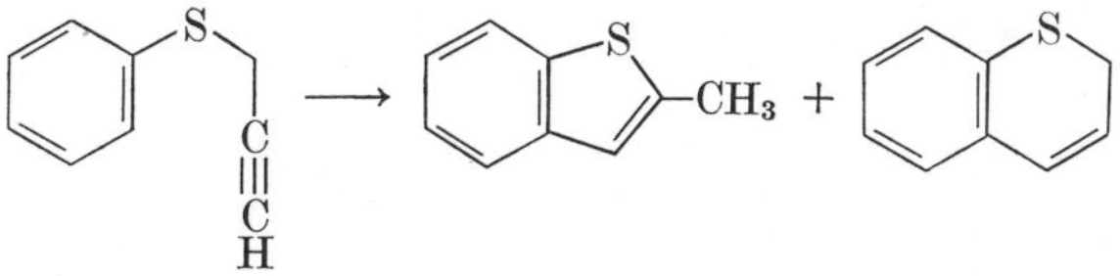

[266] H. Kwart and J. L. Schwartz, *Chem. Commun.*, **1969**, 44.

[267] Y. Makisumi and A. Murabayashi, *Tetrahedron Lett.*, **1969**, 2453.

[268] C. Y. Meyers, C. Rinaldi, and L. Bonoli, *J. Org. Chem.*, **28**, 2440 (1963).

[269] E. G. Kataev, G. A. Chmutova, A. A. Musina, and A. P. Anatas'eva, *Zh. Org. Khim.*, **3**, 597 (1967). [*C.A.*, **67**, 11354c (1967)].

[270] H. Kwart and T. J. George, *Chem. Commun.*, **1970**, 433.

side chain have been detected in this reaction as well as in the rearrangement of the corresponding 2-butynyl sulfide.[270]

Rearrangements of allyl and propargyl sulfides of heterocyclic nuclei appear to proceed more readily and with fewer complications than those of the carbocyclic aromatics. Allyl 4-quinolyl sulfides produce the cyclized products, **64**, in good yield when heated alone at 200° for 1 hour.[271] The

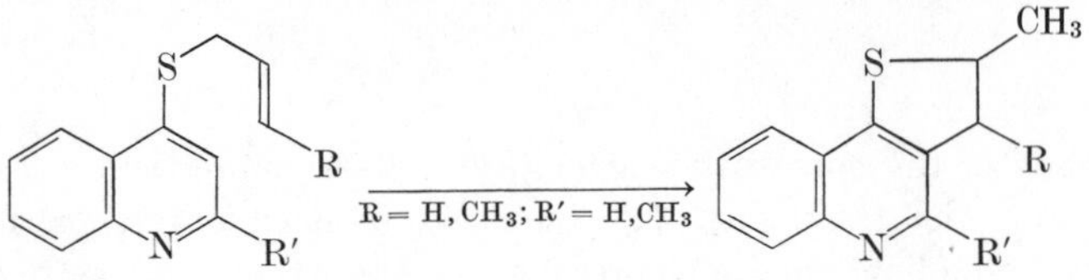

expected Claisen products, the thiones, **65**, were not detected, but when the rearrangement of allyl 4-quinolyl sulfide was conducted in the presence of butyric anhydride, the thione could be trapped as the ester, **66**, in 87% yield.[272]

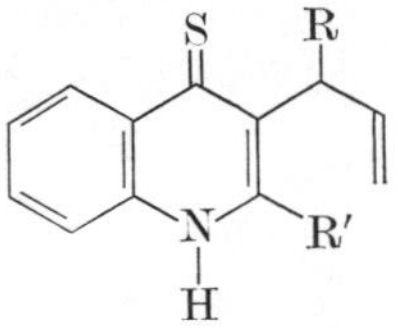

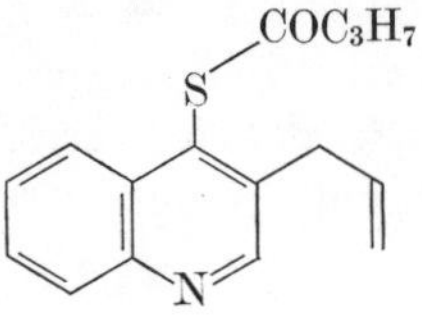

Similar results were recorded for allyl 3-quinolyl derivatives, illustrated by the accompanying example.[267, 273]

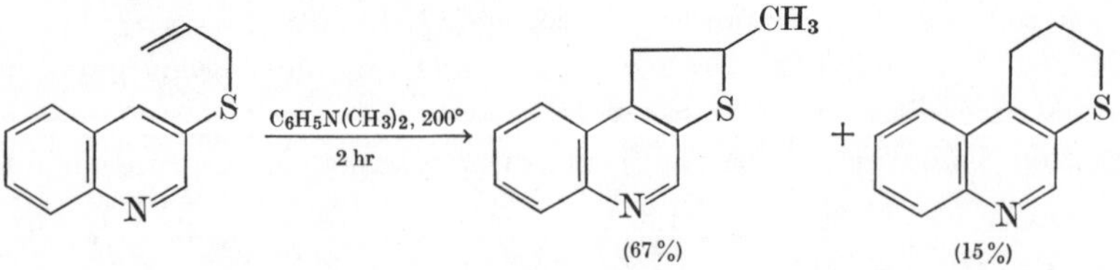

Studies of the rearrangements of 2- and 3-thienyl propargyl sulfides in a variety of solvent systems have shown that the product distribution is strongly affected by the reaction medium.[274, 275] For example, rearrange-

[271] Y. Makisumi, *Tetrahedron Lett.*, **1966**, 6399.
[272] Y. Makisumi and A. Murabayashi, *Tetrahedron Lett.*, **1969**, 1971.
[273] Y. Makisumi and A. Murabayashi, *Tetrahedron Lett.*, **1969**, 2449.
[274] L. Brandsma and H. J. T. Bos, *Rec. Trav. Chim. Pays-Bas*, **88**, 732 (1969).
[275] L. Brandsma and D. Schuijl-Laros, *Rec. Trav. Chim. Pays-Bas*, **89**, 110 (1970).

ment of the sulfide **67** in hexamethylphosphoramide proceeded in excellent yield to give only the thienothiapyran **68**.[274] In dimethyl sulfoxide con-

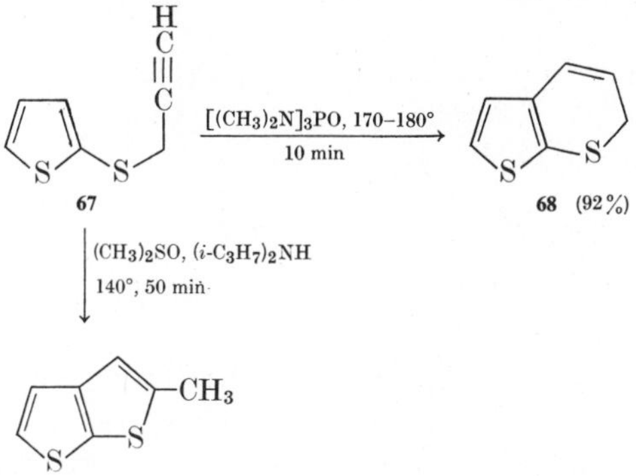

taining a small amount of diisopropylamine, rearrangement of the sulfide proceeded less well (53%) but gave a mixture consisting of three parts of the thienothiophene **69** and one part of the thienothiapyran **68**.[275]

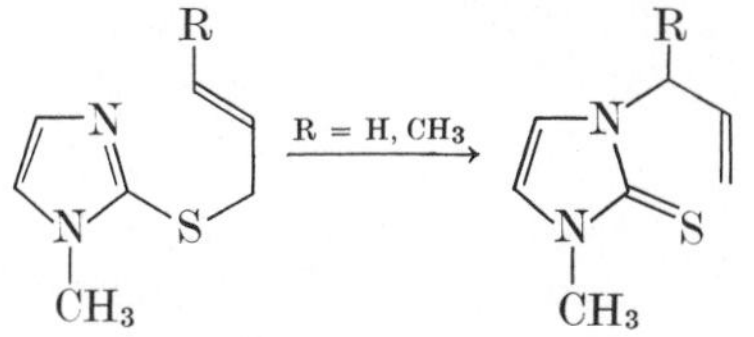

Rearrangements of 2-allylthio derivatives of imidazoles **70** have been reported to give the expected N-allyl 2-thiones.[276] Similar findings have been reported for indole derivatives of the general structures **71** and **72**.[34]

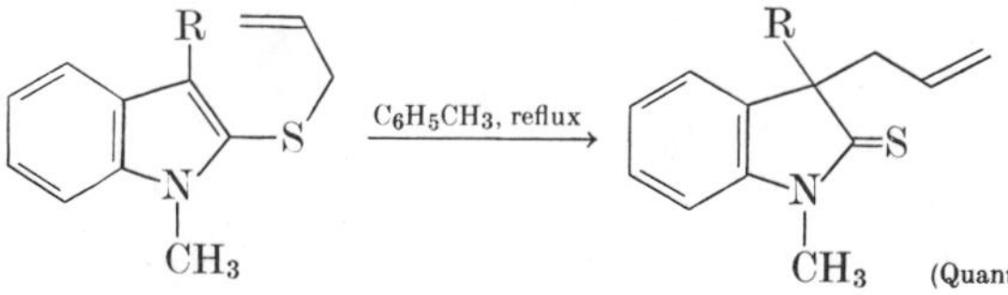

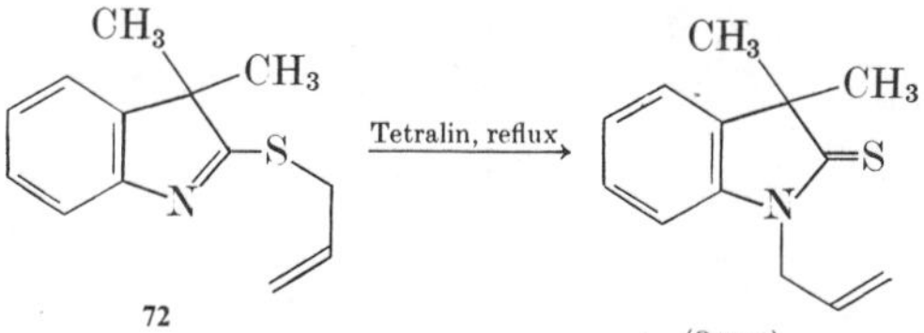

---

[276] K. M. Krivozheiko and A. V. El'tsov, *Zh. Org. Khim.*, **4**, 1114 (1968) [*C.A.*, **69**, 52070s (1968)].

Both allyl and propargyl derivatives rearranged normally in such systems. Allyl 2-indolyl sulfonium salts, exemplified by **73**, also have been rearranged successfully.[277]

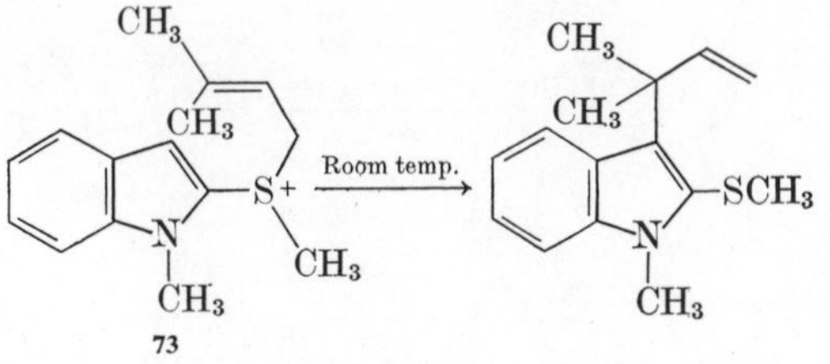

73

Aliphatic thio-Claisen rearrangements of allyl and propargyl sulfides proceed with ease. Propargyl vinyl sulfide rearranges in hexamethylphosphoramide in the presence of pyridine to give 2H-thiapyran, in 80% conversion.[278] The pyridine presumably catalyzes the ring closure of the

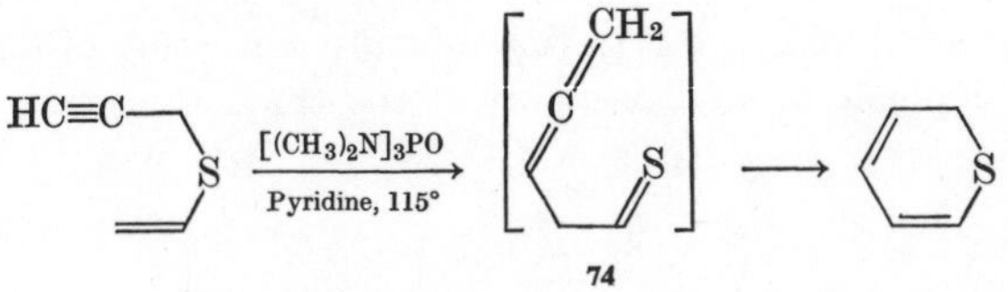

74

first-formed homoallenyl thione **74**. Thio-Claisen rearrangements of allyl vinyl sulfide derivatives of the general structure **75** readily occur *in situ*[279] in analogy to the Meerwein-Eschenmoser[135] and the Johnson[97] methods for the corresponding oxygen systems. Propargyl vinyl sulfide derivatives

75

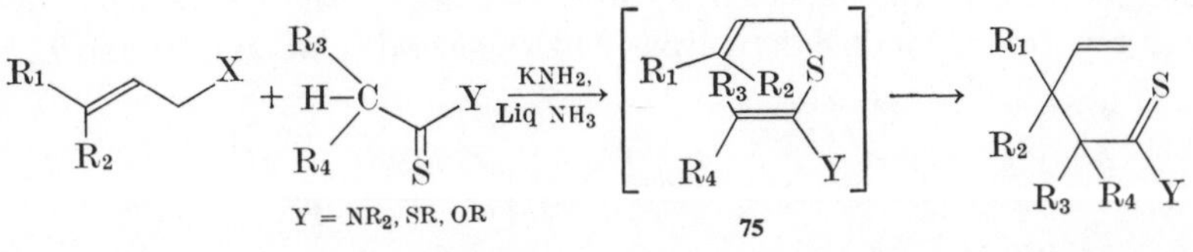

of the same type rearrange at about 100° to form allenyl dithioesters **76**.[280] On further heating or treatment with base the latter furnish 2H-thiapyrans and substituted thiophenes. The allenyl vinyl sulfide derivatives **77** also rearrange at moderate temperatures to give fair yields of the alkynyl dithioesters.[281]

[277] B. W. Bycroft and W. Landon, *Chem. Commun.*, **1970**, 967.

[278] L. Brandsma and P. J. W. Schuijl, *Rec. Trav. Chim. Pays-Bas*, **88**, 30 (1969).

[279] P. J. W. Schuijl and L. Brandsma, *Rec. Trav. Chim. Pays-Bas*, **87**, 929 (1968).

[280] P. J. W. Schuijl, H. J. T. Bos, and L. Brandsma, *Rec. Trav. Chim. Pays-Bas*, **88**, 597 (1969).

[281] P. J. W. Schuijl and L. Brandsma, *Rec. Trav. Chim. Pays-Bas*, **88**, 1201 (1969).

$$R = H, C_2H_5$$

In current studies, the usefulness of which lies in their extension to spiroannulation methods, Corey and Schulman have employed a mercuric oxide-promoted Claisen rearrangement of vinyl allyl sulfides exemplified by the reaction of the sulfide **78**. The products are isolated as the aldehydes.[282]

Cope Rearrangements

Thermal reorganizations in all-carbon bis(allylic) systems have their greatest synthetic utility when one of the isomeric components is strongly favored at equilibrium. The driving force responsible for shifting the equilibrium in a given direction may be an increase in conjugative interactions, the relief of ring strain, or, in more complex cyclic systems, subtle differences in conformational stabilities. In the oxy-Cope rearrangement, the formation of an enolic product provides the thermodynamic drive which makes this variation of the rearrangement especially attractive for preparative purposes. The examples included in the tabulation of Cope rearrangements have been selected with an eye to their synthetic usefulness; consequently, some reactions involving unfavorable equilibria achieved from difficultly accessible starting materials have been omitted. Degenerate Cope rearrangements also are omitted for the same reason,

[282] E. J. Corey and J. I. Shulman, *J. Amer. Chem. Soc.*, **92**, 5522 (1970).

although it seems appropriate here to draw attention to the fundamentally important studies of such rearrangements in fluxional molecules.[61, 283–286]

Rearrangements of 1,5-Hexadienes. In the original studies carried out by Cope and co-workers, the 1,5-hexadiene systems were constructed to provide a driving force of double-bond conjugation with cyano, carbethoxy, or phenyl groups.[2, 138–140, 143] Rearrangements in such systems occur readily in the range 150–200° in good to excellent yields when the starting materials are heated alone for short periods as illustrated for the malonic ester derivative **79**.[139] The vinyl double bond of the

vinylallylmalonic acid derivative may be part of a ring as in the indene derivative **80**,[140] but the reaction fails when the vinyl group is part of a benzene or naphthalene nucleus.[141, 142, *] Conia and co-workers have extended these principles to systems in which the activating group is ketonic;[144, 146–150] rearrangement then gives rise to $\alpha,\beta\text{-}\epsilon,\zeta$-dienic ketones,

which in turn may undergo deconjugation and enolene cyclization to 2-acyl-3-alkylcyclopentenes. For example, 3-isopropenyl-3-methyl-5-hexen-2-one (**81**) rearranges upon heating to a mixture of the $\alpha,\beta\text{-}\epsilon,\zeta$-dienic ketones, **82** and **83**, and the deconjugated isomer, **84**. Further heating of this mixture of equilibrated ketones results in quantitative and stereospecific conversion to the cyclopentene **85**.[146]

---

* Under strongly basic conditions, an "aromatic Cope" rearrangement has been realized in the equilibration of butenylbenzenes and o-propenyltoluenes. W. von E. Doering and R. A. Bragole, *Tetrahedron*, **22**, 385 (1966).

283 L. Birladeanu, D. L. Harris, and S. Winstein, *J. Amer. Chem. Soc.*, **92**, 6387 (1970) and references therein.

284 J. B. Lambert, *Tetrahedron Lett.*, **1963**, 1901.

285 L. A. Paquette, T. J. Barton, and E. P. Whipple, *J. Amer. Chem. Soc.*, **89**, 5481 (1967).

286a J. S. McKennis, L. Brener, J. S. Ward, and R. Pettit, *J. Amer. Chem. Soc.*, **93**, 4957 (1971).

286b H. Hopf, *Chem. Ber.*, **104**, 1499 (1971).

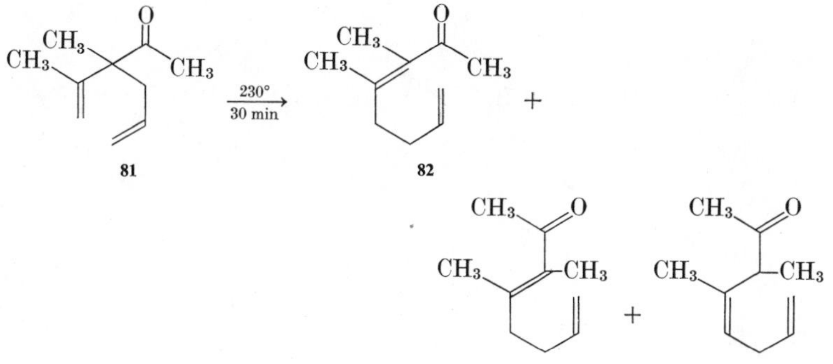

Other examples of Cope rearrangements in hexadiene systems in which the activating group is carbonyl are to be found in allylated cyclohexadienones[58, 165, 166] such as **86**[58] and **87**,[166] which represent the intermediates in the reversible *ortho* ⇌ *para* migrations of the *para*-Claisen rearrangement. Such transformations, having the additional driving force of

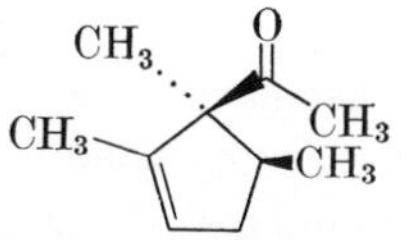

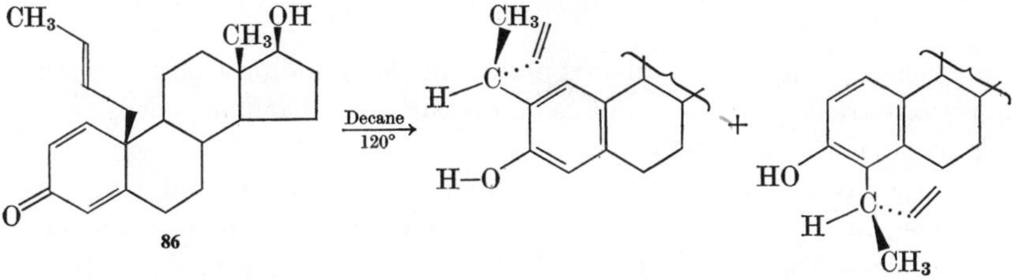

aromatization, occur with ease. Cope rearrangements in the nitrogen heterocyclic systems, **88** and **89**, occur readily over the enamine tautomers.[287]

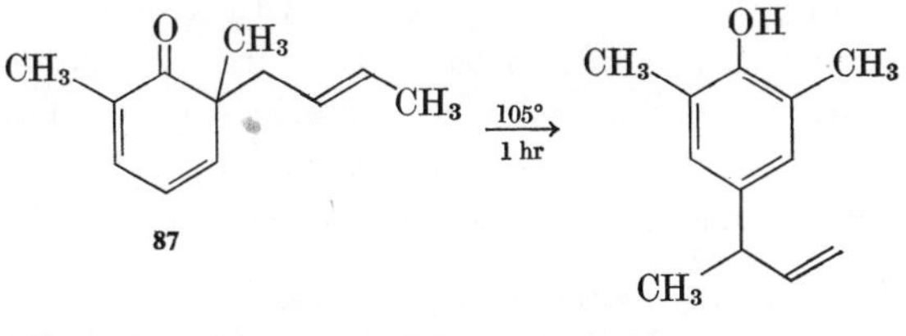

---

[287] R. K. Bramley and R. Grigg, *Chem. Commun.*, **1969,** 99.

88 (88%)

89 (89%)

When the acyclic 1,5-hexadiene system carries only alkyl substituents, rearrangement requires more vigorous conditions and may be incomplete because of the comparable stabilities of the isomers. 3-Methyl-1,5-hexadiene and 1,5-heptadiene, for example, equilibrate in the gas phase at 220–300° to give an equilibrium composition of about 85% of the more highly alkylated dienes.[39, 44] When the starting hexadiene is more highly

(*trans/cis* ~2)

substituted with alkyl or phenyl groups in the 3 and/or 4 positions, the reactions are essentially complete and occur somewhat more easily (refs. 52, 57, 154, 155, 288, 289).

cis-1,2-Divinyl derivatives of three- and four- membered rings rearrange with ease because of the attendant relief of ring strain. *cis*-Divinylcyclopropane itself· has been isolated only very recently;[67b] stable at −20°, it rearranges to 1,4-cycloheptadiene with a half-life of 90 seconds at 35°. Some derivatives of this basic structure have proved stable enough to be isolable under ordinary conditions of work-up while others have been detected as fleeting intermediates.[63–76a, 290–294] The bicyclic derivative **90** rearranges at room temperature with a half-life of

[288] H. M. Frey and R. K. Solly, *Trans. Faraday Soc.*, **65**, 1372 (1969).

[289] R. P. Lutz, S. Bernal, R. J. Boggio, R. D. Harris, and M. W. McNicholas, *J. Amer. Chem. Soc.*, **93**, 3985 (1971).

[290] P. K. Freeman and D. G. Kuper, *Chem. Ind.* (London), **1965**, 424.

[291] O. L. Chapman and J. D. Lassila, *J. Amer. Chem. Soc.*, **90**, 2449 (1968).

[292] M. S. Baird and C. B. Reese, *Chem. Commun.*, **1970**, 1519.

[293] K. Hojo, R. T. Schneider, and S. Masamune, *J. Amer. Chem. Soc.*, **92**, 6641 (1970).

[294] T. J. Katz, J. J. Cheung, and N. Acton, *J. Amer. Chem. Soc.*, **92**, 6643 (1970).

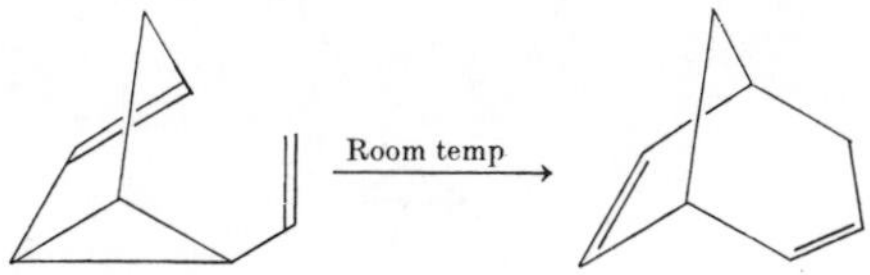

90

1 day.[63, 64] The ketene **92**, a transient intermediate in the photolytic decomposition of the diazoketone **91**, presumably accounts for the isolation of the bicyclic ketone **93** as the major product.[290]

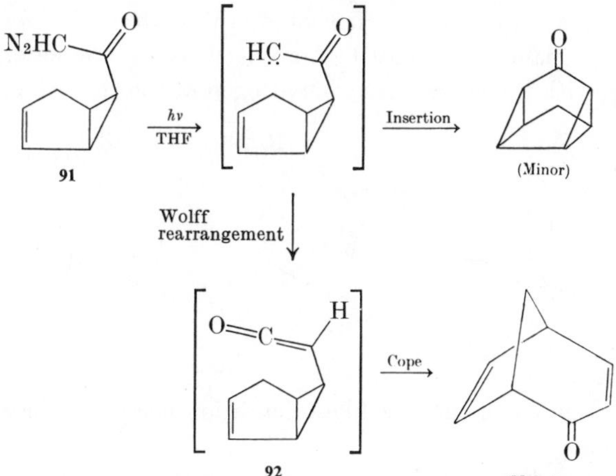

91

(Minor)

Wolff rearrangement

92

(Major)
93

Chapman and Lassila were able to detect the very labile ketene **95** in the low-temperature (−190°) irradiation of the methoxyketones, **94** or **96**.[291] On warming to −70°, the ketene spontaneously isomerizes in a Cope process to give an equilibrium mixture of the methoxy ketones in which the more stable isomer **96** predominates. Several investigations directed

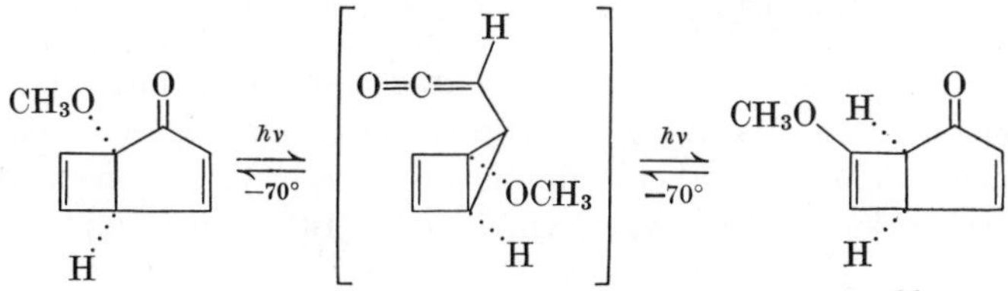

94   **95**   **96**

toward the synthesis of the odoriferous principle of the sea, dictyopterene A (**97**), have been successful in isolating or detecting the corresponding *cis*-divinylic cyclopropanes **98** and **99**,[65–67a] both of which rearrange readily

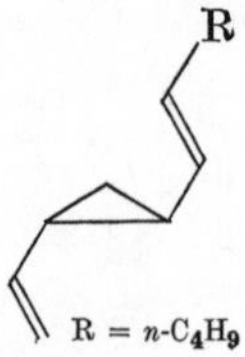

97 (*trans,trans*)

to 6-*n*-butyl-1,4-cycloheptadiene. The *cis,cis* isomer **98** can be isolated and
its rearrangement studied at 75°; the more labile *cis,trans* isomer **99**
rearranges in the process of preparation and workup.[65] The stability of
the *cis,cis* isomer in contrast to that of the *cis,trans* isomer or of *cis*-
divinylcyclopropane itself may be traced to steric interaction of the *n*-
butyl group with the *cis*-methylene hydrogen of the ring which raises the

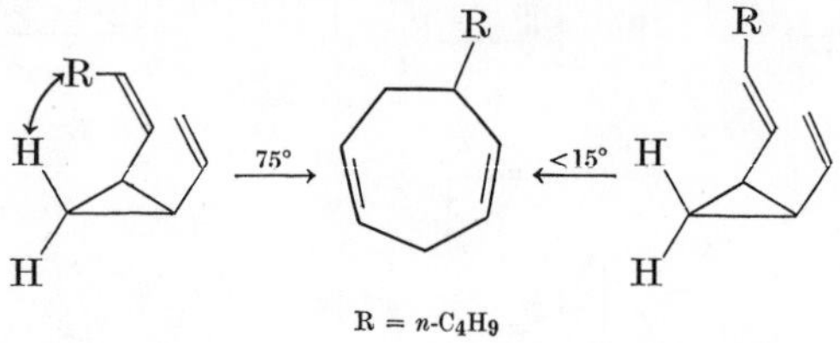

$R = n\text{-}C_4H_9$

98 (*cis,cis*)   **99** (*cis,trans*)

energy requirement of the boatlike orientation necessary for a concerted
rearrangement.

Analogs of *cis*-divinylcyclopropane in which the methylene group of the
ring is replaced by O,[178, 295, 296] N-,[297, 298] and SO$_2$[299] also undergo rearrangements. Generally, *cis*-divinyl derivatives of these systems (**100**) rearrange

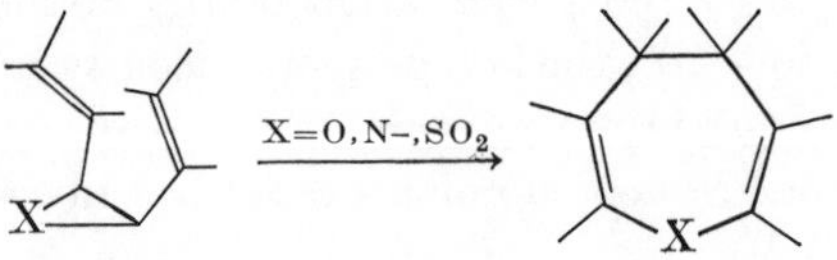

$X = O, N-, SO_2$

100

spontaneously or with very slight encouragement. Rearrangement of the
corresponding *trans*-divinyl derivatives requires more vigorous conditions
and appears to involve either a diradical process or prior isomerization to
the labile *cis* isomer. The *trans*-divinyloxiran (**101**), for example, yields a

[295] E. L. Stogryn, M. H. Gianni, and A. J. Passannante, *J. Org. Chem.*, **29**, 1275 (1964).
[296] J. C. Pommelet, N. Manisse, and J. Chuche, *C.R. Acad. Sci.*, Ser. C, **270**, 1894 (1970).
[297] E. L. Stogryn and S. J. Brois, *J. Org. Chem.*, **30**, 88 (1965).
[298] E. L. Stogryn and S. J. Brois, *J. Amer. Chem. Soc.*, **89**, 605 (1967).
[299] W. L. Mock, *Chem. Commnu.*, **1970**, 1254.

mixture of the Cope product and the vinyldihydrofuran when heated to 170–200°.[300]

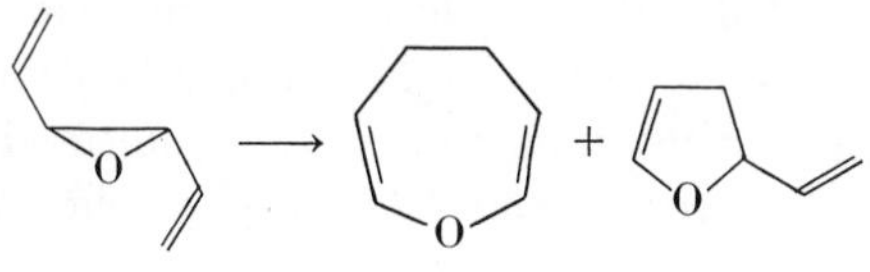

101

In contrast to the cyclopropane derivative, cis-divinylcyclobutane can be isolated without difficulty, but it too rearranges quantitatively at 120° to give a single product, cis,cis-1,5-cyclooctadiene.[38, 68]

When the 1,5-hexadiene system is incorporated in rings of 9, 10, or 11 members, the strain energy of the medium-ring systems becomes the factor which determines the position of equilibrium. Vogel and co-workers reported that cis,cis-1,5-cyclononadiene (102) and cis-1,2-divinylcyclopentane (103) equilibrate at 220° to a mixture which strongly favors the latter.[301] The more strained isomer, cis,trans-1,5-cyclononadiene (104),

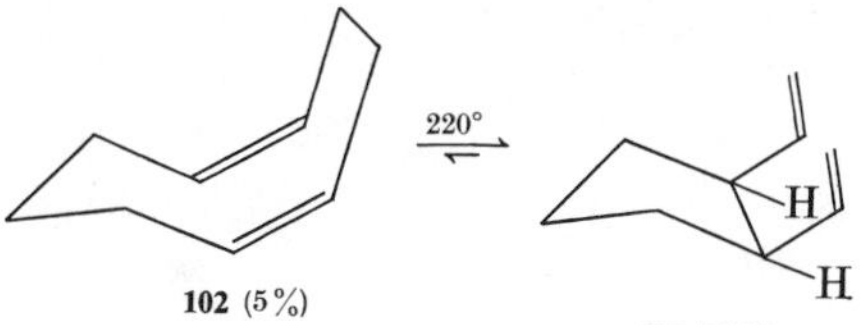

102 (5%)

103 (95%)

also isomerizes to cis-1,2-divinylcyclopentane (103) when heated in the gas phase at 130°.[301] In the first equilibration (102 ⇌ 103), a boatlike

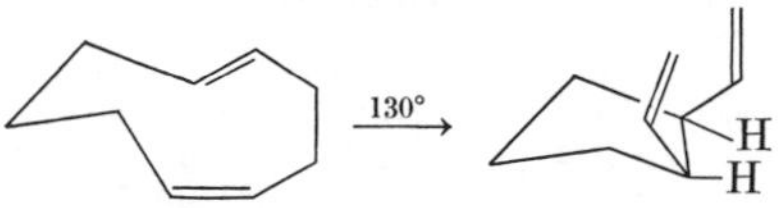

104

transition-state geometry must be involved, whereas in the rearrangement of diene 104 a chairlike arrangement can readily be achieved.

A somewhat similar situation exists in 1,5-cyclodecadiene systems. cis,trans-1,5-Cyclodecadiene (105) isomerizes quantitatively at 150° to the

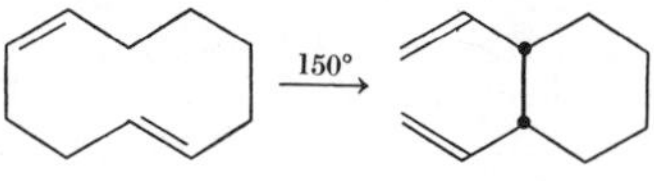

105   (Quant.)

[300] E. Vogel and H. Günther, Angew. Chem., Int. Ed. Engl., 6, 385 (1967).
[301] E. Vogel, W. Grimme, and E. Dinné, Angew. Chem., Int. Ed. Engl., 2, 739 (1963).

single product, *cis*-1,2-divinylcyclohexane, a process which, again, involves the favored chairlike geometry of the transition state.[302]

trans,trans-1,5-Cyclodecadiene (106) isomerizes with the same high degree of stereoselectivity to *trans*-1,2-divinylcyclohexane;[303] once more, a chairlike geometry of the transition state is indicated since a boatlike geometry would give rise to *cis*-1,2-divinylcyclohexane. This stereo-

106   H̤

specific ring closure of *trans,trans*-1,5-cyclodecadienes has been widely observed in sesquiterpenoid derivatives and has been used as supporting evidence for structural assignments.[304-318] Germacrone (107), for example, is quantitatively transformed to β-elemenone (108) when slowly distilled under reduced pressure at 165°.[308] In the closely related costunolide (109)-

107   108

dehydrosaussurea lactone (110) system, an equilibrium exists between the Cope isomers in which the divinylcyclohexane derivative predominates (2:1).[307] In other only slightly modified derivatives, the stabilities of the Cope isomers are even more closely balanced.[310, 316, 317]

[302] P. Heimbach, *Angew. Chem., Int. Ed. Engl.*, **3**, 702 (1964).

[303] C. A. Grob, H. Link, and P. W. Schiess, *Helv. Chim. Acta*, **46**, 483 (1963).

[304] E. D. Brown, M. D. Solomon, J. K. Sutherland, and A. Torre, *Chem. Commun.*, **1967**, 111.

[305] N. Hayashi, S. Hayashi, and T. Matsuura, *Tetrahedron Lett.*, **1968**, 4957.

[306] H. Hikino, K. Agatsuma, and T. Takemoto, *Tetrahedron Lett.*, **1968**, 931.

[307] T. C. Jain, C. M. Banks, and J. E. McCloskey, *Tetrahedron Lett.*, **1970**, 841.

[308] G. Ohloff, H. Farnow, W. Phillipp, and G. Schade, *Ann. Chem.*, **625**, 206 (1959); G. Ohloff and E. G. Hoffmann, *Z. Naturforsch.*, **16b**, 298 (1961) [*C.A.*, **55**, 25805b (1961)].

[309] A. S. Rao, A. Paul, Sadgopal, and S. C. Bhattacharyya, *Tetrahedron*, **13**, 319 (1961).

[310] K. Takeda, I. Horibe, and H. Minato, *J. Chem. Soc., C.*, **1970**, 1142.

[311] K. Takeda, I. Horibe, and H. Minato, *J. Chem. Soc., C.*, **1970**, 1547.

[312] K. Takeda, I. Horibe, and H. Minato, *J. Chem. Soc., C.*, **1970**, 2704.

[313] K. Takeda, I. Horibe, and H. Minato, *Chem. Commun.*, **1971**, 88.

[314] K. Takeda, I. Horibe, M. Teraoka, and H. Minato, *Chem. Commun.*, **1968**, 940.

[315] K. Takeda, I. Horibe, M. Teraoka, and H. Minato, *J. Chem. Soc., C.*, **1969**, 1491.

[316] K. Takeda, I. Horibe, M. Teraoka, and H. Minato, *J. Chem. Soc., C.*, **1970**, 973.

[317] K. Takeda, H. Minato, and M. Ishikawa, *J. Chem. Soc., C*, **1964**, 4578.

[318] K. Takeda, K. Tori, I. Horibe, M. Ohtsuru, and H. Minato, *J. Chem. Soc., C*, **1970**, 2697.

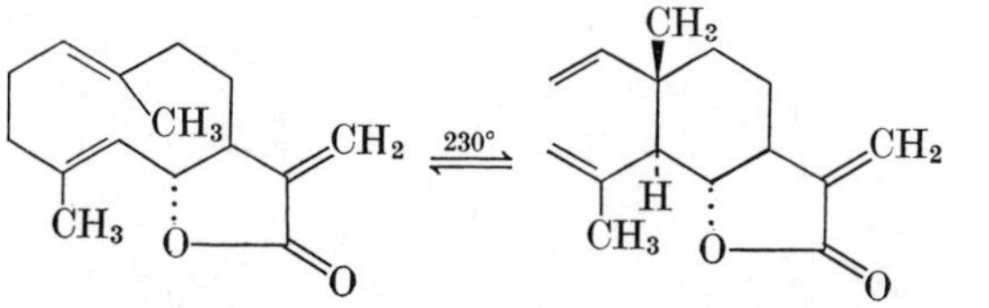

109   110

Although the simple *cis,trans*-1,5-cyclodecadiene (105) is smoothly and exclusively converted to *cis*-1,2-divinylcyclohexane through a chairlike arrangement, the same result is not always observed in the more highly substituted sesquiterpenoid derivatives of this ring system and caution must be exercised in interpreting the outcome of thermal isomerizations in such systems. Takeda and co-workers have reported that neolinderalactone (111) and sericenine (112), both of which contain a *cis,trans*-1,5-cyclodecadiene ring system, rearrange only with difficulty and in poor yield to the *trans*-substituted divinylic cyclohexane derivatives rather than the expected *cis* derivatives.[311–313]

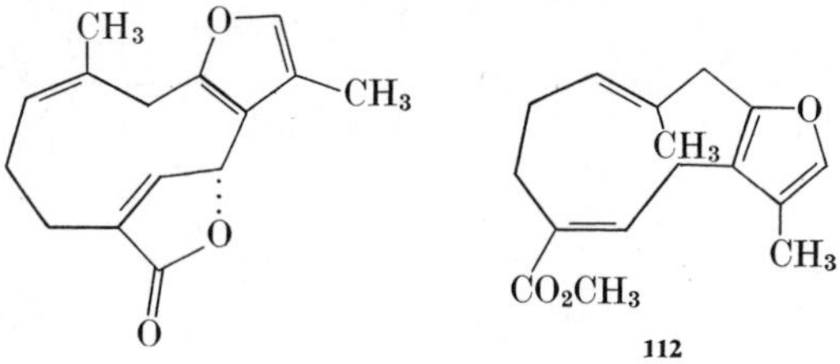

111   112

The *trans,trans*-1,5-diene system in the eleven-membered ring compound, zerumbone (113), readily rearranges to ψ-photozerumbone (114) which contains the *trans*-divinyl feature.[319]

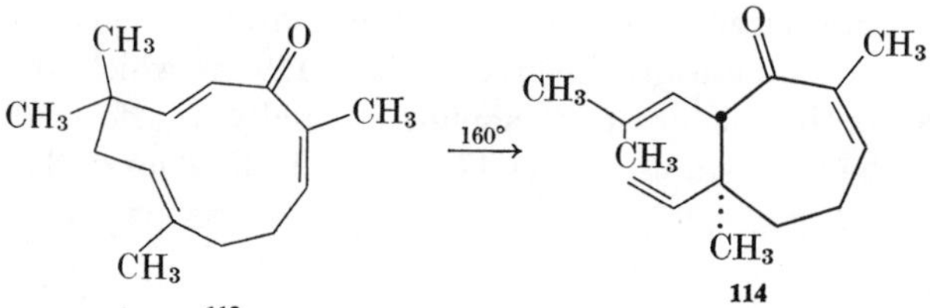

113   114

The facile Cope rearrangement of 1,5-hexadiene systems in bicyclic structures, first observed by Woodward and Katz[74] in α- and β-1-hydroxydicyclopentadiene, has been extended to a variety of bicyclic derivatives.[75–77, 320, 321] The *endo* adduct (115) of tropilidene and di-

[319] H. N. Subba Rao, N. P. Damodaran, and S. Dev, *Tetrahedron Lett.*, **1967**, 227.
[320] R. C. Cookson, J. Hudec, and R. O. Williams, *Tetrahedron Lett.*, **22**, 29 (1960).
[321] M. F. Ansell, A. F. Gosden, and V. J. Leslie, *Tetrahedron Lett.*, **1967**, 4537.

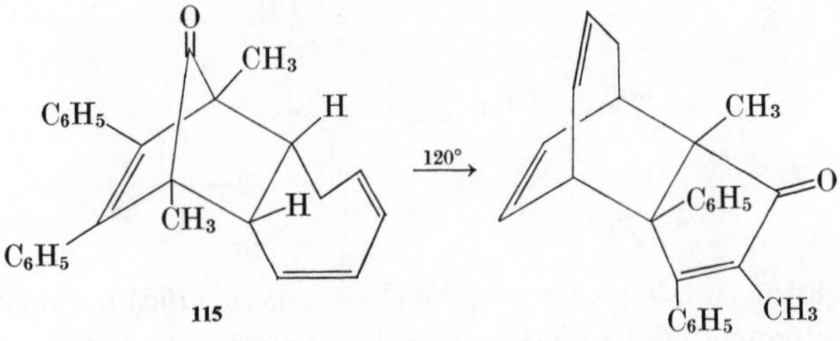

115 → 116

methyldiphenylcyclopentadienone, for example, rearranges completely to the Cope product **116** at 120°.[77] The tropone-dimethyl acetylenedicarboxylate adduct **117** thermally isomerizes to the ketene **118** which can be trapped as the ester in the presence of methanol.[322]

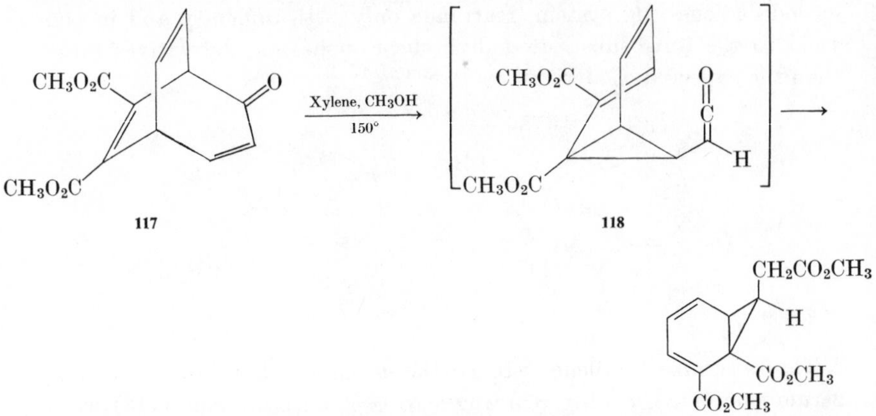

The diene adducts of *o*-benzoquinones and cyclopentadiene furnish additional interesting examples of Cope rearrangements in bicyclic systems.[321] The kinetically favored adduct **119** in which the *o*-benzoquinone plays the role of the dienophile is rapidly transformed in boiling benzene to the more stable adduct **120** in which the roles of the diene and

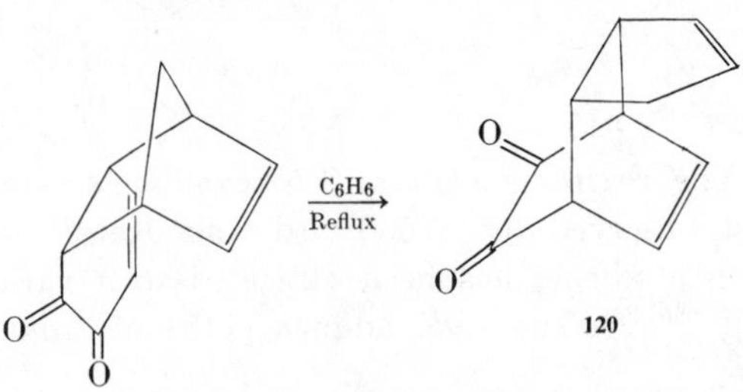

[322] T. H. Kinstle and P. D. Carpenter, *Tetrahedron Lett.*, **1969**, 3943.

dienophile have been reversed. It was established that the conversion was a true intramolecular Cope rearrangement and not a dissociation and recombination process.

Oxy-Cope Rearrangements.[5]   When the 1,5-hexadiene system bears a hydroxyl group at positions 3 and/or 4, the Cope rearrangement leads to an enolic or bis(enolic) product and the reaction becomes useful for the preparation of δ,ϵ-unsaturated carbonyl compounds and α,δ-dicarbonyl derivatives. The simple 1,5-hexadien-3-ol **(121)** gives 57 % of the rearrangement product, **122**, when distilled through a helices-packed column at 380°.[168] The major side reaction in the oxy-Cope rearrangement is a β-hydroxy olefin cleavage (a 1,5-hydrogen shift) which leads to fragmentation products.

A study of the effect of alkylation on the relative importance of cleavage *vs* rearrangement showed that methylation at C_2, C_3, or C_4 of the hexadiene system has little effect, whereas methyl groups at C_1, C_5, and C_6 enhance cleavage at the expense of rearrangement.[167]

Marvell and Whalley examined the thermal behavior of *trans*- and *cis*-1,2-divinylcyclohexanol.[171] The *trans* isomer, **123**, rearranges in excellent yield to the single product, *trans*-5-cyclodecen-1-one, when heated in solution at 220°. The *cis* isomer, **124**, under the same conditions, produces in 50 % yield a mixture containing *trans*- and *cis*-5-cyclodecen-1-one in a 3:2 ratio (Formulae on p. 62). These results, which are accommodated in terms of the chairlike arrangements available to the rearranging isomers, provide an attractive approach to functionalized ten-membered rings.

Berson and Jones have examined bicyclic systems which incorporate the structural requirements for an oxy-Cope rearrangement but in which the rigidity of the system makes a concerted [3,3]sigmatropic process

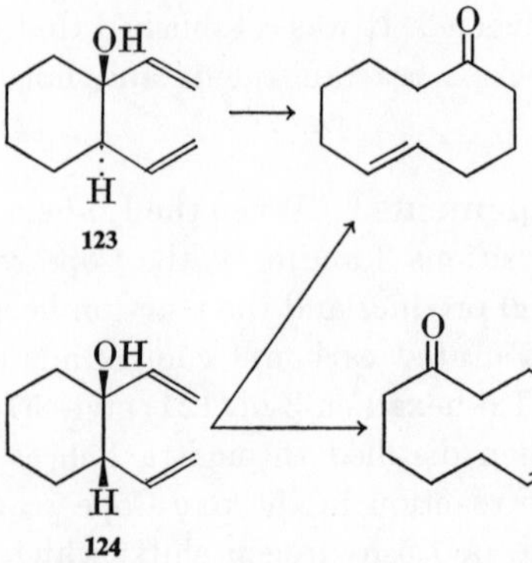

123

124

impossible or highly unlikely.[170, 323] *syn*-7-Vinyl-2-bicyclo[2.2.1]hepten-7-ol **(125)** for example, heated in the gas phase at 320°, gave only 3–5% of the Cope product, **126**; the major product **127** is formally the result of a

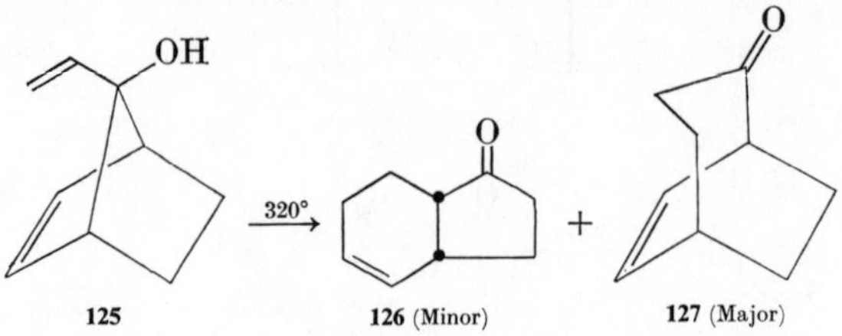

125   126 (Minor)   127 (Major)

1,3 shift. The major product of the gas-phase pyrolysis of *endo*-2-vinyl-5-bicyclo[2.2.2]octen-2-ol **(128)** is the Cope product, produced in 45% yield; in this case, too, the authors favor the intermediacy of a diradical rather than a concerted rearrangement, even though, structurally, the

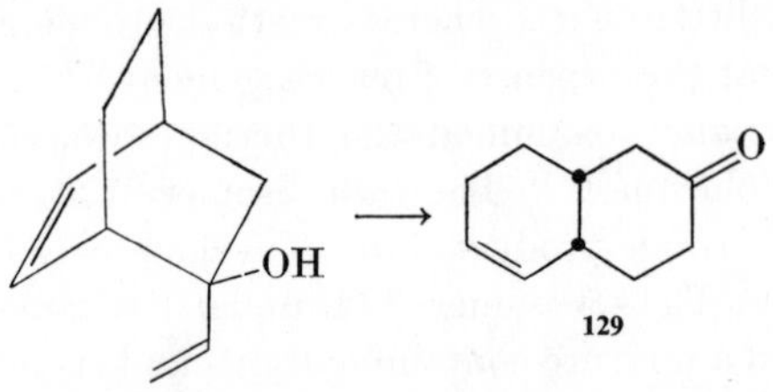

128

129

reacting molecule could achieve a six-centered transition-state geometry.[323] It is noteworthy that the methyl ether of **128** underwent rearrangement to the enol ether of **129** in much better yield (87%).[324]

[323] J. A. Berson and M. Jones, Jr., *J. Amer. Chem. Soc.*, **86**, 5017 (1964).
[324] J. A. Berson and E. J. Walsh, Jr., *J. Amer. Chem. Soc.*, **90**, 4729 (1968).

Other cyclic systems in which the oxy-Cope rearrangement appears to be inhibited by steric factors have been reported.[325]

As discussed earlier for the Cope rearrangement of dienes, the oxy-Cope process has also been observed and used in structural assignments in naturally occurring compounds containing *trans,trans*-1,5-cyclodecadiene ring systems.[326–328]

Triple bonds may replace either of the double bonds of the allyl vinyl carbinol structure.[50, 169, 329, 330] 5-Hexen-1-yn-3-ol **(130)**, pyrolyzed in a flow system at 370°, gave rise to the Cope product, **131**, as well as two cyclic products.[50] The cyclopentene derivatives appear to arise from the initially

| 130 | 131 (35%) | (42%, total) |

formed allenol. Propargyl vinyl carbinol **(132)**, pyrolyzed under the same conditions, gave rise to the Cope product in 12% yield.[169] The major product was 3-cyclopentenecarboxaldehyde. Extensive fragmentation also occurred.

When both C_3 and C_4 of the hexadiene carry hydroxy functions, the oxy-Cope rearrangement leads through the intermediate bis(enol) either to the corresponding dicarbonyl compound or to a cyclized product, depending on the structure of the starting diol and the conditions imposed (refs. 49, 172–177, 180, 181, 183). For example, the diol **133** gave rise to 2,7-octanedione when heated alone at 190° for 1 hour; above 240° the sole product was the internal aldol condensation product, **134**.[173] In many cases it appears that the rate of cyclization of the initial rearrangement product is so rapid that the dicarbonyl compound cannot be isolated. This

[325] R. W. Thies and M. T. Wills, *Tetrahedron Lett.*, **1970**, 513.
[326] N. H. Fischer and T. J. Mabry, *Chem. Commun.*, **1967**, 1235.
[327] N. H. Fischer, T. J. Mabry, and H. B. Kagan, *Tetrahedron*, **24**, 4091 (1968).
[328] W. Renold, H. Yoshioka, and T. J. Mabry, *J. Org. Chem.*, **35**, 4264 (1970).
[329] J. Chuche and N. Manisse, *C.R. Acad. Sci., Ser. C*, **267**, 78 (1968).
[330] J. W. Wilson and S. A. Sherrod, *Chem. Commun.*, **1968**, 143.

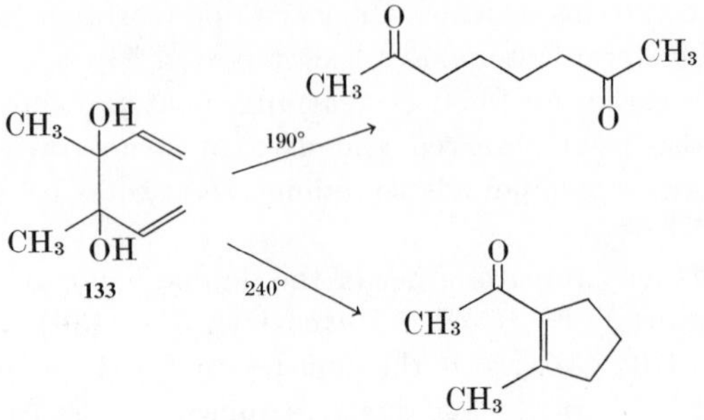

has proved to be so in the rearrangements of divinylcyclanediols of rings of 5, 6, or 7 members, e.g., **135** → **136** + **137**, and presents a limitation to what appeared to be a useful entry to the medium rings of 9, 10, or 11 members (refs. 49, 172, 174, 175, 177). With divinylcyclanediols of eight

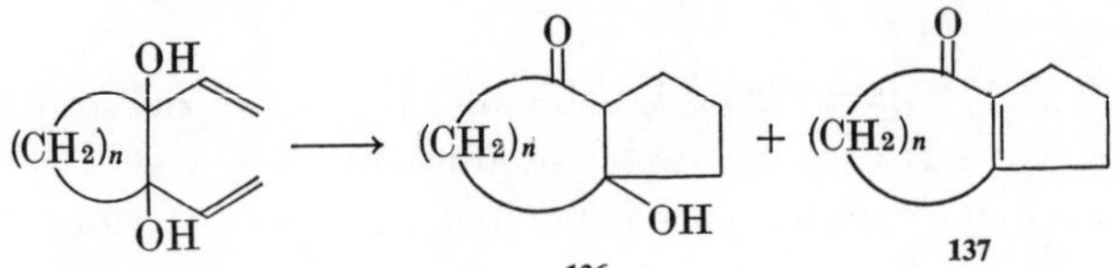

or more ring members good yields of the ring-expanded diones are realized.[49, 174, 176]

It appears that the competing reaction of hydroxy olefin cleavage which often predominates in the oxy-Cope rearrangement as well as undesired cyclization processes could be minimized by the use of esters of the alcohols or glycols in place of the hydroxy compounds themselves. The diacetate **138** is reported to rearrange to the dienol ester in 70% yield at 240°.[331] This may be contrasted with the result reported for the glycol

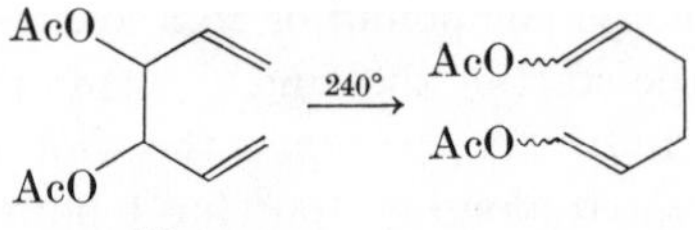

139 which gives 1-formylcyclopentene in only 40% yield when heated under reduced pressure in the same temperature range.[180, 181] Recently trimethylsilyl derivatives of alcohols have been used successfully to eliminate undesired competing and subsequent reactions (siloxy-Cope rearrangement).[332]

[331] A. Bader and F. Weiss, Fr. Pat. 1519701 (1968) [*C.A.*, **70**, 96177g (1969)].
[332] R. W. Thies, *Chem. Commun.*, **1971**, 237; *J. Amer. Chem. Soc.*, **94**, 7074 (1972).

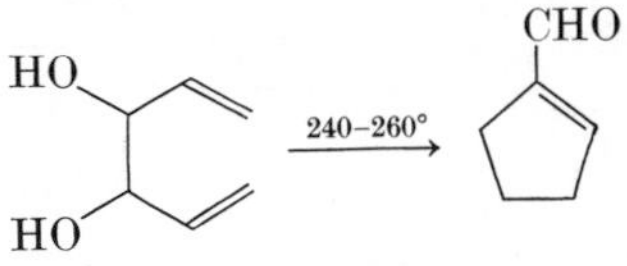

139

Miscellaneous Cope Rearrangements. Cope rearrangements in 1,5-hexenynes[151, 152, 333] and 1,5-hexadiynes[153, 334, 335] appear to proceed readily as do those in systems incorporating allenyl functions as the unsaturated linkages.[301, 336–338] Since the diynes furnish cyclized products predominantly, the intermediacy of the expected Cope product is conjectural in these systems. In a flow system at 340° the enyne **140** produced the three products shown.[152] With longer contact times the amount of the

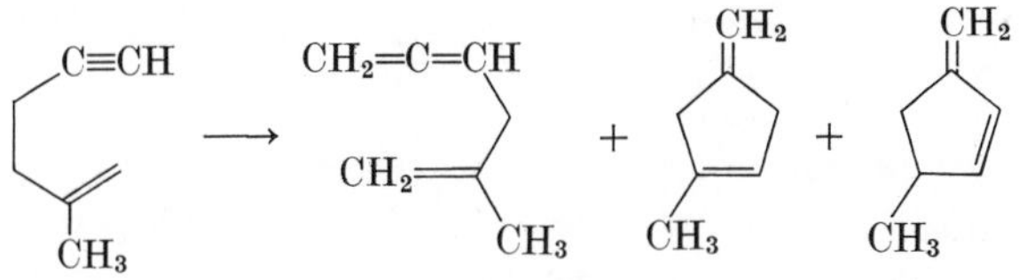

140

allenyl product decreased and the amounts of the cyclic products increased, supporting the notion that the allenyl Cope product is the precursor of the cyclic ones. Rearrangement in the heterocyclic derivative, **141** (generated by the action of methylhydrazine on 3-allyl-3-propargyl-2,4-pentanedione) proceeds through both the hexadiene and the hexenyne pathways at comparable rates to give a mixture of the rearrangement products **142** and **143** in a ratio of 1.6:1.[333] 1,5-Hexadiyne is transformed

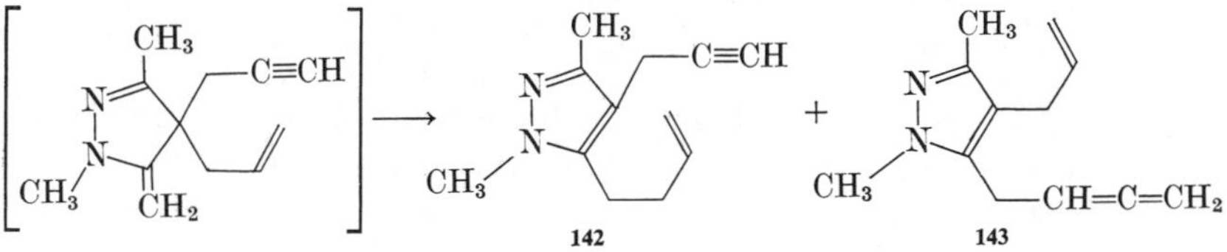

141   142   143

in 85% yield to the cyclized product **144** in a flow reactor at 350°.[335] In this system none of the normal Cope product, bis(allene), could be detected.

[333] D. T. Manning, H. A. Coleman, and R. A. Langdale-Smith, *J. Org. Chem.*, **33**, 4413 (1968).

[334] M. B. D'Amore and R. G. Bergman, *J. Amer. Chem. Soc.*, **91**, 5694 (1969).

[335] W. D. Huntsman and H. J. Wristers, *J. Amer. Chem. Soc.*, **85**, 3308 (1963).

[336] J. F. Harris, *Tetrahedron Lett.*, **1965**, 1359.

[337] L. Skattebol and S. Solomon, *J. Amer. Chem. Soc.*, **87**, 4506 (1965).

[338] K. G. Untch and D. J. Martin, *J. Amer. Chem. Soc.*, **87**, 4501 (1965).

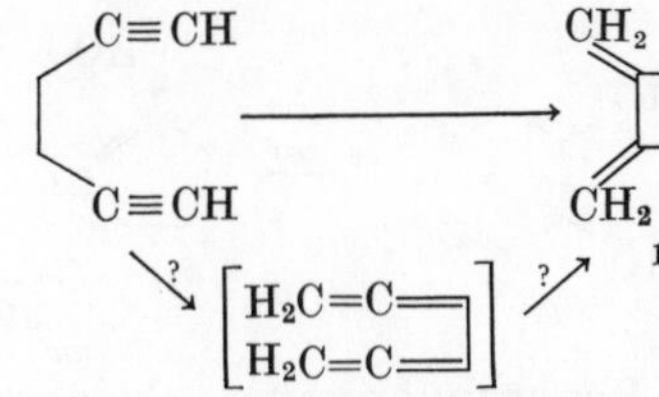

The cyclic bis(allenyl) compound **145** rearranges quantitatively in the gas phase at 300° to the Cope product.[336, 337] Other cyclic[301, 338] and open-chain[337] allenyl derivatives show the behavior expected of a Cope process.

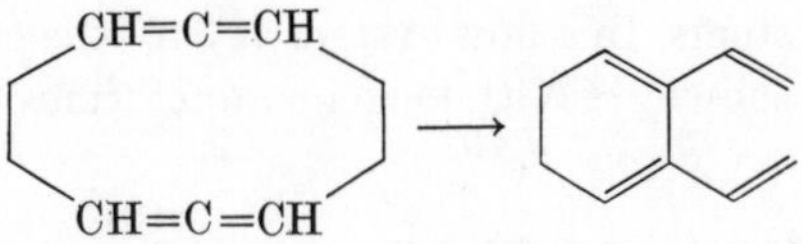

Several cyclopropane derivatives bearing *cis*-disposed vinyl and iso-cyanate groups have been reported to undergo Cope-like reorganizations leading to derivatives of 2-azepinone.[69–72] Brown and co-workers were able to isolate the isocyanate **146** and to demonstrate its conversion to the azepinone at room temperature.[71] The more heavily substituted cyclo-

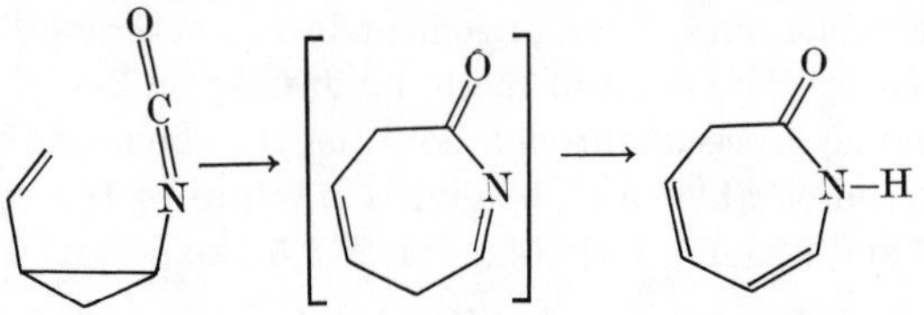

propane derivative **147** equilibrates with the azepinone only when heated in xylene to give a mixture of 1 part of **147** to 7 parts of azepinone.[72]

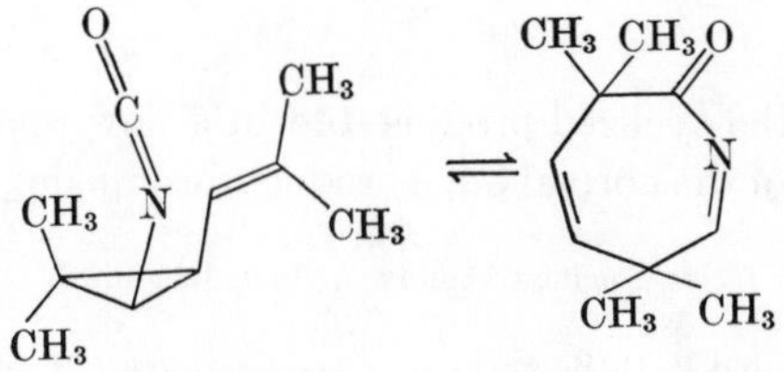

A transformation closely related to the retro-Claisen rearrangement (p. 29) has been reported for the heterocycle **148** produced by the action of benzenesulfonyl azide on norbornadiene.[339] At room temperature the conversion to the bicyclic product **149** is quantitative. The transformation is postulated to pass through the intermediate sulfonyl imine which then

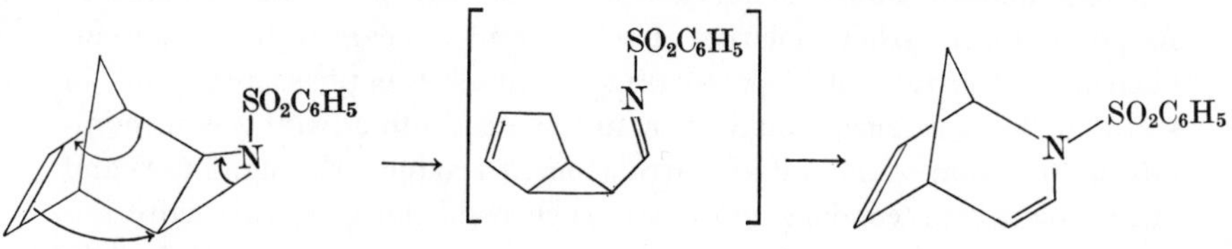

148 **149**

suffers a bis(allylic) rearrangement. This sequence of steps finds a complete parallel in the oxygen analogs.[78,340]

Schiff bases of *cis*-1,2-diaminocyclopropanes readily rearrange to 2,3-dihydro-1H-1,4-diazepins.[73] The tribenzylidene derivative **150** rearranges in the process of its formation. Schiff bases of *trans*-1,2-diaminocyclopropane also rearrange when heated to 120–140°.[73]

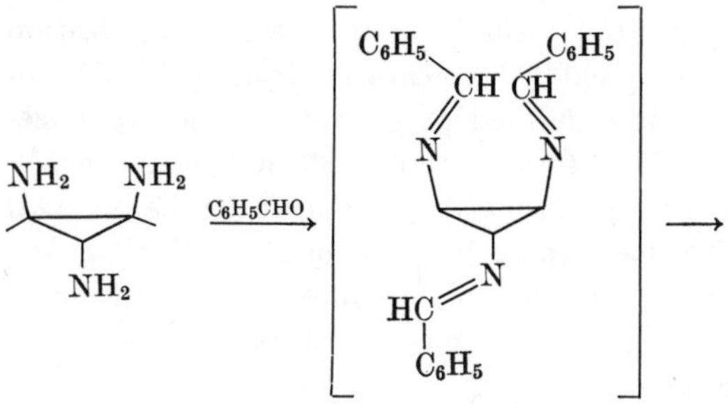

150

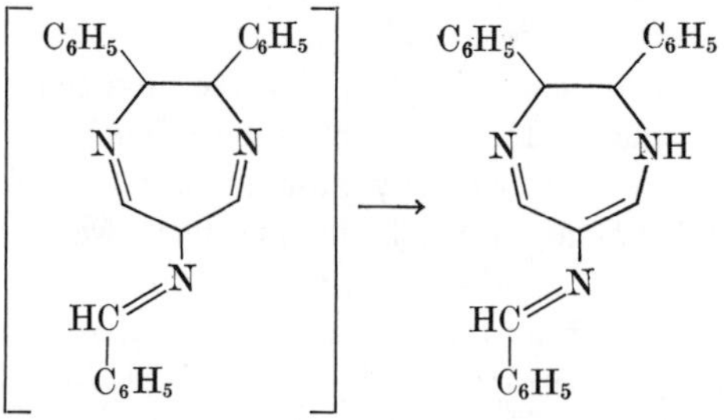

Experimental Conditions. The majority of Cope rearrangements have been conducted simply by heating neat samples of the starting materials in sealed tubes at the temperatures necessary to bring about the

[339] A. C. Oehlschlager and L. H. Zalkow, *Chem. Commun.*, **1965**, 70.

[340] J. Meinwald, S. S. Lebana, and M. S. Chadha, *J. Amer. Chem. Soc.*, **85**, 582 (1963).

desired change. In some instances inert solvents such as cyclohexane, decane, or tetralin have been included, but there has been no systematic study of the effect of solvent on yield or on product distribution when side reactions and subsequent reactions of the Cope products are possible. As pointed out earlier, polarity of the solvent appears to have little influence on the rate of Cope rearrangements.[85] Gas-phase reactions, in static or flow systems, would seem to have definite advantages in minimizing diversion of the initial rearrangement products through inter- and intra-molecular secondary processes; such conditions are especially important in oxy-Cope rearrangements and rearrangements which lead to thermally unstable products such as are encountered in 1,5-hexenyne and hexadiyne derivatives.

EXPERIMENTAL PROCEDURES

The earlier chapter in *Organic Reactions*[1] should be consulted for a procedure for the preparation of o-allylphenol in 86% yield; described there, too, are methods for conversion of o-allylphenol to the coumaran, 2-methyldihydrobenzofuran, in 51% yield by acid-catalyzed cyclization and to o-propenylphenol in 75% yield by base-catalyzed double-bond isomerization. *Organic Syntheses* has published procedures for the rearrangement of the allyl ether of guaiacol to o-eugenol in 80–90% conversion[341] and for the *in situ* dealcoholation and rearrangement of the diallyl ketal of cyclohexanone to 2-allylcyclohexanone in yields of 85–91%.[342]

The procedures collected in this section illustrate some of the more recently developed applications of Claisen and Cope processes in aromatic, heterocyclic, and aliphatic systems.

5-Methoxy-7-hydroxy-8-(α,α-dimethylallyl)coumarin. (*Aromatic Claisen Rearrangement with Trapping of Normal ortho Rearrangement Product.*)[219] Oxygen-free nitrogen was passed over a suspension of 60 mg (0.23 mmol) of 5-methoxy-7-(γ,γ-dimethylallyloxy)coumarin in 0.5 ml of N,N-diethylaniline and 0.3 ml of butyric anhydride contained in a 1-ml flask for 1 hour. The flask was then immersed in an oil bath at 185 ± 5°, shaken for 5 minutes to ensure that the melt had dissolved, and kept at that temperature under nitrogen for 8 hours. The mixture was diluted with 10 ml of ice water, set aside for 2 hours, and then extracted with ethyl acetate. The organic layer was washed with dilute hydrochloric acid (1% w/v) to pH 2, dilute potassium carbonate (5% w/v) to pH 11, saturated salt solution to neutrality, dried, and evaporated. The residue was purified by preparative tlc [2X ethyl acetate-light petroleum ether

---

[341] C. F. H. Allen and J. W. Gates, Jr., *Org. Syntheses*, **Coll. Vol. 3**, 418 (1955).
[342] W. L. Howard and N. B. Lorette, *Org. Synthsess*, **42**, 1434 (1962).

(3:7); then 1X chloroform] and sublimed at 155° (0.02 mm). This afforded the butyrate of the rearrangement product as colorless needles (70 mg, 92%), mp 162–164°. Saponification of 32 mg of the ester with ethanolic sodium hydroxide (1% w/v) and final tlc isolation (2X ethyl acetate-petroleum ether) yielded the title compound as colorless needles (22 mg, 86%), mp 161–166° dec after recrystallization from ether.

2-(3-Butenyl)-3-methyl-4(1H)-quinolone. (*Out-of-Ring Claisen Rearrangement in a Heterocyclic System.*)[212] 4-Allyloxy-2,3-dimethyl-quinoline (5 g, 2.35 mmol) was heated without solvent at 200° for 30 minutes. The reaction mixture was digested with benzene and the crystals were collected by filtration to yield 4.6 g (92%) of the title compound, mp 261–262° after recrystallization from ethanol. The benzene-soluble portion of the reaction mixture was chromatographed on alumina and eluted with benzene-chloroform (1:1) to give two fractions which proved to be 1-allyl-2,3-dimethyl-4(1H)-quinolone, mp 133–134° (2%), and 1,4-dimethyl-2,3-dihydropyrrolo[1,2-a]quinolin-5(1H)-one, mp 191–192°(2%).

2-Allyl-1-naphthylamine. (*Aromatic Amino-Claisen Rearrangement.*)[29] The free base from 3.0 g (1.37 mmol) of N-allyl-1-naphthylamine hydrochloride was heated in a sealed tube at 260° for 3 hours. The crude, pale-yellow product distilled in a short-path still as a colorless, viscous liquid (1.76 g, 70%), $n^{21.5}$D 1.6497; hydrochloride, mp 216–218°, needles after recrystallization from ethanol.

6H-Thieno[2,3-b]thiin. (*Thio-Claisen Rearrangement in a Heterocyclic System. Solvent Effect on Product Distribution.*)[274] In a 100-ml, three-necked flask, provided with a dropping funnel, a thermometer, and a gas-inlet tube, was placed 30 ml of hexamethylphosphoramide while nitrogen was passed through the flask. The solvent was heated to 170–180°. At 2-minute intervals five 5-g portions (16.25 mmol total) of 2-(propargylthio)thiophene were added with shaking. The internal temperature was kept between 170 and 180° by occasional cooling or heating. When all of the sulfide had been added and the evolution of heat had ceased, the mixture was cooled to 20° and poured into 200 ml of ice water. Extraction with ether was followed by 3 water washes of the ether extract and drying the ether solution over magnesium sulfate. Distillation of the residue remaining after evaporation of the ether afforded the thienothiin, bp 126–128° (10 mm), n^{20}D 1.678, in 92% yield, in purity of over 97% by nmr and mass spectrometry.

2-Methylthio[2,3-b]thiophene. (*Thio-Claisen Rearrangement in a Heterocyclic System. Solvent Effect on Product Distribution.*)[275] 2-(Propargylthio)thiophene (46.2 g, 0.3 mol) was dissolved in 200 ml of dimethyl

sulfoxide and heated for 50 minutes at 140–142° in the presence of 2.5 ml of diisopropylamine. The internal temperature was then gradually raised to 170° over 15 minutes. To the reaction mixture, cooled to 20°, 25 g of powdered potassium *t*-butoxide (alcohol-free) was added with shaking to remove the concurrently formed thienothiopyran and other contaminants. After 15 minutes the black reaction mixture was poured into 500 ml of ice-cold 3 *N* hydrochloric acid and extracted twice with ether. The ether extract was washed with water, shaken vigorously with potassium hydroxide pellets to remove any thiol, and concentrated under reduced pressure. Distillation of the residue afforded the thiothiophene in 53% yield, bp 114–115° (12 mm), n^{20}D 1.6394; purity better than 98% by nmr and glpc analyses.

1-Acetyl-3-methyl-3-cyclohexene. (*Aliphatic Claisen Rearrangement of a Cyclic Vinyl Ether.*)[80] A solution of 2-(2-propenyl)-6-methyl-3,4-dihydro-2H-pyran in hexane was pyrolyzed at 425° in a flow system consisting of a Pyrex tube packed with 17.5 cm of Pyrex helices and heated over this distance in an electric furnace. Nitrogen served as the carrier gas at a flow of 30 bubbles/minute in conjunction with a material drop rate of 20 drops/minute. The pyrolysis liquid was collected in a flask cooled in an ice-salt bath. Concentration of the condensate under reduced pressure followed by distillation furnished the rearrangement product in 75% yield, a colorless liquid, bp 80–83° (9.5 mm), homogeneous by glpc; semicarbazone, mp 170–172.5°. A comparable yield was obtained by heating the starting material without solvent in a sealed tube at 240° for 25 minutes.

6,10-Dimethyl-4,5,9-undecatrien-2-one and 6,10-Dimethyl-3,5,9-undecatrien-2-one (ψ-Ionone). (*Aliphatic in situ Claisen Rearrangement of a Propargylic Vinylic Ether and Isomerization of the β-Ketoallenic Product.*)[133] A solution of 152 g (1 mol) of 3,7-dimethyl-6-octen-1-yne-3-ol (dehydrolinalool), 150 mg of *p*-toluenesulfonic acid, 300 ml of ligroin (bp 150–160°), and 150 g of isopropenyl methyl ether was stirred in an autoclave for 17 hours at 92° under a nitrogen pressure of 10 atm. The reaction mixture, treated with 0.5 ml of triethylamine, was freed of solvent and other volatile components (dimethoxypropane) under reduced pressure and finally distilled at 60–100° (0.04 mm) to give 160 g (83%) of the β-keto-allene, a light yellow oil, bp 68° (0.04 mm), n^{20}D 1.4860; phenylsemicarbazone, mp 85° after recrystallization from methanol.

Direct isomerization of the crude β-ketoallene preparation to ψ-ionone was accomplished as follows. The crude reaction mixture was slowly poured into a solution of 1.5 ml of 30% sodium hydroxide in 150 ml of methanol cooled to 0°, the temperature of the mixing solutions being maintained at 0–10°. The mixture was stirred 30 minutes at 0–10° and then neutralized with 0.75 ml of acetic acid. The solvent together with the dimethoxypropane formed in the reaction was removed under water aspirator vacuum. The residue (204 g), consisting of 92% ψ-ionone by uv analysis (95% yield), distilled at 102–104° (0.05 mm), n^{20}D 1.5305. Glpc analysis (10% Apiezon on Celite) of this material showed three peaks identified as the *cis* isomer (60%), the *trans* isomer (39%), and a minor third isomer (1%).

N,N - Dimethyl - 3β - hydroxypregna - 5,20 - dien - 17α - acetamide.

(*Aliphatic Claisen Rearrangement. Meerwein-Eschenmoser in situ Method.*)[343] A solution of 10.4 g (33 mmol) of pregna-5,17(20)-dien-3β,21-diol in 86 ml of 1,1-diethoxy-1-dimethylaminoethane[134] was distilled until the vapor temperature reached 120° and then was heated under reflux in a nitrogen atmosphere for 5 hours. Concentration of the solution under reduced pressure yielded an oil which crystallized when triturated with cold methanol to give 11.7 g (89%) of product containing 0.5 mol of methanol of crystallization, mp 183–185°. A sample was recrystallized from methanol for analysis, mp 186–188°, $[\alpha]^{24}$D —60°.

N,N-Dimethyl-2-methyl-1-naphthaleneacetamide.

(*Claisen Rearrangement of a Benzyl Vinyl Ether Derivative. Meerwein-Eschenmoser in situ Method.*)[135] A solution of 1.0 g (6.3 mmol) of 2-naphthylcarbinol (mp 80–81.5°) in 10 ml of absolute dimethylformamide was treated with 1.28 g (12.7 mmol) of 1-dimethylamino-1-methoxyethene[134] and stirred for 24 hours in an oil bath maintained at 160°. The reaction product was taken up in ether-methylene chloride, extracted twice with 10-ml portions of phosphate buffer solution (pH 5) and twice with saturated salt solution, dried over anhydrous sodium sulfate, and concentrated in a rotary evaporator. The crude product (1.61 g) was chromatographed on Kieselgel (60-fold quantity); elution with benzene-ether (9:1) afforded 1.36 g (94%) of practically pure amide, mp 114–115° after recrystallization from methyl acetate-petroleum ether.

All-*trans*-2,23-dichloro-3,22-dioxo-2,6,10,15,19,23-hexamethyl-tetracosa-6,10,14,18-tetraene.

(*Aliphatic Claisen Rearrangement. Johnson in situ Chloroketal Method.*)[99] A mixture of all-*trans*-3,-4-dihydroxy-2,6,11,15-tetramethylhexadeca-1,6,10,15-tetraene (350 mg, 1.14 mmol), 3-chloro-2,2-dimethoxy-3-methylbutane (1.98 g, 11.2 mmol), and 2,4-dinitrophenol (21 mg, 0.12 mmol) in 2.7 ml of toluene was stirred at 94° for 24 hours using a heated Dean-Stark trap (70–80°) for the removal

343 D. F. Morrow, T. P. Culbertson, and R. M. Hofer, *J. Org. Chem.*, **32**, 361 (1967).

of methanol. After reaction times of 13 and 18 hours, two supplemental 500-mg portions of the chloroketal were added. The solvent was removed at room temperature under reduced pressure and the remaining yellow oil was chromatographed on 60 g of silica gel. Elution with hexane-ether (98:2) yielded 350 mg (60%) of a slightly yellow oil which, after drying at room temperature and 0.01 mm, was shown by tlc to be pure title compound; ir (film) 1725, 1670 cm^{-1}; mass spectrum m/e 510 (M$^+$).

2,6-Dimethylocta-2-*trans*-6-*trans*-dienal and 2,6-Dimethylocta-2-*cis*-6-*trans*-dienal. (*Consecutive in situ aliphatic Claisen-Cope Rearrangements. Thomas Method.*)[102] A mixture of 50 g (0.58 mol) of 2-methyl-2-butenol (tiglic alcohol), 125 g of 1-ethoxy-2-methyl-1,3-butadiene, 15 g of mercuric acetate, and 5 g of anhydrous sodium acetate was heated at 100° for 15 hours in an argon atmosphere. Filtration and distillation gave 62.5 g of a fraction, bp 62–96° (10 mm). Column chromatography (silica gel, petroleum ether) of this fraction gave 21.3 g of a fraction, bp 96–97° (10 mm), which consisted of 85% of the *trans,trans* dienal. Further purification of this fraction by glpc (Carbowax 20M on Chromosorb W) gave pure *trans,trans* isomer: semicarbazone, mp 175–176°; 2,4-dinitrophenylhydrazone, mp 155–156°. Elution of a section of the silica gel column adjacent to that which yielded the *trans,trans* isomer afforded 9.5 g of a product which, after fractionation in a spinning-band column and further purification by glpc (Carbowax 20M on Chromosorb W) was pure *cis,trans* dienal: semicarbazone, mp 121–123°.

Ethyl (2-Allyl-1-indanylidene)cyanoacetate. (*Cope Rearrangement.*)[140] Ethyl (3-indenyl)allylcyanoacetate (4.0 g, 14.5 mmol) was heated under nitrogen at 124–128° for 3 hours in a sealed Pyrex tube. The solid that separated on cooling was recrystallized from 1:1 pentane-hexane containing 15% ether to give 2.5 g (63%) of the title compound, mp 79.5–80.5°. Recrystallization furnished an analytically pure sample, mp 80.5–81°.

***cis*- and *trans*-3,4-Dimethyl-3,7-octadien-2-one, 3,4-Dimethyl-4,7-octadien-2-one, and 2,3,4-Trimethyl-3-acetylcyclopentene.** (*Cope Rearrangement, Deconjugation, and Cyclization.*)[146] One-gram portions of 3-isopropenyl-3-methyl-5-hexen-2-one were sealed in 2-ml high-pressure ampoules and kept at a temperature of 230° in a metal bath for 30 minutes. Vacuum distillation of material from several such runs resulted in an almost quantitative yield of a mixture of three ketones, *trans*-3,4-dimethyl-3,7-octadien-2-one (35%), *cis*-3,4-dimethyl-3,7-octadien-2-one (32%), and the deconjugated isomer, 3,4-dimethyl-4,7-octadien-2-one (28%). When the mixture of ketones was maintained at a temperature of 300° for 50 minutes in a sealed ampoule, a greenish liquid, bp 32° (0.25 mm), was obtained in 70 % yield. Glpc analysis indicated that the product consisted of the single ketone, 2,3,4-trimethyl-3-acetylcyclopentene, in which the acetyl and 4-methyl groups are *cis*; oxime, mp 82° after recrystallization from methanol.

2-Methyl-5-hexenal. (*Oxy-Cope Rearrangement.*)[167] 2-Methyl-1,5-hexadien-3-ol (6.12 g, 55 mmol) was rearranged in the vapor phase at 370–380° in a flow system consisting of an externally heated Pyrex tube packed with Pyrex helices for a length of 45 cm. The sample was admitted to the flow system at the rate of 4–10 drops/minute in a nitrogen atmosphere and under a pressure of 21 mm. The pyrolysis liquid (5.4 g, 88%), trapped in dry ice-acetone cooled receivers, consisted of 73 % 2-methyl-5-hexenal, 26 % methacrolein, and a small amount of a low-boiling constituent believed to be propylene. The condensate was separated by fractional distillation and the methacrolein identified as its 2,4-dintrophenylhydrazone. The higher-boiling fraction, 2-methyl-5-hexenal, bp 140–141°, n^{27}D 1.4288, formed a 2,4-dinitrophenylhydrazone, mp 87–88°.

1,6-Cyclododecanedione. (*Oxy-Cope Rearrangement.*)[49] One gram (5.1 mmol) of 1,2-divinyl-1,2-cyclooctanediol in a sealed Pyrex ampoule was heated for 1 hour at 220° in a metal bath. The product, crystalline on cooling, represented a quantitative yield of 1,6-cyclododecanedione, mp 93° after recrystallization from methanol; bis-2,4-dinitrophenylhydrazone, mp 269°.

TABULAR SURVEY

The survey that follows is a tabulation of the Claisen rearrangements reported since 1943 and of all the Cope rearrangements that have been located up to January 1972. Unsuccessful reactions have been omitted. Processes that involve a sequence of Claisen and Cope rearrangements are tabulated under the heading of the first reaction.

The first three sections of Table I, the aromatic Claisen rearrangements, are organized in terms of benzene ring and allyl or propargyl side-chain substitutions. In Section A are the ring-substituted allyl ethers arranged in alphabetical order of substituent names. Section B includes ethers containing a substituted allyl group and, in many cases, substituted phenyl groups. Here the order is based first on the position (α, β, or γ), number, and complexity of the allylic substitutions and, secondly, on the number of aryl substituents. Owing to the complexity of the structures in this section, especially, a rather arbitrary placement of certain compounds has been made. The propargyl aryl ethers of Section C may also have both side-chain and ring substituents. Section D is a short miscellaneous section of diaryl diether structures. Polycyclic and heterocyclic allyl ethers are listed in Section E in order of increasing carbon number and heteroatom(s) when they are present. All out-of-ring Claisen rearrangements are in Section F in order of increasing number of carbon atoms.

The aliphatic Claisen rearrangements have been divided into four sections in Table II, each arranged in order of complexity of molecular formula. Section A is a listing of acyclic allyl vinyl ethers; while in Section B the allylic double bond of each ether is part of a cyclic structure. The ethers in Section C have the vinyl double bond as part of a ring, and in Section D are the propargyl and allenyl vinyl ethers. In all of these sections there are certain ether structures which have not been isolated but are thought to be the *in situ* rearranging species. They have been bracketed.

Both Table III, the amino-Claisen rearrangements, and Table IV, the thio-Claisen rearrangements, have been divided, for convenience, into A, Aromatic and Heterocyclic Compounds, and B, Aliphatic Compounds. Again, in these tables, brackets indicate *in situ* prepared starting materials, *i.e.*, ammonium ions, sulfides, or sulfonium ions.

In Table V for the Cope rearrangements the 1,5-hexadiene structures comprise Section A, the oxy-Cope rearrangements comprise Section B, and all the others are tabulated in Section C as miscellaneous Cope rearrangements. The compounds in all three of these sections are listed in order of increasing complexity of molecular formula.

Throughout all of the tables, summaries of reaction conditions have been given when they were available. Product ratios and/or percentage yields are recorded. Data for reactions at equilibrium are so marked. Consultation of the references cited will reveal useful additional information on these reactions which was not readily tabulated.

TABLE I. AROMATIC CLAISEN REARRANGEMENTS OF ALLYL AND PROPARGYL ETHERS

A. Allyl Ethers of Benzene Derivatives

Ring Substituents in $CH_2=CHCH_2OC_6H_5$	Conditions	Product(s) and Ratio ( ), Substituents in Phenol Ring	Yield(s), %	Refs.
None	Reflux, 6 hr, neat	2-Allyl	77	344
2-Acetamido	190°, 30 min, $(CH_3)_2NC_6H_5$	2-Acetamido-6-allyl,	50	191
		2-acetamido-4-allyl	7.6	
3-Acetamido	Reflux, 6 hr, $(CH_3)_2NC_6H_5$, N_2 atm	2-Allyl-5-acetamido,	40	345
		2-allyl-3-acetamido	46	
4-Acetamido	Reflux, 0.1 hr, $(C_6H_5)_2O$	2-Allyl-4-acetamido	63	344
3-Acetoxy	Reflux, 50 min, $(C_2H_5)_2NC_6H_5$, N_2 atm	2-Allyl-3-acetoxy, 2-allyl-5-acetoxy	73 (total)	360
4-Acetyl	Reflux, 1 hr, $(C_6H_5)_2O$	2-Allyl-4-acetyl	76	344
2-Acetyl-3-methoxy	215-220°, 24 hr, sealed tube	2-Acetyl-3-methoxy-6-allyl	35.4	346
2-Allyl-6-(α-phenylallyl)	210°, 3 hr, $(C_2H_5)_2NC_6H_5$	2,6-Diallyl-4-(γ-phenylallyl), 2,4-diallyl-6-(α-phenylallyl)	42 (total)	347
2-Amino	190-195°, 30 min, $(C_6H_5)_2O$	2-Amino-6-allyl,	42	191
		2-amino-4-allyl	21	
4-Amino	Reflux, 0.1 hr, $(C_6H_5)_2O$	2-Allyl-4-amino	53	344
4-t-Amyl	210-220°, 2 hr, CO_2 atm	2-Allyl-4-t-amyl	85	348
3-Benzoyl	200°, 1-3 half-lives, Carbitol, sealed tube	2-Allyl-5-benzoyl (1), 2-allyl-3-benzoyl (3.4)	—	349
4-Benzoyl	Reflux, 1 hr, $(C_6H_5)_2O$	2-Allyl-4-benzoyl	73	344
2-Benzoyl-3-methoxy	Heat	2-Benzoyl-3-methoxy-6-allyl, 2-benzoyl-3-methoxy-4-allyl	—	350

Note: References 344-439 are on pp. 251-252.

75

TABLE I. AROMATIC CLAISEN REARRANGEMENTS OF ALLYL AND PROPARGYL ETHERS (Continued)

A. Allyl Ethers of Benzene Derivatives (Continued)

Ring Substituents in $CH_2=CHCH_2OC_6H_5$	Conditions	Product(s) and Ratio ( ), Substituents in Phenol Ring	Yield(s), %	Refs.
3-Bromo	200°, 1–3 half-lives, Carbitol, sealed tube	2-Allyl-3-bromo (1.9), 2-allyl-5-bromo (1)	—	349
4-Bromo	Reflux, 0.1 hr, neat	2-Allyl-4-bromo	51	344
2-t-Butyl	195–200°, neat, N_2 atm	2-t-Butyl-4-allyl (1), 2-t-butyl-6-allyl (15)	—	186
4-t-Butyl	210–220°, 1.5 hr, CO_2 atm	2-Allyl-4-t-butyl	80	351
3-Chloro	200°, 1–3 half-lives, Carbitol, sealed tube	2-Allyl-3-chloro (2), 2-allyl-5-chloro (1)	—	349
4-Chloro	Reflux, 0.3 hr, neat	2-Allyl-4-chloro	55	344
2-(p-Chlorobenzoyl)-3-methoxy	Heat, $(C_2H_5)_2NC_6H_5$	2-(p-Chlorobenzoyl)-3-methoxy-6-allyl	—	352
3-Cyano	200°, 1–3 half-lives, Carbitol, sealed tube	2-Allyl-3-cyano (2.3), 2-allyl-5-cyano (1)	—	349
4-Cyano	Reflux, 0.1 hr, neat	2-Allyl-4-cyano	56	344
2,6-Dichloro	193–200°, 90 min, neat	2-Allyl-6-chloro (3.35), 2-allyl-4,6-dichloro (1), 2,6-dichloro-4-allyl (45.7), 2-methyl-5,7-dichlorocoumaran (trace)	—	354
	180–185°, 5.5 hr, $(C_6H_5)_2O$	2-Allyl-6-chloro (1), 2,6-dichloro-4-allyl (33)	—	354
	180–185°, 3 hr, $C_6H_5NO_2$	2-Allyl-6-chloro (1.4), 2-allyl-4,6-dichloro (1), 2,6-dichloro-4-allyl (10.6)	—	354

Substituent	Conditions	Product	Yield (%)	Refs.
2,4-Dichloro-5-methy	Reflux, 8 hr, $(CH_3)_2NC_6H_5$	2-Allyl-3-methyl-4,6-dichloro	77	355
2,6-Diisobutyl	244–288°, 4.75 hr, neat	2-Allyl-3-methyl-4,6-dichloro	23	355
2,6-Di(methallyl)	230–260°, 45 min, $(C_2H_5)_2NC_6H_5$	2,6-Diisobutyl-4-allyl	61	353
	230°, 3.5 min, $(C_2H_5)_2NC_6H_5$	2,6-Di(methallyl)-4-allyl (1.2), 2,4-di(methallyl)-6-allyl (1)	61 (total)	353
2,4-Di(methoxycarbonyl)	180–200°, 10 hr, neat	2,4-Di(methoxycarbonyl)-6-allyl	81	356
2,5-Dimethoxy-3-tosyloxy-4-acetyl	190–195°, 1.75 hr, reduced pres.	2,5-Dimethoxy-3-tosyloxy-4-acetyl-6-allyl	60	206
3,4-Dimethyl	245°, 30 min, $(C_2H_5)_2NC_6H_5$	2-Allyl-3,4-dimethyl (1), 2-allyl-4,5-dimethyl (2.3)	71 (total)	357
4-Dimethylamino	220°, 9 hr, $(C_6H_5)_2O$, N_2 atm	2-Allyl-4-dimethylamino	—	83
2-Ethoxy	200°	2-Allyl-6-ethoxy	81–85	358
2-Ethyl	195–200°, neat, N_2 atm	2-Ethyl-6-allyl (9), 2-ethyl-4-allyl (1)	—	186
4-Ethyl	220°, 9 hr, $(C_6H_5)_2O$, N_2 atm	2-Allyl-4-ethyl	—	83
2-Hydroxy	200–205°, 5 min, N_2 atm	2-Hydroxy-3-allyl, 2-hydroxy-4-allyl	45 / 39	190
3-Hydroxy	Reflux, 1 hr, $(C_2H_5)_2NC_6H_5$	2-Allyl-5-hydroxy (1), 2-allyl-3-hydroxy (1.3)	94 (total)	359
3-Hydroxy-4-methoxy-methylcarbonyl	190–195°, 2 hr, reduced pres.	2-Allyl-3-hydroxy-4-methoxy-methylcarbonyl	66.7	205
2-Isopropyl	195–200°, neat, N_2 atm	2-Isopropyl-6-allyl (9), 2-isopropyl-4-allyl (1)	—	186
2-Methallyl	240–250°, 30 min, $(C_2H_5)_2NC_6H_5$	2-Allyl-6-methallyl	71	353
3-Methoxy	200°, 1–3 half-lives, diethylene glycol, sealed tube	2-Allyl-3-methoxy (1), 2-allyl-5-methoxy (2)	—	349

Note: References 344–439 are in pp. 251–252.

77

TABLE I. Aromatic Claisen Rearrangements of Allyl and Propargyl Ethers (Continued)

A. Allyl Ethers of Benzene Derivatives (Continued)

Ring Substituents in $CH_2=CHCH_2OC_6H_5$	Conditions	Product(s) and Ratio ( ), Substituents in Phenol Ring	Yields(s) %	Refs.
4-Methoxy	Reflux, 1 hr, $(C_6H_5)_2O$	2-Allyl-4-methoxy	62	344
3-Methoxycarbonyl	220°, 9 hr, $(C_6H_5)_2O$, N_2 atm	2-Allyl-5-methoxycarbonyl, 2-allyl-3-methoxycarbonyl	98 (total)	83
2-Methoxycarbonyl-6-methyl	Reflux, 3 hr, $(C_2H_5)_2NC_6H_5$, N_2 atm	2-Methoxycarbonyl-4-allyl-6-methyl	70	103
2-Methyl	195–200°, neat, N_2 atm	2-Methyl-4-allyl (1), 2-methyl-6-allyl (5.7)	— —	186
3-Methyl	Reflux, 8.5 hr, $(CH_3)_2NC_6H_5$	2-Allyl-3-methyl (1.1), 2-allyl-5-methyl (1)	57 (total)	355
3-Methyl	200°, 1–3 half-lives, Carbitol, sealed tube	2-Allyl-3-methyl (1.5), 2-allyl-5-methyl (1)	— —	349
4-Methyl	Reflux, 1 hr, $(C_6H_5)_2O$	2-Allyl-4-methyl	55	344
4-Methylsulfinyl	Reflux, 1 hr, $(C_6H_5)_2O$	2-Allyl-4-methylsulfinyl	50	344
2-(3-Morpholinooxycarbonyl)-4-methoxycarbonyl	180–200°, 10 hr, neat	2-Allyl-4-methoxycarbonyl-6-(3-morpholinooxycarbonyl)	64	356
4-Nitro	Reflux, 6 hr, o-$C_6H_4Cl_2$	2-Allyl-4-nitro	59	344
2,3,4,5,6-Pentafluoro	365°, vapor phase	(cyclohexadienone structure with F substituents and $CH_2CH{=}CH_2$)	32	433
4-Phenyl	Reflux, 1 hr, $(C_6H_5)_2O$	2-Allyl-4-phenyl	60	344
2-Propionyl-3-methoxy	215–220°, 24 hr, sealed tube	2-Propionyl-3-methoxy-6-allyl	30	346
3-Trifluoromethyl	Reflux, 15 hr, neat, CO_2 atm	2-Allyl-5-trifluoromethyl	75	361

78

$CH_2=CHCH_2O$ — (benzene ring, positions 1–6, with γ β α on the allyl chain)

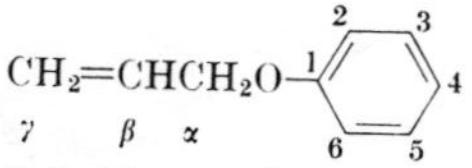

Substituents in

Allyl Group	Ring	Product(s) and Ratio ( )	Conditions	Yield(s), %	Refs.
α-Carboxy-β-methyl	None	2-HOC₆H₄CH=C(CH₃)CH₂CO₂H	270°, 12 hr, Na in diethylene glycol	65	362
α-Carboxy-β-methyl	4-Methyl	CH=C(CH₃)CH₂CO₂H, OH, CH₃	270°, 12 hr, Na in diethylene glycol	74	362
α-Carboxy-γ-methyl	None	2-HOC₆H₄C(CH₃)=CHCH₂CO₂H	Reflux, 12 hr, Na in diethylene glycol	64	362
α-Ethyl	2-Methoxycarbonyl-6-methyl	OH, CO₂CH₃, CH₃, CH(C₂H₅)CH=CH₂	120°, 18 hr, neat	17	103
α-Methoxy-α-methyl[a]	None	OCH₃, CH₃	150°, 11.5 hr, C₆H₆, trace (C₂H₅)₂NC₆H₅, sealed under vacuum	82	128b
α-Methoxy-α-methyl[a]	2-Methyl	OCH₃, CH₃, CH₃	160°, 21 hr, C₆H₆, sealed tube	76	128b

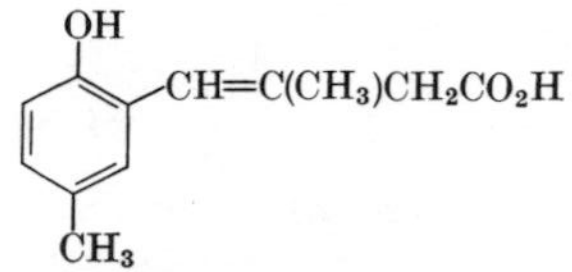

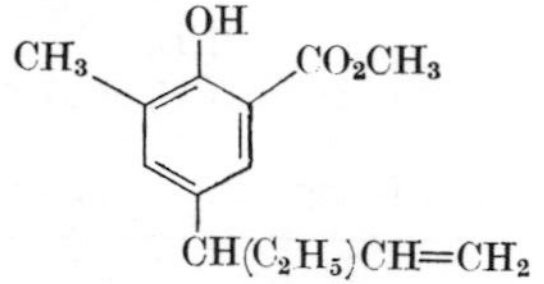

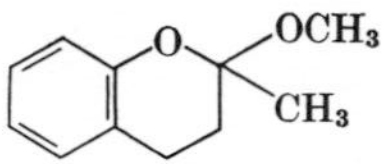

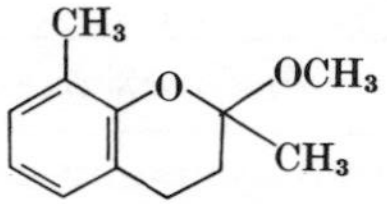

Note: References 344–439 are on pp. 251–252.
[a] The ether was prepared *in situ* (see pp. 14–15).

79

TABLE I. AROMATIC CLAISEN REARRANGEMENTS OF ALLYL AND PROPARGYL ETHERS (*Continued*)

B. *Substituted Allyl Ethers of Benzene Derivatives* (*Continued*)

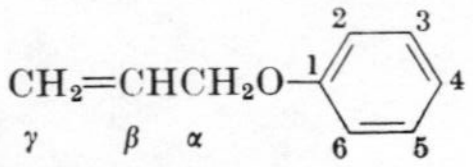

CH_2=$CHCH_2O$ — (benzene ring with positions 1, 2, 3, 4, 5, 6; allyl group labeled γ β α)

Substituents in		Conditions	Product(s) and Ratio ( )	Yield(s), %	Refs.
Allyl Group	Ring				
α-Methyl	None	169°, 2 hr, $(C_2H_5)_2NC_6H_5$	2-$HOC_6H_4CH_2CH$=$CHCH_3$ (*trans/cis* 14/1)	—	106
		220°, 9 hr, $(C_6H_5)_2O$, N_2 atm	2-$HOC_6H_4CH_2CH$=$CHCH_3$	94	83
α-Methyl	2-Methyl	169°, 2 hr, $(C_2H_5)_2NC_6H_5$	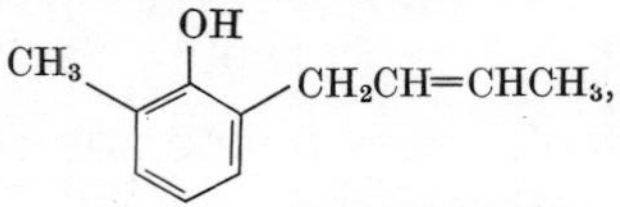 2,6-dimethylphenol with CH_2CH=$CHCH_3$, (*trans/cis* 38/1) (22.8); methylphenol with $CH(CH_3)CH$=CH_2 (1)	—	106
α-Methyl	2,6-Dimethyl	200°, 6 hr, sealed tube	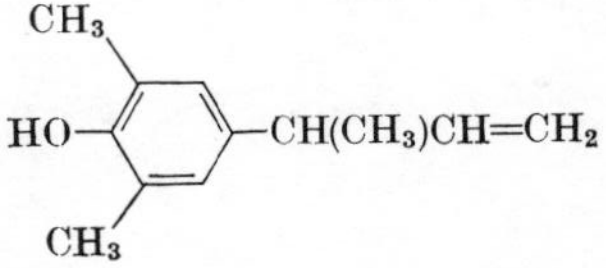 2,6-dimethyl-4-[$CH(CH_3)CH$=CH_2]phenol (HO, CH_3, CH_3)	80	105

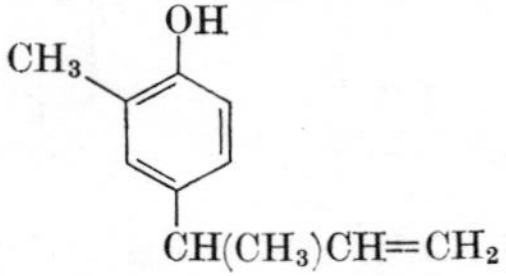

α-Methyl	2-Methoxycarbonyl-6-methyl	120°, 27 hr, neat	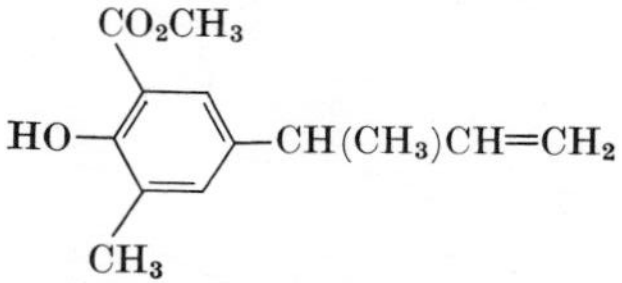	23	104
α,α-Dimethyl	None	Reflux, 0.5 hr, $(C_2H_5)_2NC_6H_5$	$2\text{-}HOC_6H_4CH_2CH=C(CH_3)_2$	>90	363
α,α-Dimethyl	3-Methoxy	Reflux, 1 hr, $(C_2H_5)_2NC_6H_5$	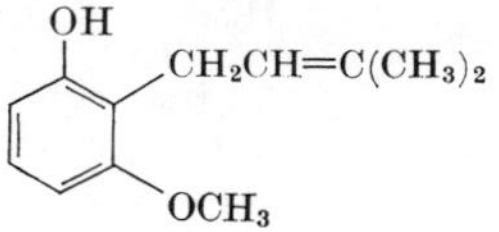 (1), 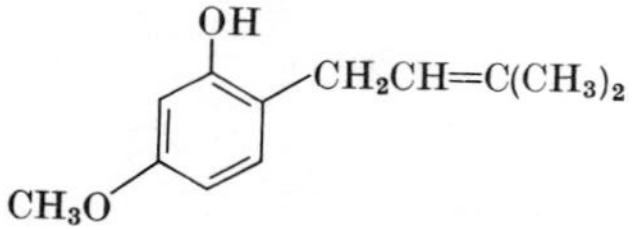	90	363
α,α-Dimethyl	4-Methoxy	Reflux, 1 hr, $(C_2H_5)_2NC_6H_5$	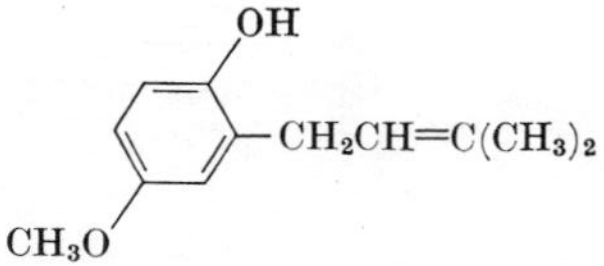	>90	363

Note: References 344–439 are on pp. 251–252.

81

TABLE I. AROMATIC CLAISEN REARRANGEMENTS OF ALLYL AND PROPARGYL ETHERS (Continued)

B. Substituted Allyl Ethers of Benzene Derivatives (Continued)

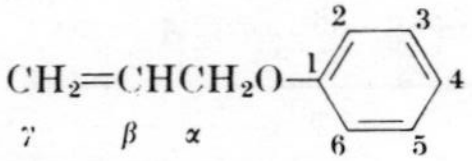

$CH_2=CHCH_2O$

Substituents in

Allyl Group	Ring	Conditions	Product(s) and Ratio ( )	Yield(s), %	Refs.
α,α-Dimethyl	3,5-Dimethoxy-4-acetyl	Reflux, 1 hr, $(C_2H_5)_2NC_6H_5$	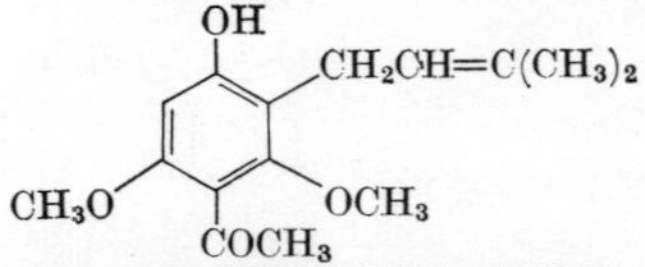	90	363
cis-α,γ-Dimethyl	None	165°, 24 hr, mesitylene	2-$HOC_6H_4CH(CH_3)CH=CHCH_3$ ($trans/cis$ 98/cis trace)	—	364
$trans$-α,γ-Dimethyl	None	165°, 24 hr, mesitylene	($trans/cis$ 9/1)	—	364
$trans$-α,γ-Dimethyl	None	195–200°, 1 hr, sealed tube	($trans$, only)	60	365
$R(+)$-$trans$-α,γ-Dimethyl	None	200°, 1 hr, neat	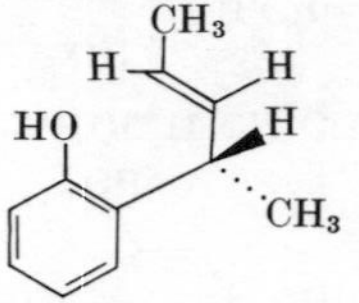 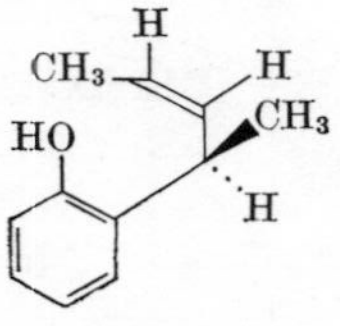	50 (total)	60

β-t-Butyl	None	195–197°, 24 hr, (C₆H₅)₂O, sealed tube	2-HOC₆H₄CH₂C(C₄H₉-t)=CH₂	44	366
β-t-Butyl	4-Methoxy	195–197°, 24 hr, (C₆H₅)₂O, sealed tube	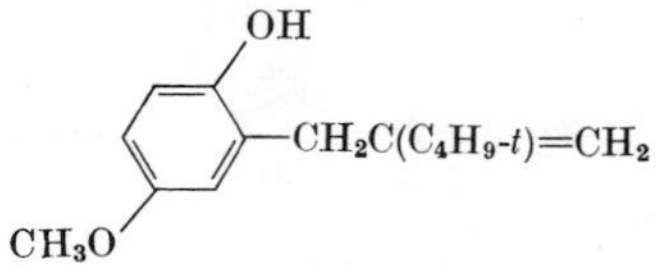 [structure: OH, CH₃O ring, CH₂C(C₄H₉-t)=CH₂]	56	366
β-Methyl	None	205–216°, 3.3 hr, neat	2-HOC₆H₄CH₂C(CH₃)=CH₂, I (4.4); 2-HOC₆H₄CH=C(CH₃)₂, II (1); 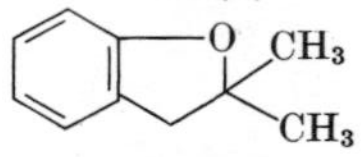 [benzofuran, 2,2-diCH₃], III (2.3)	—	202
		200–206°, 4.2 hr, C₆H₅NO₂	I (7.3), II (1.1), III (1)	—	202
		198–199°, 3.5 hr, 2,6-xylenol	I (1), II (1.7), III (12.2)	—	202
		212–215°, 3.7 hr, 2,6-xylidine	I (1), II (5.5), III (4)	—	202
		208–216°, 3.5 hr, m-CH₃C₆H₄N(CH₃)₂	I (81), II (4), III (1)	—	202
		205–215°, 7.8 hr, (n-C₄H₉)₃N	I (42.5), II (1.5), III (1)	—	202
		208–218°, 3.0 hr, p-CH₃C₆H₄CN	I (13.3), II (1.5), III (1)	—	202
		203–210°, 5.5 hr, dodecane	I (8.4), II (1), III (1.1)	—	202
		188–200°, 10 hr, o-CH₃C₆H₄N(CH₃)₂	I (81), II (3), III (1)	—	202
		199–205°, 4.8 hr, (CH₃)₂NC₆H₅	I (90), II (2), III (1)	—	202
		207–218°, 2.8 hr, (C₂H₅)₂NC₆H₅	I (43), II (2), III (1)	—	202
		214–225°, 1.9 hr, p-CH₃C₆H₄N(C₂H₅)₂	I (43.5), II (1.5), III (1)	—	202

Note: References 344–439 are on pp. 251–252.

TABLE I. AROMATIC CLAISEN REARRANGEMENTS OF ALLYL AND PROPARGYL ETHERS (*Continued*)

B. *Substituted Allyl Ethers of Benzene Derivatives* (*Continued*)

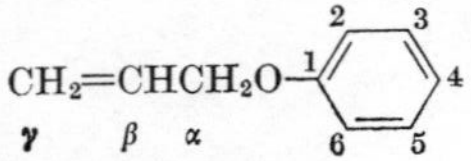

CH$_2$=CHCH$_2$O

Substituents in		Conditions	Product(s) and Ratio ( )	Yield(s), %	Refs.
Allyl Group	Ring				
β-Methyl	4-Methoxy	195–197°, 24 hr, (C$_6$H$_5$)$_2$O, sealed tube	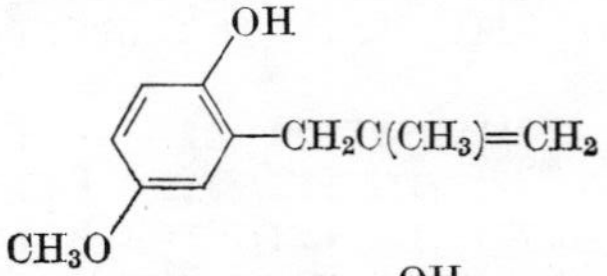	45	366
β-Methyl	2-Allyl-6-methallyl	240°, 48 min, (C$_2$H$_5$)$_2$NC$_6$H$_5$	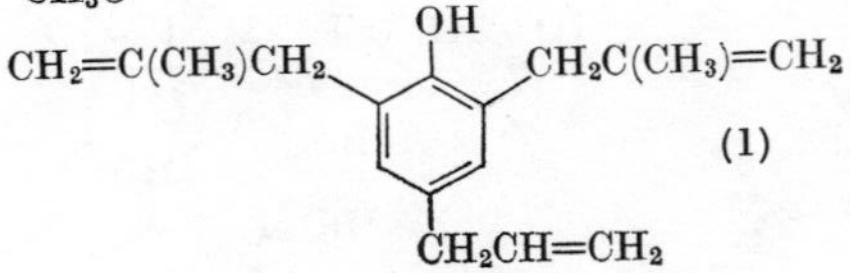 (1) 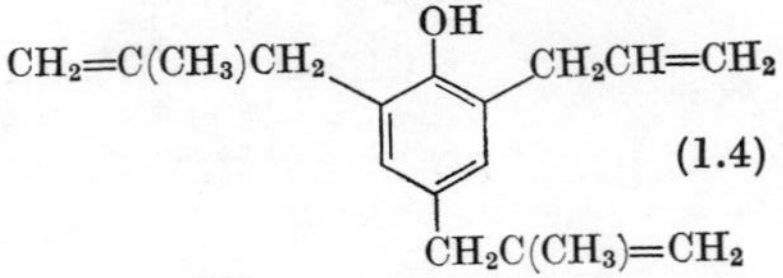 (1.4)	63 (total)	353
β-Methyl	2-Isobutyl-6-*n*-propyl	250°, 30 min, (C$_2$H$_5$)$_2$NC$_6$H$_5$	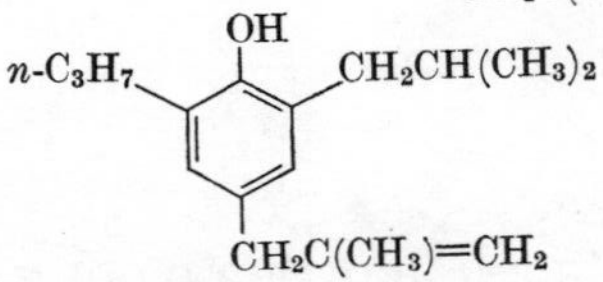	63	353

Substituent	Substituent	Conditions	Product	Yield (%)	Refs.
β-Methyl-γ-methoxy-carbonyl	None	270–280°, 3 hr	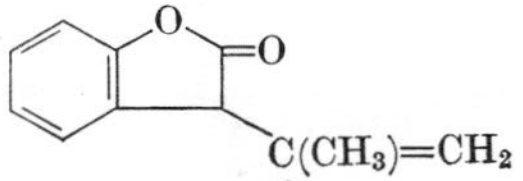	74	362
β-Methyl-γ-methoxy-carbonyl	4-Methyl	300°, 3 hr	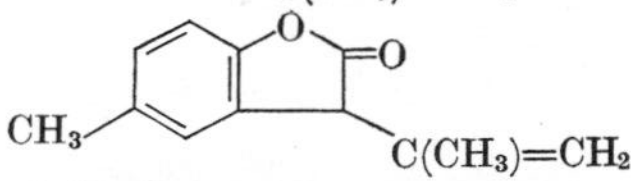	70	362
γ-Ethyl	None	165°, 175 hr, mesitylene 195°, 48 hr, $(C_2H_5)_2NC_6H_5$	2-$HOC_6H_4CH(C_2H_5)CH=CH_2$ 2-$HOC_6H_4CH(C_2H_5)CH=CH_2$, (1), 2-$HOC_6H_4CH(CH_3)CH=CH(CH_3)$, (1.35)	— —	195 195
γ-Ethyl	2,6-Dimethyl	Reflux, 3 hr, $(C_2H_5)_2NC_6H_5$	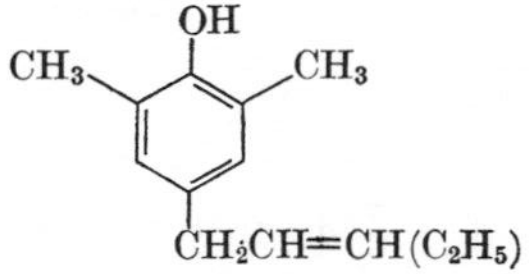	79	368
γ-Ethyl	2-Methoxycarbonyl-6-methyl	Reflux, 3 hr, $(C_2H_5)_2NC_6H_5$, N_2 atm	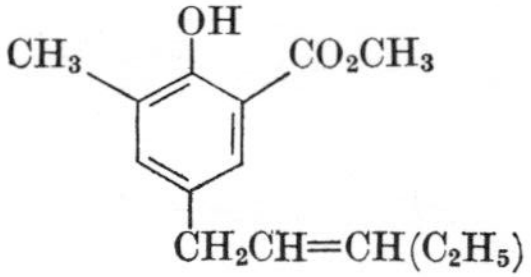	60	103
γ-Methyl (^{14}C)	4-Ethoxycarbonyl	220–235°, 80 min, reduced pres.	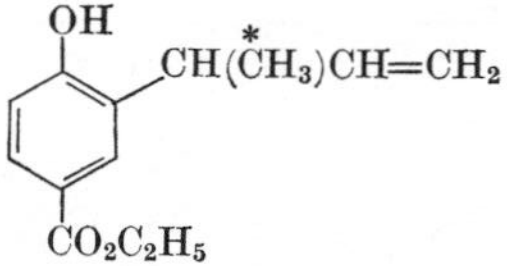 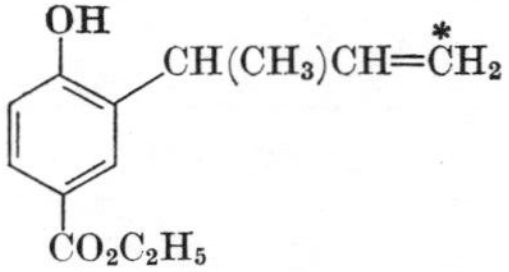	91 (total)	196, 197

Note: References 344–439 are on pp. 251–252.

85

B. *Substituted Allyl Ethers of Benzene Derivatives* (*Continued*)

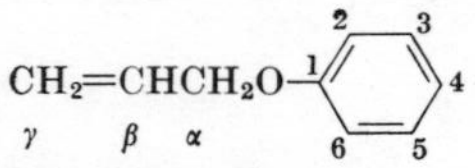

Substituents in

$CH_2=CHCH_2O$
$\gamma \quad \beta \quad \alpha$

Allyl Group	Ring	Conditions	Product(s) and Ratio ( )	Yield(s), %	Refs.
γ-Methyl *cis* or *trans*	4-Methoxy	195–197°, 24 hr, $(C_6H_5)_2O$, sealed tube	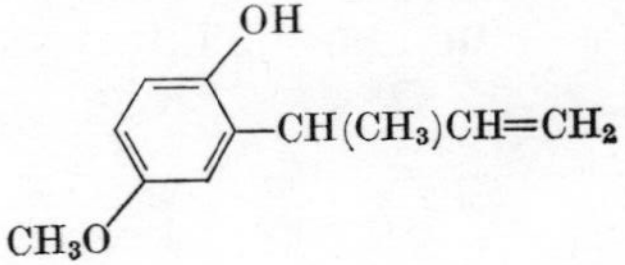	54	107
γ-Methyl	3-Methyl	186.5°, 16 hr, $(C_2H_5)_2NC_6H_5$	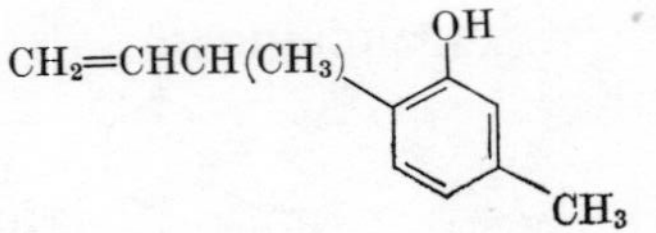 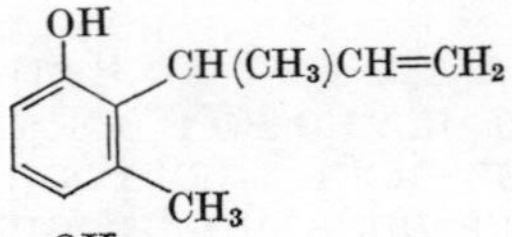 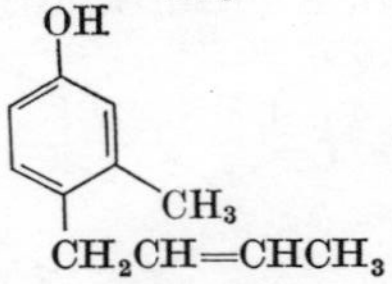	—	184

86

γ-Methyl (¹⁴C) (cis-crotyl)	4-Methyl	230°, 3 hr, (C₂H₅)NC₆H₅, reduced pres.	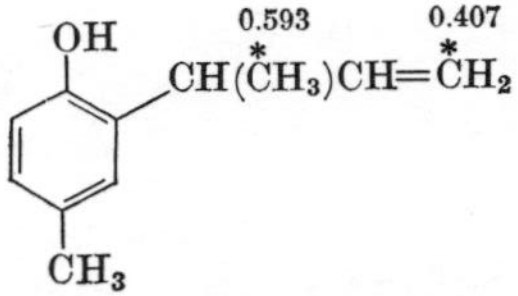 OH, CH(CH₃)CH=CH₂ (0.593*, 0.407*), CH₃	98	192
		197°, 24 hr, (C₂H₅)₂NC₆H₅, reduced pres.	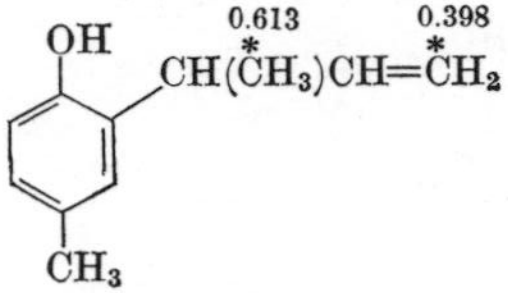 OH, CH(CH₃)CH=CH₂ (0.613*, 0.398*), CH₃	84	192
γ-Methyl	3,5-Diethyl	186.5°, 16 hr, (C₂H₅)₂NC₆H₅	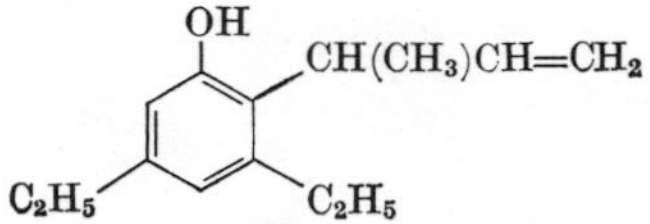 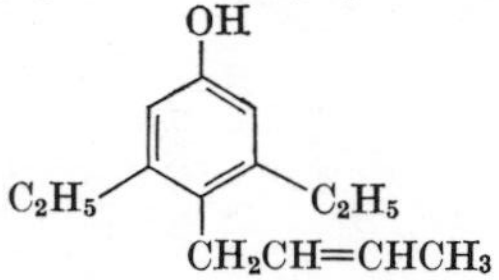 OH, C₂H₅, C₂H₅, CH(CH₃)CH=CH₂, C₂H₅ (2.4); OH, C₂H₅, C₂H₅, CH₂CH=CHCH₃ (1)	—	184

Note: References 344–439 are on pp. 251–252.

TABLE I. Aromatic Claisen Rearrangements of Allyl and Propargyl Ethers (*Continued*)

B. Substituted Allyl Ethers of Benzene Derivatives (*Continued*)

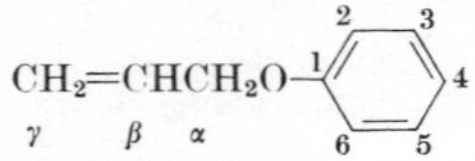

Substituents in Allyl Group	Substituents in Ring	Conditions	Product(s) and Ratio ( )	Yield(s), %	Refs.
γ-Methyl	3,5-Dimethoxy	186.5°, 16 hr, $(C_2H_5)_2NC_6H_5$	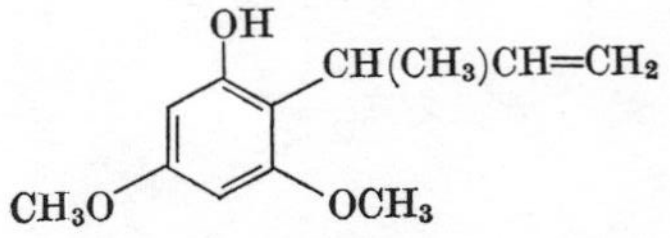 (5), 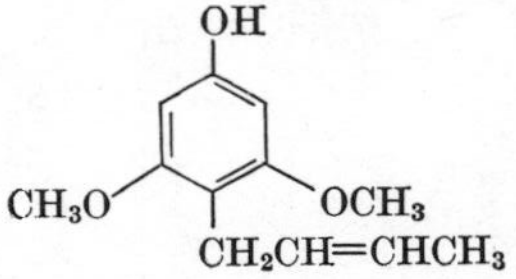 (1)	—	184
γ-Methyl	2,4-Di(methoxy-carbonyl)	180–200°, 10 hr, neat	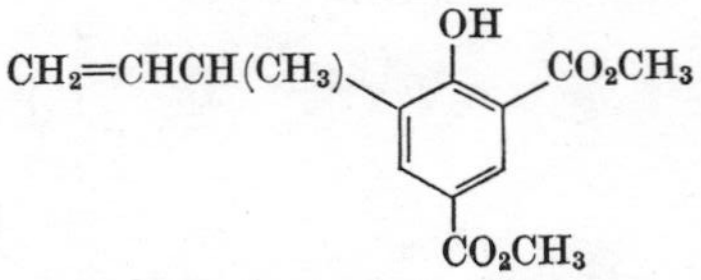	74	356
γ-Methyl	2,4-Dimethyl	190–200°, 2.5 hr, decalin	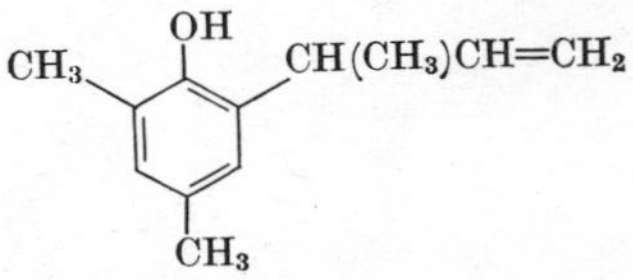	78	367

			Products	Yield (%)	Ref.

Reactant		Conditions	Products	Yield (%)	Ref.
γ-Methyl, cis and trans	3,5-Dimethyl	195–197°, 24 hr, (C₆H₅)₂O, sealed tube	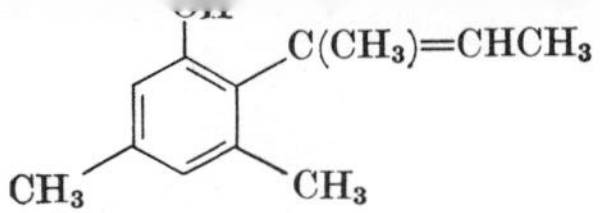 [structure: C(CH₃)=CHCH₃; CH₃, CH₃]	35	107
cis-γ-Methyl	3,5-Dimethyl	186°, 26.5 hr, (C₂H₅)₂NC₆H₅, reduced pres.	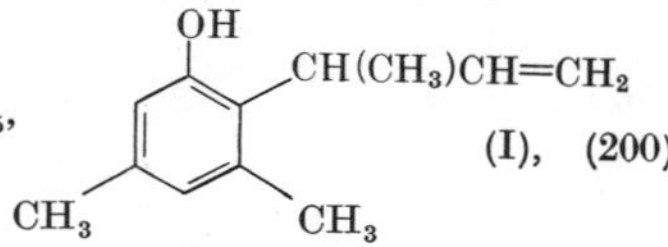 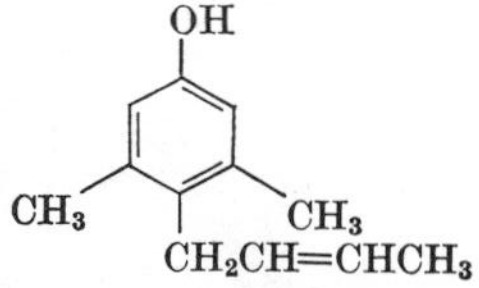 CH(CH₃)CH=CH₂, OH, CH₃, CH₃ (I), (200); CH₃, OH, CH₃	92 (total)	106
trans-γ-Methyl	3,5-Dimethyl	186°, 16 hr, (C₂H₅)₂NC₆H₅, reduced pres.	II (168); I (113.5), II	77 (total)	106
trans-γ-Methyl	3,5-Dimethyl	186°, 16 hr, (C₂H₅)₂NC₆H₅, reduced pres.	[OH, CH₃, CH₂CH=CHCH₃, CH₃]; II (cis 1) (trans 8.1); II (cis 1) (trans 5.6); II (cis 1) (trans 32.4)	88 (total)	106
γ-Methyl	3,5-Dimethyl	186°, 30 hr, neat	I (2.6), II (4.7), side products (III, 1)	—	184
		186°, 30 hr, decalin	I (3.5), II (4.5), III (1)	—	184
		186°, 30 hr, (C₂H₅)₂NC₆H₅	I (2.4), II (1)	—	184
		186°, 30 hr, C₆H₅CN	I (5), II (1)	—	184
		186°, 30 hr, HCON(CH₃)₂	I (17.3), II (1)	—	184
trans-γ-Methyl	2,6-Dimethyl	186°, 3.5 hr, decane	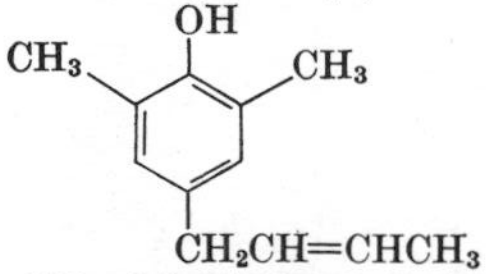 OH, CH₃, CH₃, CH₂CH=CHCH₃; IV (cis 1) (trans 6.7); 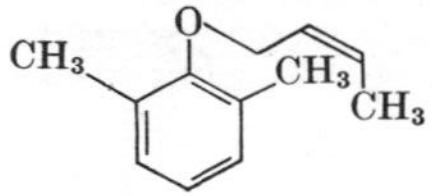 [CH₃, O, CH₃ structure] (cis 1)	91 (total)	106

Note: References 344–439 are on pp. 251–252.

89

TABLE I. AROMATIC CLAISEN REARRANGEMENTS OF ALLYL AND PROPARGYL ETHERS (Continued)

B. Substituted Allyl Ethers of Benzene Derivatives (Continued)

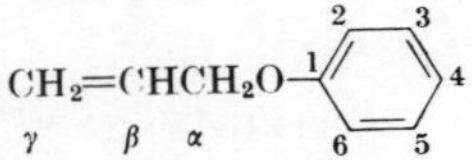

$$CH_2{=}CHCH_2O \quad (\gamma\ \beta\ \alpha)$$

Substituents in		Conditions	Product(s) and Ratio ( )	Yield(s), %	Refs.
Allyl Group	Ring				
cis-γ-Methyl	2,6-Dimethyl	186°, 20 hr, decane	IV (cis 1) (trans 2.5), 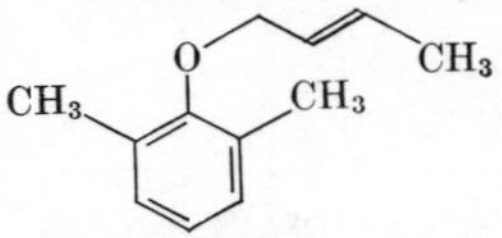 trans (3.5)	88 (total)	106
γ-Methyl	2,6-Dimethyl	Reflux, 3 hr, (C₂H₅)₂NC₆H₅	IV	67	368
γ-Methyl cis or trans	2,6-Dimethyl	186°, 0.5–1.5 hr, N₂ atm, sealed tube	IV (largely trans)	—	369
γ-Methyl	2,6-Dimethyl	200°, 6 hr, sealed tube	IV (largely trans)	79	105
γ-Methyl	2-Methoxycarbonyl-6-methyl	Reflux, 3 hr, (C₂H₅)₂NC₆H₅		73	104
γ-Methyl	2-Methyl-4-methoxy	Heat, (CH₃)₂NC₆H₅	(Major)	—	367

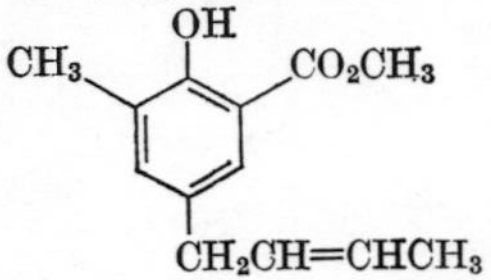

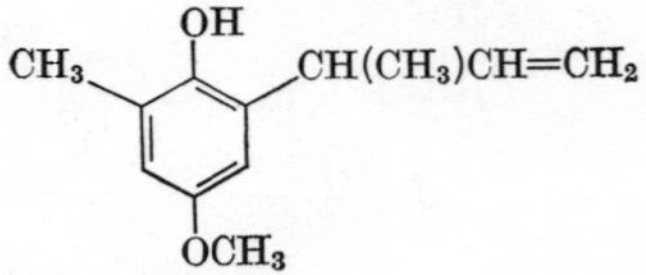

γ-Methyl	3-Methyl-4-methoxy	Heat, $(CH_3)_2NC_6H_5$	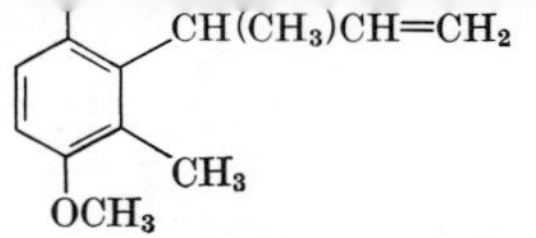 (1), 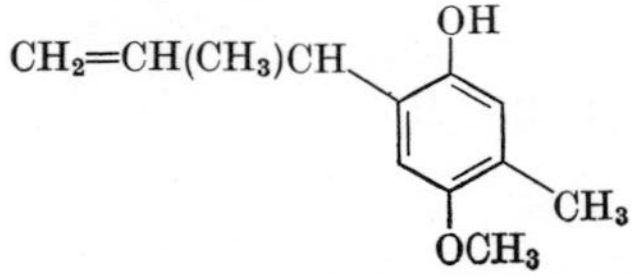 (3.8)	—	367
γ-Methyl	2-(3-Piperidinyloxy-carbonyl)-4-methoxy-carbonyl	180–200°, 10 hr, neat	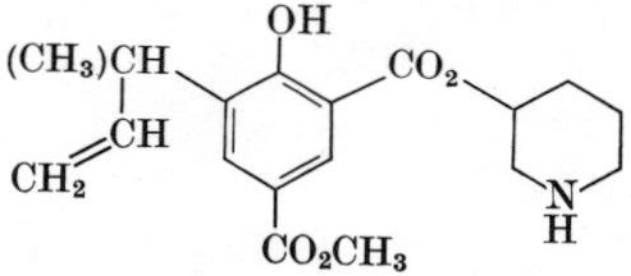	65	356
γ-Methyl	2,3-Dimethyl-4-allyl	Heat, $(CH_3)_2NC_6H_5$	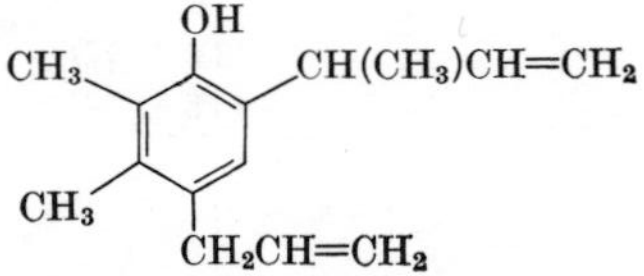	75	367
γ-Methyl	2,3-Dimethyl-4-methoxy	Heat, $(CH_3)_2NC_6H_5$	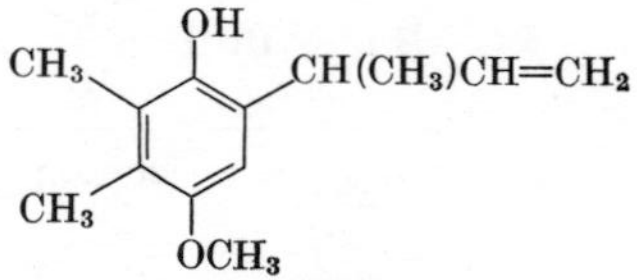 (Major)	—	367
γ-Methyl	2,5-Dimethyl-4-methoxy	Heat, $(CH_3)_2NC_6H_5$	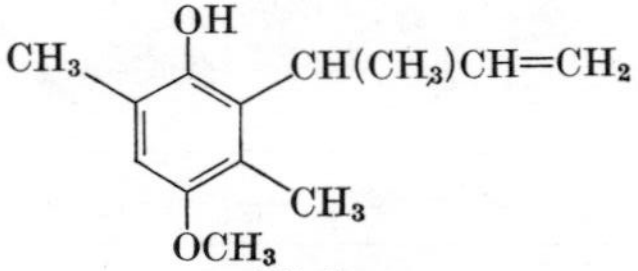 (Major)	—	367

Note: References 344–439 are on pp. 251–252.

TABLE I. AROMATIC CLAISEN REARRANGEMENTS OF ALLYL AND PROPARGYL ETHERS (Continued)

B. Substituted Allyl Ethers of Benzene Derivatives (Continued)

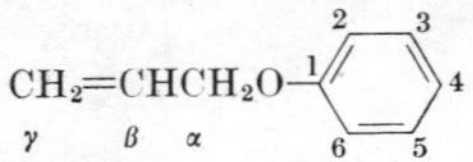

$CH_2=CHCH_2O$
γ β α

| Substituents in | | | | Yield(s), | |
Allyl Group	Ring	Conditions	Product(s) and Ratio ( )	%	Refs.
γ-Phenyl (cis or trans)	None	Reflux, 4–13 hr, $(C_2H_5)_2NC_6H_5$	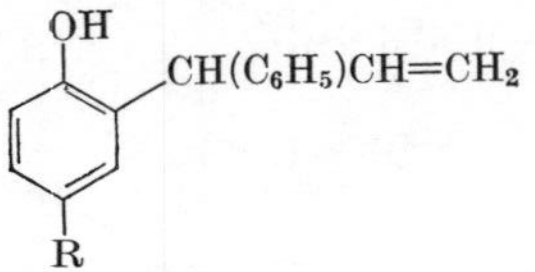 V, R = H	29 from cis	369
		150°, 10 half-lives, Carbitol, sealed tube	V, R = H	90	370
γ-Phenyl	4-Methoxy	195–197°, 24 hr, $(C_6H_5)_2O$, sealed tube	V, R = OCH$_3$	55	107
γ-Phenyl	4-Methyl	150°, 10 half-lives, Carbitol, sealed tube	V, R = CH$_3$	80	370
γ-Phenyl	2,6-Diallyl	201–205°, 1–2 mm, distil	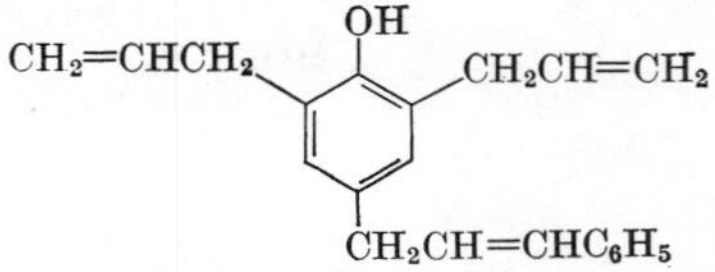	31	347
γ-Phenyl	3,5-Dimethyl	195–197°, 24 hr, $(C_6H_5)_2O$, sealed tube	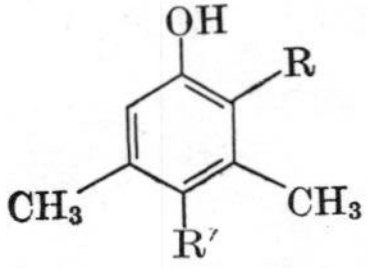 VI R = $C(C_6H_5)$=CHCH$_2$, R' = H	83	107

γ-Phenyl	2,4-Di(methoxy-carbonyl)	186°, 16 hr, $(C_2H_5)_2NC_6H_5$	VI, R = $CH(C_6H_5)CH=CH_2$, R′ = H (1) VI, R = H, R′ = $CH_2CH=CH=CH(C_6H_5)$ (7.9) [structure: CH_3O_2C / OH / $CH(C_6H_5)CH=CH_2$ / CO_2CH_3]	—	184
γ-(m-Chlorophenyl)	None	180–200°, 10 hr, neat	2-$HOC_6H_4CH(CH=CH_2)C_6H_4Cl$-m	62	356
γ-(m-Chlorophenyl)	4-Methyl	150°, 10 half-lives, Carbitol, sealed tube	[structure: OH / CHC_6H_4Cl-m / $CH=CH_2$ / CH_3]	75	370
γ-(p-Chlorophenyl)	4-Methyl	150°, 10 half-lives, Carbitol, sealed tube	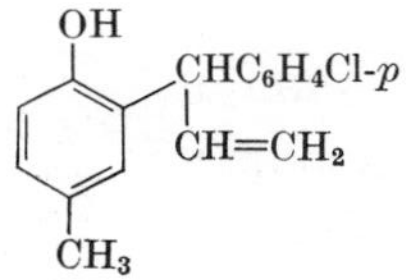 [structure: OH / CHC_6H_4Cl-p / $CH=CH_2$ / CH_3]	90	370
γ-(m-Cyanophenyl)	4-Methyl	150°, 10 half-lives, Carbitol, sealed tube	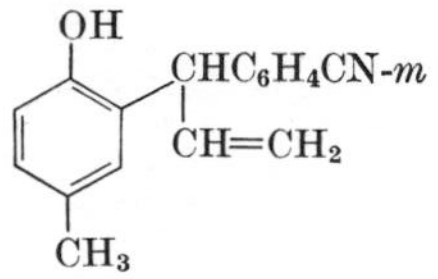 [structure: OH / CHC_6H_4CN-m / $CH=CH_2$ / CH_3]	80	370

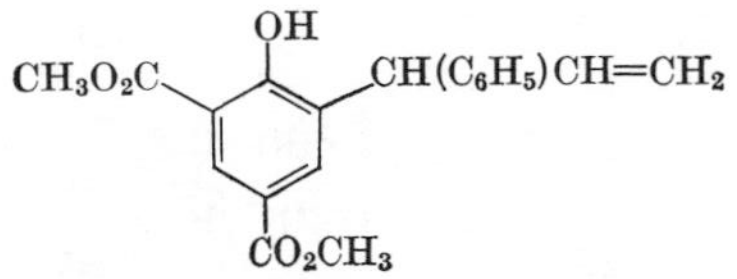

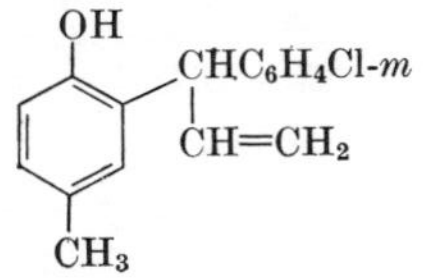

Note: References 344–439 are on pp. 251–252.

TABLE I. AROMATIC CLAISEN REARRANGEMENTS OF ALLYL AND PROPARGYL ETHERS (Continued)

B. Substituted Allyl Ethers of Benzene Derivatives (Continued)

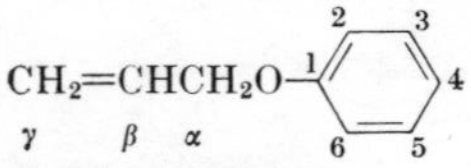

$$CH_2=CHCH_2O \quad \gamma \ \beta \ \alpha$$

| Substituents in | | | | Yield(s), | |
Allyl Group	Ring	Conditions	Product(s) and Ratio ( )	%	Refs.
γ-(m-Methoxyphenyl)	4-Methyl	150°, 10 half-lives, Carbitol, sealed tube	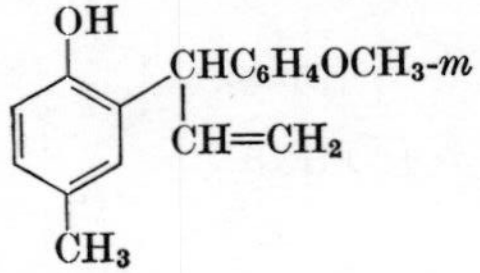	95	370
γ-(p-Methoxyphenyl)	4-Methyl	150°, 10 half-lives, Carbitol, sealed tube	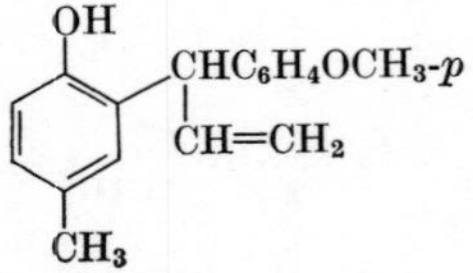	95	370
γ-(m-Nitrophenyl)	4-Methyl	150°, 10 half-lives, Carbitol, sealed tube	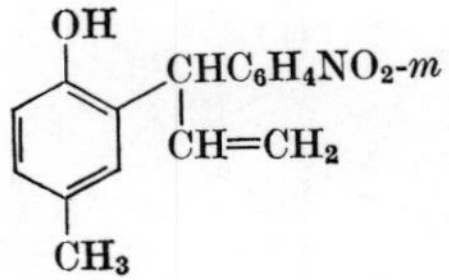	75	370
γ-(m-Tolyl)	4-Methyl	150°, 10 half-lives, Carbitol, sealed tube	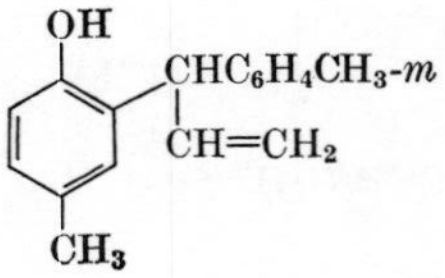	95	370

94

γ-(p-Tolyl)	4-Methyl	(structure: benzene ring bearing OH; ring substituent –CH(C6H4CH3-p)–CH=CH2; ring CH3)	150°, 10 half-lives, Carbitol, sealed tube	95	370
γ,γ-Dimethyl	None	4-HOC6H4CH2CH=C(CH3)2 (35.4), 2-HOC6H4CH2CH=C(CH3)2 (20.6), 4-HOC6H4CH2CH2C(CH3)=CH2 (8.5), 2-HOC6H4CH(CH3)C(CH3)=CH2 (1)	220°, neat	49.2 (total)	371
		2-HOC6H4C(CH3)2CH=CH2 (16.6), 2-HOC6H4CH2CH=C(CH3)2 (1), 2-HOC6H4CH(CH3)C(CH3)=CH2 (14)	205–220°, +Na2CO3	—	371
		2-HOC6H4CH(CH3)C(CH3)=CH2 (VII, 29.6), 4-HOC6H4CH2CH=C(CH3)2 (VIII, 1.3), (benzofuran structure with two CH3 groups) (IX, 1), C6H5OH X (Trace)	184°, 90 hr, HCON(CH3)2, bomb tube	—	187
		VII (1), VIII (2.4), IX (Trace), X (Trace)	184°, 90 hr, (C2H5)2NC6H5, bomb tube	—	187
		VII (1), VIII (21.5), IX (Trace), X (3.5)	184°, 90 hr, neat, bomb tube	—	187
		VII (1.9), VIII (1), IX (1.3), X (Trace)	184°, 90 hr, +Na2CO3, bomb tube	—	187
		VII (1), VIII (4.1), IX (Trace), X (Trace)	214°, 4.5 hr, (C2H5)2NC6H5	—	187

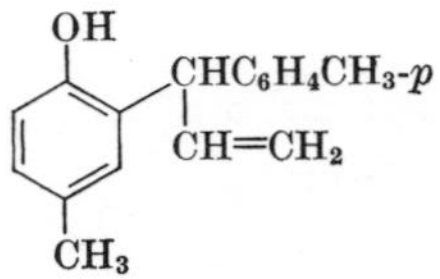

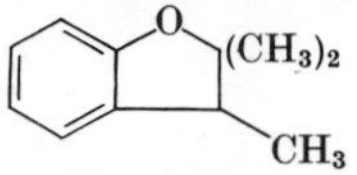

Note: References 344–439 are on pp. 251–252.

TABLE I. AROMATIC CLAISEN REARRANGEMENTS OF ALLYL AND PROPARGYL ETHERS (*Continued*)

B. *Substituted Allyl Ethers of Benzene Derivatives* (*Continued*)

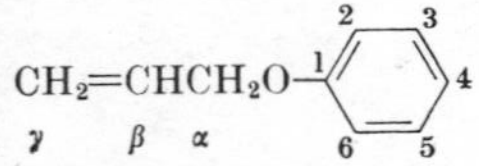

$$CH_2=CHCH_2O$$
$$\gamma \quad \beta \quad \alpha$$

Substituents in				
Allyl Group	Ring	Conditions	Product(s) and Ratio ( )	Yield(s), %	Refs.

Allyl Group	Ring	Conditions	Product(s) and Ratio ( )	Yield(s), %	Refs.
γ,γ-Dimethyl	4-Methyl	$220°$, $(CH_3)_2NC_6H_5$	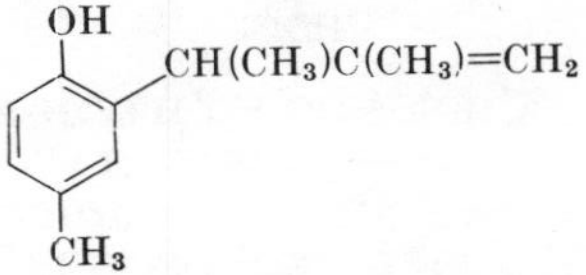	—	372
γ,γ-Dimethyl	2-Acetyl-3,5-dimethoxy	Reflux, 5 hr, $(CH_3)_2NC_6H_5$	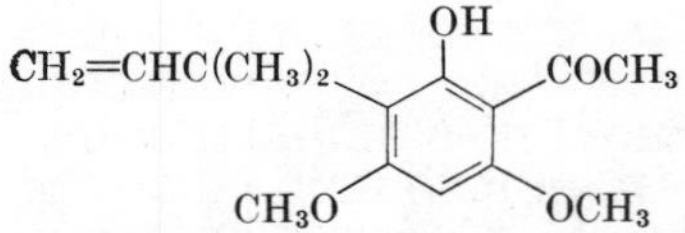 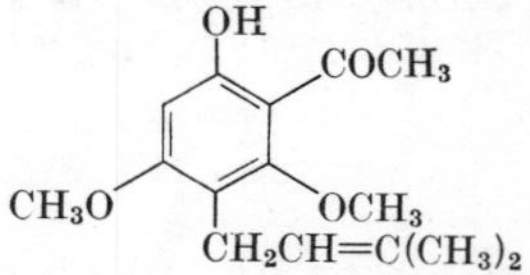	~70 (total) crude	188

C. *Propargyl Aryl Ethers*

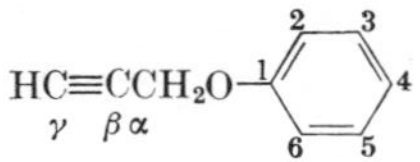

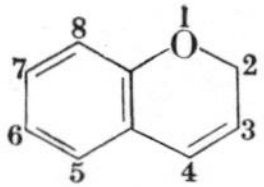

HC≡CCH₂O
$\gamma\ \beta\ \alpha$

Substituents in

Substituents in Product(s) and Yield(s) (%)

Propargyl Group	Ring	Conditions	Substituents in Product(s) and Yield(s) (%)	Refs.
None	None	220–230°, 12 hr, (C₂H₅)₂NC₆H₅	None (22)	109
None	4-Methoxy	''	6-Methoxy (30)	109
None	3-Methoxy	''	7-Methoxy (12.5)	109
None	2-Methoxy	''	8-Methoxy (11.9)	109
None	4-Chloro	''	6-Chloro (16.6)	109
None	3-Chloro	''	7-Chloro (48)	109
None	2-Chloro	''	8-Chloro (16)	109
γ-Phenyl	None	''	4-Phenyl (70)	109
γ-Phenyl	4-Methoxy	''	4-Phenyl-6-methoxy (46.4)	109
γ-Phenyl	3-Methoxy	''	4-Phenyl-7-methoxy (56.7)	109
γ-Phenyl	2-Methoxy	''	4-Phenyl-8-methoxy (26.6)	109
γ-Phenyl	4-Chloro	''	4-Phenyl-6-chloro (30)	109
γ-Phenyl	3-Chloro	''	4-Phenyl-7-chloro (43)	109
γ-Phenyl	2-Chloro	''	4-Phenyl-8-chloro (30)	109
γ-Phenyl	4-Nitro	''	4-Phenyl-6-nitro (15)	109

Note: References 344–439 are on pp. 251–252.

97

TABLE I. AROMATIC CLAISEN REARRANGEMENTS OF ALLYL AND PROPARGYL ETHERS (*Continued*)

C. Propargyl Aryl Ethers (Continued)

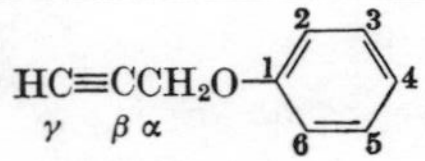

HC≡CCH₂O
γ β α

Substituents in

Propargyl Group	Ring	Conditions	Product(s) and Yield(s) (%),	Refs.
None	2,6-Dimethyl	200°	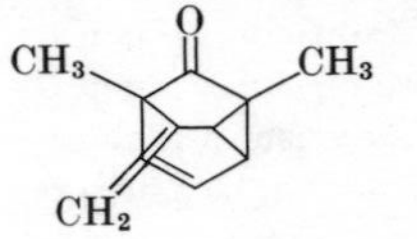 —	110
None	2,4,6-Trimethyl	200°	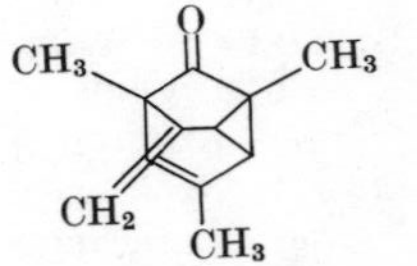 —	110
α-Methyl	2,6-Dimethyl	200°	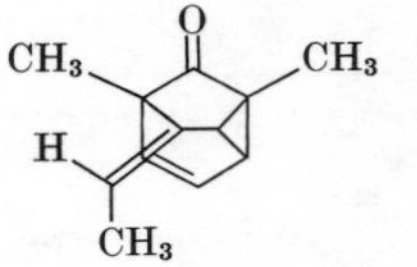 —	110
γ-Methyl	2,6-Dimethyl	200°	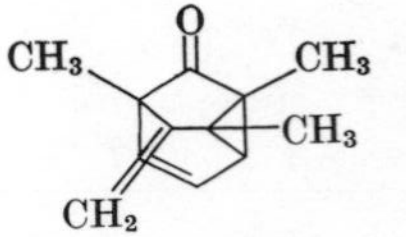 (Major), 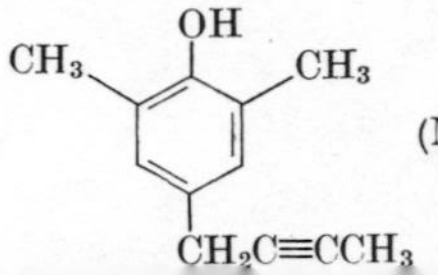 (Minor)	110

Propargyl Aryl Ether	Conditions	Product and Yield (%)	Refs.
2-C$_{10}$H$_7$OCH$_2$C≡CH	Reflux, 40 min, (C$_2$H$_5$)$_2$NC$_6$H$_5$	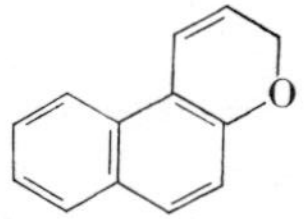 (40)	108
2-C$_{10}$H$_7$OCH$_2$C≡CCH$_3$	Reflux, 4 hr, (C$_2$H$_5$)$_2$NC$_6$H$_5$	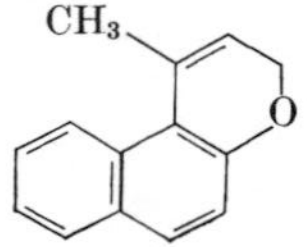 (49)	108
1-C$_{10}$H$_7$OCH$_2$C≡CC$_6$H$_5$	Reflux, 4.5 hr, (C$_2$H$_5$)$_2$NC$_6$H$_5$	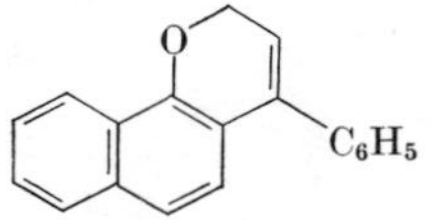 (50)	108
2-C$_{10}$H$_7$OCH$_2$C≡CC$_6$H$_5$	Reflux, 40 min, (C$_2$H$_5$)$_2$NC$_6$H$_5$	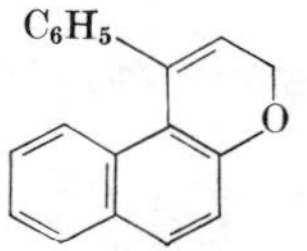 (95)	108

D. Miscellaneous Diaryl Diether Rearrangements

Diaryl Ether	Conditions	Product(s) and Ratio ( )	Yield, %	Refs.
C$_6$H$_5$OCH$_2$C≡CCH$_2$OC$_6$H$_5$	Reflux, 10–12 hr, (C$_2$H$_5$)$_2$NC$_6$H$_5$	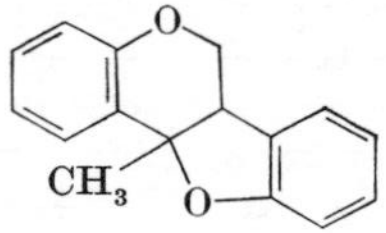	60	218

Note: References 344–439 are on pp. 251–252.

99

TABLE I. AROMATIC CLAISEN REARRANGEMENTS OF ALLYL AND PROPARGYL ETHERS (Continued)

D. Miscellaneous Diaryl Diether Rearrangements (Continued)

Diaryl Ether	Conditions	Product(s) and Ratio ( )	Yield, %	Refs.
$p\text{-ClC}_6\text{H}_4\text{OCH}_2\text{C}{\equiv}\text{CCH}_2\text{OC}_6\text{H}_4\text{Cl-}p$	Reflux, 10 hr, $(\text{C}_2\text{H}_5)_2\text{NC}_6\text{H}_5$	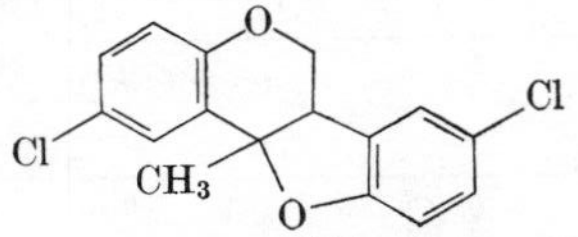	40	373
$\text{C}_6\text{H}_5\text{OCH}_2\text{CCH}_2\text{OC}_6\text{H}_5$ $\overset{\|}{}\ \text{CH}_2$	195–200°, 6 hr, $(\text{CH}_3)_2\text{NC}_6\text{H}_5$		66	374
$o\text{-CH}_3\text{C}_6\text{H}_4\text{OCH}_2\text{CCH}_2\text{OC}_6\text{H}_4\text{CH}_3\text{-}o$ $\overset{\|}{}\ \text{CH}_2$	200–210°, 8 hr, $(\text{CH}_3)_2\text{NC}_6\text{H}_5$	$2\text{-HOC}_6\text{H}_4\text{CH}_2\text{CCH}_2\text{C}_6\text{H}_4\text{OH-2}$ $=$	55 (total)	374
	200–210°, 8 hr, $(\text{CH}_3)_2\text{NC}_6\text{H}_5$	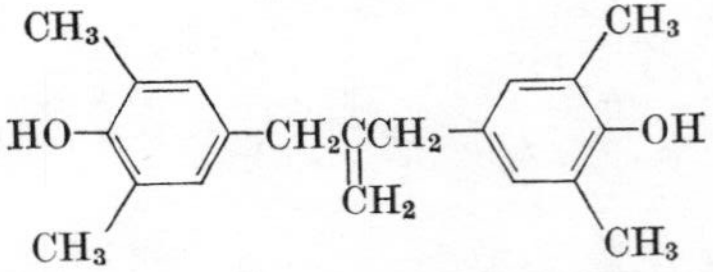	73	374

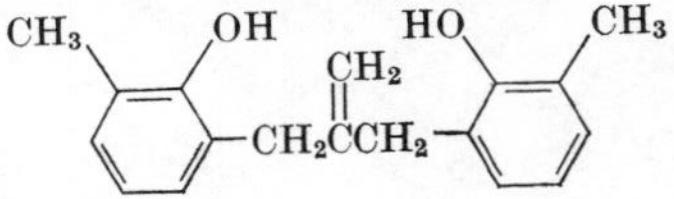

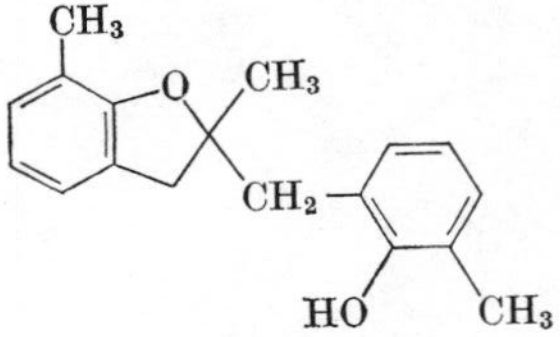

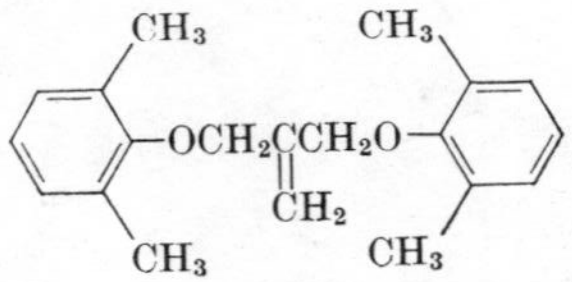

Reactant	Conditions	Product	Yield (%)	Reference
$C_6H_5OCH_2CH\overset{t}{=}CHCH_2OC_6H_5$	Reflux, 10 hr, $(C_2H_5)_2NC_6H_5$	$o\text{-}HOC_6H_4CHCH_2OC_6H_5$, $CH=CH_2$	76	375
$o\text{-}ClC_6H_4OCH_2CH\overset{t}{=}CHCH_2OC_6H_4Cl\text{-}o$	Reflux, 10 hr, $(C_2H_5)_2NC_6H_5$	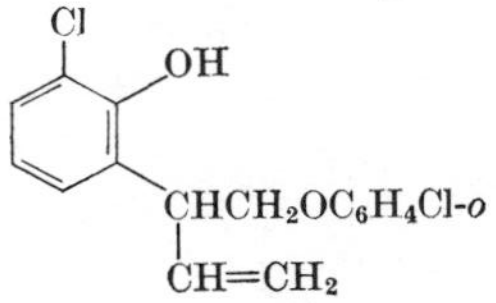	62	375
$p\text{-}ClC_6H_4OCH_2CH\overset{t}{=}CHCH_2OC_6H_4Cl\text{-}p$	Reflux, 10 hr, $(C_2H_5)_2NC_6H_5$	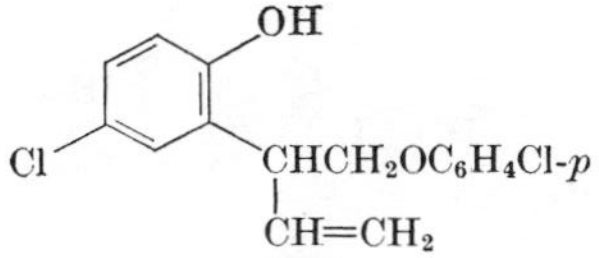	55	375
$p\text{-}BrC_6H_4OCH_2CH\overset{t}{=}CHCH_2OC_6H_4Br\text{-}p$	Reflux, 10 hr, $(C_2H_5)_2NC_6H_5$	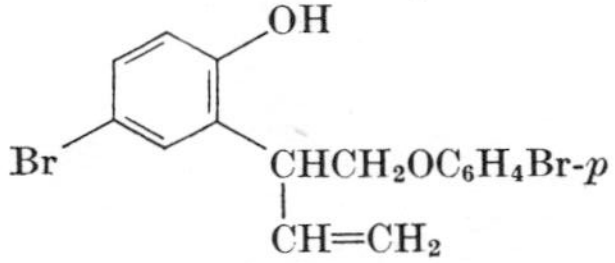	50	375
$p\text{-}ClC_6H_4OCH_2CH\overset{t}{=}CHCH_2OC_6H_4OCH_3\text{-}o$	Reflux, 10–12 hr, $(C_2H_5)_2NC_6H_5$	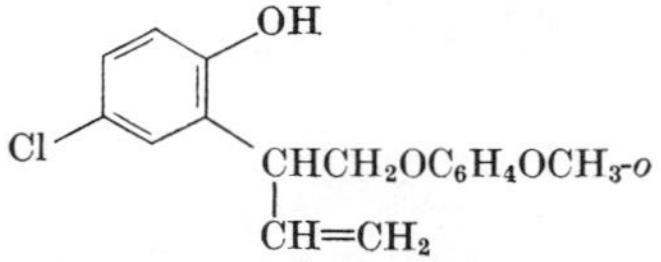 (1), 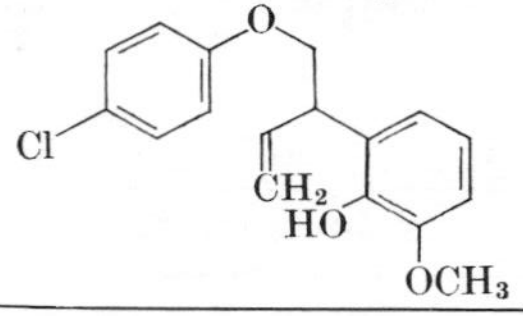 (10)	55–82 (total)	376

Note: References 344–439 are on pp. 251–252.

101

TABLE I. Aromatic Claisen Rearrangements of Allyl and Propargyl Ethers (*Continued*)

D. *Miscellaneous Diaryl Diether Rearrangements* (*Continued*)

Diaryl Ether	Conditions	Product(s) and Ratio ( )	Yield, %	Refs.
$p\text{-CH}_3\text{C}_6\text{H}_4\text{OCH}_2\text{CH}\overset{t}{=}\text{CHCH}_2\text{OC}_6\text{H}_4\text{CH}_3\text{-}o$	Reflux, 10–12 hr, $(\text{C}_2\text{H}_5)_2\text{NC}_6\text{H}_5$	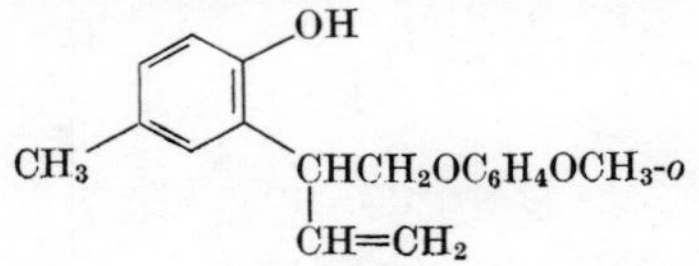 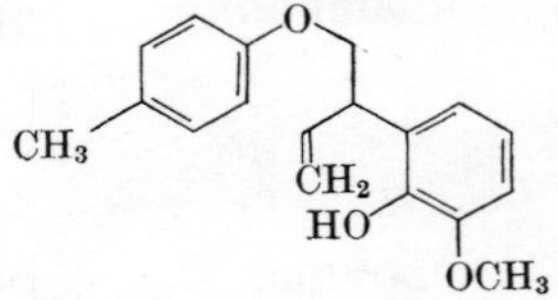 (1), and (11)	55–82 (total)	376
$m\text{-CH}_3\text{C}_6\text{H}_4\text{OCH}_2\text{CH}\overset{t}{=}\text{CHCH}_2\text{OC}_6\text{H}_4\text{CH}_3\text{-}o$	Reflux, 10–12 hr, $(\text{C}_2\text{H}_5)_2\text{NC}_6\text{H}_5$	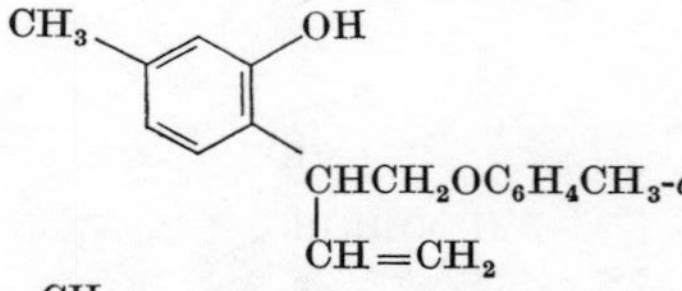 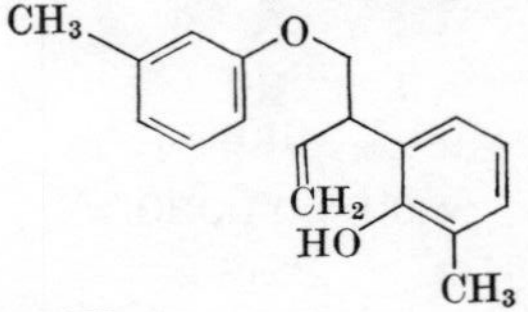 (1), and (4)	—	376
$o\text{-CH}_3\text{OC}_6\text{H}_4\text{OCH}_2\text{CH}\overset{t}{=}\text{CHCH}_2\text{OC}_6\text{H}_4\text{OCH}_3\text{-}o$	Reflux, 10 hr, diethylene glycol	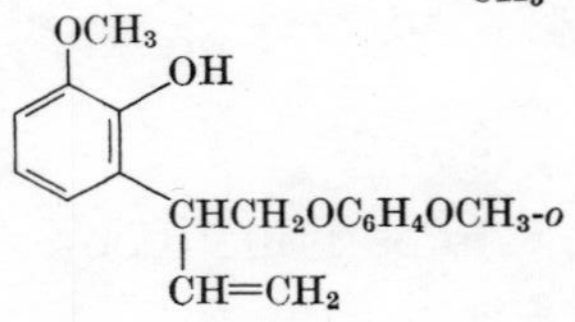	80	375

102

Reactant	Conditions	Product	Yield (%)	Reference
$o\text{-}CH_3C_6H_4OCH_2CH\overset{t}{=}CHCH_2OC_6H_4CH_3\text{-}o$	Reflux, 10 hr, $(C_2H_5)_2NC_6H_5$	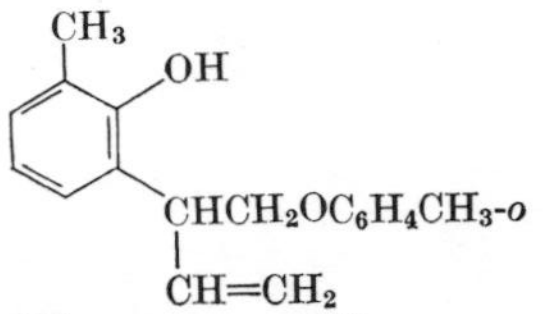 [structure]	60	375
$m\text{-}CH_3C_6H_4OCH_2CH\overset{t}{=}CHCH_2OC_6H_4CH_3\text{-}m$	Reflux, 10 hr, $(C_2H_5)_2NC_6H_5$	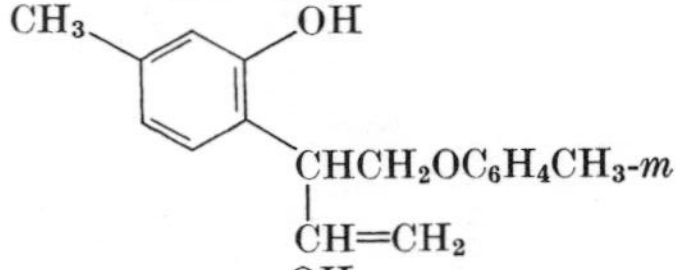 [structure]	64	375
$p\text{-}CH_3C_6H_4OCH_2CH\overset{t}{=}CHCH_2OC_6H_4CH_3\text{-}p$	Reflux, 10 hr, $(C_2H_5)_2NC_6H_5$	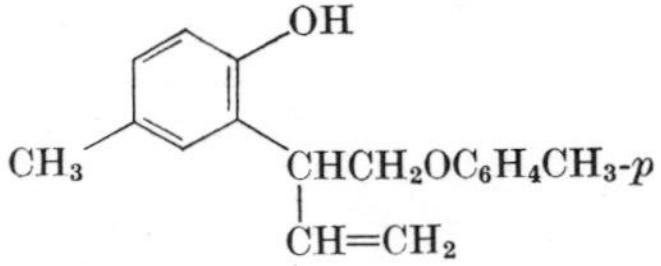 [structure]	65	375
$2,6\text{-}(CH_3)_2C_6H_3OCH_2CH=CHCH_2OC_6H_3(CH_3)_2\text{-}2,6$	Reflux, $(C_2H_5)_2NC_6H_5$	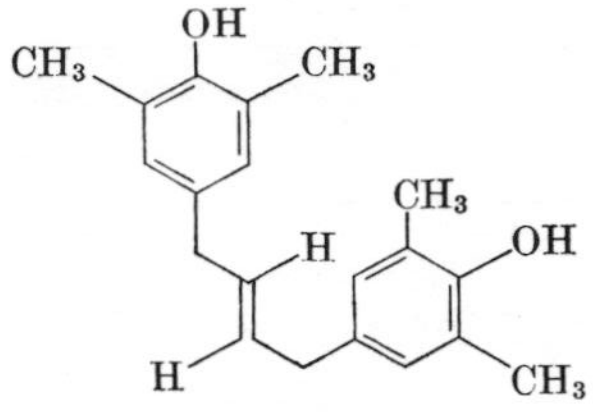 [structure] *trans* only	50	377
trans and *cis*				
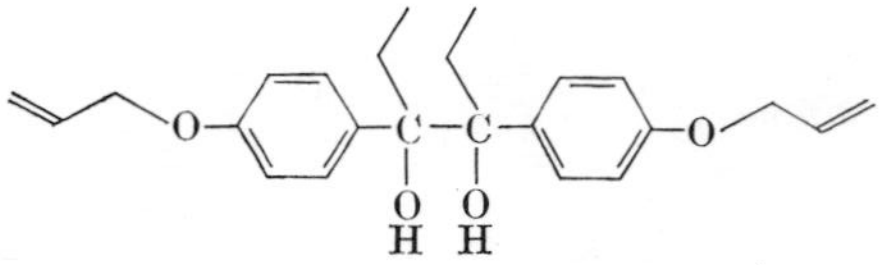 [diallyl ether structure]	Reflux, 6 hr, $(C_2H_5)_2NC_6H_5$	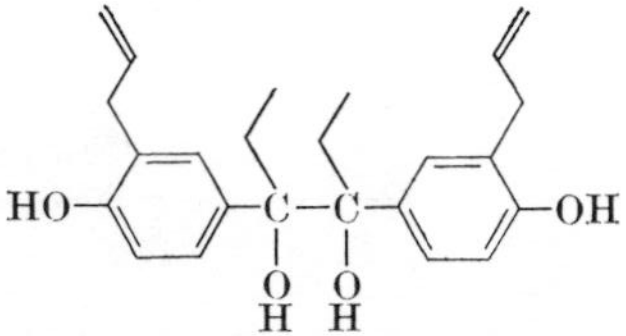 [structure]	—	378

Note: References 344–439 are on pp. 251–252.

103

TABLE I. AROMATIC CLAISEN REARRANGEMENTS OF ALLYL AND PROPARGYL ETHERS (Continued)

D. Miscellaneous Diaryl Diether Rearrangements (Continued)

Compound	Conditions	Product(s) and Ratio ( )	Yield, %	Refs.
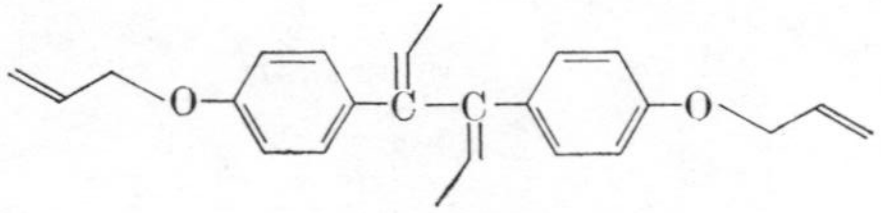	Reflux, 6 hr, $(C_2H_5)_2NC_6H_5$	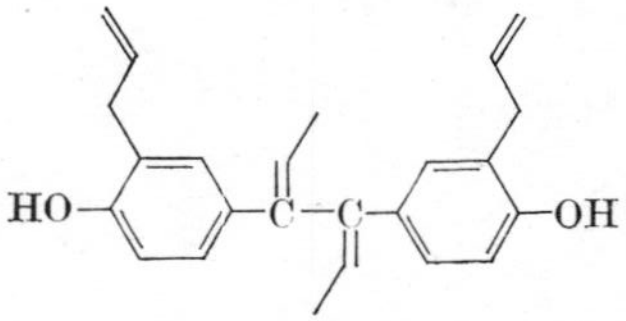	—	378

E. Allyl and Substituted Allyl Ethers of Polycyclic and Heterocyclic Systems

Carbocyclic Systems

C No.	Compound	Conditions	Product(s) and Ratio ( )	Yield(s), %	Refs.
C_{13}	$2\text{-}C_{10}H_7O$ 	Reflux, 2 hr, $(CH_3)_2NC_6H_5$	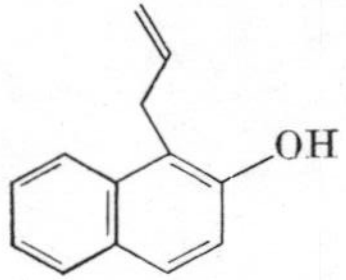 (9) 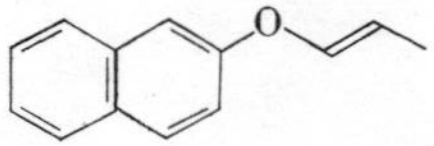 (1)	—	214
C_{14}	$2\text{-}C_{10}H_7O$ 	194°, 2.5 hr, $(CH_3)_2NC_6H_5$	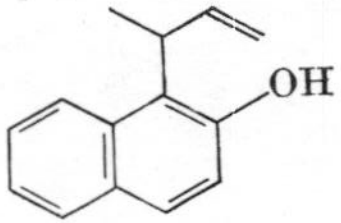	52	367, 217

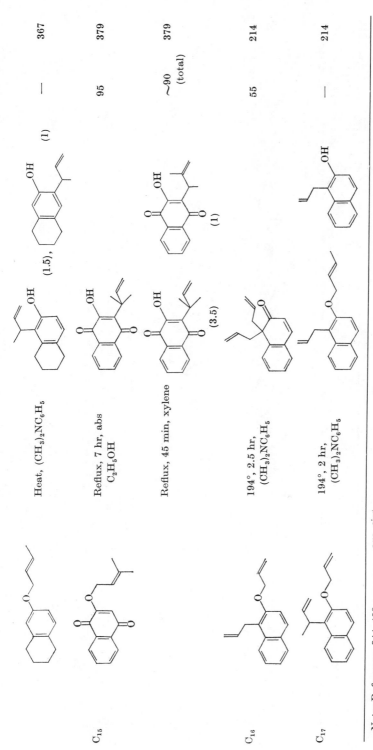

Heat, $(CH_3)_2NC_6H_5$

Reflux, 7 hr, abs C_2H_5OH

Reflux, 45 min, xylene

194°, 2.5 hr, $(CH_3)_2NC_6H_5$

194°, 2 hr, $(CH_3)_2NC_6H_5$

C_{15}

C_{16}

C_{17}

(1.5),

(1)

(3.5)

(1)

—

95

~90 (total)

55

—

367

379

379

214

214

Note: References 344–439 are on pp. 251–252.

E. Allyl and Substituted Allyl Ethers of Polycyclic and Heterocyclic Systems (Continued)

Carbocyclic Systems

C No.	Compound	Conditions	Product(s) and Ratio ( )	Yield(s), %	Refs.
C_{21}	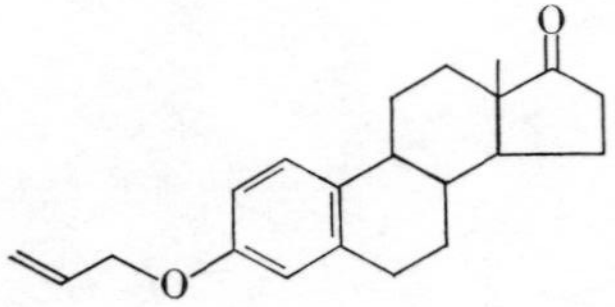	Reflux, 10 hr, $(C_2H_5)_2NC_6H_5$, N_2 atm	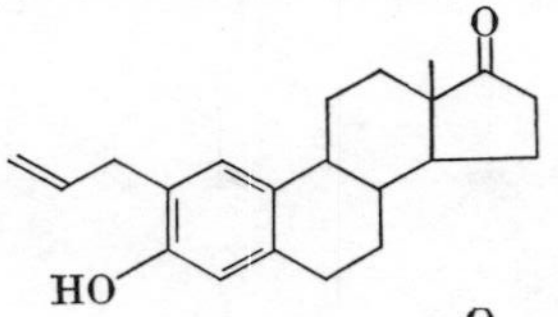 (1), 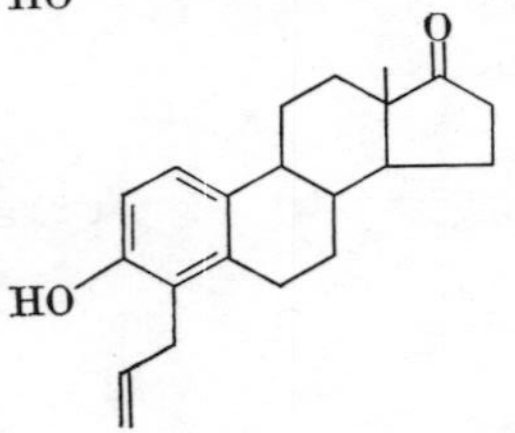 (3)	86 (total)	380, 381
C_{22}	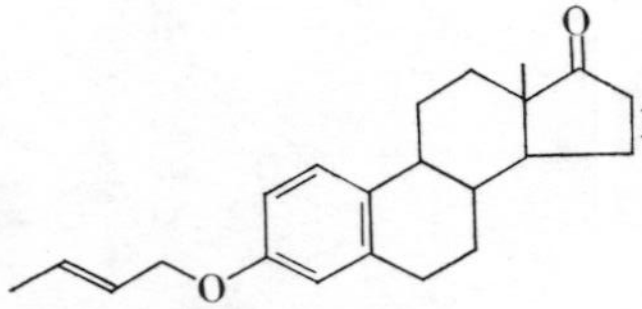	Reflux, 12 hr, $(C_2H_5)_2NC_6H_5$, N_2 atm	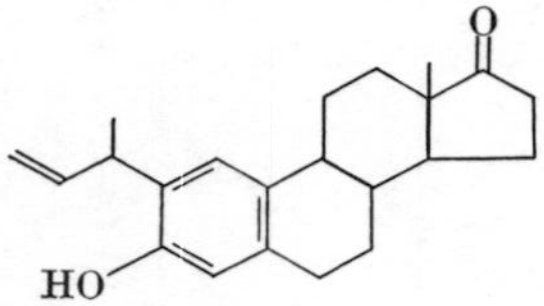	58	381
C_{23}	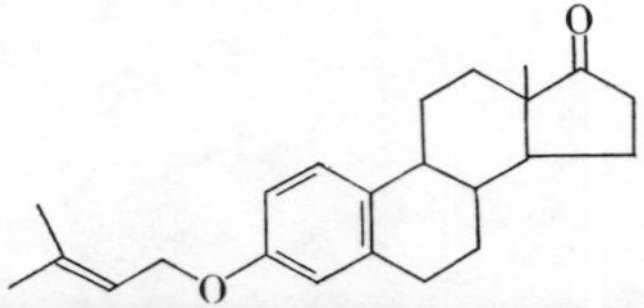	Reflux, 20 hr, $(C_2H_5)_2NC_6H_5$, N_2 atm	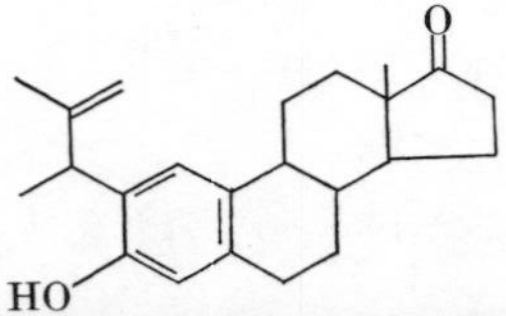	36 (total)	201

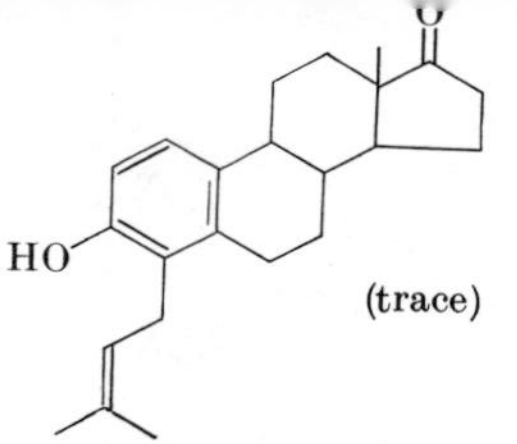

Heterocyclic Systems

C No.	Ring Hetero- atom(s)	Compound	Conditions	Product(s) and Ratio ( )	Yield(s), %	Refs.

(trace)

C$_7$	N$_2$	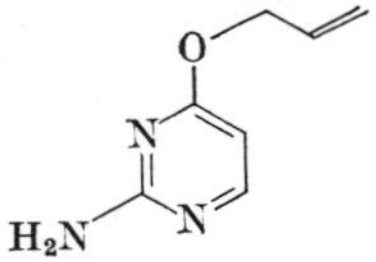	240°, 4 hr, (C$_2$H$_5$)$_2$NC$_6$H$_5$	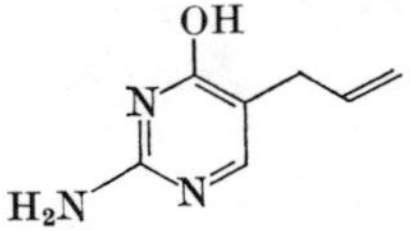	2–14	382
C$_8$	N	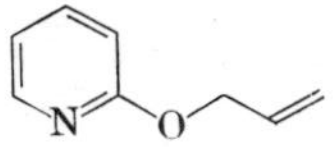	250°, 8 hr, (C$_2$H$_5$)$_2$NC$_6$H$_5$, autoclave	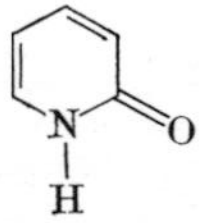 (Major), I II	24 (total) (crude)	383

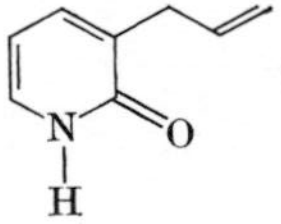

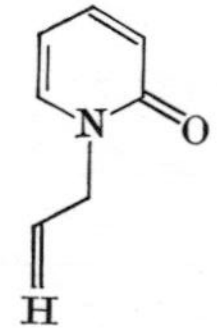

Note: References 344–439 are on pp. 251–252.

TABLE I. AROMATIC CLAISEN REARRANGEMENTS OF ALLYL AND PROPARGYL ETHERS (*Continued*)

E. *Allyl and Substituted Allyl Ethers of Polycyclic and Heterocyclic Systems* (*Continued*)

Heterocyclic Systems (*Continued*)

C No.	Ring Hetero-atom(s)	Compound	Conditions	Product(s) and Ratio ( )	Yield(s), %	Refs.
C$_8$	N (*Contd.*)		90–150°, 9 hr, H$_2$PtCl$_6$, i-C$_3$H$_7$OH, N$_2$ atm		85	88
			255°, 12 hr, (CH$_3$)$_2$NC$_6$H$_5$, sealed tube	I (1.1), II (1)	55	89
		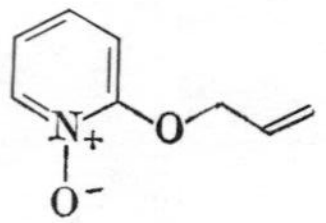	137°, (CH$_3$OCH$_2$CH$_2$)$_2$O	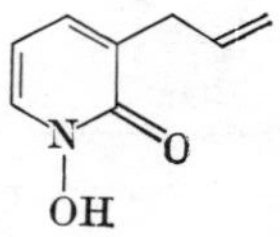 II (1.4), 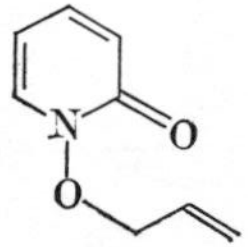 (1)	94	384
C$_8$	N$_2$	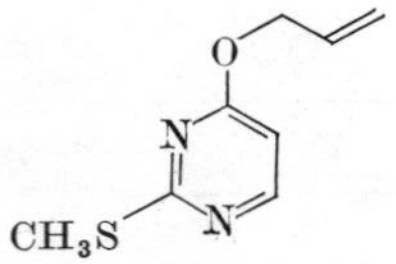	240°, 7–8 hr, m-CH$_3$C$_6$H$_4$N(C$_2$H$_5$)$_2$	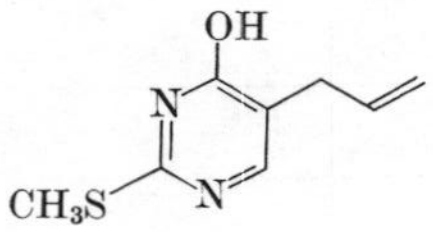 (1.7), 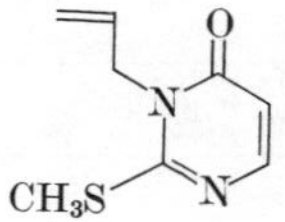 (1)	51	385
		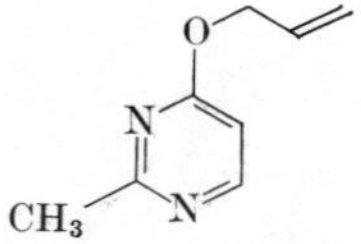	240°, 20–22 hr, (C$_2$H$_5$)$_2$NC$_6$H$_5$	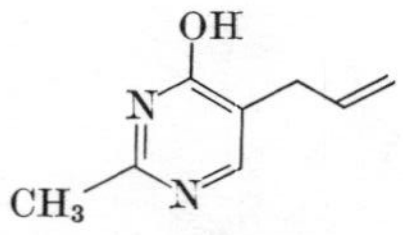 (1.6), 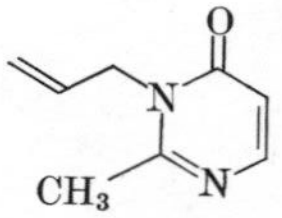 (1)	41	385

108

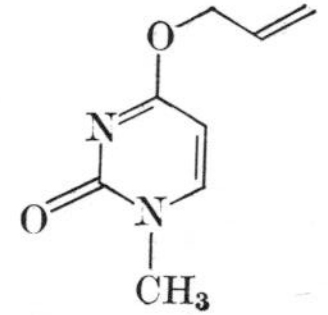

	Conditions		Yield (%)	Refs.
C₉ N				
(4-allyloxy-1-methylpyridin-2(1H)-one)	240°, 1.75 hr, neat	(1-methyl-3-allyluracil-type product)	70	385
(2-(but-2-enyloxy)pyridine)	245°, 4 hr, (CH₃)₂NC₆H₅, sealed tube	(3-(but-2-enyl)-1H-pyridin-2-one) (1.1), and (1-(but-2-enyl)pyridin-2-one) (1)	64	89
(2-(but-3-en-2-yloxy)pyridine)	250°, 7 hr, (CH₃)₂NC₆H₅, sealed tube	(3-(but-3-en-2-yl)-1H-pyridin-2-one) (1.2), and (1-(but-3-en-2-yl)pyridin-2-one) (1)	67	89
(2-(but-2-enyloxy)pyridine N-oxide)	125°, 36 hr, H₂PtCl₆, i-C₃H₇OH, sealed tube	(1-hydroxy-3-(but-3-en-2-yl)pyridin-2-one)	∼Quant.	88
	137°, (CH₃OCH₂CH₂)₂O		40	384
C₉ N₂				
(4-(but-2-enyloxy)-2-methylthiopyrimidine)	245°, 6 hr, m-CH₃C₆H₄N(C₂H₅)₂	(5-(but-3-en-2-yl)-4-hydroxy-2-methylthiopyrimidine)	28	386

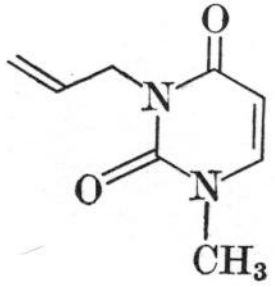

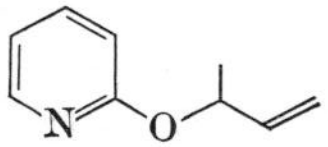

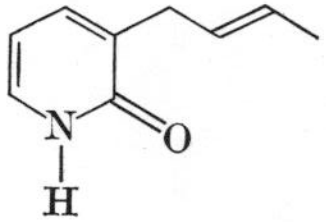

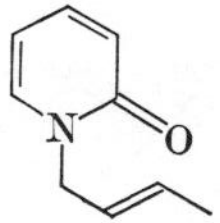

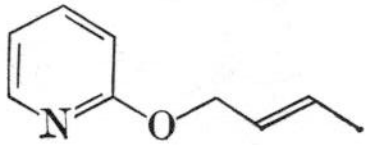

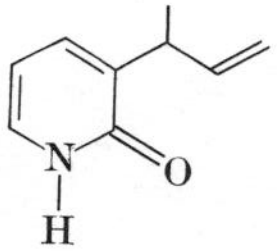

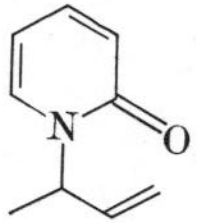

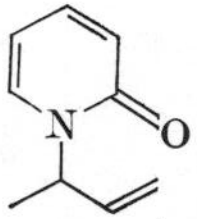

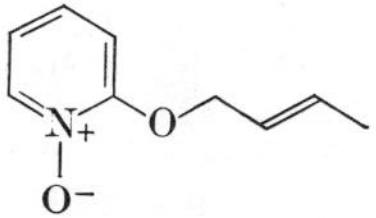

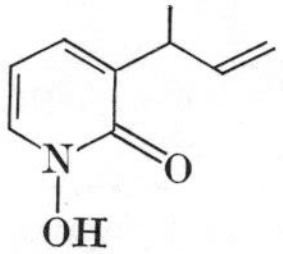

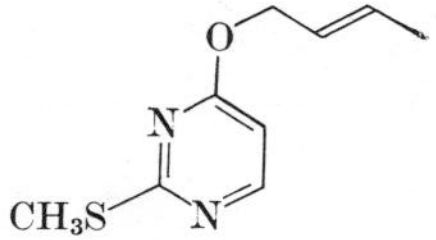

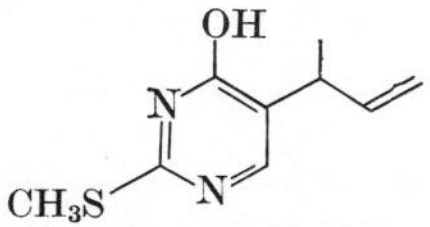

Note: References 344–439 are on pp. 251–252.

TABLE I. AROMATIC CLAISEN REARRANGEMENTS OF ALLYL AND PROPARGYL ETHERS (*Continued*)

E. Allyl and Substituted Allyl Ethers of Polycyclic and Heterocyclic Systems (Continued)

Heterocyclic Systems (Continued)

C No.	Ring Hetero-atom(s)	Compound	Conditions	Product(s) and Ratio ( )	Yield(s), %	Refs.
C$_9$ (*Contd.*)	N$_2$	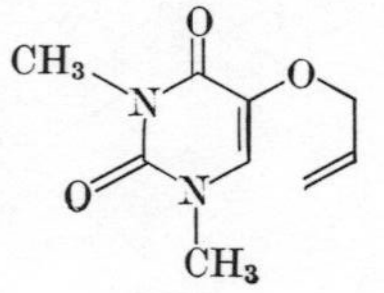	120° (melted), 10 min	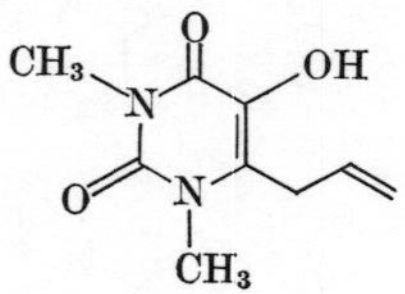	96	256
C$_9$	N$_4$	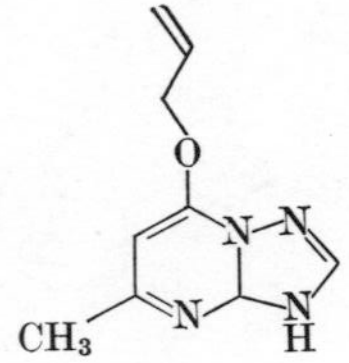	150°, 30 min	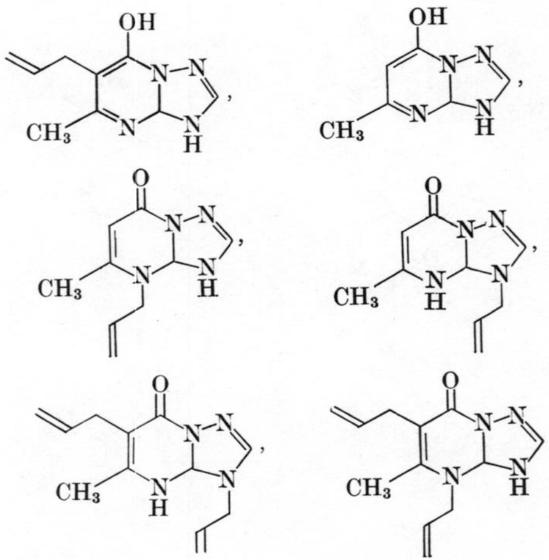	—	387

	Substrate	Conditions	Product(s)	Yield (%)	Refs.
C$_{10}$ N	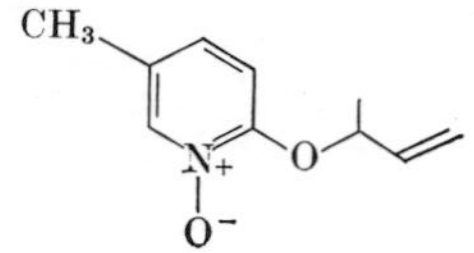	137°, (CH$_3$OCH$_2$CH$_2$)$_2$O	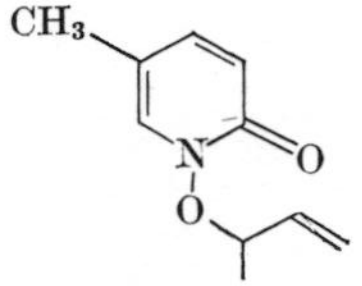 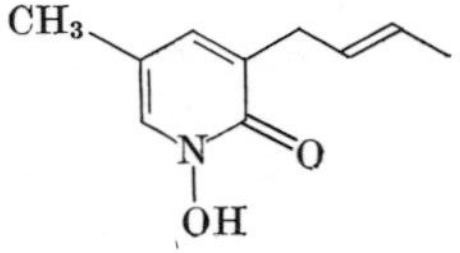 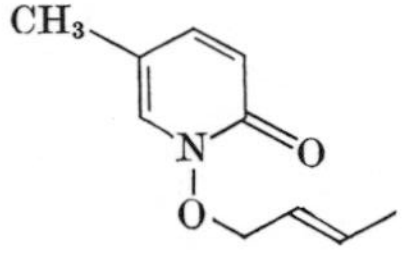 (4), (1), (15), —		384
C$_{10}$ N$_2$	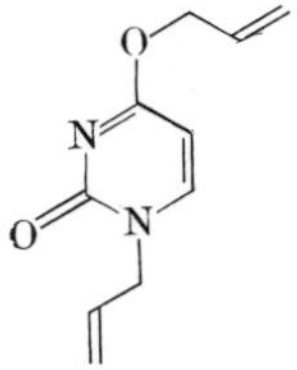	240°, 1.75 hr, neat	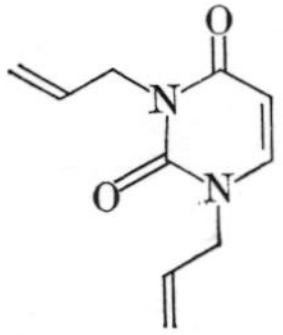	62	385
C$_{10}$ N$_4$	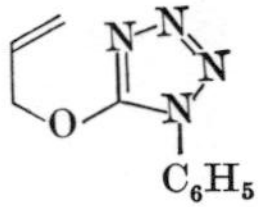	200°, 1 hr, neat	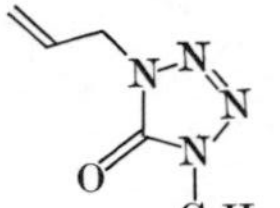	—	388
C$_{10}$ N, O	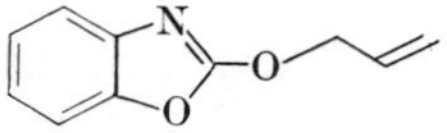	200°, 1 hr	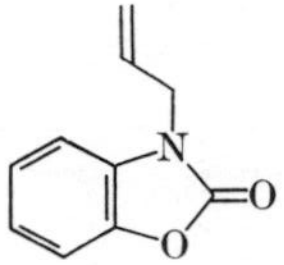	—	388
C$_{10}$ N, S	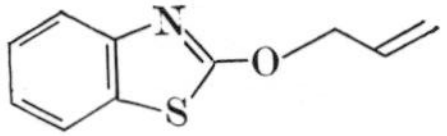	230°, 2 hr	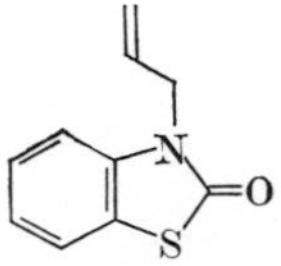	—	388

Note: References 344–439 are on pp. 251–252.

TABLE I. AROMATIC CLAISEN REARRANGEMENTS OF ALLYL AND PROPARGYL ETHERS (*Continued*)

E. *Allyl and Substituted Allyl Ethers of Polycyclic and Heterocyclic Systems* (*Continued*)

Heterocyclic Systems (*Continued*)

Ring C No.	Hetero- atom(s)	Compound	Conditions	Product(s) and Ratio ( )	Yield(s), %	Refs.
C_{11}	N_4	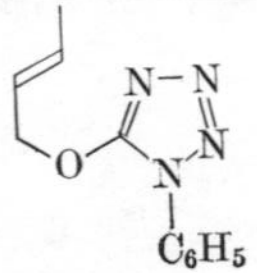	120°, 2–3 hr	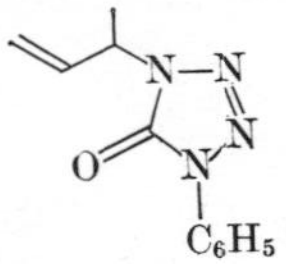	—	388
		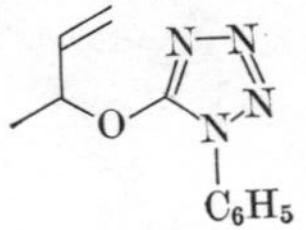	120°, several hr	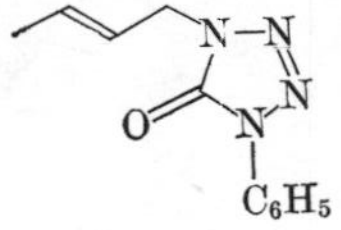	—	388
C_{12}	N	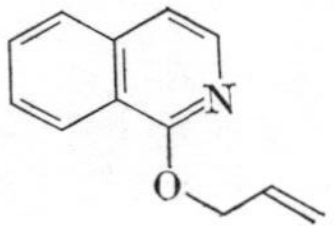	250°, 5 hr, neat, sealed tube	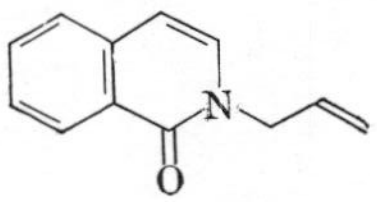	94	389
		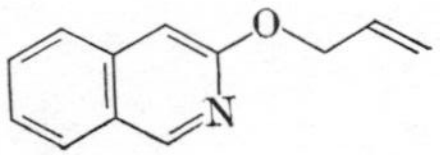	190°, 1 hr, 2—$C_{10}H_7CH_3$, sealed tube	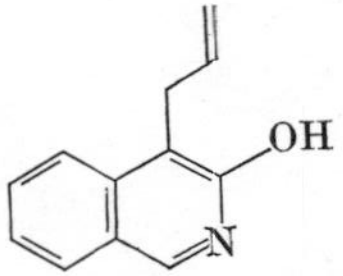	56	389
		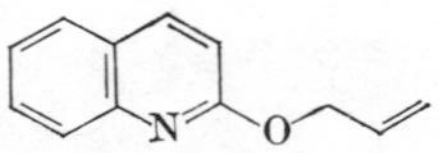	250°, 1 hr, neat	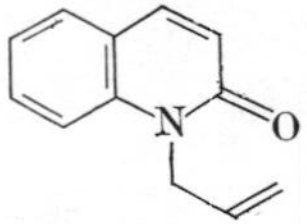 (1)	18	390

		300°, 1 hr, neat	T(1.3), 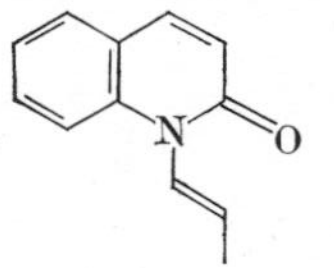 (1)	87	390
	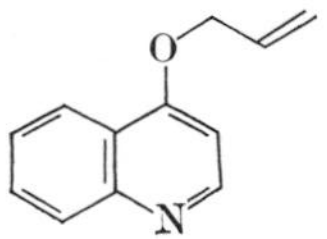	200°, 1.5 hr, 1—$C_{10}H_7CH_3$	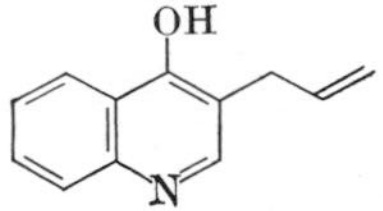	93	391
C_{12} N_2	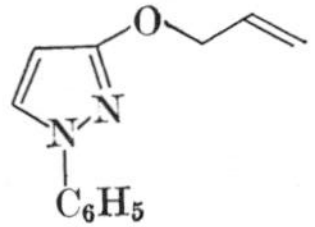	250°, m-$CH_3C_6H_4N(C_2H_5)_2$	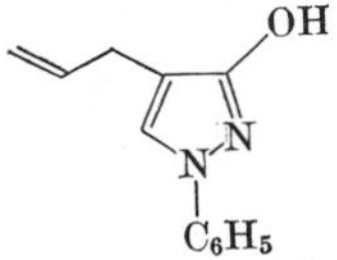	49	392
C_{12} N_4	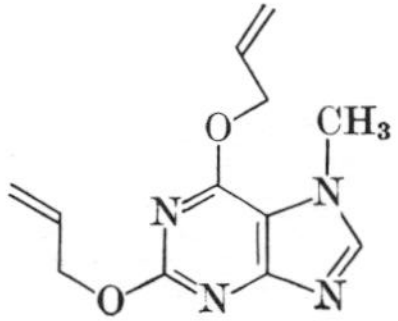	150°, 2 hr	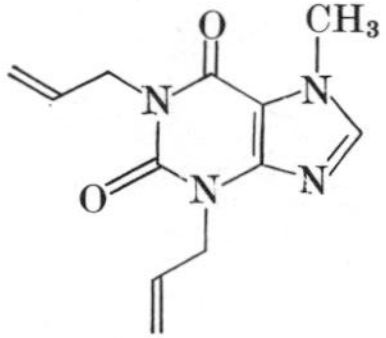	—	393
C_{12} N_2 O	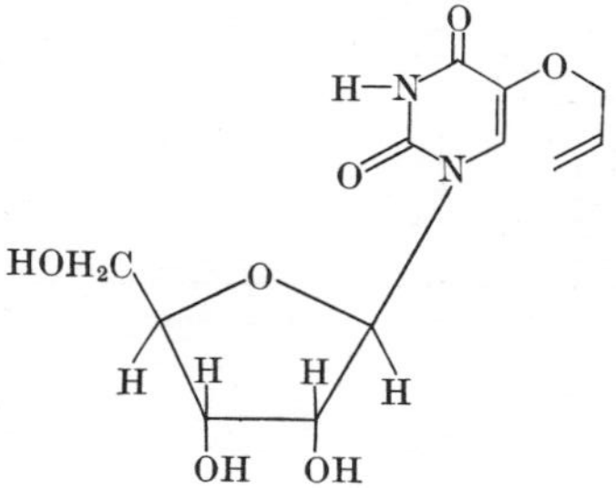	153°, 10 min, $HCON(CH_3)_2$	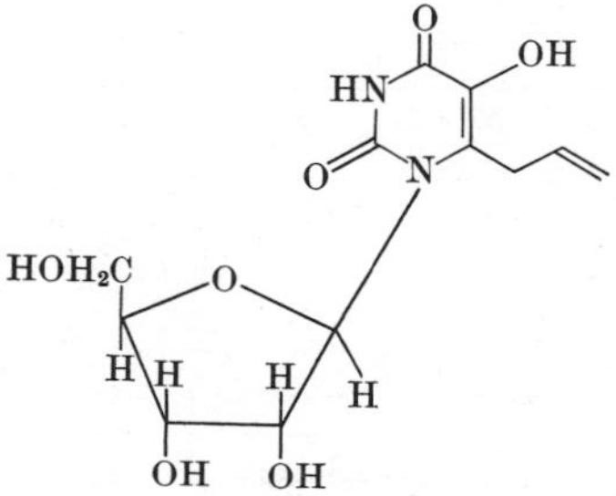	79	256

Note: References 344–439 are on pp. 251–252.

TABLE I. Aromatic Claisen Rearrangements of Allyl and Propargyl Ethers (*Continued*)

E. Allyl and Substituted Allyl Ethers of Polycyclic and Heterocyclic Systems (*Continued*)

Heterocyclic Systems (*Continued*)

C No.	Ring Hetero-atom(s)	Compound	Conditions	Product(s) and Ratio ( )	Yield(s), %	Refs.
C_{13}	N	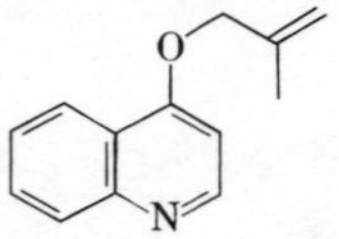	200°, 1.5 hr, 1-$C_{10}H_7CH_3$	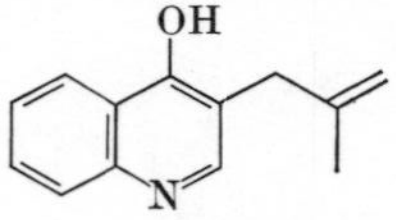 (271), 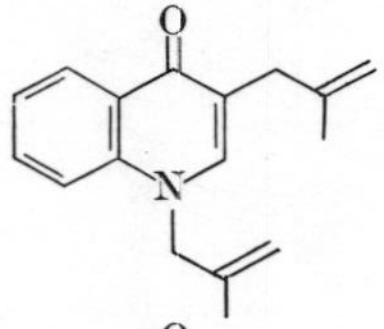 (1), 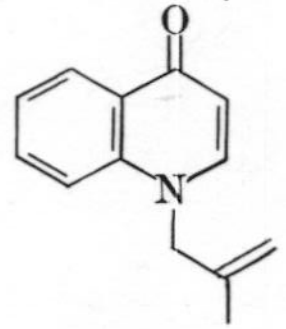 (1.3)	65 (total)	391
		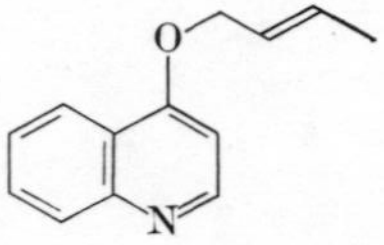	200°, 1.5 hr, 1-$C_{10}H_7CH_3$	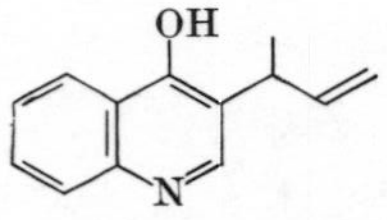	91	391

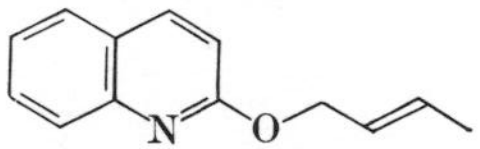

C13 N2

C13 O

300°, 1 hr

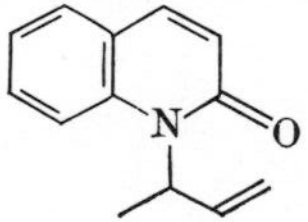

(37), (31), (1), (14) 81 (total) 390

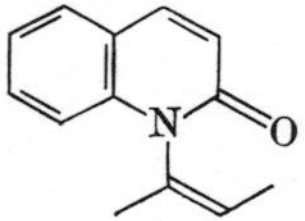

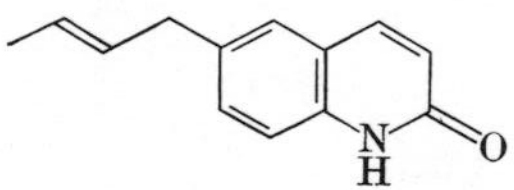

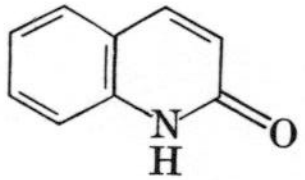

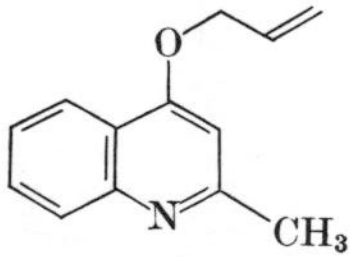

200°, neat, 30 min

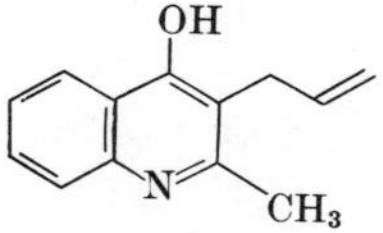

(9.3), (1), — 394

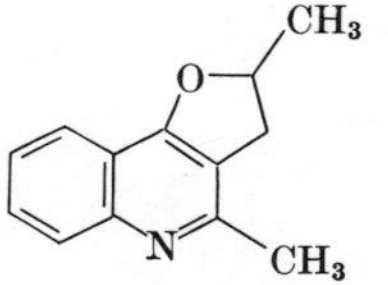

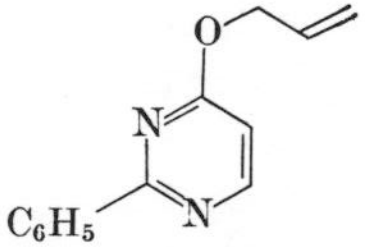

240°, 20–22 hr, m-CH$_3$C$_6$H$_4$N(C$_2$H$_5$)$_2$

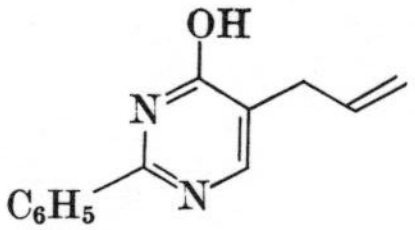

(2.6), (1), 50 385

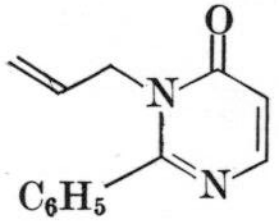

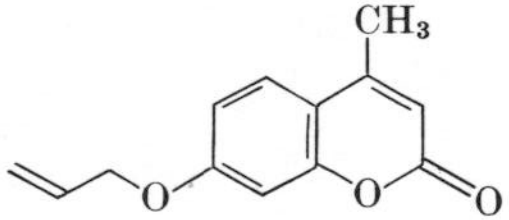

215–220°, 2 hr

~94 360

220°, 75 min

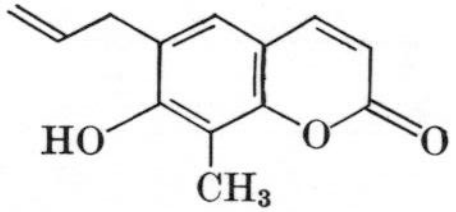

80 360

CH$_3$ CH$_3$ CH$_3$ C$_6$H$_5$ OH   OH   CH$_3$ HO   HO   CH$_3$

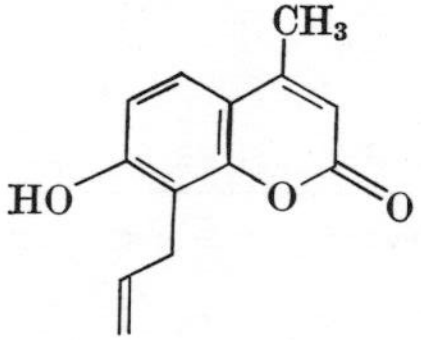

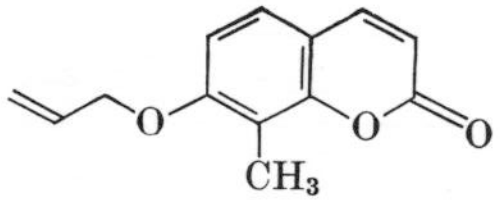

Note: References 344–439 are on pp. 251–252.

115

TABLE I. AROMATIC CLAISEN REARRANGEMENTS OF ALLYL AND PROPARGYL ETHERS (Continued)

E. Allyl and Substituted Allyl Ethers of Polycyclic and Heterocyclic Systems (Continued)

Heterocyclic Systems (Continued)

Ring C No.	Hetero- Atom(s)	Compound	Conditions	Product(s) and Ratio ( )	Yield(s), %	Refs.
C_{13} (Contd.)	O	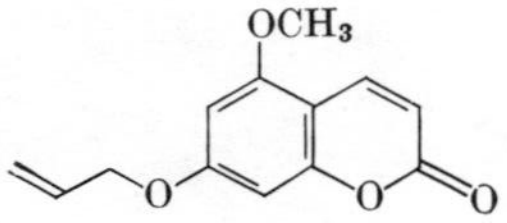	200°, 2 hr, in vacuum	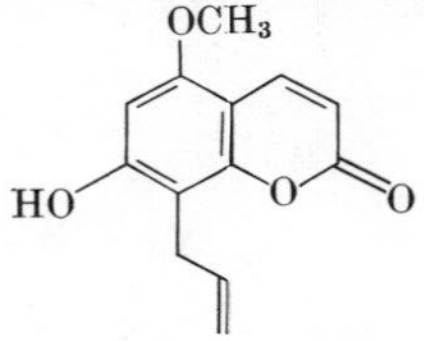	62	395
C_{14}	N	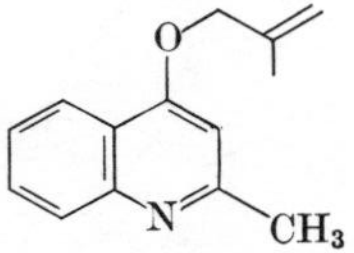	200°, neat	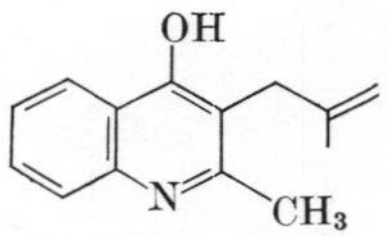 (2.2), 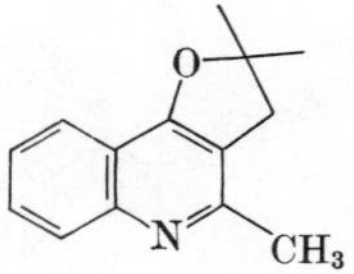	(1)	394
		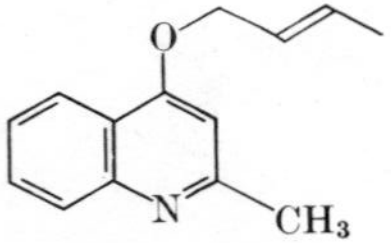	200°, neat	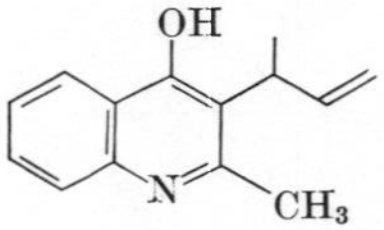 (10.3), 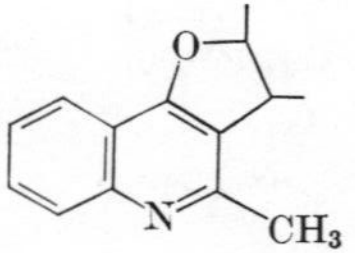	(1)	394
		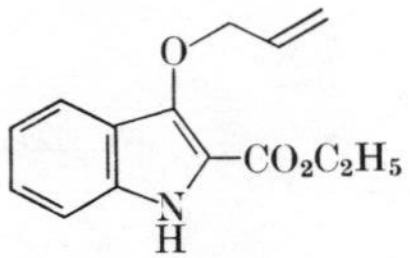	Heat, cyclohexane	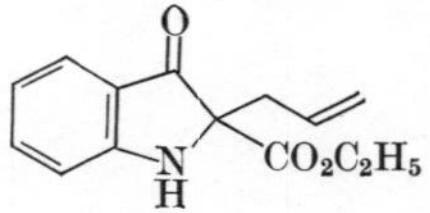	Quant.	396

C_{14}	N_2		240°, 7–8 hr, m-CH$_3$C$_6$H$_4$N(C$_2$H$_5$)$_2$	(1), $C_6H_5CH_2$—S	(2.4), $C_6H_5CH_2S$	48 (total)	385, 89
C_{14}	O		Reflux, 0.5–1 hr, (C$_2$H$_5$)$_2$NC$_6$H$_5$		(5.3), HO	>90	363
			130°, 1.5 hr		(1), HO	88 (total)	111
C_{15}	N	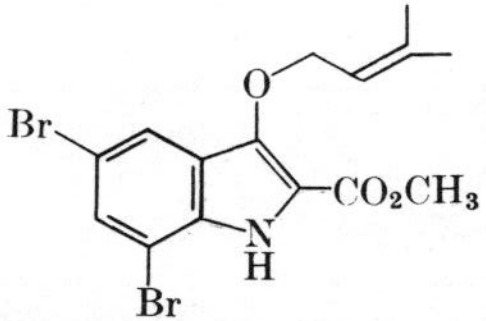	Heat, 24 hr, cyclohexane		CO$_2$CH$_3$, Br, Br	Quant.	396

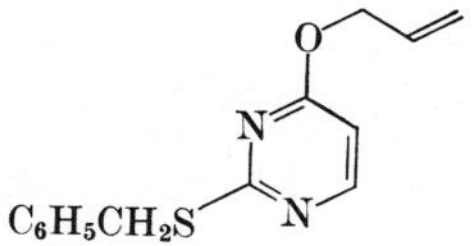

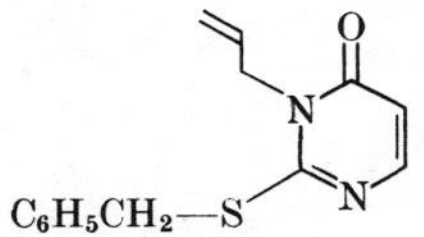

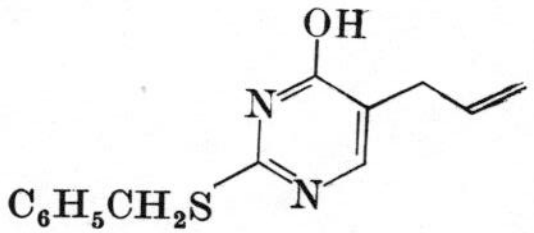

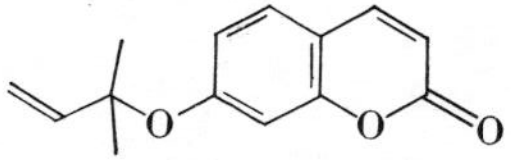

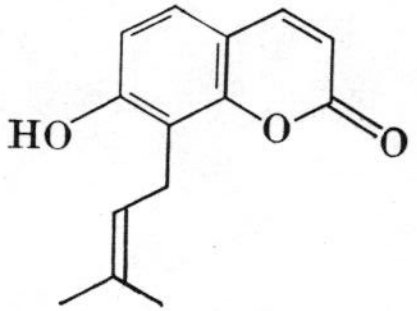

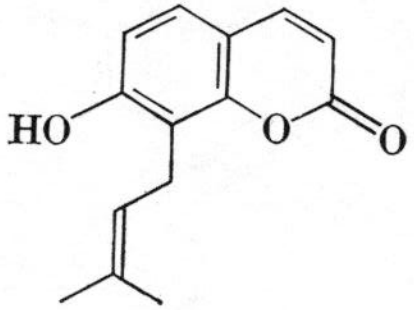

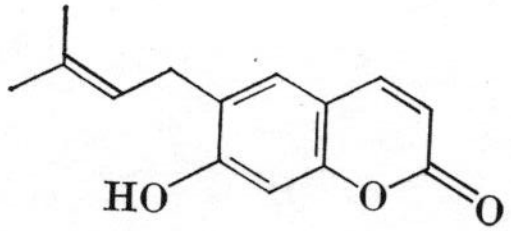

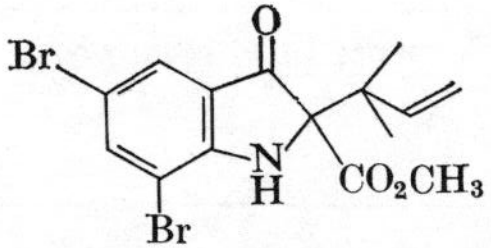

Note: References 344–439 are on pp. 251–252.

117

TABLE I. AROMATIC CLAISEN REARRANGEMENTS OF ALLYL AND PROPARGYL ETHERS (Continued)

E. Allyl and Substituted Allyl Ethers of Polycyclic and Heterocyclic Systems (Continued)

Heterocyclic Systems (Continued)

C No.	Ring Hetero- Atom(s)	Compound	Conditions	Product(s) and Ratio ( )	Yield(s), %	Refs.
C_{15} (Contd.)	N	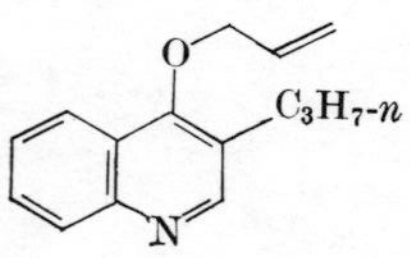	200°, 30 min, neat	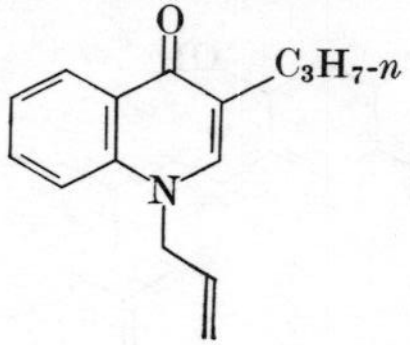	96	397
		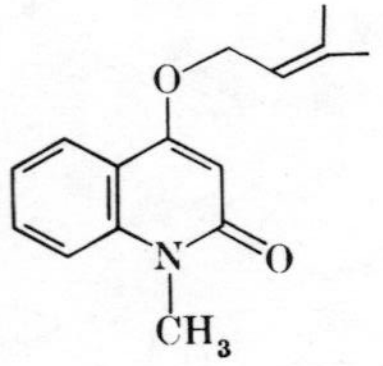	130–140°, 5 hr, N_2 atm	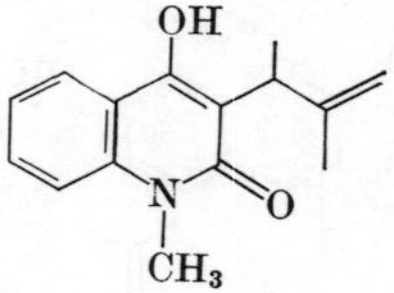 (1), 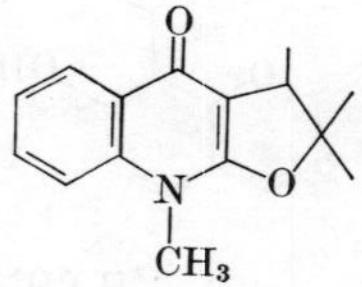 (1.8), 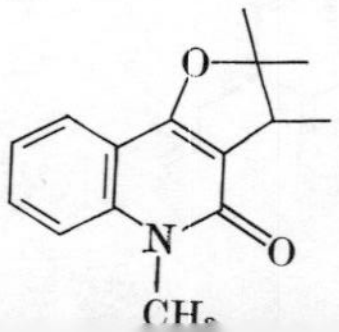 (2.2)	94 (total)	398

C$_{15}$ N$_2$

C$_6$H$_5$CH$_2$S

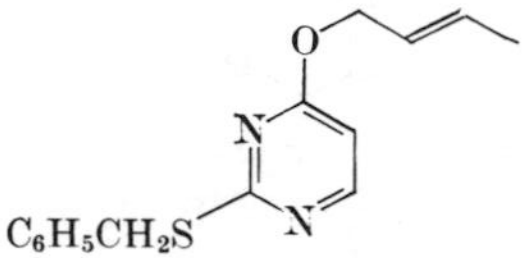

245°, 6 hr,
m-CH$_3$C$_6$H$_4$N(C$_2$H$_5$)$_2$

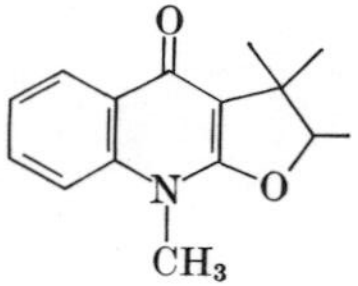

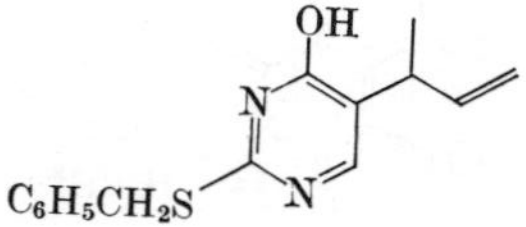

OH

C$_6$H$_5$CH$_2$S

32

386

C$_{15}$ O

OCH$_3$

160°, 1 hr

OCH$_3$

HO

88

111

C$_{15}$

OCH$_3$

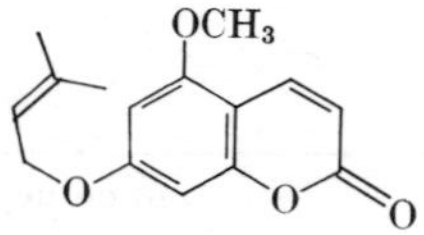

185°, 8 hr,
(CH$_3$)$_2$NC$_6$H$_5$,
(n-C$_3$H$_7$CO)$_2$O

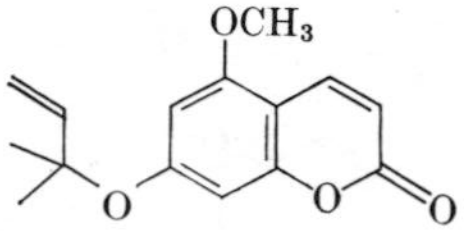

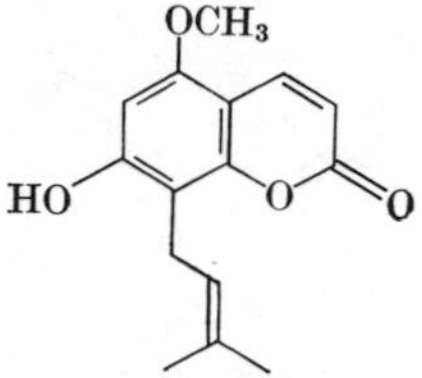

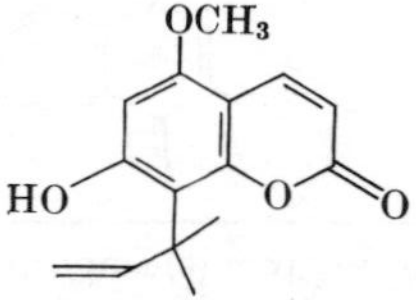

OCH$_3$

HO

(as butyrate)

92

219

C$_{16}$ N

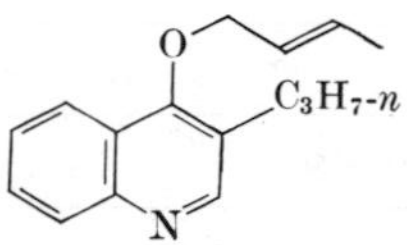

C$_3$H$_{7}$-n

200°, 30 min, neat

C$_3$H$_{7}$-n

97

397

140–145°, 4.5 hr,
Na$_2$CO$_3$, N$_2$ atm

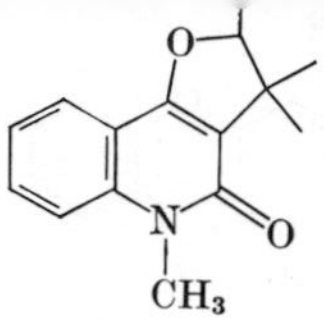

CH$_3$

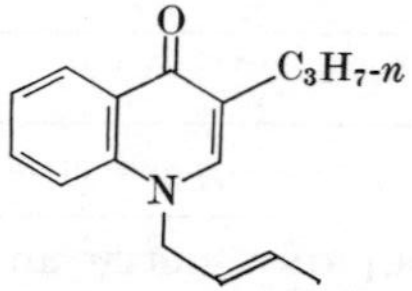

(15),

CH$_3$

(1)

88 (total)

398

Note: References 344–439 are on pp. 251–252.

TABLE I. AROMATIC CLAISEN REARRANGEMENTS OF ALLYL AND PROPARGYL ETHERS (*Continued*)

E. *Allyl and Substituted Allyl Ethers of Polycyclic and Heterocyclic Systems* (*Continued*)

Heterocyclic Systems (*Continued*)

C No.	Ring Hetero-atom(s)	Compound	Conditions	Product(s) and Ratio ( )	Yield(s), %	Refs.
C_{16}	N (*Contd.*)	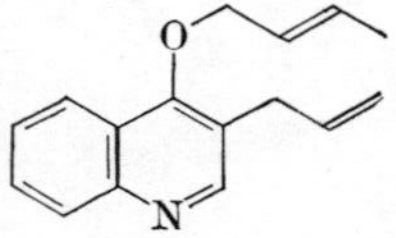	200°, 30 min, neat	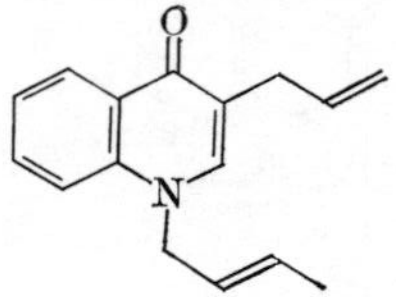	Quant.	397
		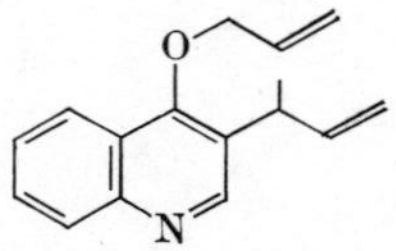	200°, 30 min, neat	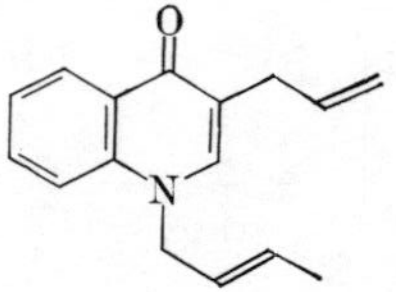	Quant.	397
C_{16}	N_2	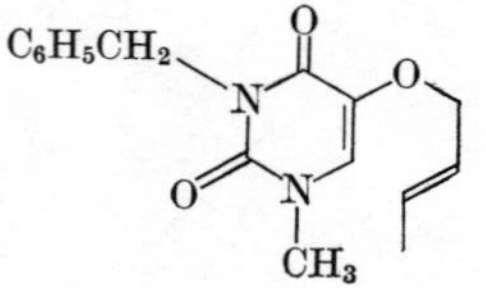	125°, 20 min	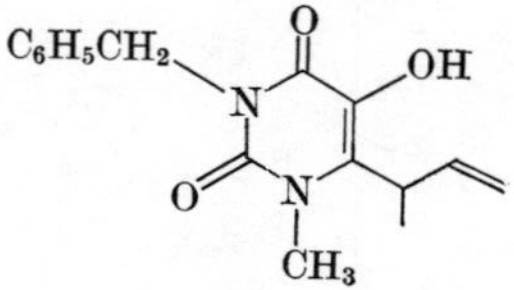	97	256
C_{16}	O	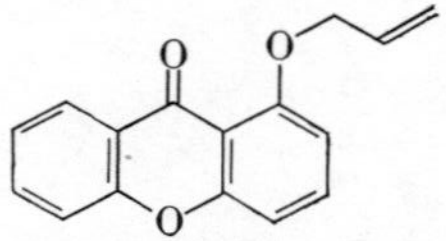	Reflux, 3 hr, $(CH_3)_2NC_6H_5$	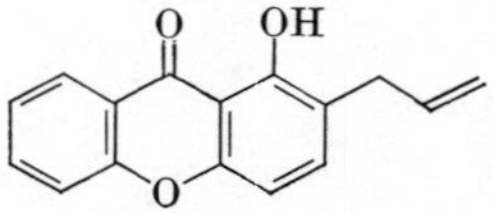	Quant.	399

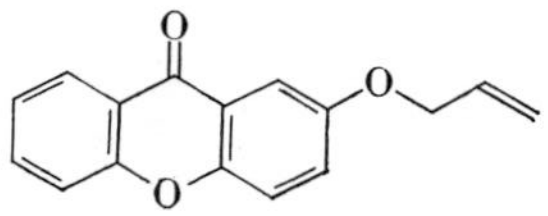

79   400

80   400

97 (total)   401

78   400

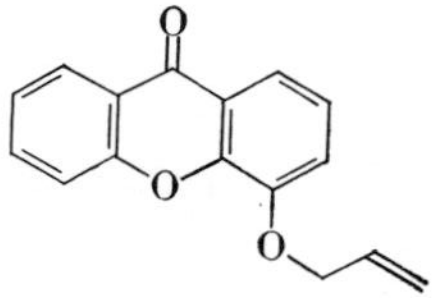

(6.5)

(1)

Reflux, 3 hr,
(CH$_3$)$_2$NC$_6$H$_5$

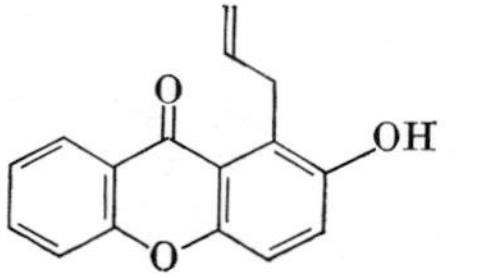

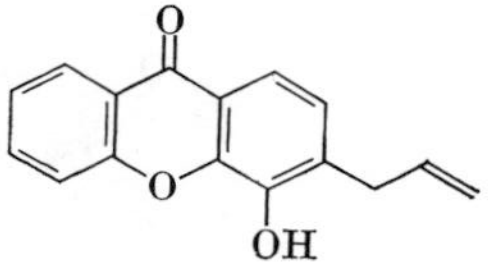

Reflux, 3 hr,
(CH$_3$)$_2$NC$_6$H$_5$

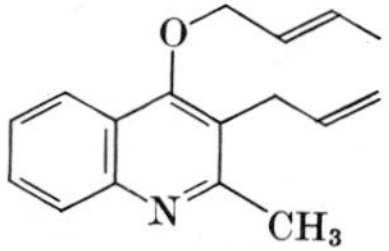

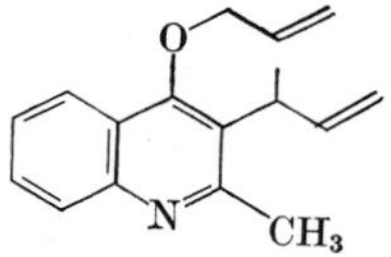

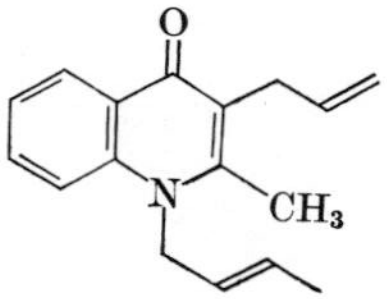

200°, 30 min, neat

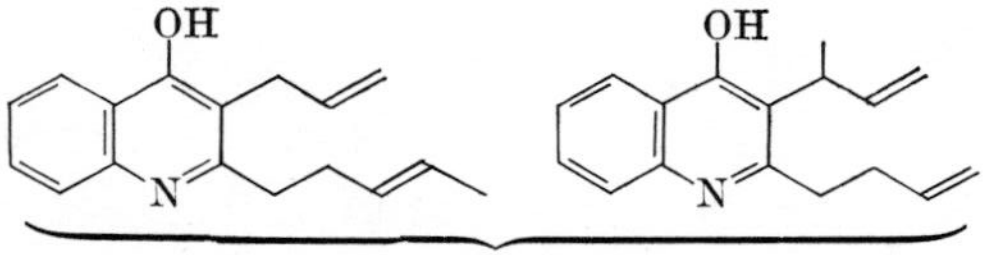

Reflux, 3 hr,
(CH$_3$)$_2$NC$_6$H$_5$

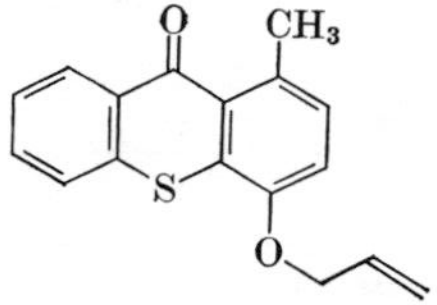

C$_{17}$ N

or

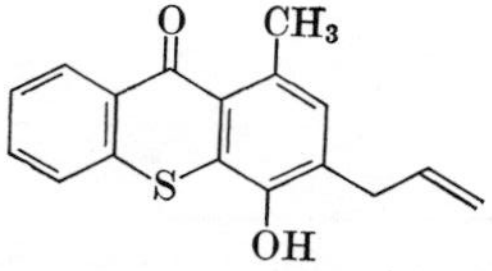

C$_{17}$ S

Note: References 344–439 are on pp. 251–252.

TABLE I. AROMATIC CLAISEN REARRANGEMENTS OF ALLYL AND PROPARGYL ETHERS (*Continued*)

E. Allyl and Substituted Allyl Ethers of Polycyclic and Heterocyclic Systems (Continued)

Heterocyclic Systems (Continued)

C No.	Ring Hetero-atom(s)	Compound	Conditions	Product(s) and Ratio ( )	Yield(s), %	Refs.
C_{18}	N_4	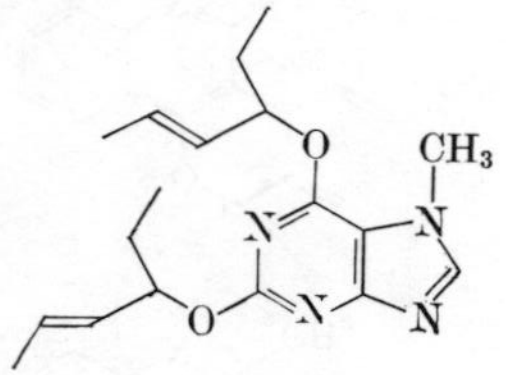	165°, 4 hr, sealed tube	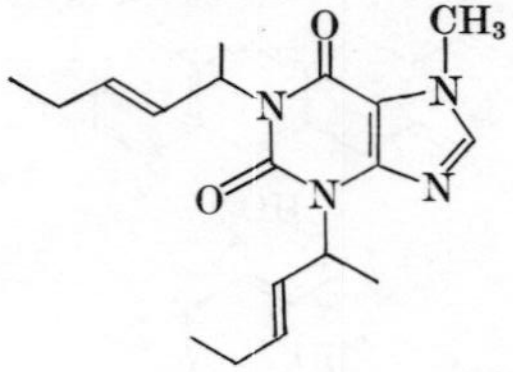	—	393
C_{19}	O	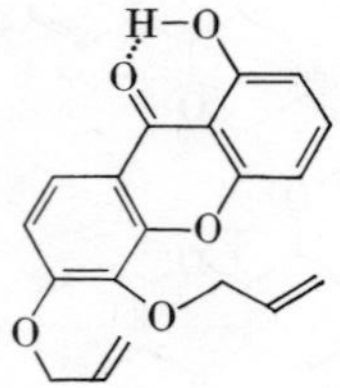	Reflux, 14 hr, decalin	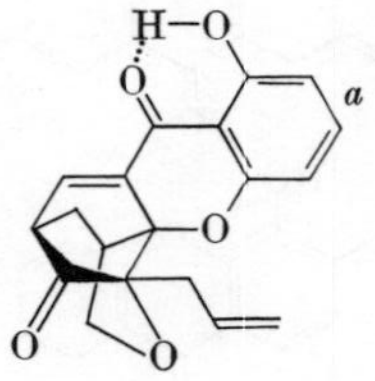	—	402
		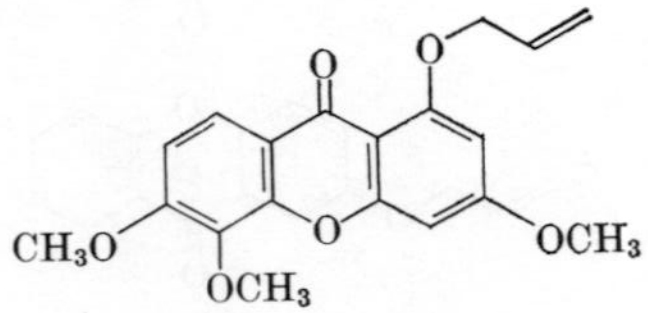	Reflux, 4 hr, $(CH_3)_2NC_6H_5$	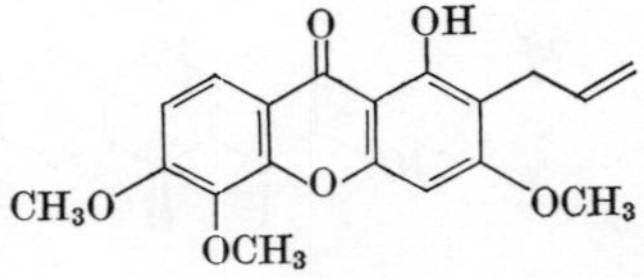	57	188

C_{20} O

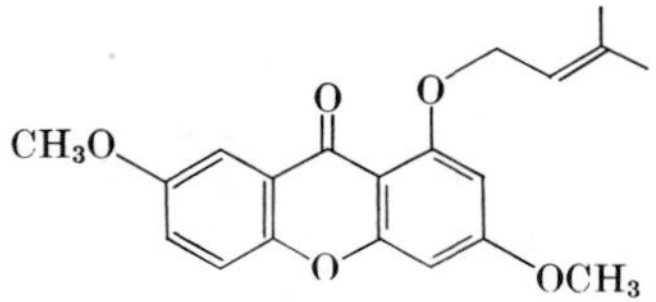

Reflux, 3 hr,
$(CH_3)_2NC_6H_5$, N_2 atm

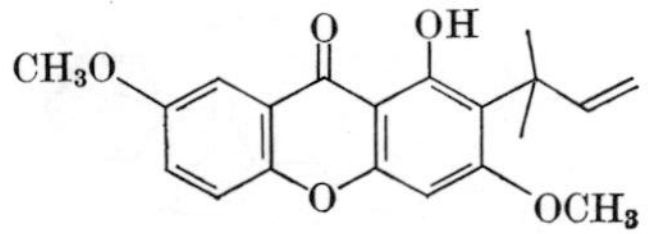

(1), (1.7),

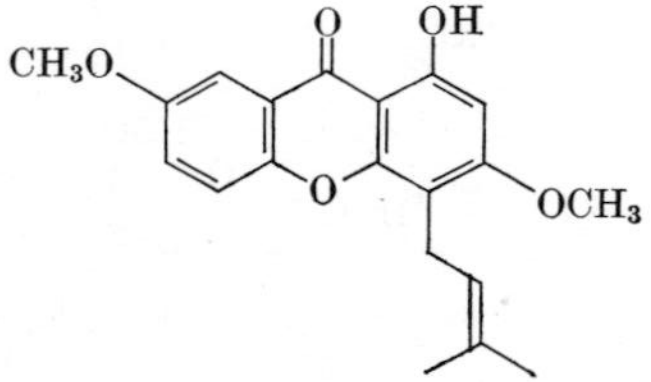

63 (total)   403

C_{21} O

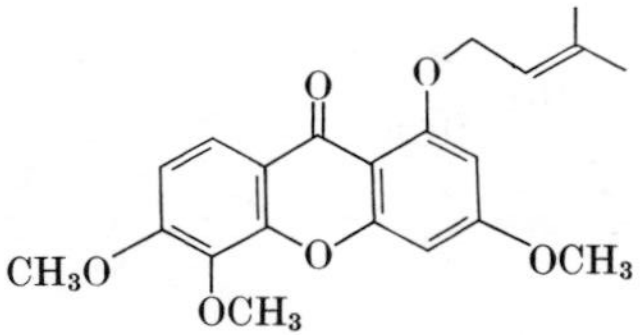

Reflux, 8 hr,
$(CH_3)_2NC_6H_5$

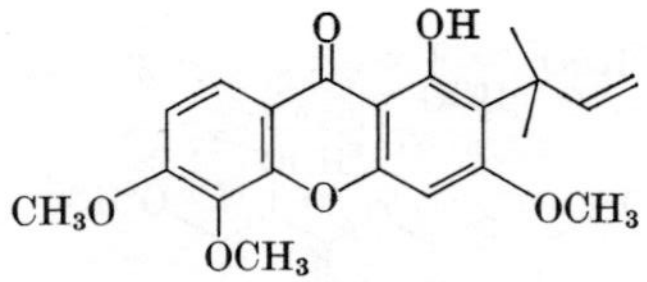

(1), (2.2),

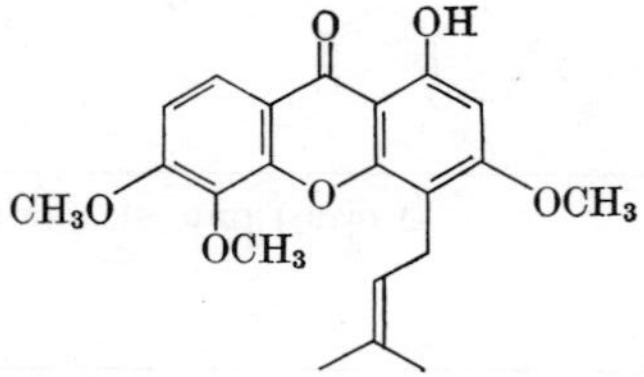

73 (total)   188

Note: References 344–439 are on pp. 251–252.

[a] The *o*-Claisen rearrangement is followed by a Diels-Alder reaction.

123

TABLE I. AROMATIC CLAISEN REARRANGEMENTS OF ALLYL AND PROPARGYL ETHERS (*Continued*)

E. *Allyl and Substituted Allyl Ethers of Polycyclic and Heterocyclic Systems* (*Continued*)

Heterocyclic Systems (*Continued*)

Ring C No.	Hetero-atom(s)	Compound	Conditions	Product(s) and Ratio ( )	Yield(s), %	Refs.
C22	O	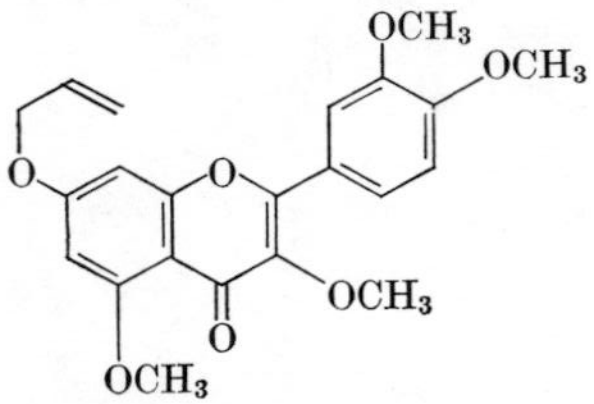	235°, 6 hr, (C2H5)2NC6H5	Trace	50	404
C24	O	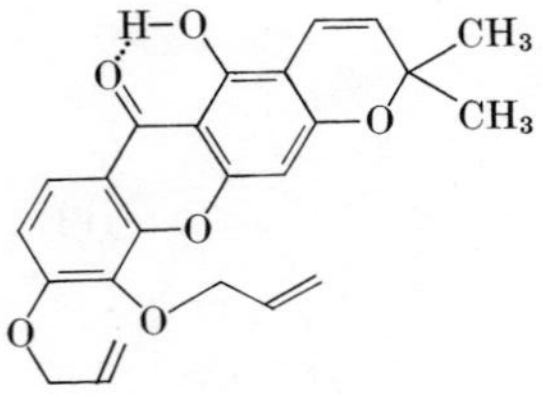	Reflux, 14 hr, decalin	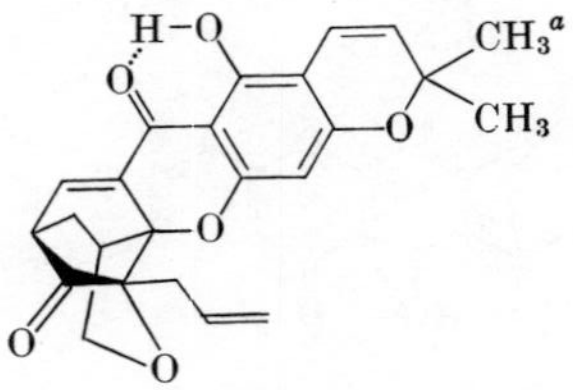	—	402

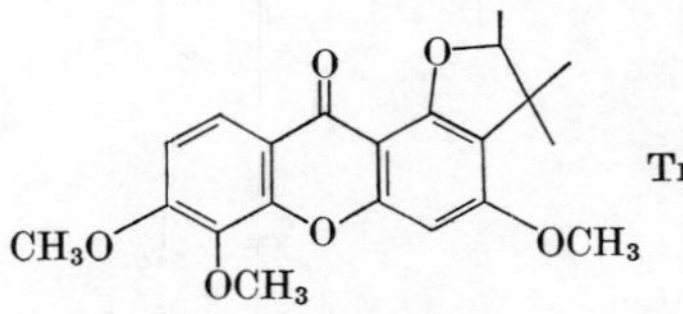

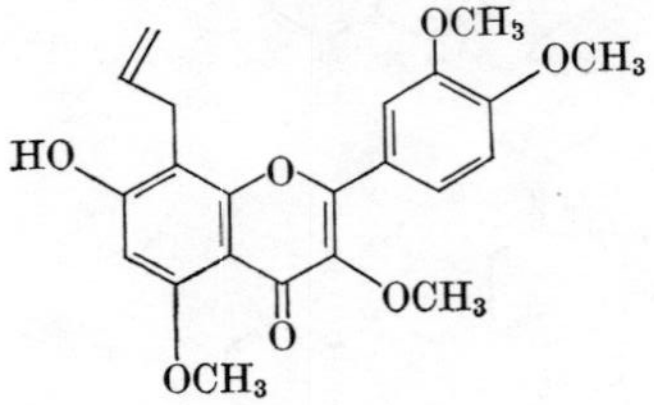

C$_{25}$ O

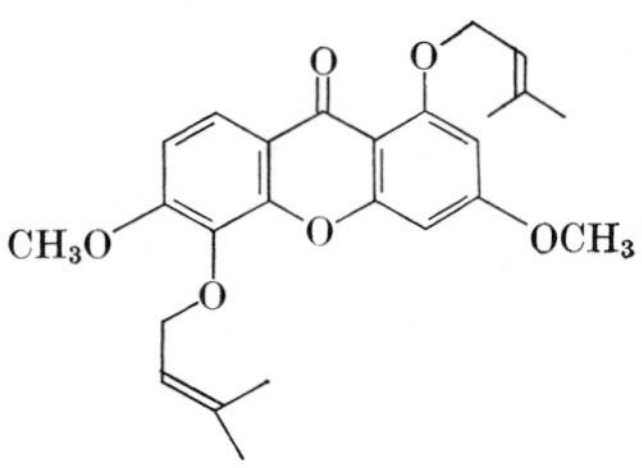

Reflux, 5 hr,
(CH$_3$)$_2$NC$_6$H$_5$

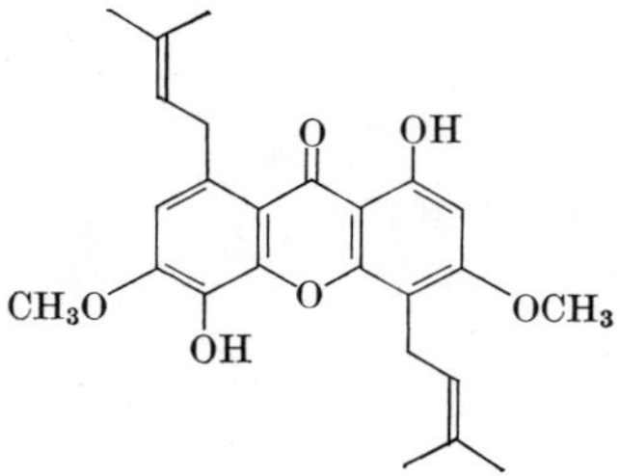

(1),

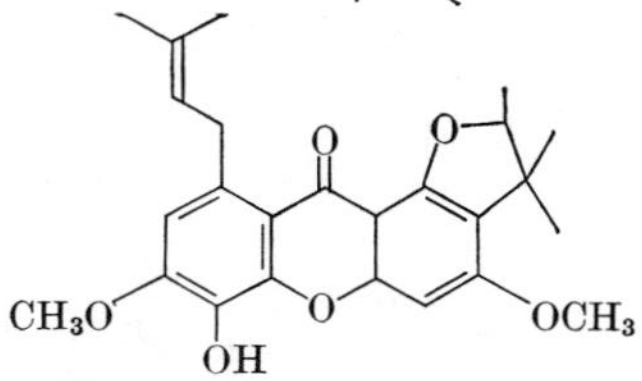

(1)

65 (total) 188

C$_{29}$ O

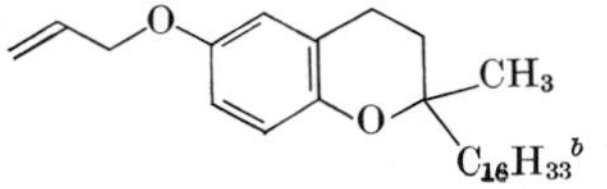

Heat, (CH$_3$)$_2$NC$_6$H$_5$

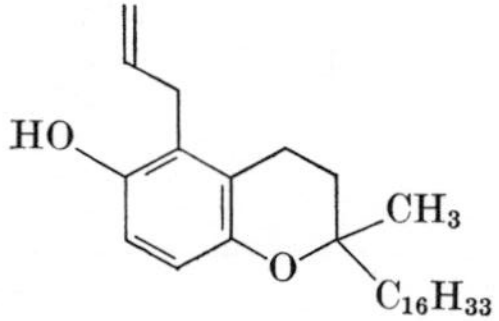

(3),

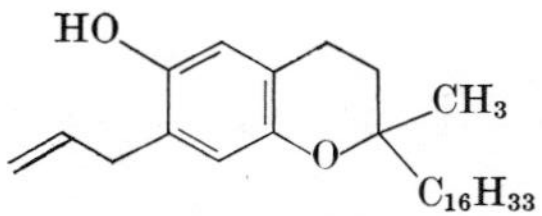

(1)

217 —

Note: References 344–439 are on pp. 251–252.

[a] The *o*-Claisen rearrangement is followed by a Diels-Alder reaction.

[b] C$_{16}$H$_{33}$ = (CH$_3$)$_2$CH(CH$_2$)$_3$CH(CH$_3$)(CH$_2$)$_3$CH(CH$_3$)(CH$_2$)$_3$—

125

TABLE I. Aromatic Claisen Rearrangements of Allyl and Propargyl Ethers (*Continued*)

E. *Allyl and Substituted Allyl Ethers of Polycyclic and Heterocyclic Systems* (*Continued*)

Heterocyclic Systems (*Continued*)

| Ring | | | | Yield(s) | |
C No.	Hetero-atom(s)	Compound	Conditions	Product(s) and Ratio ( )	%	Refs.
C_{30}	O	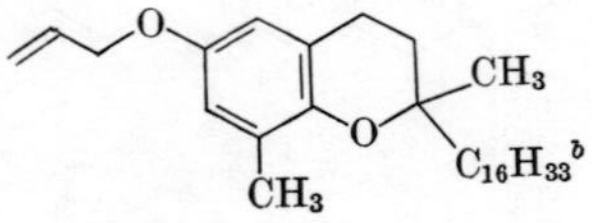	Heat, $(CH_3)_2NC_6H_5$	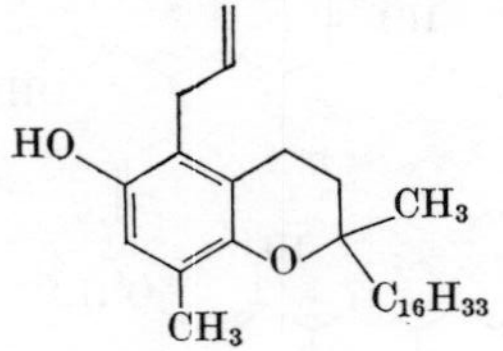 (4), 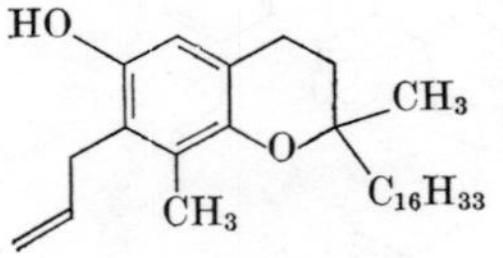 (1)	—	217
		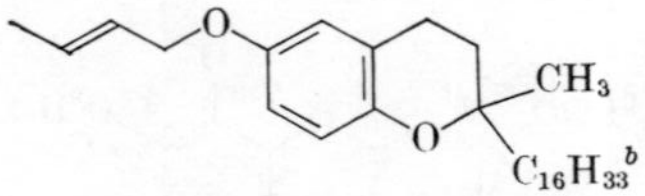	190–200°, 2 hr, $(CH_3)_2NC_6H_5$	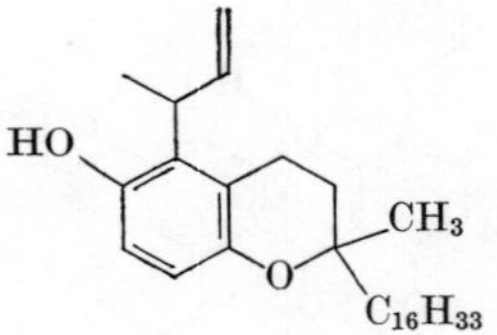 (1.7),	—	367

C$_{31}$ O

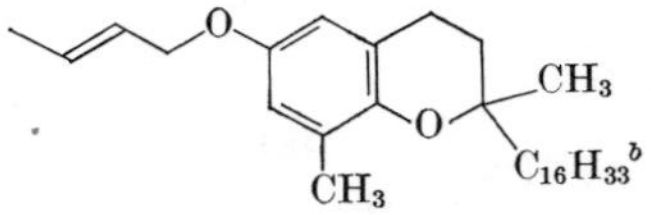

190–200°, 2 hr,
(CH$_3$)$_2$NC$_6$H$_5$

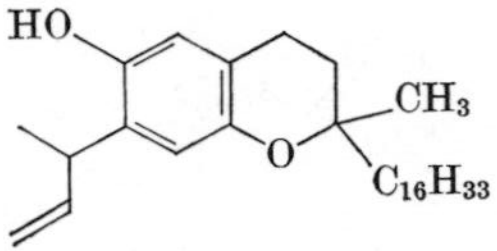

(1),

Trace of cleavage and other products

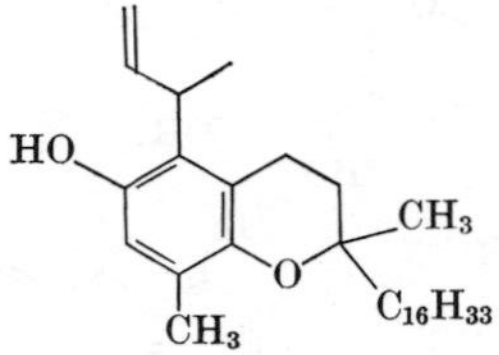

(4)$_1$

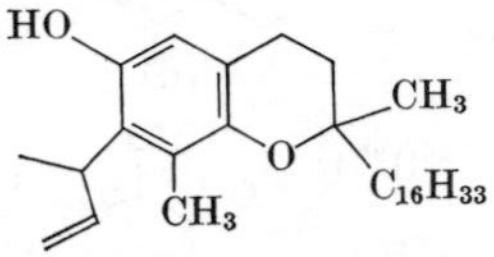

(1),

45 (total)   367

180–200°, 2 hr,
(CH$_3$)$_2$NC$_6$H$_5$

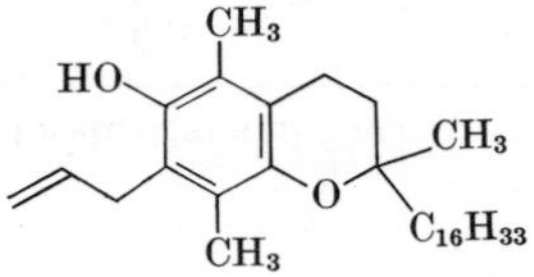

Trace of phenol and unrearranged
side-chain products

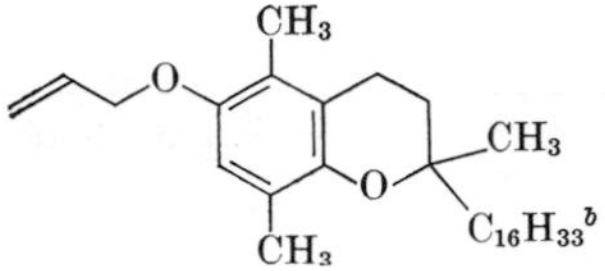

70   217

Note: References 344–439 are on pp. 251–252.

[b] C$_{16}$H$_{33}$ = (CH$_3$)$_2$CH(CH$_2$)$_3$CH(CH$_3$)(CH$_2$)$_3$CH(CH$_3$)(CH$_2$)$_3$–.

TABLE I. AROMATIC CLAISEN REARRANGEMENTS OF ALLYL AND PROPARGYL ETHERS *(Continued)*

E. *Allyl and Substituted Allyl Ethers of Polycyclic and Heterocyclic Systems (Continued)*

Heterocyclic Systems (Continued)

C No.	Ring Hetero-atom(s)	Compound	Conditions	Product(s) and Ratio ( )	Yield(s), %	Refs.
C_{31} *(Contd.)*	O	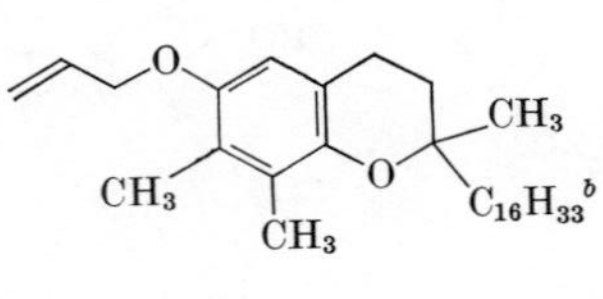 (allyl ether chroman structure with CH_3, $C_{16}H_{33}$, CH_3)	180–200°, 2 hr, $(CH_3)_2NC_6H_5$	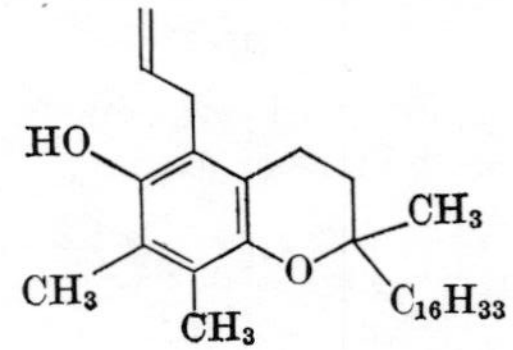 (allyl phenol chroman structure with CH_3, $C_{16}H_{33}$, HO, CH_3)	90	217
C_{32}	O	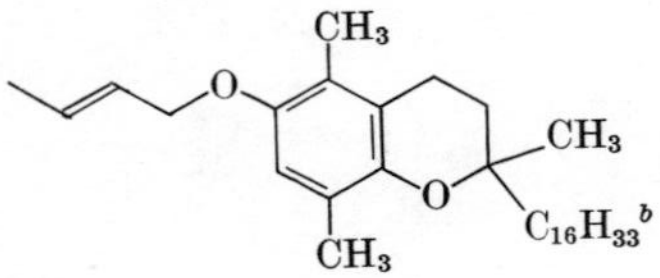 (butenyl ether chroman structure with CH_3, $C_{16}H_3^b$, CH_3)	180–200°, 3 hr, neat	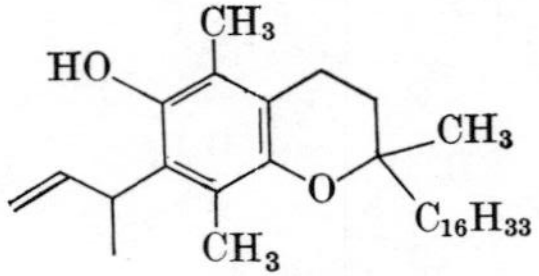 I (Major); 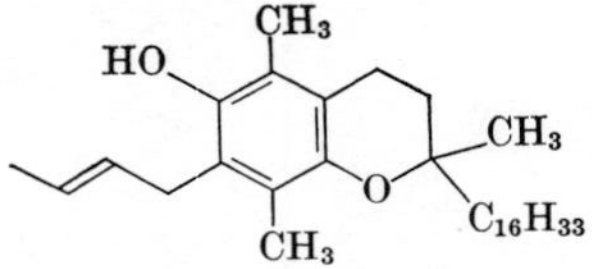 II (Minor); 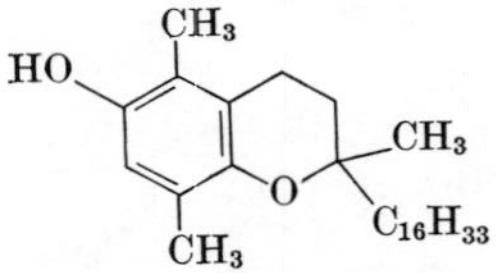 III	37 (total)	367
			195°, 3 hr, decalin	Mostly II and III; little normal product	—	405

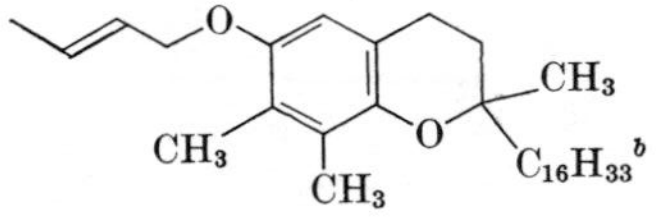

50 (total) 367

180–200°, 3 hr, neat

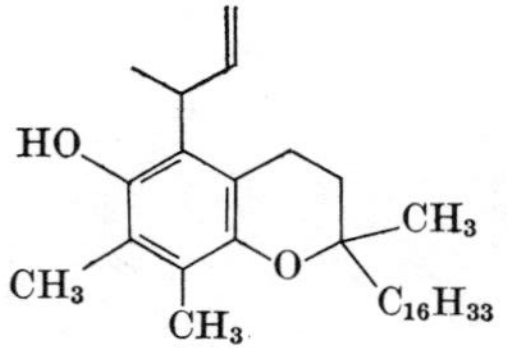

(Major),

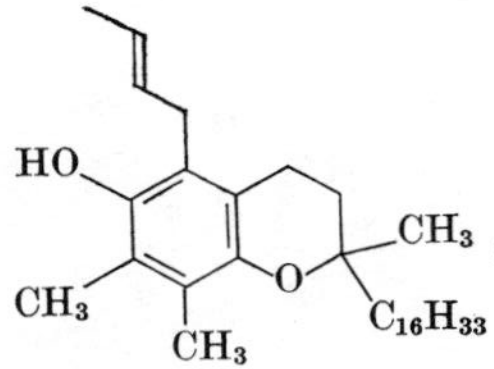

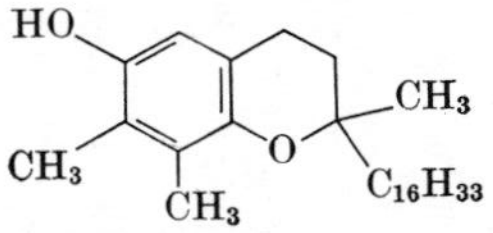

F. Out-of-Ring Migrations

C No.	Hetero-atom(s)	Compound	Conditions	Product(s) and Ratio ( )	Yield(s), %	Refs.
C$_{10}$	N$_4$	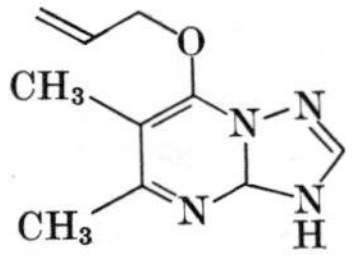	180°, 1 hr, neat	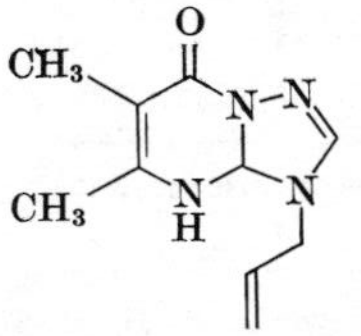 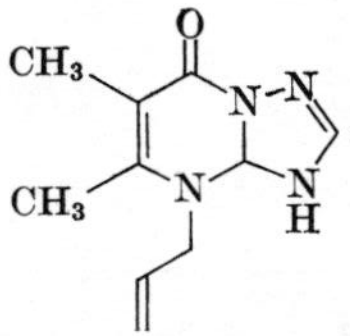 ,	—	406

Note: References 344–439 are on pp. 251–252.

^{b} C$_{16}$H$_{33}$ = (CH$_2$)$_2$CH(CH$_2$)$_3$CH(CH$_3$)(CH$_2$)$_3$CH(CH$_3$)(CH$_2$)$_3$–.

TABLE I. AROMATIC CLAISEN REARRANGEMENTS OF ALLYL AND PROPARGYL ETHERS (*Continued*)

F. *Out-of-Ring Migrations* (*Continued*)

C No.	Hetero-atom(s)	Compound	Conditions	Product(s) and Ratio ( )	Yield(s), %	Refs.
C$_{14}$	—	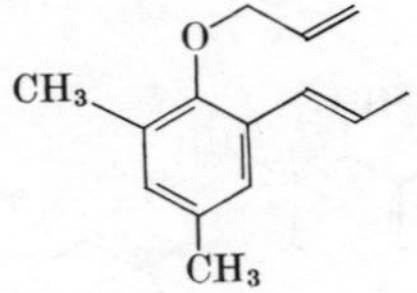	184°, 13 hr, 34 mm, N$_2$ atm	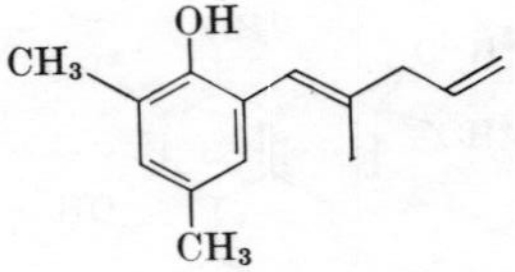 three out-of-ring products:	36.1 (total)	207

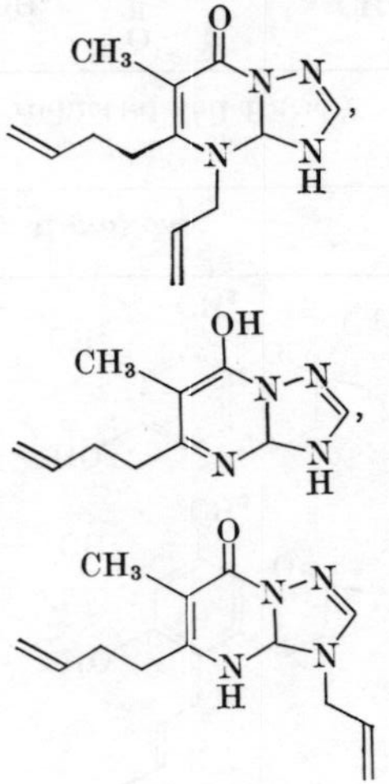

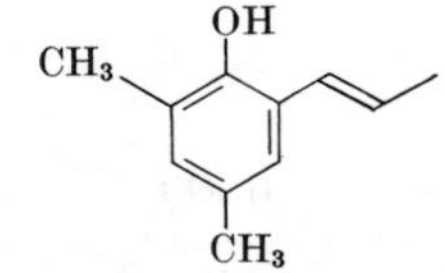

	44	208
	41 (total)	209
	96 (total)	212

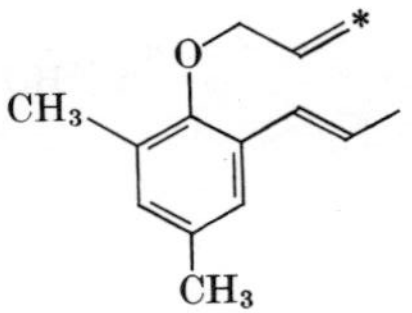

170°, 20 hr,
$(C_2H_5)_2NC_6H_5$,
reduced pres.

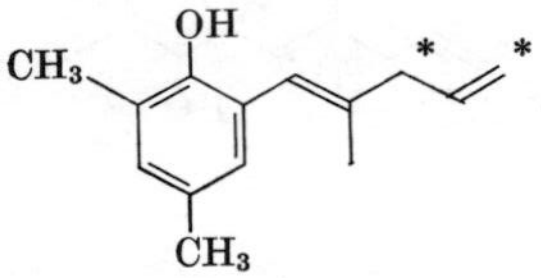

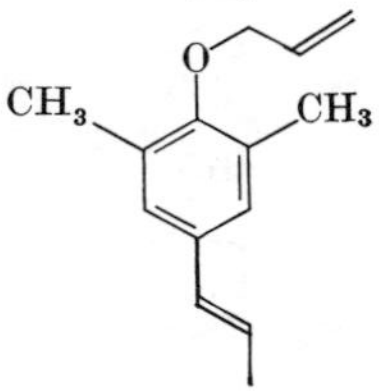

200°, 8 hr, 95%
C_2H_5OH, sealed tube

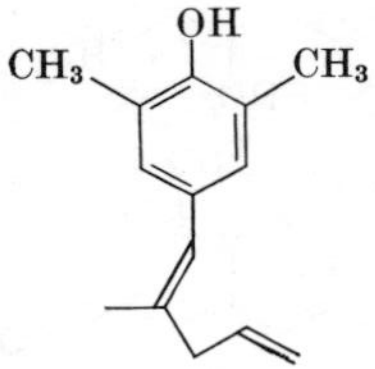

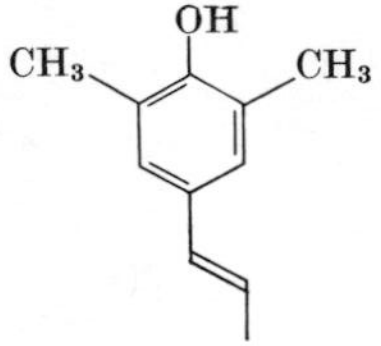

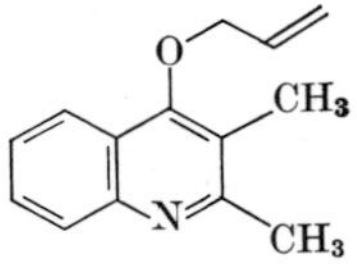

200°, 30 min, neat

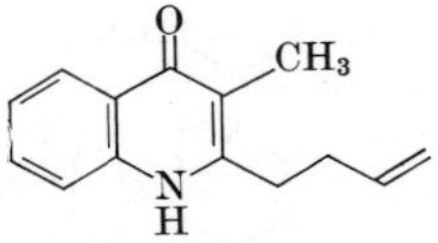

(1)

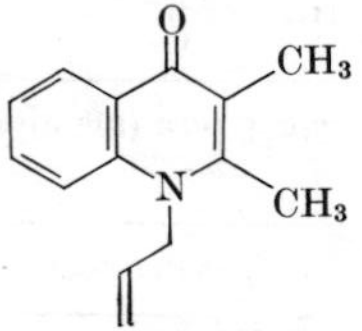

(4),

(1)

(46),

(1),

C_{14} N

Note: References 344–439 are on pp. 251–252.

* ^{14}C.

131

TABLE I. AROMATIC CLAISEN REARRANGEMENTS OF ALLYL AND PROPARGYL ETHERS (Continued)

F. Out-of-Ring Migrations (Continued)

C No.	Hetero-atom(s)	Compound	Conditions	Product(s) and Ratio ( )	Yield(s), %	Refs.
C_{15}	—	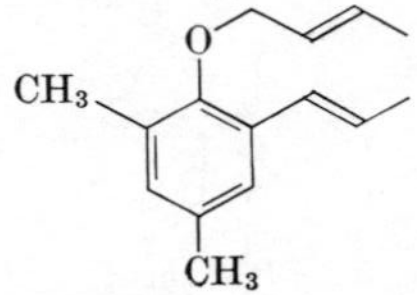	$174°$, 5 hr, 44 mm, N_2 atm	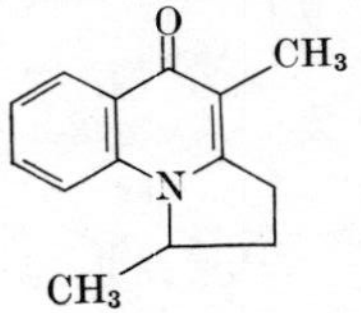 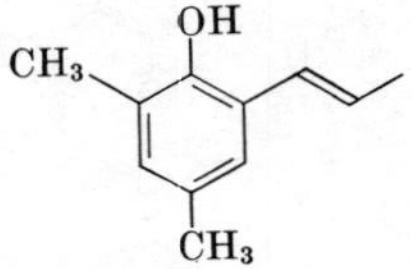 (1) 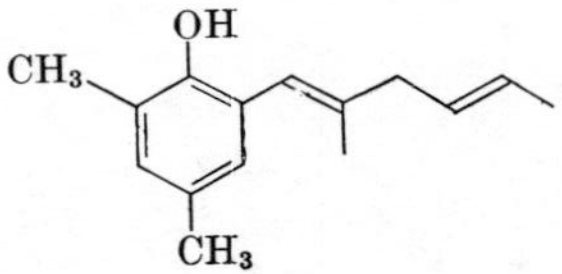	48 (total)	207
		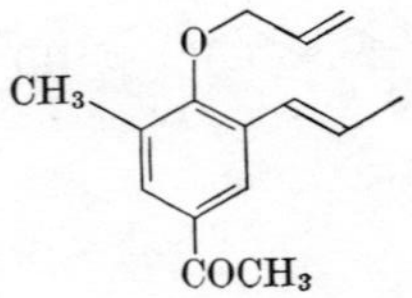	$170°$, 20 hr, $(C_2H_5)_2NC_6H_5$, reduced pres.	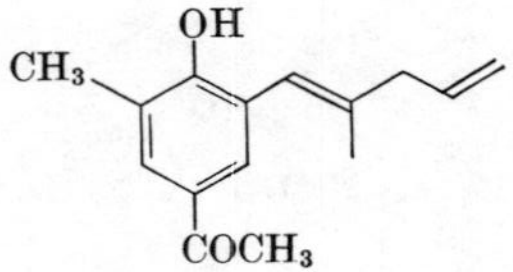 (1.1)	76.5	208

C$_{15}$ O

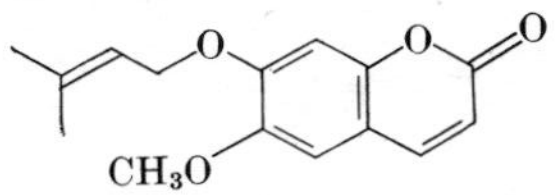

195°, 2 hr

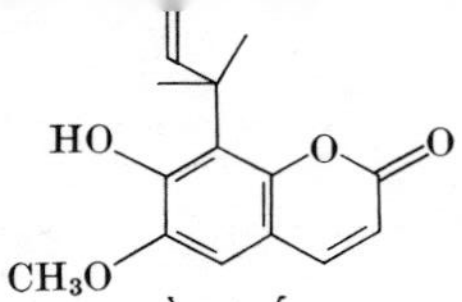

(1),

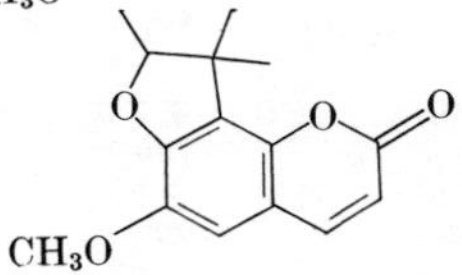

(2.3),

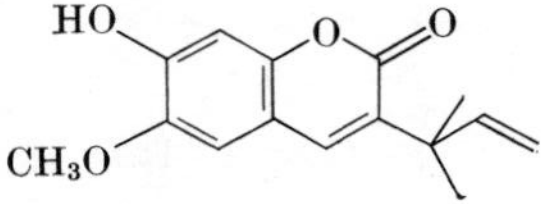

(1.5)

44 (total) 210, 211

C$_{16}$ N

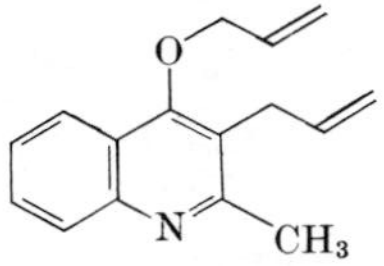

200°, 30 min, neat

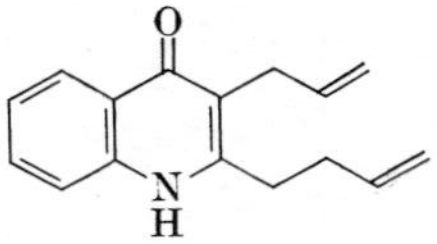

(30),

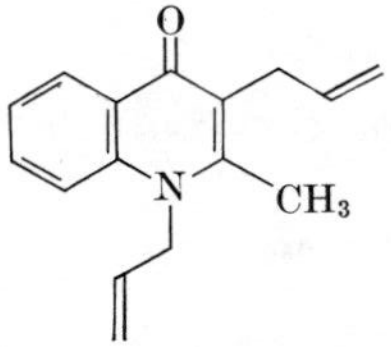

(1),

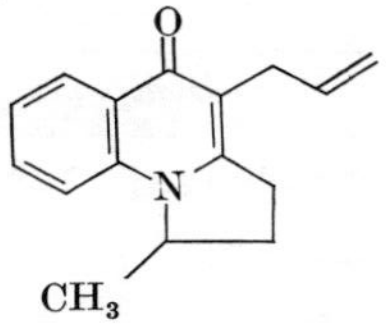

(1.3)

97 (total) 212

Note: References 344–439 are on pp. 251–252.

133

TABLE II. ALIPHATIC CLAISEN REARRANGEMENTS

A. Acyclic Allyl Vinyl Ethers

Molecular Formula	Ether	Conditions	Product(s) and Ratio ( )	Yield(s), %	Refs.
C_5H_8O	$CH_2{=}CHCH_2OCH{=}CH_2$	100°, 72 hr, $(C_6H_5CO_2)_2$, sealed tube	$CH_2{=}CHCH_2CH_2CHO$	36	123
$C_7H_5OF_7$	$[CH_2{=}CHCH_2OCF{=}C(CF_3)_2]$	<50° allyl alcohol, distil	$(CF_3)_2CCO_2CH_2CH{=}CH_2$ $\quad\ CH_2CH{=}CH_2$	15	128a
$C_7H_6OF_6$	$[CH_2{=}CHCH_2OC(CF_3){=}CHCF_3]$	<90°, distil	$CF_3COCH(CF_3)CH_2CH{=}CH_2$	74	128a
$C_7H_{10}O$	$[CH_2{=}CHCHOCH{=}CH_2$ $\qquad\quad CH{=}CH_2]$	Reflux, 3 hr, $Hg(OAc)_2$, NaOAc	$CH_2{=}CHCH{=}CHCH_2CH_2CHO$	72	233
	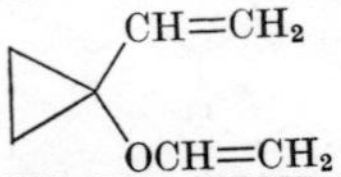 (cyclopropyl)$OCH{=}CH_2$	150°, 30 min, sealed tube	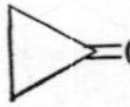 $={=}CHCH_2CH_2CHO$ (cyclopropyl)	—	231
$C_7H_{12}O$	$CH_3CH{=}CHCH(CH_3)OCH{=}CH_2$ *cis* or *trans*	170–180°, 15–20 min, sealed tube	$CH_3CH{=}CHCH(CH_3)CH_2CHO$ *trans*	80–90	365
	$CH_3CH{=}CHCH_2OCH{=}CHCH_3$ *trans, cis*	142.5 ± 0.1°, 120 min, 5% in heptane, bomb tube	$CH_2{=}CHCH(CH_3)CH(CH_3)CHO$ *erythro* (97 ± 1) *threo* (3 ± 1)	67.7	54
	cis,cis	", 240 min	*erythro* (2.2 ± 0.1) *threo* (97.8 ± 0.1)	52	54
	trans,trans	", 90 min	*erythro* (2.2 ± 0.7) *threo* (97.8 ± 0.7)	91	54
	$(CH_3)_2C{=}CHCH_2OCH{=}CH_2$	200°, 30 min, sealed tube, N_2 atm	$CH_2{=}CHC(CH_3)_2CH_2CHO$	Quant	118
	$CH_2{=}CHCH(C_2H_5)OCH{=}CH_2$	200°, sealed tube	$C_2H_5CH{=}CHCH_2CH_2CHO$ *trans*	—	118

$C_7H_{12}O$ (Contd.)	[CH$_2$=CHCH$_2$OC(CH$_3$)=CHCH$_3$ + CH$_2$=CHCH$_2$OC(C$_2$H$_5$)=CH$_2$ / CH$_2$=CHC(CH$_3$)$_2$OCH=CH$_2$]	Heat, p-TsOH, C$_6$H$_5$CH$_3$, distil	CH$_2$=CHCH$_2$CH(CH$_3$)COCH$_3$ and CH$_2$=CHCH$_2$CH$_2$COC$_2$H$_5$	79	117
		150°, 2 hr, H$_3$PO$_4$, pres. reactor, N$_2$	(CH$_3$)$_2$C=CHCH$_2$CH$_2$CHO	4 / 81	132
	[CH$_2$=CHCH$_2$OCH=CH(C$_2$H$_5$)]	Heat, 85% H$_3$PO$_4$, N$_2$	CH$_2$=CHCH$_2$CH(C$_2$H$_5$)CHO	37	115
	[CH$_2$=CHCH$_2$OCH=C(CH$_3$)$_2$]	Heat, p-TsOH, allyl alcohol	(CH$_2$=CHCH$_2$)$_2$C(C$_2$H$_5$)CHO	36	115
		Heat, 85% H$_3$PO$_4$	CH$_2$=CHCH$_2$C(CH$_3$)$_2$CHO I	77	115
		Heat, p-TsOH, cymene		89	115
$C_7H_{13}NO$	[CH$_2$=CHCH$_2$OC(N(CH$_3$)$_2$)=CH$_2$]	130°	CH$_2$=CHCH$_2$CH$_2$CON(CH$_3$)$_2$	90	134
$C_8H_8O_2F_6$	[CH$_2$=CHCH$_2$OC(OCH$_3$)=C(CF$_3$)$_2$]	<50°, distil	CH$_2$=CHCH$_2$C(CF$_3$)$_2$CO$_2$CH$_3$	39	128a
$C_8H_{10}O$	[CH$_2$=CHCHOCH=CH$_2$ (CH$_3$C≡C–)]	35°, Hg^{2+}	CH$_3$C≡CCH=CHCH$_2$CH$_2$CHO (cis/trans 1/2)	~40	228, 232
$C_8H_{12}O$	[CH$_3$CH=CHCHOCH=CH$_2$ (CH=CH$_2$)] *trans*	′′	CH$_2$=CHCH=CHCH(CH$_3$)CH$_2$CHO II, (1), CH$_3$CH=CHCH=CHCH$_2$CH$_2$CHO III, (2)	~65 (total)	229, 227
	[CH$_3$CH=CHCHOCH=CH$_2$ (CH=CH$_2$)] *cis*	′′	II (1), III (19)	~65	229
$C_8H_{12}O_2$	[(CH$_2$=CHCH$_2$O)$_2$C=CH$_2$]	Reflux, 5 hr, t-BuOH, t-BuOK	CH$_2$=CHCH$_2$CH$_2$CO$_2$CH$_2$CH=CH$_2$	43	119
$C_8H_{14}O$	(CH$_3$)$_2$C=CHCH(CH$_3$)OCH=CH$_2$	170–180°, 15–20 min, sealed tube	CH$_3$CH=CHC(CH$_3$)$_2$OCH$_2$CH$_2$CHO *trans*	80–90	365
	[CH$_2$=CHCH$_2$OC(C$_2$H$_5$)=CHCH$_3$]	Heat, p-TsOH, toluene, distil	CH$_2$=CHCH$_2$CH(CH$_3$)COCH$_2$CH$_3$	91	117
	[CH$_2$=CHC(CH$_3$)OC(CH$_3$)=CH$_2$]	125°, 13–15 hr, H$_3$PO$_4$, pres. reactor, N$_2$ atm	(CH$_3$)$_2$C=CHCH$_2$CH$_2$COCH$_3$	94	131

Note: References 344–439 are on pp. 251–252.

135

TABLE II. ALIPHATIC CLAISEN REARRANGEMENTS (Continued)

A. Acyclic Allyl Vinyl Ethers (Continued)

Molecular Formula	Ether	Conditions	Product(s) and Ratio ( )	Yield(s), %	Refs.
$C_8H_{14}O$ (contd.)	$[CH_2=CHC(CH_3)_2OC(CH_3)=CH_2]$	130–150°, 24 hr, ligroin, reflux	$(CH_3)_2C=CHCH_2CH_2COCH_3$	41	131
	$[CH_2=CHC(CH_3)(C_2H_5)OCH=CH_2]$	Heat, trace H_3PO_4	$C_2H_5C(CH_3)=CHCH_2CH_2CHO$ cis and trans	54.5	407
	$[CH_2=C(CH_3)C(CH_3)_2OCH=CH_2]$	150°, 1 hr, pres. reactor, N_2 atm	$(CH_3)_2C=C(CH_3)CH_2CH_2CHO$	89	132
	$[CH_2=CHC(CH_3)_2OCH=CH(CH_3)]$	100°, 48 hr, pres. reactor, N_2 atm	$(CH_3)_2C=CHCH_2CH(CH_3)CHO$	88	132
	$CH_2=C(CH_3)CH(C_2H_5)OCH=CH_2$	110°, sealed tube	$C_2H_5\overset{CH_3}{\underset{H}{C=C}}CH_2CH_2CHO$ (9), $H\overset{CH_3}{\underset{}{C=C}}CH_2CH_2CHO$ (1)	Quant	55
	$[CH_2=CHCH_2OCH=C(CH_3)(C_2H_5)]$	Heat, 85% H_3PO_4 Heat, p-TsOH, cymene	CH_2CH_2CHO $CH_2=CHCH_2C(CH_3)(C_2H_5)CHO$ ''	82 78	115 115
	$[CH_2=C(CH_3)CH_2OCH=CH(C_2H_5)]$ $[CH_2=C(CH_3)CH_2OCH=C(CH_3)_2]$	Heat, 85% H_3PO_4 Heat, 85% H_3PO_4	$CH_2=C(CH_3)CH_2CH(C_2H_5)CHO$ $CH_2=C(CH_3)CH_2C(CH_3)_2CHO$	63 73	115 115
$C_9H_{12}O$	$CH_3CH=CHCHOCH=CH_2$ $CH_3C\equiv C$	35°, Hg^{2+}	$CH_3C\equiv CCH=CHCH(CH_3)CH_2CHO$ (cis/trans 1/2)	~40	228, 232
	$CH_2=C(CH_3)OCH=CH_2$ $CH_3C\equiv C$	35°, 12 hr	$CH_3C\equiv CC(CH_3)=CHCH_2CH_2CHO$ (Z only)	—	408
	$CH_2=CHCHOCH=CH_2$ $C_2H_5C\equiv C$	35°, Hg^{2+}	$C_2H_5C\equiv CCH=CHCH_2CH_2CHO$	40	228

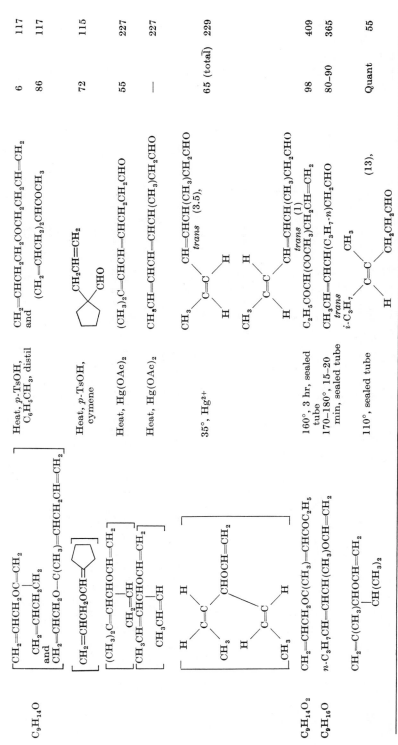

	Conditions	Products	Yield (%)	Ref.
$C_9H_{14}O$ $\big[CH_2=CHCH_2OC(CH_3)=CH_2$ and $CH_2=CHCH_2O-C(CH_3)=CHCH_2CH=CH_2\big]$	Heat, p-TsOH, $C_6H_5CH_3$, distil	$CH_2=CHCH_2CH_2COCH_2CH_2CH=CH_2$ and $(CH_2=CHCH_2)_2CHCOCH_3$	6, 86	117, 117
$\big[CH_2=CHCH_2OCH-$ cyclopentane$\big]$	Heat, p-TsOH, cymene	cyclopentane($CH_2CH=CH_2$)(CHO)	72	115
$\big[(CH_3)_2C=CHCHOCH=CH_2$, $CH_2=CH\big]$	Heat, $Hg(OAc)_2$	$(CH_3)_2C=CHCH=CHCH_2CH_2CHO$	55	227
$\big[CH_3CH=CHCHOCH=CH_2$, $CH_3CH=CH\big]$	Heat, $Hg(OAc)_2$	$CH_3CH=CHCH=CHCH(CH_3)CH_2CHO$	—	227
$\big[$H, CH$_3$ C=C CHOCH=CH$_2$ ... $\big]$	$35°$, Hg^{2+}	$CH=CHCH(CH_3)CH_2CHO$ *trans* (3.5), *trans* (1)	65 (total)	229
$C_9H_{14}O_2$ $CH_2=CHCH_2OC(CH_3)=CHCOC_2H_5$	$160°$, 3 hr, sealed tube	$C_2H_5COCH(COCH_3)CH_2CH=CH_2$	98	409
$C_9H_{16}O$ n-$C_3H_7CH=CHCH(CH_3)OCH=CH_2$	$170-180°$, 15–20 min, sealed tube	$CH_3CH=CHCH(C_3H_7\text{-}n)CH_2CHO$ *trans*	80–90	365
$CH_2=C(CH_3)CHOCH=CH_2$, $CH(CH_3)_2$	$110°$, sealed tube	(13), i-C_3H_7 ... CH_2CH_2CHO	Quant	55

Note: References 344–439 are on pp. 251–252.

137

TABLE II. ALIPHATIC CLAISEN REARRANGEMENTS (Continued)

A. Acyclic Allyl Vinyl Ethers (Continued)

Molecular Formula	Ether	Conditions	Product(s) and Ratio ( )	Yield(s), %	Refs.
$C_9H_{16}O$ (contd.)	$CH_2=C(C_2H_5)CHOCH=CH_2$ with C_2H_5	110°, sealed tube	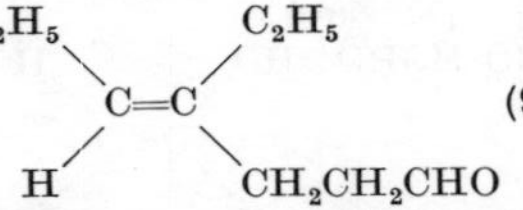 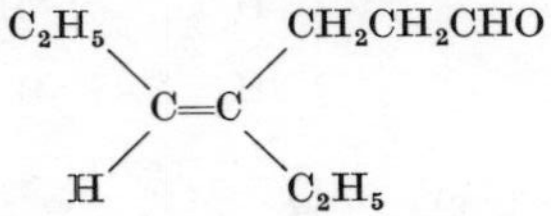 structure: $i\text{-}C_3H_7$, H, CH_2CH_2CHO, CH_3 on $C=C$ (1)	Quant	55
	$\left[CH_2=C(CH_3)CHOC(CH_3)=CH_2\right]$ with C_2H_5	110°, 24 hr, sealed tube, oxalic acid, hydroquinone	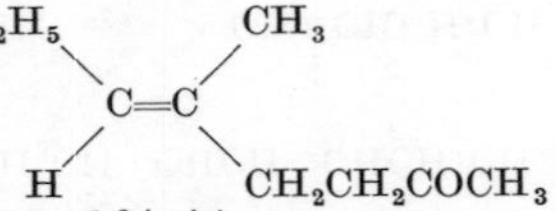 structure: C_2H_5, H, C_2H_5, CH_2CH_2CHO on $C=C$ (9), and structure: C_2H_5, C_2H_5, CH_2CH_2CHO, CH_2CH_2CHO (1)	76	55
	$[CH_3{=}CH(CH_3)CH_2OCH{=}C(CH_3)C_2H_5]$	Heat, 85% H_3PO_4	$CH_2=C(CH_3)CH_2C(CH_3)(C_2H_5)CHO$	87	115
	$[CH_2{=}CHCH(C_2H_5)OCH{=}C(CH_3)_2]$	Heat, p-TsOH, cymene	$C_2H_5CH=CHCH_2C(CH_3)_2CHO$ *trans* (+ <1% *cis*)	89	115
	$[CH_2{=}C'(CH_3)C'(CH_3)_2OC'(CH_3)=CH_2]$	125°, 13–15 hr, H_3PO_4, pres. reactor, N_2 atm	$(CH_3)_2C=C(CH_3)CH_2CH_2COCH_3$	94	131
	$[CH_2{=}C(CH_3)C(CH_3)_2OCH{=}CHCH_3]$	150°, 1 hr, H_3PO_4, pres. reactor, N_2 atm	$(CH_3)_2C=C(CH_3)CH_2CH(CH_3)CHO$	60	132
	$[CH_2{=}CHC(CH_3)_2OCH{=}CH(C_2H_5)]$	180°, 2 hr, H_3PO_4, pres. reactor, N_2 atm	$(CH_3)_2C=CHCH_2CH(C_2H_5)CHO$	56	132

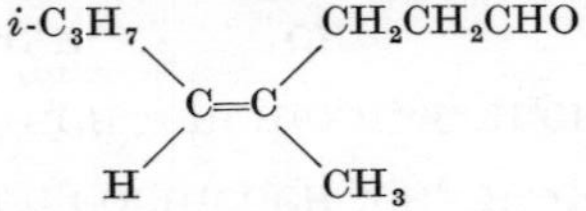

Formula	Reactant	Conditions	Product	Yield (%)	Refs.
	$[(CH_3)_2C=CHCH_2OCH=C(CH_3)_2]$	115–120°, 20 hr, hydroquinone, p-TsOH	$CH_2=CHC(CH_3)_2C(CH_3)_2CHO$	50	118
	$[CH_2=CHC(CH_3)_2OCH=C(CH_3)_2]$	135°, 30 hr, p-TsOH, hydroquinone	$(CH_3)_2C=CHCH_2C(CH_3)_2CHO$	—	118
$C_9H_{17}NO$	$[CH_2=CHCH(C_2H_5)OCH=C(CH_3)_2]$ *trans*	35°, Hg^{2+}	$C_2H_5CH=CHCH_2C(CH_3)_2CHO$	~75	118
	$\left[\begin{array}{c}CH_3CH=CHCH_2OC=CHCH_3\\ \mid \\ N(CH_3)_2\end{array}\right]$ *cis*	Reflux, 15 hr, xylene	$CH_2=CHCH(CH_3)CH(CH_3)CON(CH_3)_2$ (*erythro/threo* 19/1)	76	410
$C_{10}H_{14}O$	$\left[\begin{array}{c}CH_3CH=CHCHOCH=CH_2\\ \mid \\ C_2H_5C\!\equiv\!C\end{array}\right]$ *trans*	Reflux, 7 hr, xylene	(*erythro/threo* 1/32)	65	410
$C_{10}H_{14}O$	$\left[\begin{array}{c}CH_3CH=CHCHOHOC=CH_2\\ \mid \\ C_2H_5C\!\equiv\!C\end{array}\right]$ *trans*	35°, $Hg(OAc)_2$	$C_2H_5C\!\equiv\!CCH=CHCH(CH_3)CH_2CHO$ *trans*	~40	228
$C_{10}H_{14}O_2$	$\left[\begin{array}{c}CH_2=CHCHOC=CH_2\\ \mid \qquad \mid \\ CH_3C\!\equiv\!C \quad OC_2H_5\end{array}\right]$ *trans*	140°, 2 hr	$CH_3C\!\equiv\!CCH=CHCH_2CH_2CO_2C_2H_5$ (*cis/trans* 1/8)	25	408
$C_{10}H_{14}O_3$	$\left[\begin{array}{c}\qquad CO_2CH_2CH=CH_2\\ \mid \\ CH_3C=C\\ \mid \qquad\quad H\\ CH_2=CHCH_2O\end{array}\right]$	150°, 1 hr, neat	$CH_3COCH(CH_2CH=CH_2)CO_2CH_2CH=CH_2$	—	124
$C_{10}H_{16}N_2O$	$\left[\begin{array}{c}\quad CH=CH_2\\ \mid \quad\quad CN\\ (C_2H_5)_2N\end{array}\right]$ *	$BF_3{\cdot}(C_2H_5)_2O$, trace	(C₂H₅)₂N–CO–CH(CN)–CH₂CH=CH₂	18	434
$C_{10}H_{16}O$	$\left[\begin{array}{c}CH_3\quad CH=CH_2\\ \quad C==C\\ \quad H \qquad CH_3\\ CH_3\\ \quad C==C\\ H \qquad CH_2\end{array}\right]$	100°, 15 hr, $Hg(OAc)_2$, NaOAc, Ar atm	(Major) and (Minor) CHO isomers (CH₃-substituted unsaturated aldehydes)	—	102

Note: References 344–439 are on pp. 251–252.

* This was obtained from the reaction of allyl alcohol and $(C_2H_5)_2NC\!\equiv\!CCN$.

TABLE II. ALIPHATIC CLAISEN REARRANGEMENTS (Continued)

A. Acyclic Allyl Vinyl Ethers (Continued)

Molecular Formula	Ether	Conditions	Product(s) and Ratio ( )	Yield(s), %	Refs.
$C_{10}H_{16}O$ (contd.)	$CH_2=CHCH_2OC(CH_3)=CHCO(C_3H_7-n)$	160°, 3 hr, sealed tube	$(n\text{-}C_3H_7)COCH(COCH_3)CH_2CH=CH_2$	98	409
	$CH_2=CHCH_2OC(CH_3)=CHCO(C_3H_7-i)$	160°, 3 hr, sealed tube	$(i\text{-}C_3H_7)COCH(COCH_3)CH_2CH=CH_2$	98	409
	$[CH_2=CHCH_2OCH=C(C_2H_5)CH_2CH=CH_2]$	Heat, p-TsOH, cymene	$(CH_2=CHCH_2)_2C(C_2H_5)CHO$	36	115
	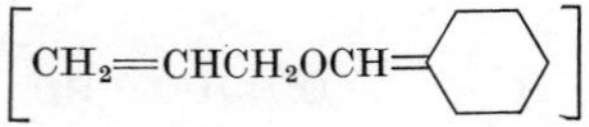	Heat, p-TsOH	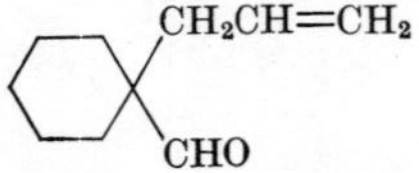	84	115
	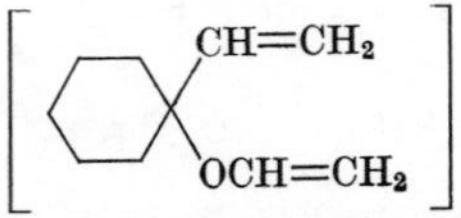	150°, 3 hr, H_3PO_4, pres. reactor, N_2 atm	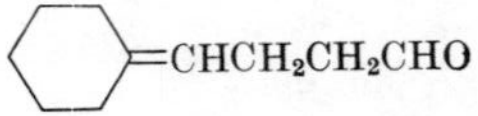	40	132
$C_{10}H_{16}OSi$		35°, Hg^{2+}	$(CH_3)_3SiC\equiv CCH=CHCH_2CH_2CHO$ (cis/trans 1/1.5)	—	232
$C_{10}H_{17}NO$		Heat	$CH_2=CHCH_2CH_2CON(C_2H_5)CH_2CH=CH_2$	—	247
$C_{10}H_{18}O$	$[CH_2=CHCH_2OCH=C(CH_3)C_4H_9\text{-}n]$ $[CH_3CH_2C(CH_3)=CHCH_2OCH=C(CH_3)_2]$	Heat, 85% H_3PO_4 115°, 20 hr, trace p-TsOH, hydroquinone	$CH_2=CHCH_2C(CH_3)(C_4H_9\text{-}n)CHO$ $CH_3CH_2C(CH_3)=CHCH_2C(CH_3)_2CHO$	73 —	115 118
	$[CH_2=CHCH(C_3H_7\text{-}i)OCH=C(CH_3)_2]$	120°, 20 hr, dry C_6H_6, trace p-TsOH, hydroquinone	$i\text{-}C_3H_7CH=CHCH_2C(CH_3)_2CHO$	~75	118

Molecular formula	Reactant	Conditions	Product	Yield (%)	Ref.
	$[CH_2=CHC(CH_3)(C_4H_9\text{-}i)OCH=CH_2]$	$200°$, 0.5 hr, H_3PO_4, pres. reactor, N_2 atm	$i\text{-}C_4H_9C(CH_3)=CHCH_2CH_2CHO$ (cis/trans 1/2.3)	65	132
$C_{10}H_{19}NO$	$\left[CH_2=C(CH_3)CH(C_2H_5)OC(N(CH_3)_2)=CH_2\right]$	$140°$, xylene	$C_2H_5CH=C(CH_3)CH_2CH_2CON(CH_3)_2$ trans, trace of cis	High	55
$C_{11}H_{12}O$	$C_6H_5CH=CHCH_2OCH=CH_2$	$200°$, sealed tube	$CH_2=CHCH(C_6H_5)CH_2CHO$	—	118
$C_{11}H_{16}O_2$	$\left[CH_2=C(CH_3)OC=CH_2\;/\;CH_3C\equiv C(OC_2H_5)(CH_2)_2C(CH_3)=CH_2\right]$	$140°$, 2 hr	$CH_3C\equiv CC(CH_3)=CHCH_2CH_2CO_2C_2H_5$ (E/Z 1/10)	~25	408
$C_{11}H_{18}O$	$CH_2=C(CH_3)CHOCH=CH_2$	$83-98°$, 61 hr	$CH_2=C(CH_3)CH_2CH_2CH=C(CH_3)CH_2CH_2CHO$ (cis/trans 1/6)	98	97
	$\left[CH_2=C(CH_3)CH(C_2H_5)OC=CH_2\;/\;C(CH_3)=CH_2\right]$	$110°$, 24 hr, oxalic acid, hydroquinone, sealed tube	$C_2H_5CH=C(CH_3)CH_2CH_2COC(CH_3)=CH_2$ trans only	—	55
$C_{11}H_{20}O$	$[CH_2=CHCH(C_4H_9\text{-}n)OCH=C(CH_3)_2]$	$120°$, 20 hr, dry C_6H_6, trace p-TsOH, hydroquinone	$n\text{-}C_4H_9CH=CHCH_2C(CH_3)_2CHO$	~75	118
	$[CH_2=CHC(CH_3)(C_4H_9\text{-}i)OC(CH_3)=CH_2]$	$125°$, 15 hr, H_3PO_4, pres. N_2 atm	$i\text{-}C_4H_9C(CH_3)=CHCH_2CH_2COCH_3$ (cis/trans 1/1.2)	80	131
	$[CH_2=CHC(CH_3)(C_4H_9\text{-}i)OCH=CHCH_3]$	$200°$, 1 hr, H_3PO_4, pres. N_2 atm	$i\text{-}C_4H_9C(CH_3)=CHCH_2CH(CH_3)CHO$ (cis/trans 1/2.3)	83	132
$C_{12}H_{14}O$	$[CH_2=CHC(CH_3)(C_6H_5)OCH=CH_2]$	$200°$, 1.5 hr, H_3PO_4, pres. N_2 atm	$C_6H_5C(CH_3)=CHCH_2CH_2CHO$	25	132
$C_{12}H_{18}O_3$	$CH_3CH=CHCH_2OC(CH_3)=CH\;/\;CH_3CH=CHCH_2O_2C\;/\;CH_3O_2C$	$150°$, 1 hr, neat	$CH_2=CHCH(CH_3)CHCOCH_3\;/\;CO_2CH_2CH=CHCH_3$	—	124
	$\left[CH_2=C(CH_3)CHOC=CH_2\;/\;C(C_2H_5)=CH_2\right]$	$100°$, 8 hr, $C_6H_5CH_3$, $2,4\text{-}(O_2N)_2C_6H_3OH$	$C_2H_5COCH_2CH_2C(CH_3)=CHCO_2CH_2CH_3$	—	96

Note: References 344–439 are on pp. 251–252.

141

TABLE II. ALIPHATIC CLAISEN REARRANGEMENTS (Continued)

A. Acyclic Allyl Vinyl Ethers (Continued)

Molecular Formula	Ether	Conditions	Product(s) and Ratio ( )	Yield(s), %	Refs.
$C_{12}H_{20}O$	$\begin{bmatrix} CH_2=CHC(CH_3)OCH=CH_2 \\ (CH_2)_2CH=C(CH_3)_2 \end{bmatrix}$	120°, 17 hr, H_3PO_4, pres., N_2 atm	$(CH_3)_2C=CH(CH_2)_2C(CH_3)=CHCH_2CH_2CHO$ (cis/trans 1/1.5)	73	132
	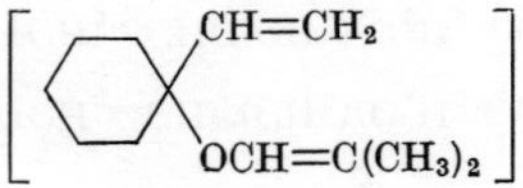 (structure: cyclohexane with $CH=CH_2$ and $OCH=C(CH_3)_2$)	120°, 20 hr, C_6H_6, hydroquinone, trace p-TsOH	 (structure: cyclohexylidene $=CHCH_2C(CH_3)_2CHO$)	50	118
	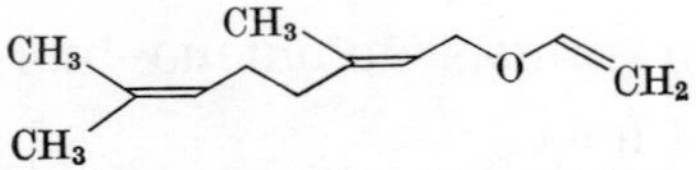 (structure with CH_3, CH_3 groups and $O-CH_2$ vinyl ether)	188°, 90 min	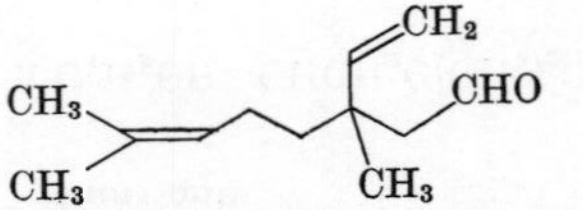 (branched structure with CH_2, CHO, CH_3, CH_3, CH_3)	70	130
$C_{12}H_{21}NO$	$\begin{bmatrix} (CH_3)_2C=CHCH=CHCH_2OC=CHCH_3 \\ N(CH_3)_2 \end{bmatrix}$	Heat, xylene	$(CH_3)_2C=CHCHCH(CH_3)CON(CH_3)_2$ $CH=CH_2$	78	411
	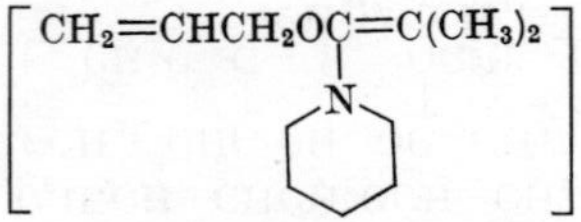 $\begin{bmatrix} CH_2=CHCH_2OC=C(CH_3)_2 \\ N \text{ (piperidine)} \end{bmatrix}$	Heat	 $CH_2=CHCH_2C(CH_3)_2CON$ (piperidine)	Quant	247
$C_{12}H_{22}O$	$\begin{bmatrix} CH_2=CHC(CH_3)OCH=CH_2 \\ (CH_2)_3CH(CH_3)_2 \end{bmatrix}$	150°, 3 hr, H_3PO_4, pres., N_2 atm	$(CH_3)_2CH(CH_2)_3C(CH_3)=CH(CH_2)_2CHO$ (cis/trans 1/1.5)	85	132

142

$C_{12}H_{23}NO$	$\left[\begin{array}{c} C_2H_5CH=CHCH(C_2H_5)OC-CHCH_3 \\ \underset{N(CH_3)_2}{\mid} \end{array}\right]$ $trans$ cis	Reflux, 15 hr, xylene	$C_2H_5CH=CHCH(C_2H_5)CH(CH_3)CON(CH_3)_2$ (*erythro/threo* 11.5/1)	92	410
		,,	(*erythro/threo* 1/9)	93	410
$C_{12}H_{23}NO_2$	$\left[\begin{array}{c} (CH_3)_2C(OH)CH=CHC(CH_3)_2OC=CH_2 \\ \underset{N(CH_3)_2}{\mid} \end{array}\right]$	Heat, xylene	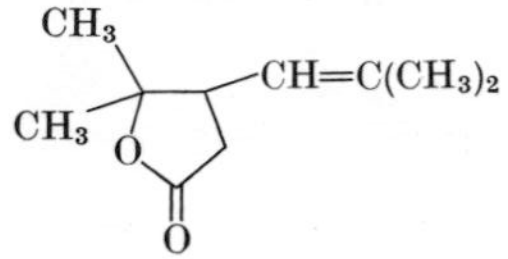	—	226
$C_{13}H_{11}OCl$	$\left[\begin{array}{c} CH_2=CHCHOCH=CH_2 \\ \underset{C\equiv CC_6H_4Cl\text{-}p}{\mid} \end{array}\right]$	$35°$, Hg^{2+}	$p\text{-}ClC_6H_4C\equiv CCH=CHCH_2CH_2CHO$ (*cis/trans* 1/1.5)	—	232
$C_{13}H_{12}O$	$\left[\begin{array}{c} CH_2=CHCHOCH=CH_2 \\ \underset{C\equiv CC_6H_5}{\mid} \end{array}\right]$	$35°$, Hg^{2+}	$C_6H_5C\equiv CCH=CHCH_2CH_2CHO$ (*cis/trans* 1/1.5)	—	232
$C_{13}H_{16}O$	$[CH_2=CHC(CH_3)C_6H_5OC(CH_3)=CH_2]$	$125°$, 13–15 hr, H_3PO_4, N_2 pres.	$C_6H_5C(CH_3)=CHCH_2CH_2COCH_3$ (*cis/trans* 1/1.6)	73	131
	$[CH_2=CHCH(C_6H_5)OCH=C(CH_3)_2]$	115–120°, 25 hr, p-TsOH, hydroquinone	$CH_2=CHCH(C_6H_5)C(CH_3)_2CHO$	75	118
$C_{13}H_{17}NO$	$\left[\begin{array}{c} C_6H_5CH=CHCH_2OCH=C(CH_3)_2 \\ \text{or} \\ CH_2=CHCH_2OC=CHCH_3 \\ \underset{N(CH_3)C_6H_5}{\mid} \end{array}\right]$	$30°$, 1 hr, trace $BF_3 \cdot (C_2H_5)_2O$	,,	75	118
			$CH_2=CHCH_2CH(CH_3)CON(CH_3)C_6H_5$	75	125
$C_{13}H_{20}O$	$[CH_2=CHCH_2OCH\text{ (bicyclic)}]$ 	Heat, p-TsOH	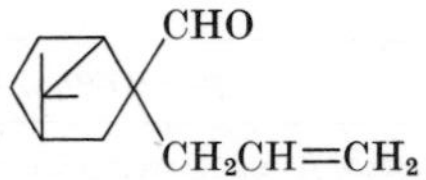	91	115

Note: References 344–439 are on pp. 251–252.

143

TABLE II. ALIPHATIC CLAISEN REARRANGEMENTS (Continued)

A. Acyclic Allyl Vinyl Ethers (Continued)

Molecular Formula	Ether	Conditions	Product(s) and Ratio ( )	Yield(s), %	Refs.
$C_{13}H_{22}O$	$\left[\begin{array}{l} CH_2{=}CHC(CH_3)OC(CH_3){=}CH_2 \\ (CH_2)_2CH{=}C(CH_3)_2 \end{array}\right]$	190°, 35 min	$(CH_3)_2C{=}CH(CH_2)_2C(CH_3){=}CH(CH_2)_2COCH_3$	37	130
		130–150°, 24 hr, H_3PO_4, ligroin, reflux	″	62	131
		125°, 13–15 hr, H_3PO_4, N_2 pres.	″ (cis/trans 1/1.5)	82	131
	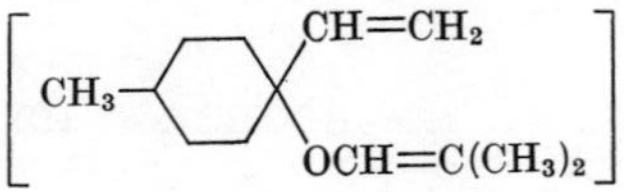 (cyclohexane ring bearing $CH{=}CH_2$ and $OCH{=}C(CH_3)_2$; CH_3 substituent)	120°, 20 hr, p-TsOH, hydroquinone	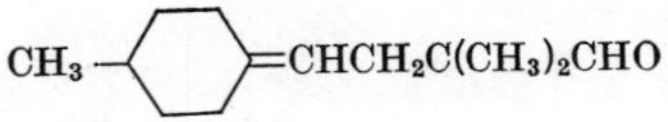 (cyclohexylidene)$=CHCH_2C(CH_3)_2CHO$; CH_3 substituent	40	118
	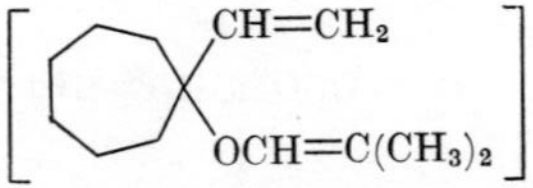 (cycloheptane ring bearing $CH{=}CH_2$ and $OCH{=}C(CH_3)_2$)	120°, 20 hr, p-TsOH, hydroquinone	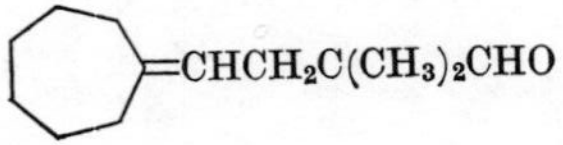 (cycloheptylidene)$=CHCH_2C(CH_3)_2CHO$	18	118
$C_{13}H_{22}O_2$	$\left[\begin{array}{l} CH_2{=}C(CH_3)CHOC{=}CH_2 \\ CH_2{=}C(CH_3)(CH_2)_2OC_2H_5 \end{array}\right]$	138°, 1 hr, trace $C_2H_5CO_2H$	$CH_2{=}C(CH_3)(CH_2)_2CH{=}C(CH_3)(CH_2)_2CO_2C_2H_5$ >98% trans	92	97
$C_{13}H_{24}O$	$\left[\begin{array}{l} CH_2{=}CHC(CH_3)OC(CH_3){=}CH_2 \\ (CH_2)_3C_3H_7{\text{-}}i \end{array}\right]$	125°, 13–15 hr, H_3PO_4, pres. reactor, N_2	$i{\text{-}}C_3H_7(CH_2)_3C(CH_3){=}CH(CH_2)_2COCH_3$ (cis/trans 1/1.7)	83	131

Formula	Reactant	Conditions	Product(s)	Yield (%)	Refs.
$C_{14}H_{14}O$	$\begin{bmatrix}CH_3CH=CHCHOCH=CH_2\\ \quad\quad\quad \mid \\ \quad\quad\quad C\equiv CC_6H_5\end{bmatrix}$ *trans cis*	35°, Hg^{2+}	$C_6H_5C\equiv CCH=CHCH(CH_3)CH_2CHO$ (*cis/trans* 1/1.9)	—	232
$C_{14}H_{18}O$	$[(CH_3)_2C=CHCH_2OCH=C(CH_3)C_6H_5]$	115°, 20 hr, C_6H_6, *p*-TsOH, hydroquinone	$CH_2=CHC(CH_3)_2C(CH_3)=C(CH_3)C_6H_5CHO$	70	118
$C_{14}H_{18}O$	$[CH_2=CHC(CH_3)_2OCH=C(CH_3)C_6H_5]$	115°, 20 hr, C_6H_6, *p*-TsOH, hydroquinone	$(CH_3)_2C=CHCH_2C(CH_3)=C(CH_3)C_6H_5CHO$	71	118
$C_{14}H_{19}NO$	$\begin{bmatrix}C_6H_5CH=CHCH_2OC=CHCH_3\\ \quad\quad\quad\quad\quad \mid \\ \quad\quad\quad\quad\quad N(CH_3)_2\end{bmatrix}$ *trans cis*	Reflux, 13 hr, xylene	$CH_2=CHCH(C_6H_5)CH(CH_3)CON(CH_3)_2$ (*erythro/threo* 11.5/1)	85	410
		Reflux, 7 hr, xylene	(*erythro/threo* 1/13.3)	79	410
$C_{14}H_{22}O$	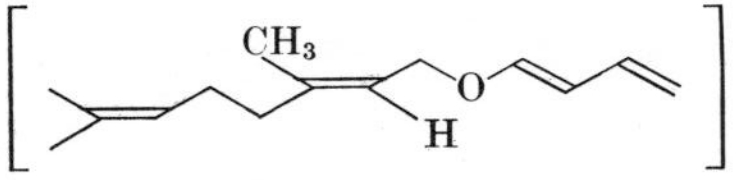 (structure, CH_3, *trans* / *cis*)	100°, 18 hr, $Hg(OAc)_2$, NaOAc, Ar atm	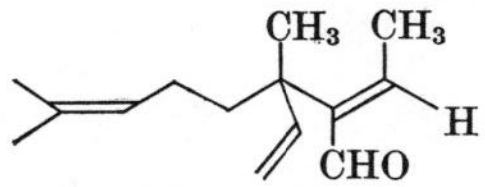 IV (4), 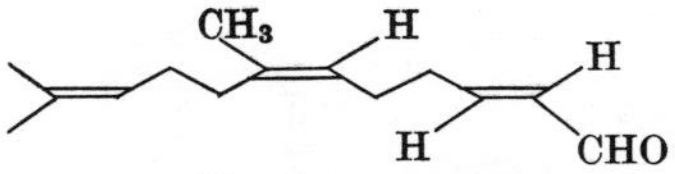 V *cis*, V *trans* (1)	—	102
	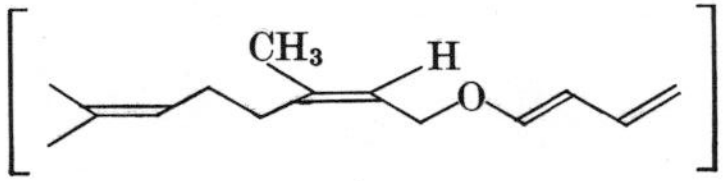	″	IV (4) V *trans trans* (1)		102
$C_{14}H_{23}NO_3$	$CH_3C=CHCH=CHCH_2OC=CHCH_3$; $\mid CH_2CO_2CH_3$; $N(CH_3)_2$	″	$CH_3C=CHCHCH=CH_2$; $CH(CH_3)CON(CH_3)_2$; $CH_2CO_2CH_3$	—	412

Note: References 344–439 are on pp. 251–252.

145

TABLE II. ALIPHATIC CLAISEN REARRANGEMENTS (Continued)

A. Acyclic Allyl Vinyl Ethers (Continued)

Molecular Formula	Ether	Conditions	Product(s) and Ratio ( )	Yield(s), %	Refs.
$C_{14}H_{24}O$	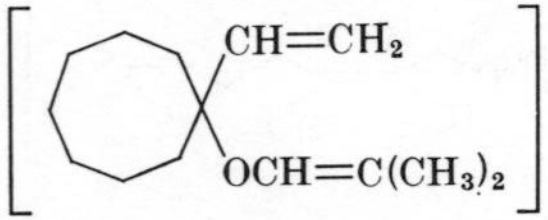 (structure)	120°, 20 hr, C_6H_6, p-TsOH, hydroquinone	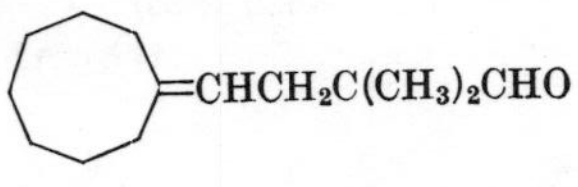 $=CHCH_2C(CH_3)_2CHO$	13	118
$C_{14}H_{25}NO$	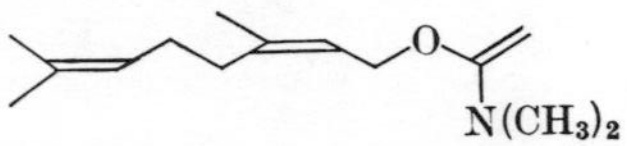 (structure) $N(CH_3)_2$	150°, 3.5 hr, xylene	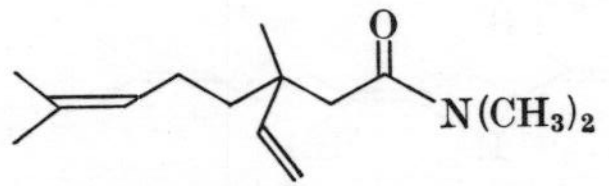 (structure) $N(CH_3)_2$	80	135
$C_{15}H_{20}O$	$[C_2H_5C(CH_3)=CHCH_2OCH=C(CH_3)C_6H_5]$	120°, 20 hr, C_6H_6, p-TsOH, hydroquinone	$C_2H_5C(CH_3)=CHCH_2C(CH_3)CHO$, C_6H_5	—	118
$C_{15}H_{20}O_2$	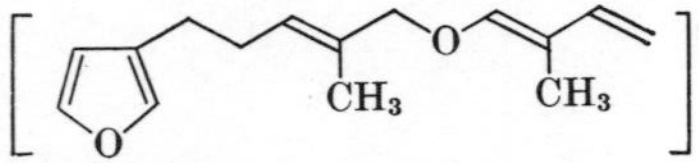 (structure)	100°, 18 hr, $Hg(OAc)_2$, NaOAc	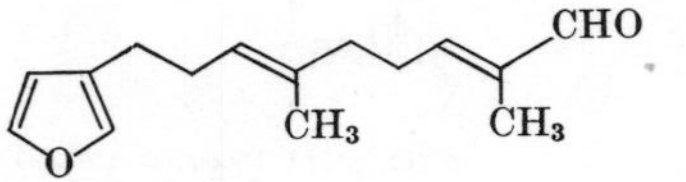 (structure) CHO	—	225
$C_{15}H_{21}NO$	$CH_2=CHCH_2OC=CHC_6H_5$, $N(C_2H_5)_2$	30°, 1 hr, trace $BF_3\cdot(C_2H_5)_2O$	$CH_2=CHCH_2CH(C_6H_5)CON(C_2H_5)_2$	85	125
$C_{15}H_{22}O$	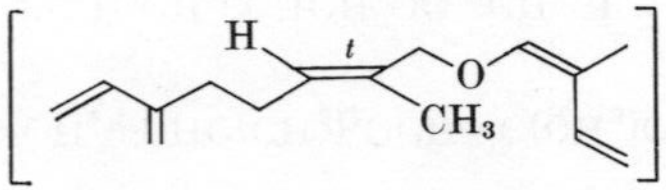 (structure) CH_3	100°, 48 hr, $Hg(OAc)_2$, NaOAc, Ar atm	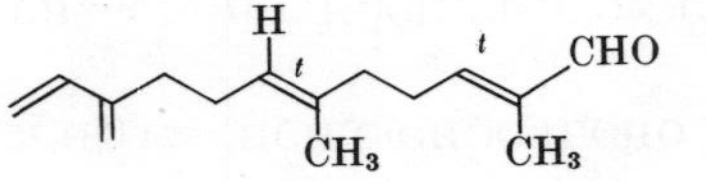 (structure) CHO, CH_3	43	102
$C_{15}H_{24}O$	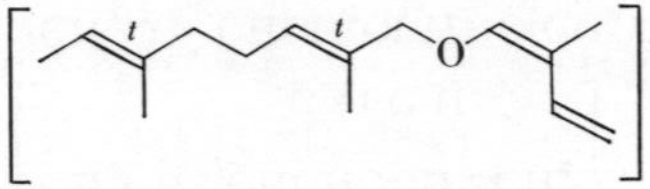 (structure)	100°, 18 hr, $Hg(OAc)_2$, NaOAc, Ar atm	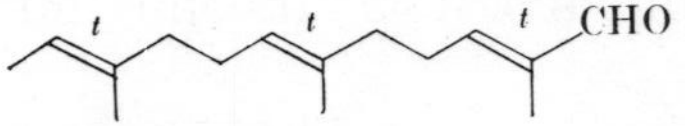 (structure) CHO (3)	—	102

			102
			102
			132
			125
			125
			125

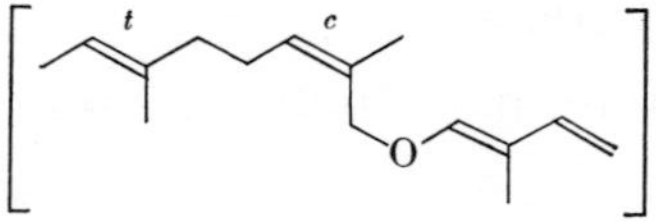

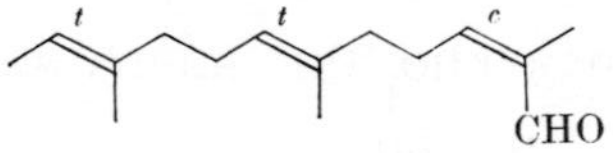

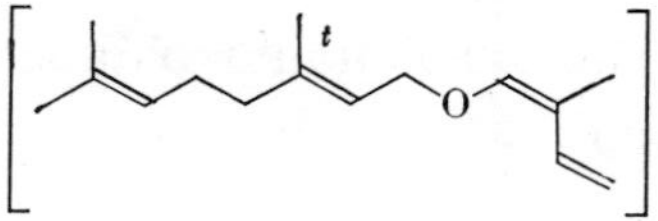

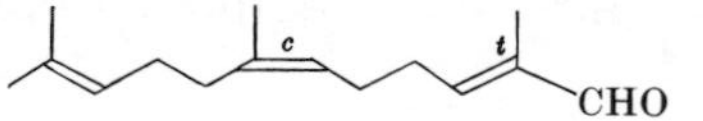

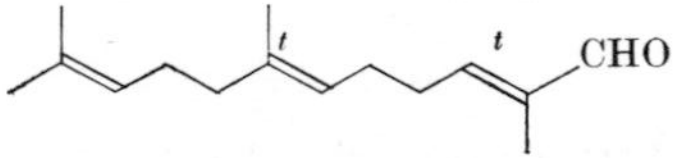

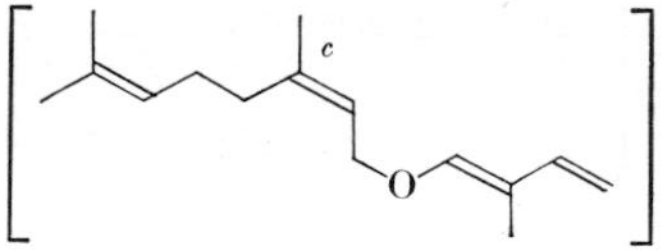

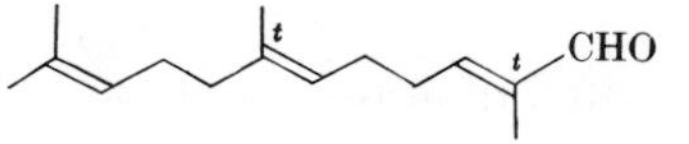

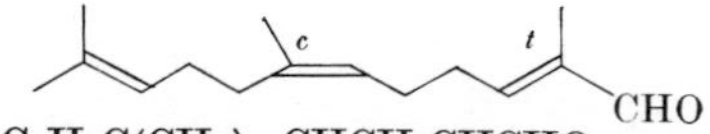

i-C$_4$H$_9$C(CH$_3$)=CHCH$_2$CHCHO

C$_5$H$_{11}$-n

CH$_2$=C(CH$_3$)CH$_2$CH(C$_6$H$_5$)CON(C$_2$H$_5$)$_2$

CH$_2$=CHCH(CH$_3$)CH(C$_6$H$_5$)CON(C$_2$H$_5$)$_2$

CH$_2$=CHCH$_2$CHCON(CH$_3$)(C$_6$H$_5$)

C$_4$H$_9$-n

150–180°, 2 hr, H$_3$PO$_4$

30°, 1 hr, trace BF$_3$·(C$_2$H$_5$)$_2$O

C$_{15}$H$_{28}$O [CH$_2$=CHC(CH$_3$)OCH=CHC$_5$H$_{11}$-n]

C$_4$H$_9$-i

C$_{16}$H$_{23}$NO [CH$_2$=C(CH$_3$)CH$_2$OC=CHC$_6$H$_5$]

N(C$_2$H$_5$)$_2$

[CH$_3$CH=CHCH$_2$OC=CHC$_6$H$_5$]

N(C$_2$H$_5$)$_2$

[CH$_2$=CHCH$_2$OC=CHC$_4$H$_9$-n]

CH$_3$NC$_6$H$_5$

or

Note: References 344–439 are on pp. 251–252.

147

TABLE II. ALIPHATIC CLAISEN REARRANGEMENTS (*Continued*)

A. *Acyclic Allyl Vinyl Ethers* (*Continued*)

Molecular Formula	Ether	Conditions	Product(s) and Ratio ( )	Yield(s), %	Refs.
$C_{16}H_{26}O$	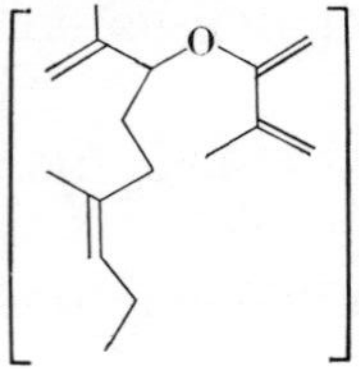	110°, 24 hr, oxalic acid, hydroquinone, sealed tube	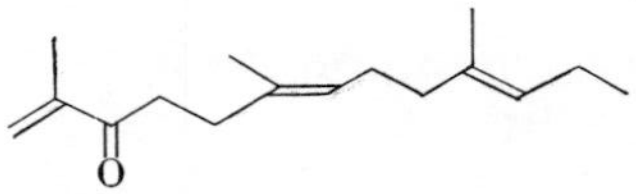	Good	55
$C_{17}H_{22}O$	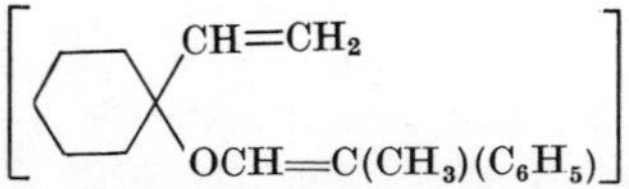	120°, 20 hr, C_6H_6, p-TsOH, hydroquinone	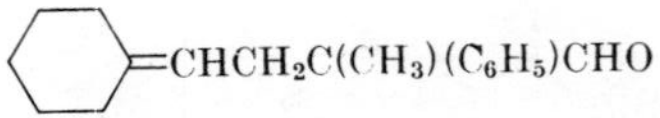 $\text{—CHCH}_2\text{C}^\cdot(\text{CH}_3)(\text{C}_6\text{H}_5)\text{CHO}$	80	118
$C_{17}H_{25}NO$	$CH_3CH=CHCH(CH_3)OC=CHC_6H_5$ $N(C_2H_5)_2$	30°, 1 hr, trace $BF_3 \cdot (C_2H_5)_2O$	$CH_3CH=CHCH(CH_3)CHCON(C_2H_5)_2$ C_6H_5	10	125
$C_{18}H_{16}O_2$	$CH_2=CHCH_2OC=CHC_6H_5$ COC_6H_5	Reflux, 20 hr, $(CH_3)_2NC_6H_5$, N_2 atm	$CH_2=CHCH_2CH(C_6H_5)COCOC_6H_5$	18	413
$C_{18}H_{24}O$	$(CH_3)_2C=CH(CH_2)_2C(CH_3)OC=CH_2$ C_6H_5 $CH=CH_2$	180°, 35 min	$(CH_3)_2C=CH(CH_2)_2C(CH_3)=CH$ $C_6H_5CO(CH_2)_2$ $CH=CH_2$	37	130
	$(CH_3)_2C=CH(CH_2)_2C(CH_3)=CHCH_2$ $CH_2=(C_6H_5)CO$	185°, 45 min	$(CH_3)_2C=CH(CH_2)_2CH_2COC_6H_5$ CH_3	48	130

148

Reactant	Conditions	Yield (%)	Product	Ref.
$C_{18}H_{28}O_3$	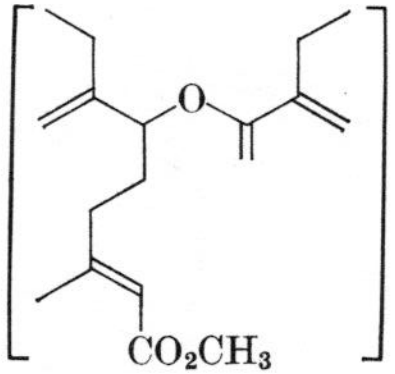 100°, 8 hr, $C_6H_5CH_3$, 2,4-$(O_2N)_2C_6H_3OH$	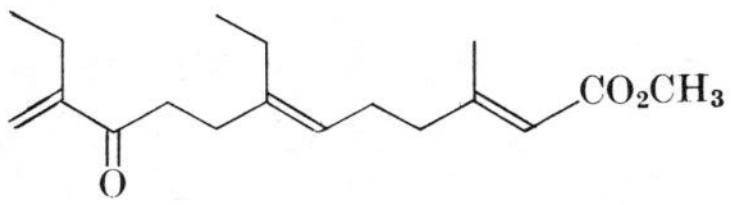 —	(CO_2CH_3)	96
$C_{18}H_{30}O$	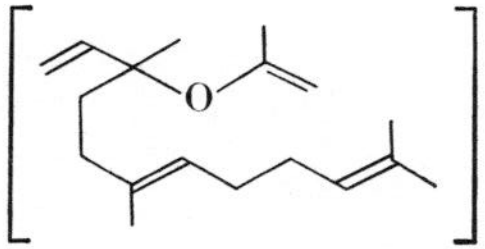 125°, 13–15 hr, H_3PO_4, N_2 pres.	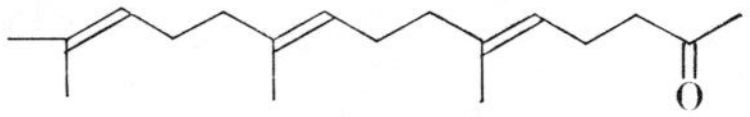 87		131
$C_{18}H_{30}O_4$	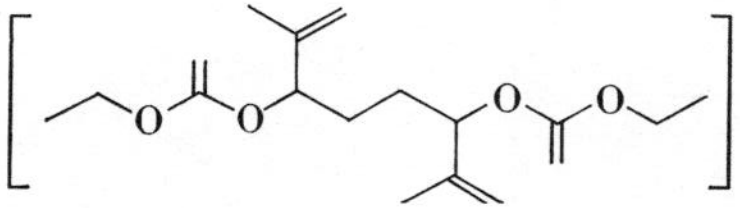 138°, 3 hr, trace $C_2H_5CO_2H$	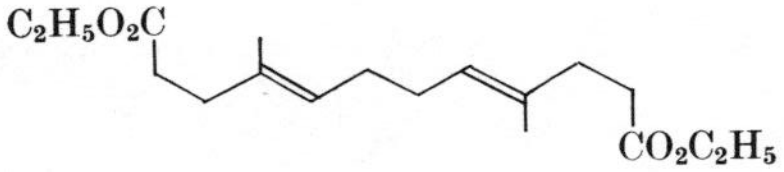 87	($CO_2C_2H_5$, $C_2H_5O_2C$)	97
$C_{20}H_{30}O_2$	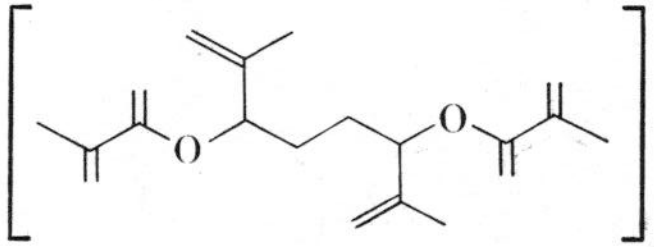 100°, 8 hr, $C_6H_5CH_3$, 2,4-$(O_2N)_2C_6H_3OH$	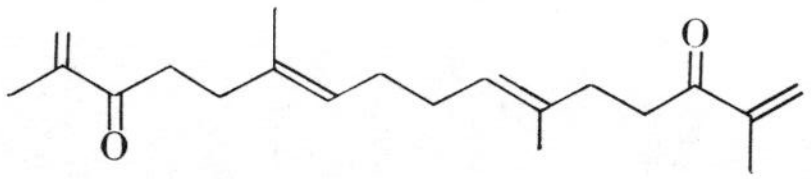 62		96
$C_{21}H_{21}NO_4$	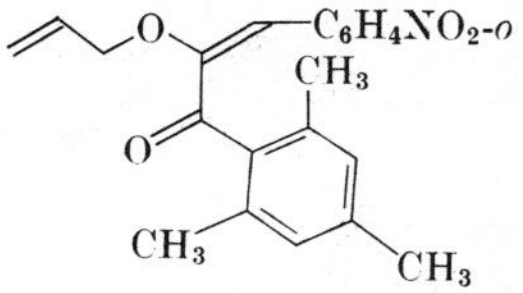 Reflux, 20 hr, $(C_2H_5)_2NC_6H_5$, N_2 atm	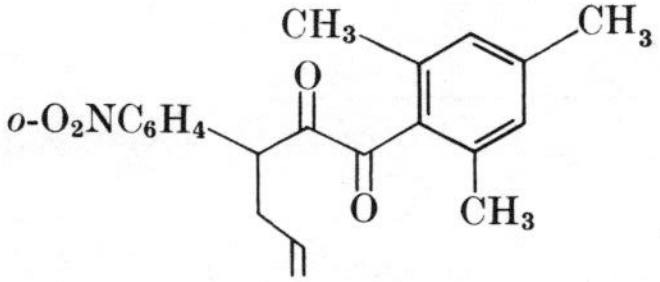 5	(CH_3, o-$O_2NC_6H_4$)	413

Note: References 344–439 are on pp. 251–252.

TABLE II. ALIPHATIC CLAISEN REARRANGEMENTS (*Continued*)

A. *Acyclic Allyl Vinyl Ethers* (*Continued*)

Molecular Formula	Ether	Conditions	Product(s) and Ratio ( )	Yield(s), %	Refs.
$C_{21}H_{21}NO_4$ (*contd.*)	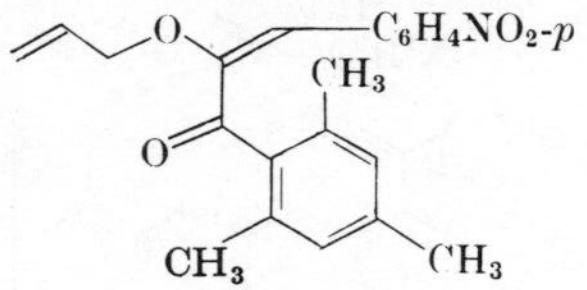	Reflux, 20 hr, $(C_2H_5)_2NC_6H_5$, N_2 atm	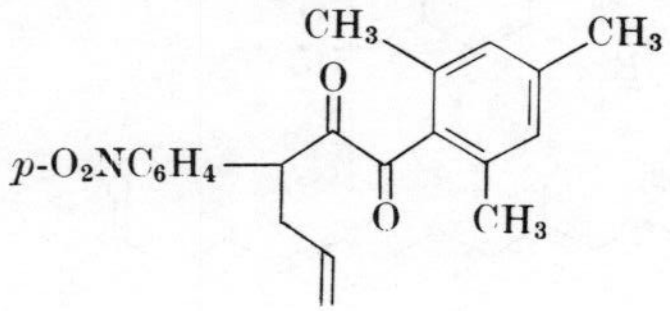 p-$O_2NC_6H_4$	6	413
$C_{21}H_{22}O_2$	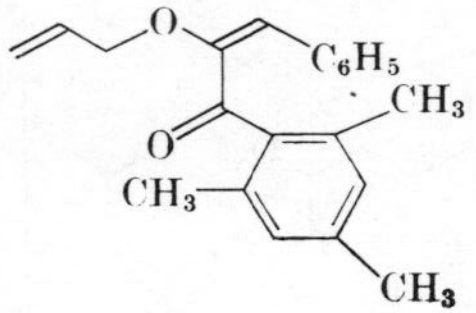	Reflux, 20 hr, $(CH_3)_2NC_6H_5$, N_2 atm	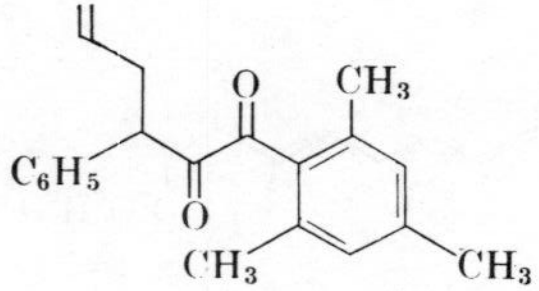	20	413
	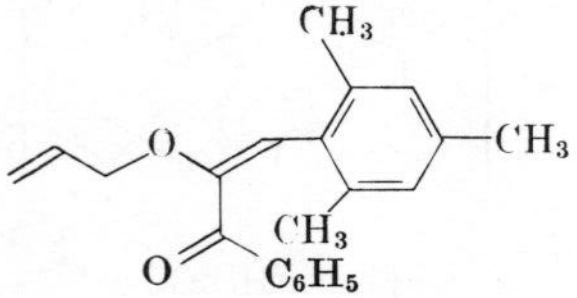	''	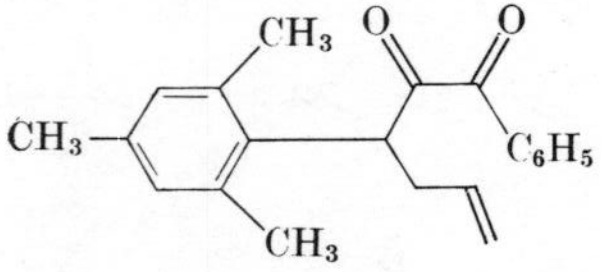	60	413
$C_{22}H_{24}O_3$	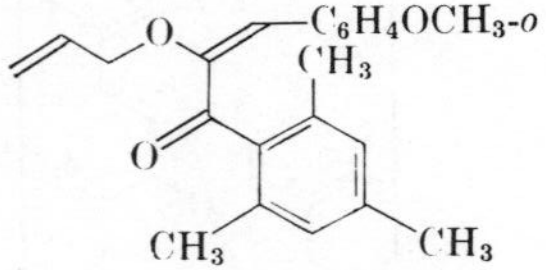	''	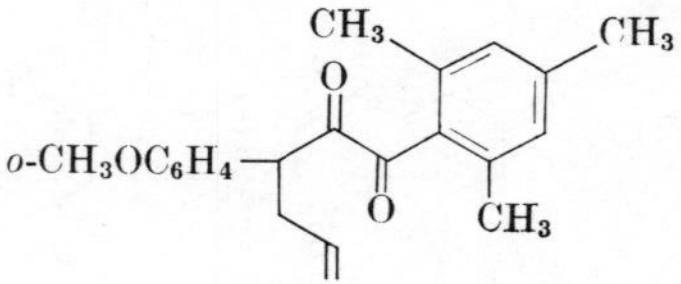 o-$CH_3OC_6H_4$	75	413

150

Formula	Reactant	Conditions	Product	Yield (%)	Refs.
	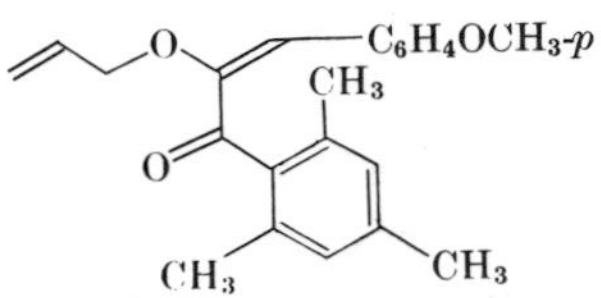	″	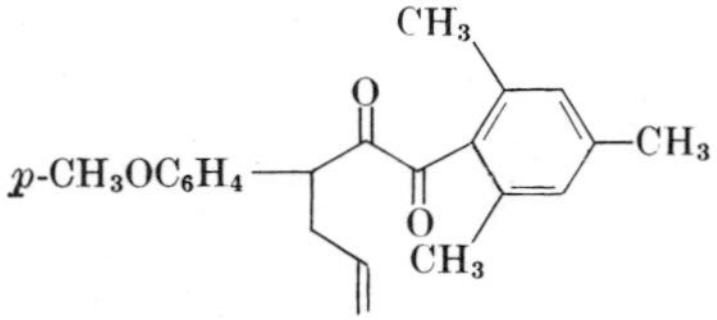 p-CH$_3$OC$_6$H$_4$ (mesityl diketone structure)	80	413
C$_{24}$H$_{27}$·NO	CH$_3$CH=CHCH=C(C$_6$H$_4$OCH$_3$-p)—O—C=CHC$_6$H$_5$ · C$_6$H$_5$ N(C$_2$H$_5$)$_2$	30°, 1 hr, trace BF$_3$·(C$_2$H$_5$)$_2$O	C$_6$H$_5$CH=CHCH(CH$_3$)CHCON(C$_2$H$_5$)$_2$ C$_6$H$_5$	50	125
C$_{22}$H$_{42}$O	[CH$_3$(CH$_2$R)C=CHCH$_2$OCH=CH$_2$] $trans$	Reflux, 48 hr, Hg(OAc)$_2$, N$_2$ atm	CH$_2$=CHC(CH$_3$)CH$_2$CHO CH$_2$R (28), phytadienes (1), rac phytol (3.3)	~95	414
	CH$_2$=CHC(CH$_3$)OCH=CH$_2$ CH$_2$R	190°, 2 hr, Hg(OAc)$_2$	RCH$_2$C(CH$_3$)=CHCH$_2$CH$_2$CHO (cis/$trans$ 2/3), (5) phytadienes (1), rac isophytol (2)	~70	414
C$_{24}$H$_{47}$·NO	[CH$_3$(CH$_2$R)C=CHCH$_2$OC=CH$_2$ N(CH$_3$)$_2$]	140°, 15 hr, xylene	CH$_2$=CHC(CH$_3$)CH$_2$CH$_2$CON(CH$_3$)$_2$ CH$_2$R	95	414
	[CH$_2$=CHC(CH$_3$)OC=CH$_2$ CH$_2$R N(CH$_3$)$_2$]	″	RCH$_2$C(CH$_3$)=CHCH$_2$CH$_2$CON(CH$_3$)$_2$ (cis/$trans$ 1/1.5),	—	414
C$_{25}$H$_{39}$NO$_2$	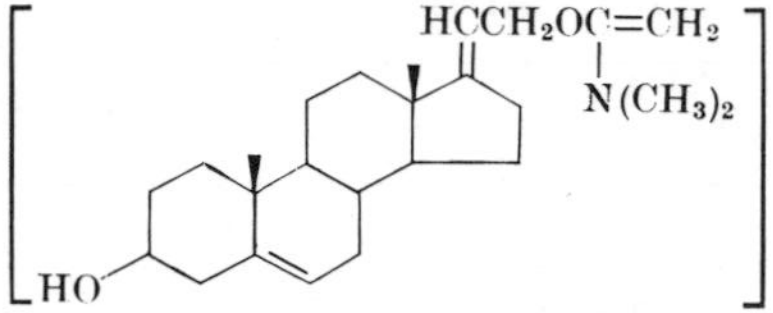 (steroid structure) HCCH$_2$OC=CH$_2$ N(CH$_3$)$_2$ HO	Reflux, 5 hr, N$_2$ atm	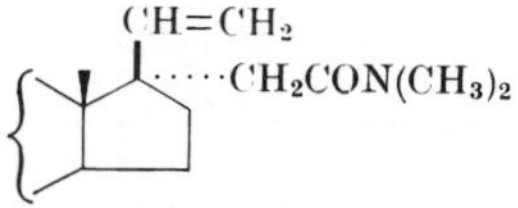 (steroid structure) (H=CH$_2$ ····CH$_2$CON(CH$_3$)$_2$)	89	343

Note: References 344–439 are on pp. 251–252.

[a] R is H[CH$_2$CH(CH$_3$)CH$_2$CH$_2$—]$_3$.

151

TABLE II. ALIPHATIC CLAISEN REARRANGEMENTS (Continued)

A. Acyclic Allyl Vinyl Ethers (Continued)

Molecular Formula	Ether	Conditions	Product(s) and Ratio ( )	Yield(s), %	Refs.
$C_{25}H_{41}NO_2$	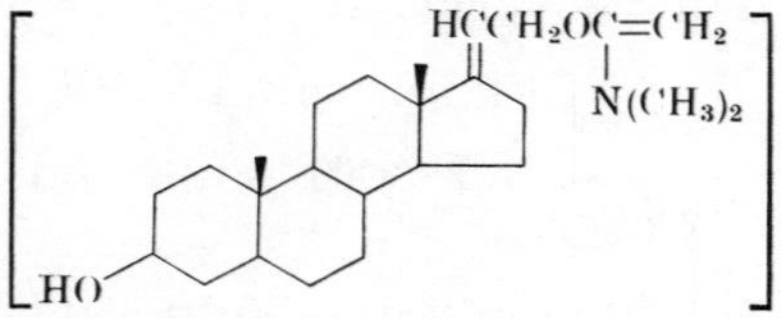	Reflux, 5 hr, N_2 atm	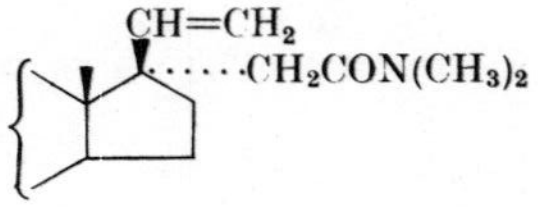	81	343
$C_{26}H_{38}O_3$	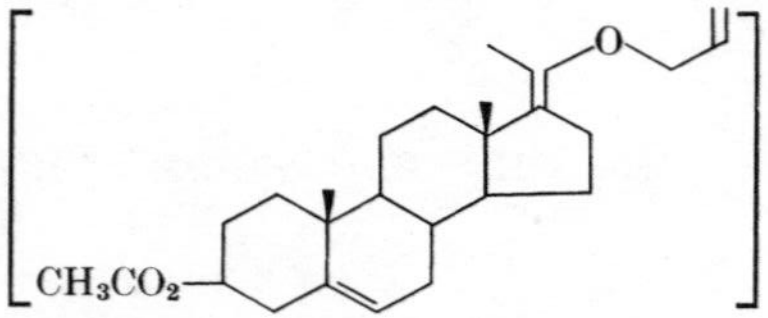	115°, 15 hr	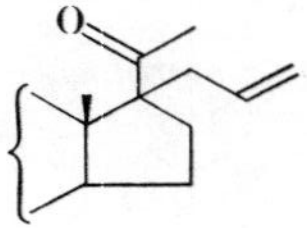	26	222
$C_{30}H_{48}O_2Cl_2$	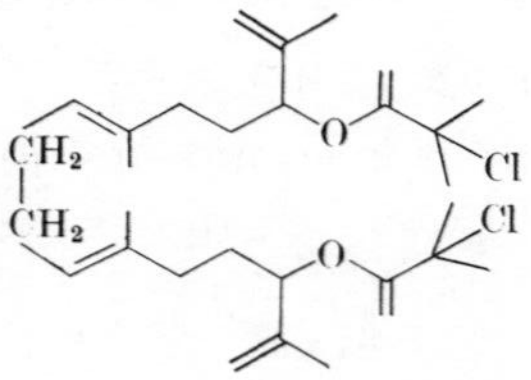	94°, 24 hr, $C_6H_5CH_3$, 2,4-$(O_2N)_2C_6H_3OH$	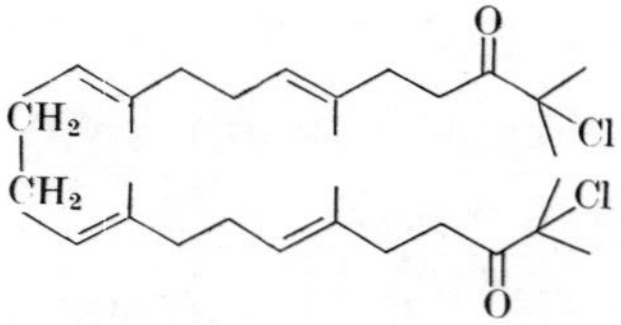	60	99, 98

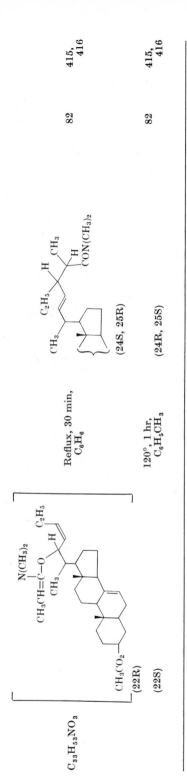

Molecular Formula	Ether	Conditions	Product(s) and Ratio ( )	Yield(s), %	Refs.
$C_{33}H_{53}NO_3$	(22R)	Reflux, 30 min, C_6H_6	(24S, 25R)	82	415, 416
	(22S)	120°, 1 hr, $C_6H_5CH_3$	(24R, 25S)	82	415, 416

B. Ethers in Which the Allyl Double Bond Is Part of a Ring

Molecular Formula	Ether	Conditions	Product(s) and Ratio ( )	Yield(s), %	Refs.
$C_7H_8O_2$		370°, C_6H_6, through packed column, N_2 atm	(11), (2.6), (6), (1)	70 (total)	417
$C_7H_{10}O$	R(+)	180–185°, sealed tube, N_2 atm	R(−)	81	59

Note: References 344–439 are on pp. 251–252.

153

TABLE II. ALIPHATIC CLAISEN REARRANGEMENTS (Continued)

B. Ethers in Which the Allyl Double Bond Is Part of a Ring (Continued)

Molecular Formula	Ether	Conditions	Product(s) and Ratio ( )	Yield(s), %	Refs.
$C_8H_{12}O$	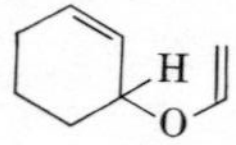	190–195°, 15 min, sealed tube, N_2 atm	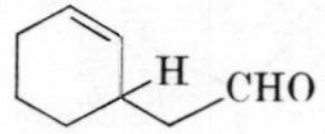	94	90
$C_9H_{10}O_2$	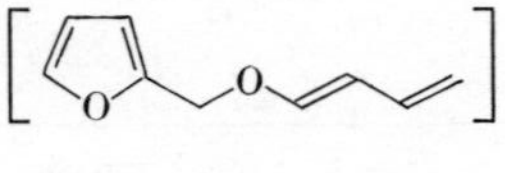	150°, 4 hr, $Hg(OAc)_2$, NaOAc, sealed tube	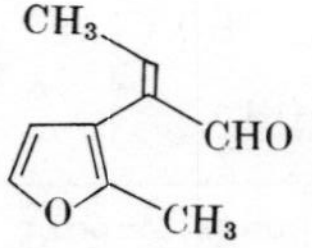 (1); 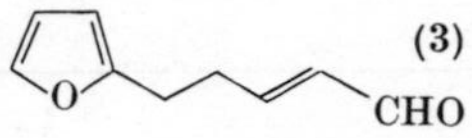 (3)	15	225
$C_9H_{13}NO_2$	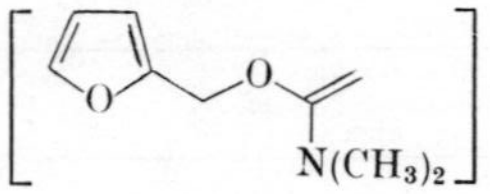	160°, 24 hr, $HCON(CH_3)_2$	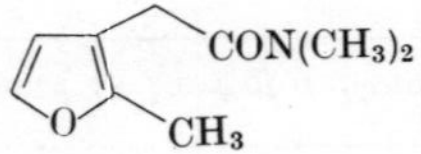 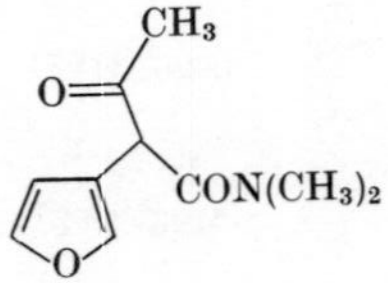 (Minor)	70–80 (total)	135
$C_9H_{14}O$	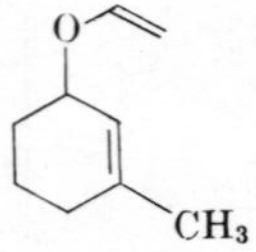	190–195°, 15 min, sealed tube, N_2 atm	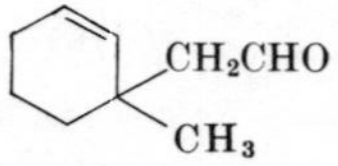	93	90

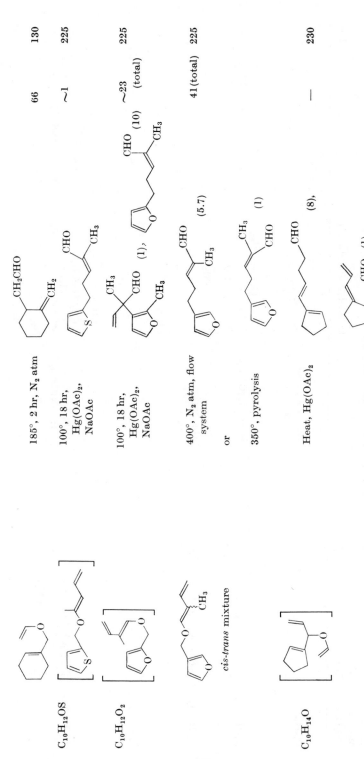

C$_{10}$H$_{12}$OS

185°, 2 hr, N$_2$ atm   66   130

100°, 18 hr,
Hg(OAc)$_2$,
NaOAc   ~1   225

C$_{10}$H$_{12}$O$_2$

100°, 18 hr,
Hg(OAc)$_2$,
NaOAc   ~23 (total)   225

cis-trans mixture

400°, N$_2$ atm, flow
system

or

350°, pyrolysis   41 (total)   225

C$_{10}$H$_{14}$O

Heat, Hg(OAc)$_2$ —   230

Note: References 344–439 are on pp. 251–252.

155

TABLE II. ALIPHATIC CLAISEN REARRANGEMENTS (*Continued*)

B. Ethers in Which the Allyl Double Bond Is Part of a Ring (*Continued*)

Molecular Formula	Ether	Conditions	Product(s) and Ratio ( )	Yield(s), %	Refs.
$C_{10}H_{16}O$	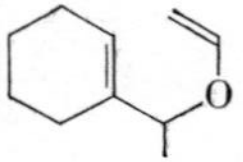 (1-(1-cyclohexenyl)ethyl vinyl ether)	195°, 2 hr, sealed tube	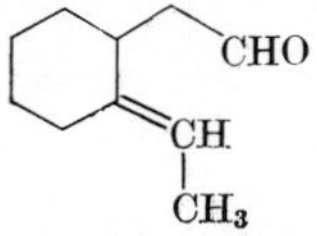 (2-(cyclohexylidene)... CH–CH₃, CHO)	—	130
$C_{11}H_{14}O$	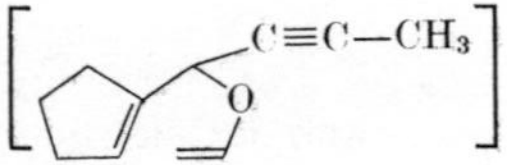 (1-(but-2-ynyl)cyclopentenyl vinyl ether)	Heat, Hg(OAc)₂	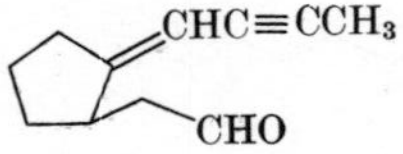 ($CHC\equiv CCH_3$, CHO)	~50	231
$C_{11}H_{14}O_2$	$m\text{-}CH_3OC_6H_4CH_2CH_2OC(CH_3)\!=\!CH_2$	240°, 1 hr, sealed tube	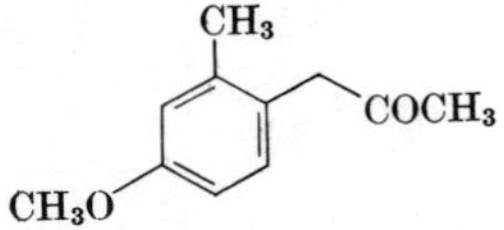 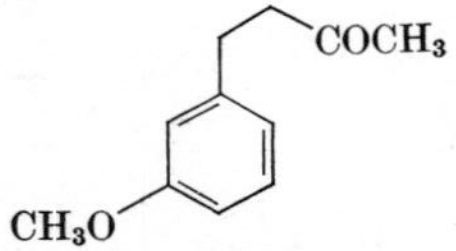 (CH_3, CH_2COCH_3, CH_3O) (1), ($COCH_3$, CH_3O) (1)	—	224
$C_{11}H_{15}NO$	$\left[\begin{array}{c} C_6H_5CH_2OC\!=\!CH_2 \\ \mid \\ N(CH_3)_2 \end{array}\right]$	160°, 20 hr, $HCON(CH_3)_2$	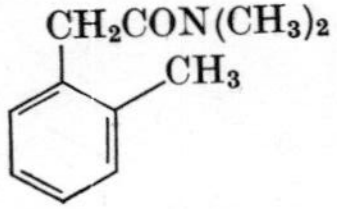 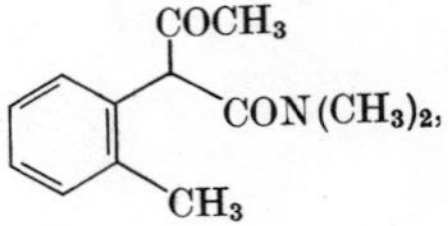 ($CH_2CON(CH_3)_2$, CH_3) (15), ($COCH_3$, $CON(CH_3)_2$, CH_3) 87 (total), Minor products (1) (1)	87 (total)	135

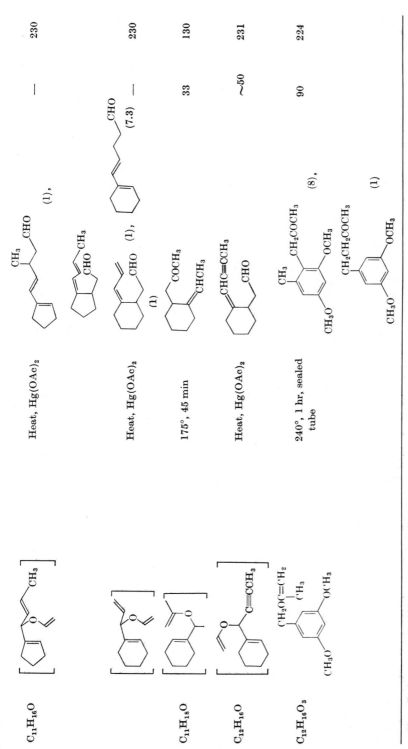

C₁₁H₁₆O	Heat, Hg(OAc)₂	(1),	—	230
C₁₁H₁₈O	Heat, Hg(OAc)₂	(1), (7.3)	—	230
	175°, 45 min	(1)	33	130
C₁₂H₁₆O	Heat, Hg(OAc)₂		~50	231
C₁₂H₁₆O₃	240°, 1 hr, sealed tube	(8), (1)	90	224

Note: References 344–439 are on pp. 251–252.

157

TABLE II. ALIPHATIC CLAISEN REARRANGEMENTS (Continued)

B. Ethers in Which the Allyl Double Bond Is Part of a Ring (Continued)

Molecular Formula	Ether	Conditions	Product(s) and Ratio ( )	Yield(s), %	Refs.
$C_{12}H_{18}O$		190–195°, 15 min, sealed tube, N_2 atm	CH_2CHO	87	90
		Heat, $Hg(OAc)_2$	CH_3 CHO (1), CHO (1.9)	—	230
		180°, 75 min, N_2 atm	CHO CH_2	40	130
$C_{12}H_{19}NO_2$		Reflux, 6 hr, C_6H_6	$CH(CH_3)CON(C_2H_5)_2$ CH_2	70	127
$C_{12}H_{20}O$		320°, flow system	CH_3 CHO CH_2 CH_3	85	236
$C_{13}H_{20}O$		180–190°, 45 min	CH_3 O	75	130

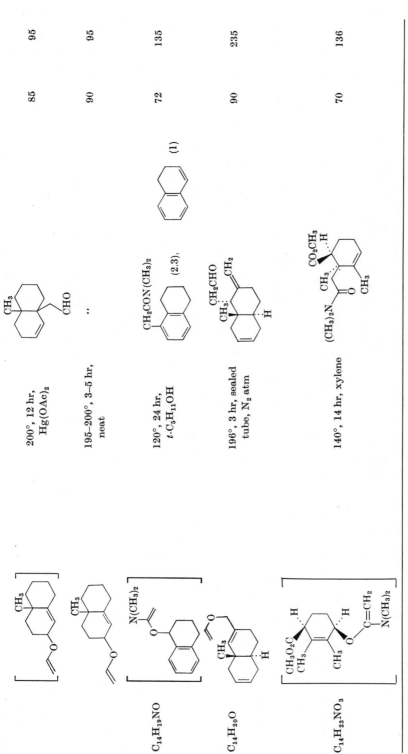

C₁₄H₁₈O	200°, 12 hr, Hg(OAc)₂		85	95
	195–200°, 3–5 hr, neat	"	90	95
C₁₄H₁₉NO	120°, 24 hr, t-C₅H₁₁OH	(2.3);	72	135
C₁₄H₂₀O	196°, 3 hr, sealed tube, N₂ atm		90	235
C₁₄H₂₃NO₃	140°, 14 hr, xylene		70	136

(1)

Note: References 344–439 are on pp. 251–252.

159

TABLE II. Aliphatic Claisen Rearrangements (*Continued*)

B. *Ethers in Which the Allyl Double Bond Is Part of a Ring (Continued)*

Molecular Formula	Ether	Conditions	Product(s) and Ratio ( )	Yield(s), %	Refs.
$C_{14}H_{23}NO_3$ (*Contd.*)	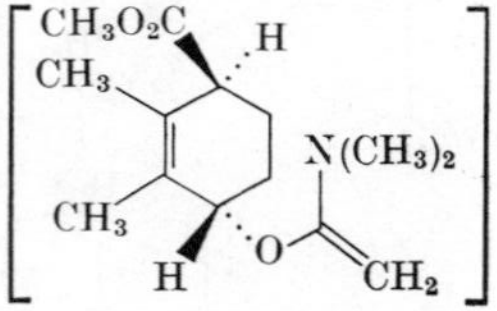	140°, 14 hr, xylene	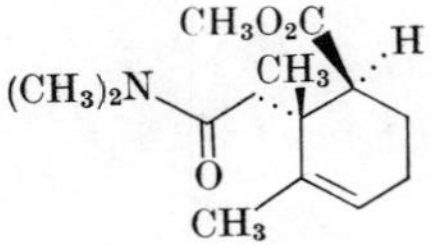	76	136
$C_{15}H_{17}NO$	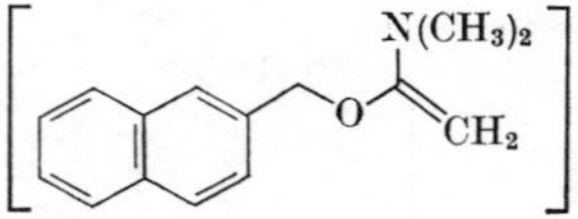	160°, 24 hr, $HCON(CH_3)_2$	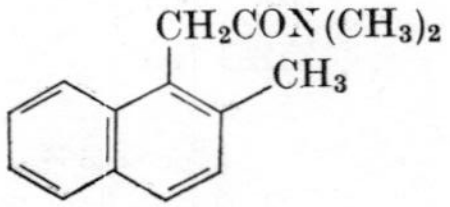	94	135
$C_{15}H_{18}O$	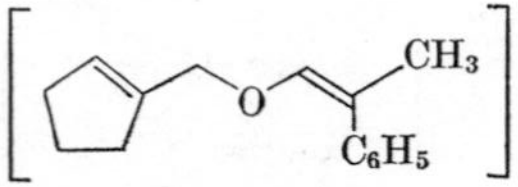	120°, 20 hr, p-TsOH, hydroquinone	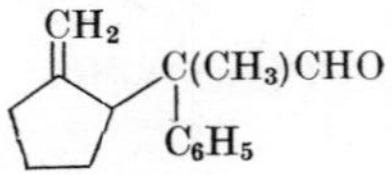	~75	118
$C_{15}H_{22}O$	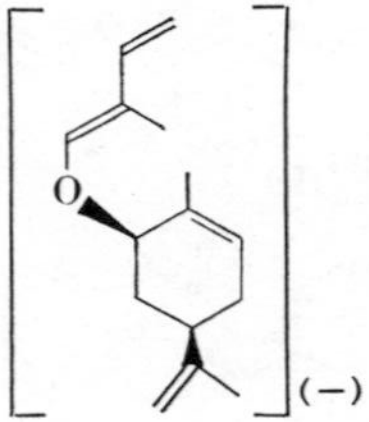	100°, 17 hr, $Hg(OAc)_2$, NaOAc	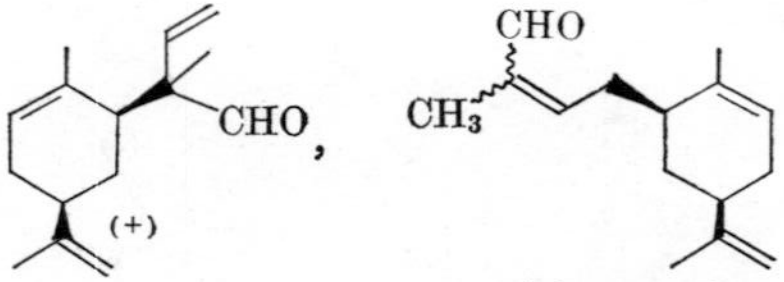 (*cis/trans* 1/4)	—	418

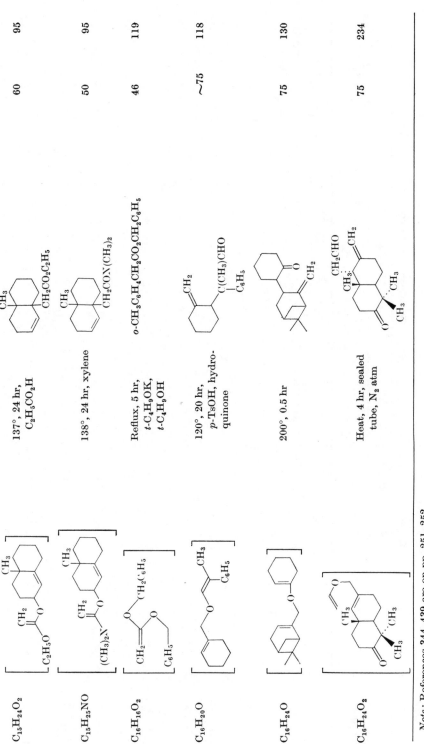

	Conditions	Product	Yield (%)	Ref.
$C_{15}H_{24}O_2$	137°, 24 hr, $C_2H_5CO_2H$		60	95
$C_{15}H_{25}NO$	138°, 24 hr, xylene		50	95
$C_{16}H_{16}O_2$	Reflux, 5 hr, t-C_4H_9OK, t-C_4H_9OH	o-$CH_3C_6H_4CH_2CO_2CH_2C_6H_5$	46	119
$C_{16}H_{20}O$	120°, 20 hr, p-TsOH, hydroquinone		~75	118
$C_{16}H_{24}O$	200°, 0.5 hr		75	130
$C_{16}H_{24}O_2$	Heat, 4 hr, sealed tube, N_2 atm		75	234

Note: References 344–439 are on pp. 251–252.

161

TABLE II. ALIPHATIC CLAISEN REARRANGEMENTS (Continued)

B. Ethers in Which the Allyl Double Bond Is Part of a Ring (Continued)

Molecular Formula	Ether	Conditions	Product(s) and Ratio ( )	Yield(s), %	Refs.
$C_{16}H_{26}O_2$	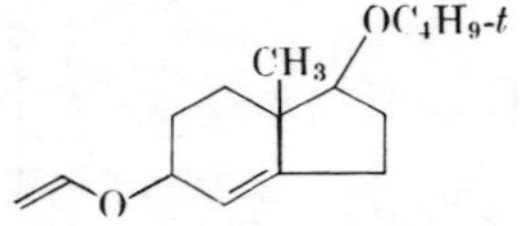	195–200°, 3–5 hr, decalin	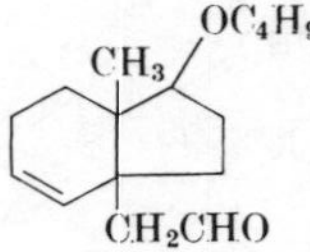	90	95
$C_{17}H_{19}NO_6$	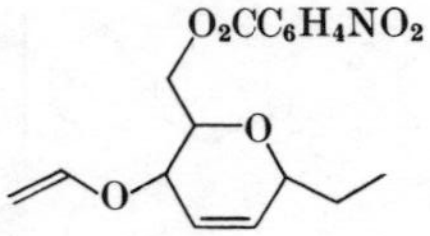	185°, 3.5 hr, $C_6H_5NO_2$	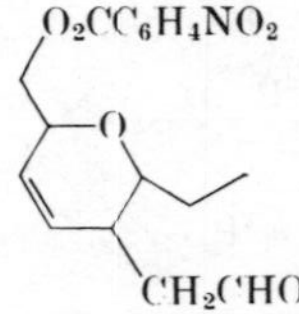	75	419
$C_{17}H_{21}NO_2$	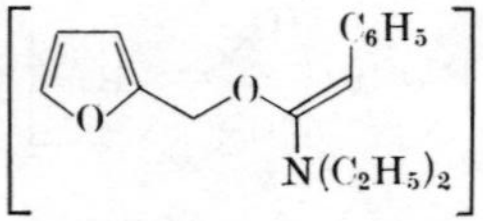	Reflux, 20 hr, C_6H_6	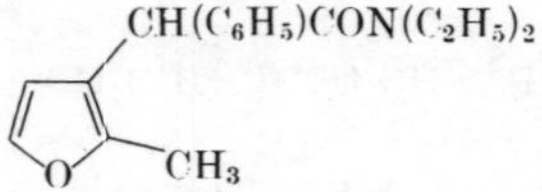	70	127
$C_{17}H_{22}O_3$	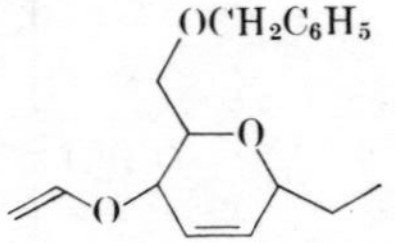	180°, 0.6 hr, $C_6H_5NO_2$	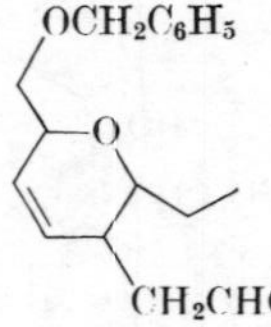	70	419

Reactant	Product	Conditions	Yield (%)	Refs.
$C_{17}H_{28}O$	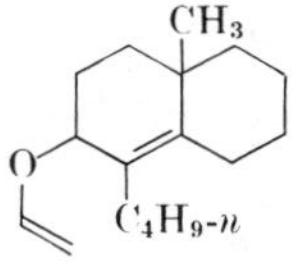	195–200°, 3–5 hr, decalin	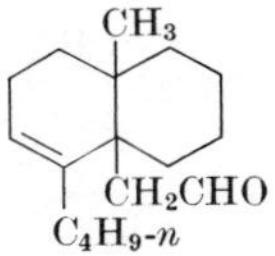 80	95
$C_{18}H_{22}O$	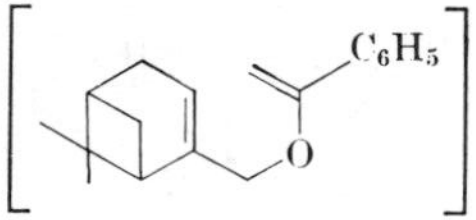	200°, 0.5 hr, N_2 atm	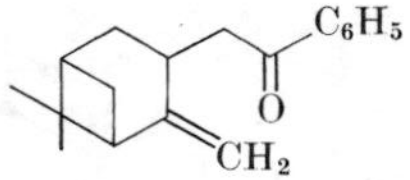 60	130
$C_{18}H_{26}N_2O$	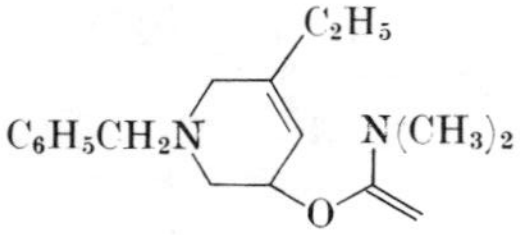	Reflux in $(CH_3OCH_2CH_2)_2O$	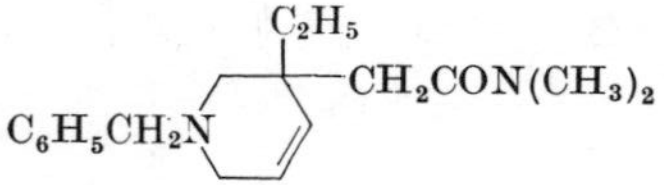 45	238
$C_{18}H_{26}O_3$	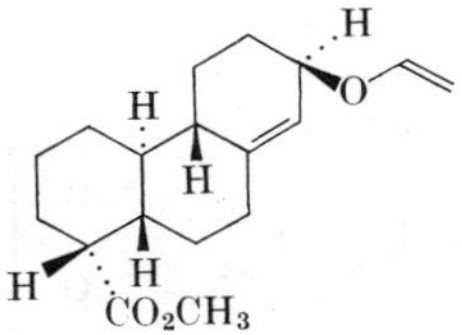	200°, several hr, decalin	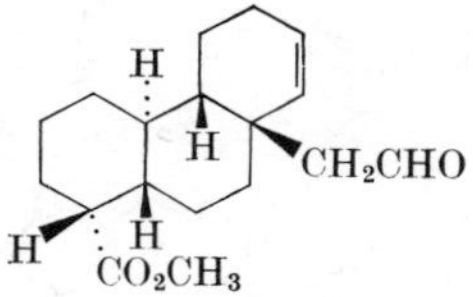 —	92
$C_{18}H_{31}NO$	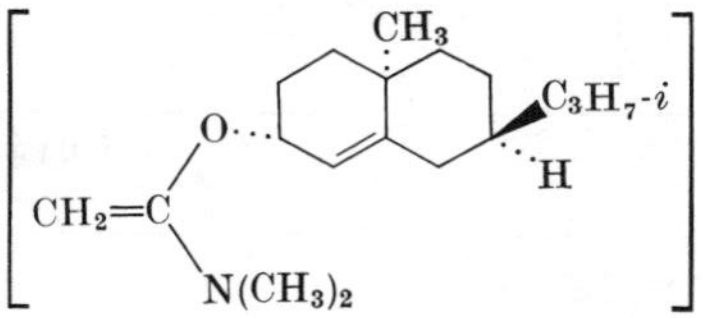	Heat, xylene	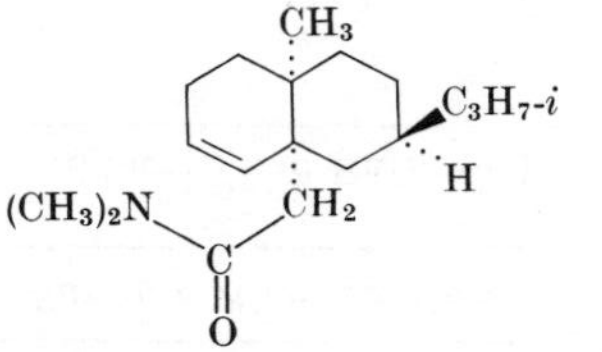 30	94

Note: References 344–439 are on pp. 251–252.

163

TABLE II. ALIPHATIC CLAISEN REARRANGEMENTS (Continued)

B. Ethers in Which the Allyl Double Bond Is Part of a Ring (Continued)

Molecular Formula	Ether	Conditions	Product(s) and Ratio ( )	Yield(s), %	Refs.
$C_{19}H_{24}N_2O_4$	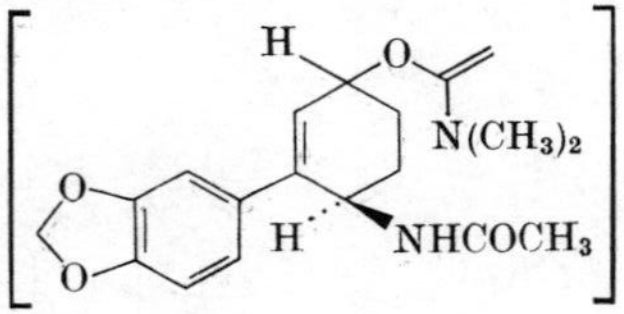	Reflux in $C_6H_5CH_3$	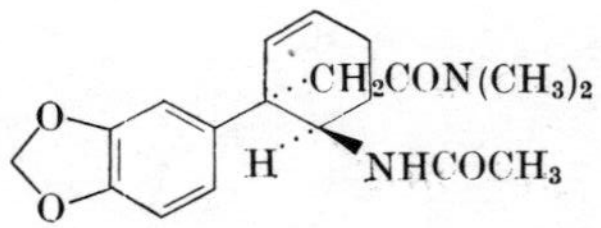 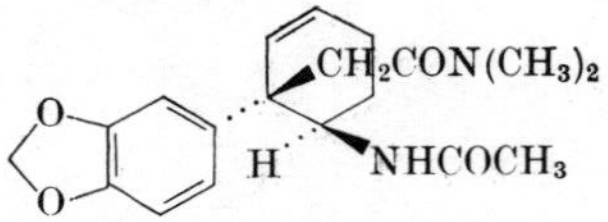	45 (total)	237
$C_{19}H_{30}O$	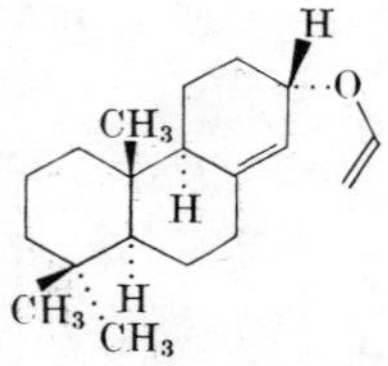	190°, 3 hr, N_2 atm	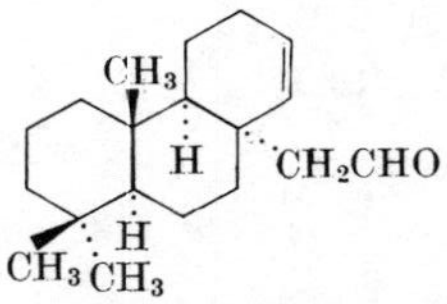	95	93
$C_{20}H_{34}O_2$	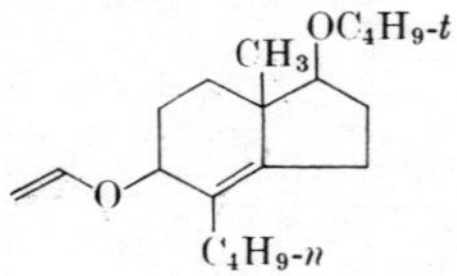	195–200°, 3–5 hr, decalin	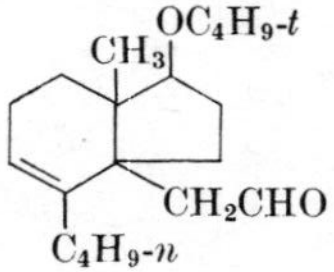	65	95

164

$C_{21}H_{27}N_3O$	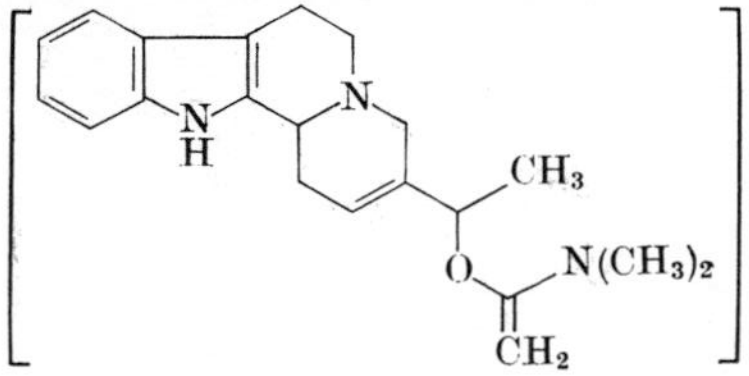 Reflux in dioxane	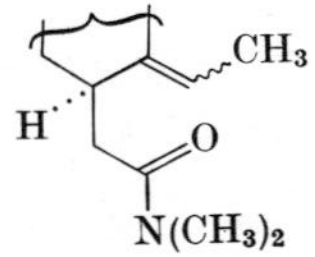	73	239
$C_{23}H_{34}O_3$	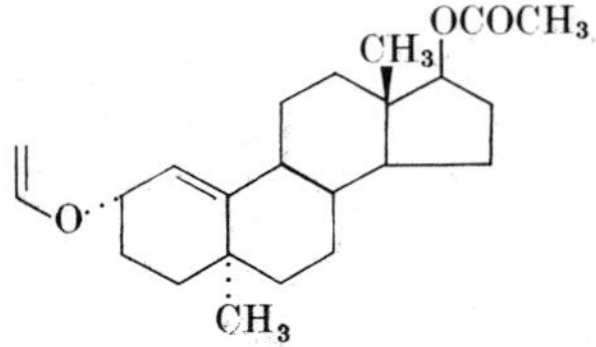 195°, decalin	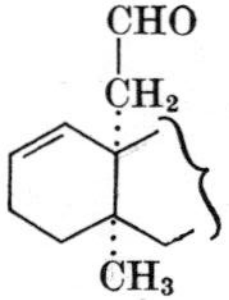	Low	91
$C_{29}H_{48}O$	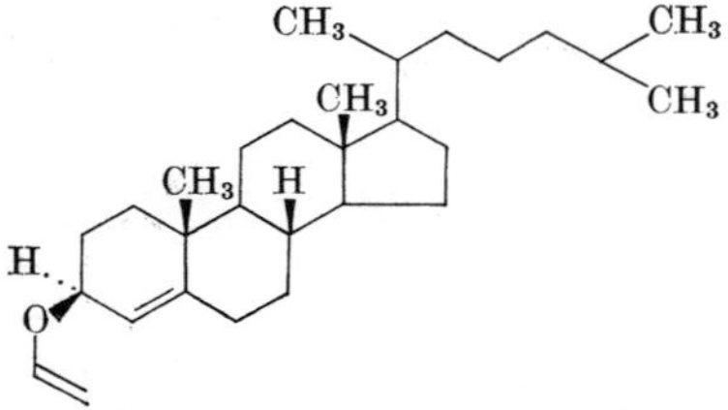 190–195°, 4 hr, decalin, sealed tube, N_2 atm	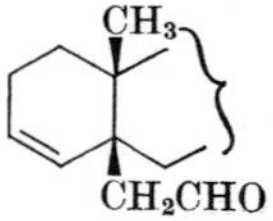	83	90
	"		—	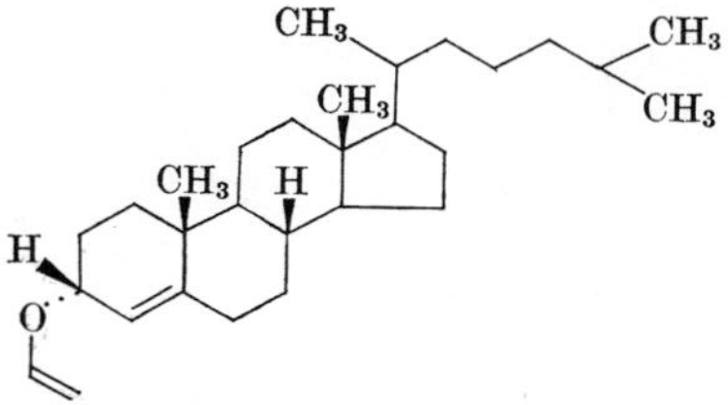 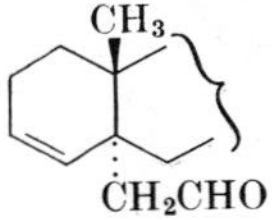 90

Note: References 344–439 are on pp. 251–252.

TABLE II. ALIPHATIC CLAISEN REARRANGEMENTS (Continued)

C. Ethers in Which the Vinyl Double Bond Is Part of a Ring

Molecular Formula	Ether	Conditions	Product(s) and Ratio ( )	Yield(s), %	Refs.
$C_7H_5OF_4Cl$	(cyclobutene structure with F_2, F_2, Cl, $OCH_2CH=CH_2$)	132°, 8 hr, reflux	(cyclobutanone structure with F_2, F_2, Cl, $CH_2CH=CH_2$, O)	69	128a
$C_7H_{10}O$	(pyran ring with $CH=CH_2$)	410°, hexane, flow system	(cyclohexene with CHO)	68	80
$C_8H_5OF_7$	(cyclopentene structure with F_2, F_2, F_2, F, O, allyl)	<105°, distil	(cyclopentanone with F_2, F_2, F_2, F, allyl, O)	35	128a
$C_8H_8O_3$	(pyranone with O–allyl)	137–140°, 15 mm, distil	(pyranone with OH and allyl)	54	420
		120–130°, neat; then HCl, C_2H_5OH	(pyranone with OH and propenyl)	63	421
$C_8H_{12}O$	(cyclopentene with O–allyl)	$C_6H_5CH_3$, p-TsOH, azeotropic distillation	(cyclopentanone with allyl)	93	117

Reactant	Conditions	Product(s)	Yield	Refs.
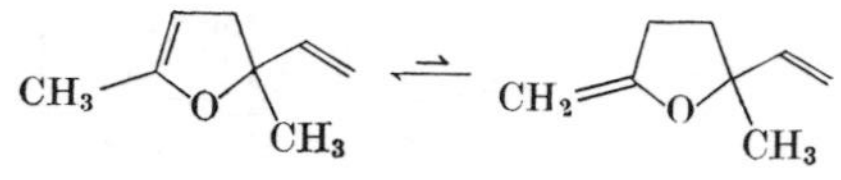	150–175°, neat or in n-decane	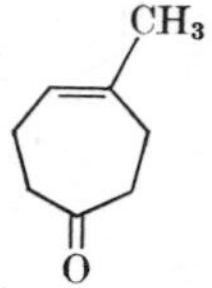	Quant.	223
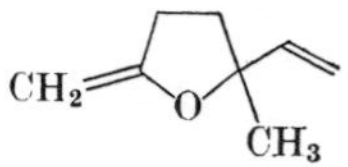	150–175°, gas phase	"	Quant.	122
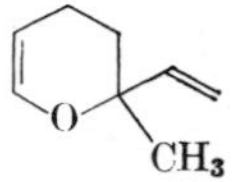	400°, hexane, flow system	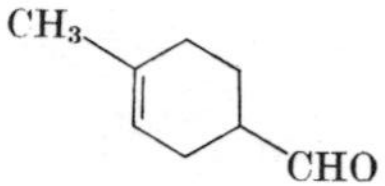 CHO	71	80
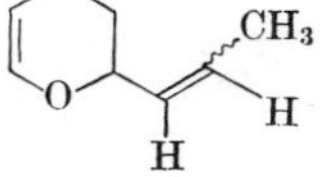 Mostly *cis*	230°, 18 hr, sealed tube	At equilibrium: CHO/CH₃ (1), CHO/CH₃ (5)	8 (total)	80
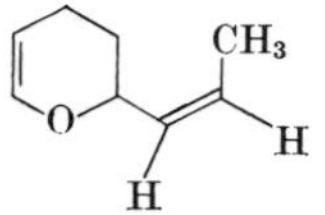 *cis*	225°, 1 hr, (C₆H₅)₂S, sealed tube, Ar atm	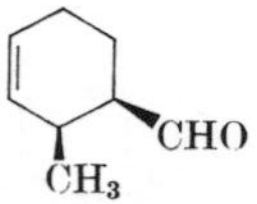 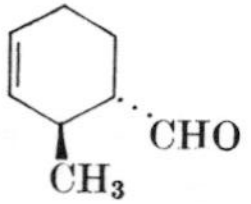 CHO/CH₃ (1) *cis*, CHO/CH₃ (2.8) *trans*, octa-4,6-dienal (1)	22 (total)	80
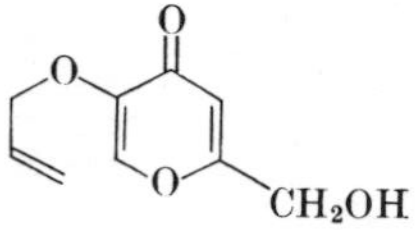 C₉H₁₀O₄	180–200°, 10–15 min, 1 mm	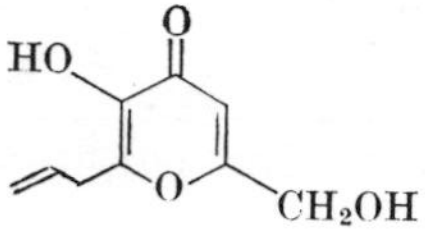 CH₂OH ... HO ...	Quant.	422

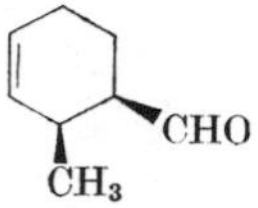

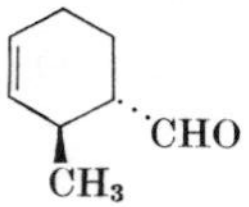

Note: References 344–439 are on pp. 251–252.

167

TABLE II. ALIPHATIC CLAISEN REARRANGEMENTS (Continued)

C. Ethers in Which the Vinyl Double Bond Is Part of a Ring (Continued)

Molecular Formula	Ether	Conditions	Product(s) and Ratio ( )	Yield(s), %	Refs.
$C_9H_{12}O_2$	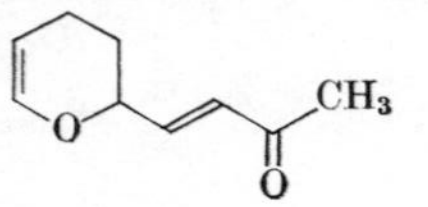 (ring ether, CH$_3$C=O substituent)	190–195°, 4.25 hr, sealed tube, Ar atm	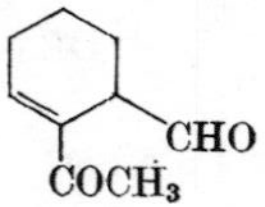 (CHO, COCH$_3$ ring) (4), unidentified components (1)	31 (total)	80
	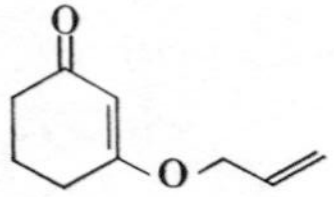 (allyloxy cyclohexenone)	Heat	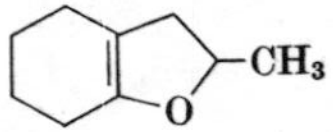 (CH$_3$, O bicyclic)	75	435
$C_9H_{12}O_3$	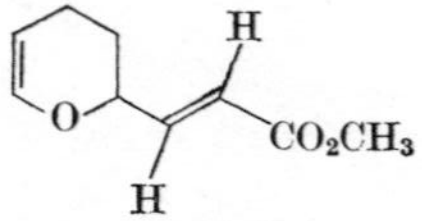 (CO$_2$CH$_3$, trans)	250°, 10 min, sealed tube, Ar atm	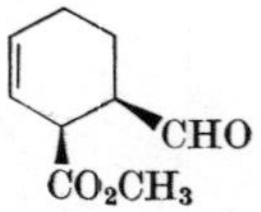 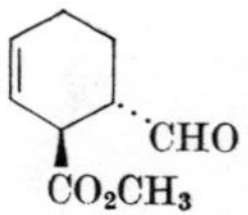 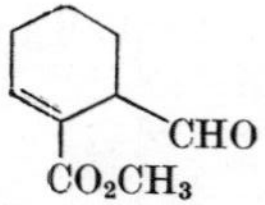 VI (19), (CHO, CO$_2$CH$_3$ ring) VII (5), (CHO, CO$_2$CH$_3$ ring)	87 (total)	80
			VIII (1), (CHO, CO$_2$CH$_3$ ring)		
	(cis)	250°, 3 hr, sealed tube	VI (1), VII (3.4), VIII (1.7)	18 (total)	80
$C_9H_{14}O$	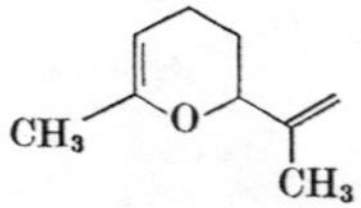 (CH$_3$, vinyl, O, CH$_3$ pyran)	425°, hexane, flow system	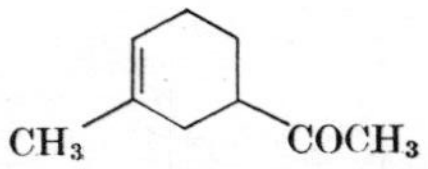 (CH$_3$, COCH$_3$ ring)	75	80
		240°, 25 min, sealed tube	"	73	80
	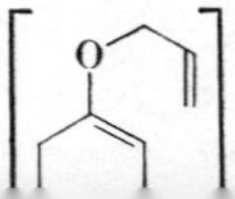 (allyloxy diene)	$C_6H_5CH_3$, p-TsOH, azeotropic distillation	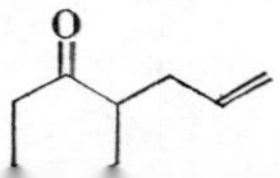 (allyl ketone)	98	117

168

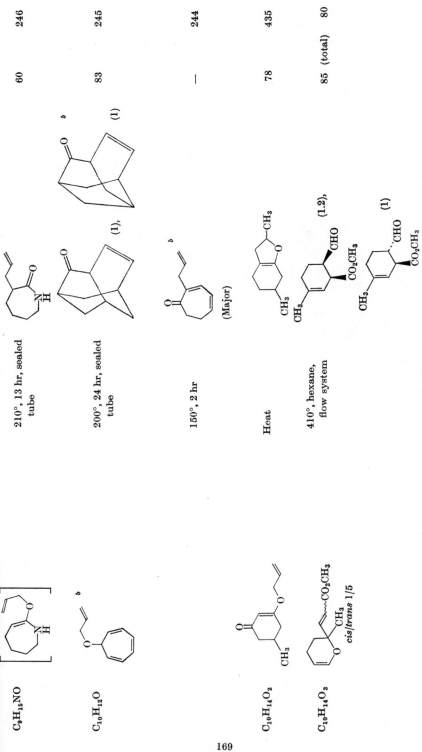

C₉H₁₅NO

C₁₀H₁₂O

C₁₀H₁₄O₂

C₁₀H₁₄O₃ *cis/trans* 1/5

210°, 13 hr, sealed tube

200°, 24 hr, sealed tube

150°, 2 hr

Heat

410°, hexane, flow system

(1), b

(1)

(Major) b

(1,2), b

(1)

60   246

83   245

—   244

78   435

85 (total)   80

Note: References 344–439 are on pp. 251–252.
b See the discussion on pp. 39, 40.

169

TABLE II. ALIPHATIC CLAISEN REARRANGEMENTS (*Continued*)

C. *Ethers in Which the Vinyl Double Bond Is Part of a Ring* (*Continued*)

Molecular Formula	Ether	Conditions	Product(s) and Ratio ( )	Yield(s), %	Refs.
$C_{10}H_{16}O$		Heat		62	121
	cis + *trans*	460°, hexane, flow system	IX (1.2), X (1)	14 (total)	80
		240°, 2.5 hr, sealed tube, Ar atm	IX (1.3), X (1)	30 (total)	80
$C_{11}H_{14}O$		150°, 2 hr	(Major)	—	244

170

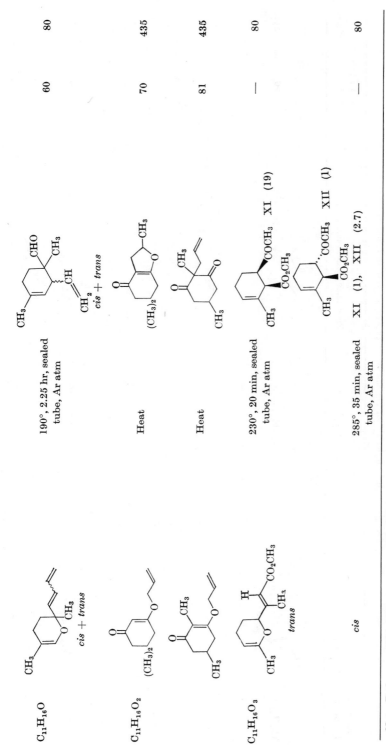

	Conditions	Product	Yield (%)	Refs.
C₁₁H₁₆O	190°, 2.25 hr, sealed tube, Ar atm	*cis + trans*	60	80
C₁₁H₁₆O₂	Heat		70	435
	Heat		81	435
C₁₁H₁₆O₃	230°, 20 min, sealed tube, Ar atm	XI (19)	—	80
	285°, 35 min, sealed tube, Ar atm *cis*	XI (1), XII (2.7)	—	80

Note: References 344–439 are on pp. 251–252.

[b] See the discussion on pp. 39, 40.

TABLE II. ALIPHATIC CLAISEN REARRANGEMENTS (Continued)

C. Ethers in Which the Vinyl Double Bond Is Part of a Ring (Continued)

Molecular Formula	Ether	Conditions	Product(s) and Ratio ( )	Yield(s), %	Refs.
C$_{13}$H$_{16}$O$_2$	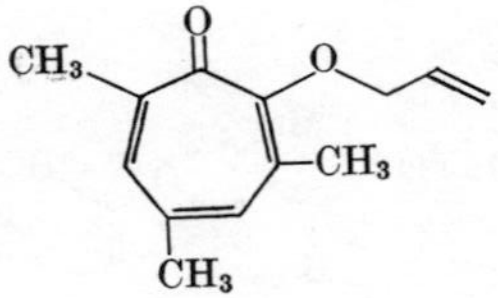	160°, neat	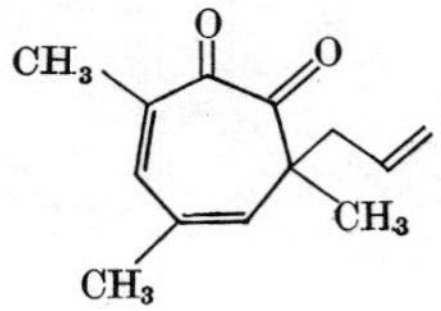 (5), 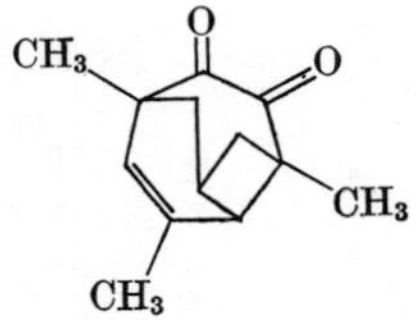 (1), 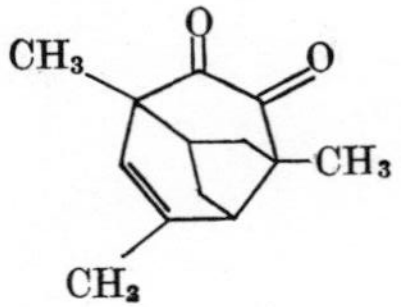 (trace)	90 (total)	242, 436
C$_{13}$H$_{20}$O$_3$	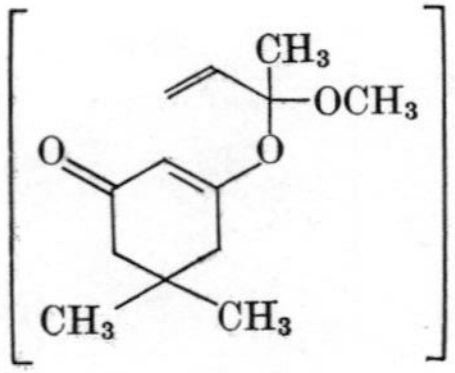	Reflux, 16.5 hr, C$_6$H$_6$	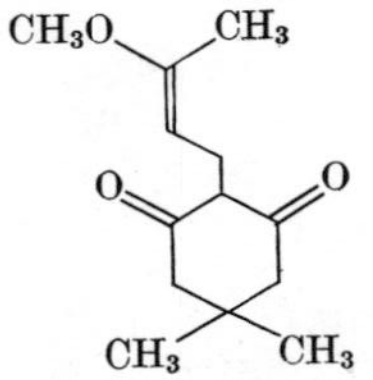	92	128b

172

$C_{15}H_{19}NO$	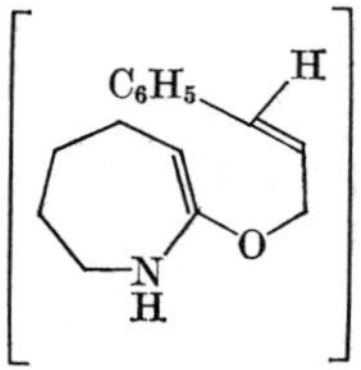 210°, 13 hr, neat sealed tube	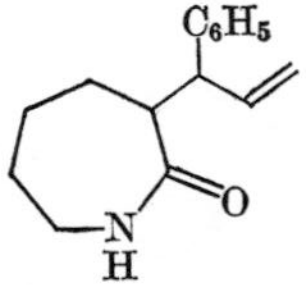 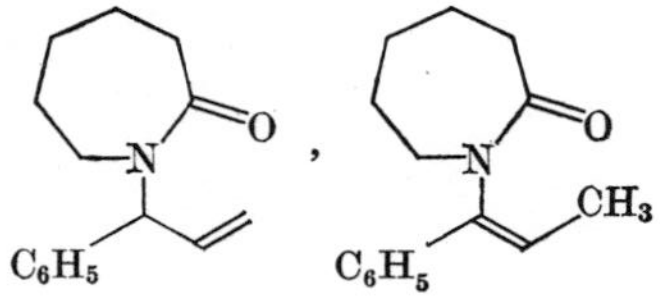 C_6H_5 ... , traces of C_6H_5		246
$C_{15}H_{20}O_2$	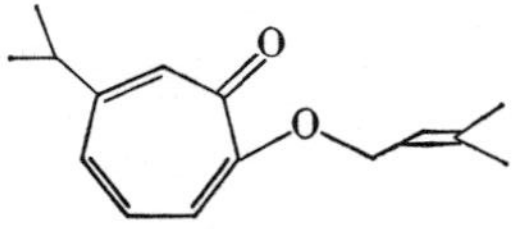 Reflux, 3 hr in xylene (Impure)	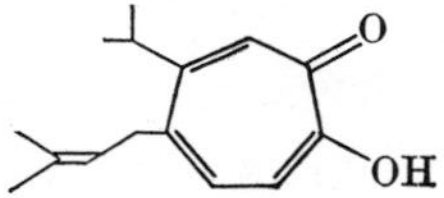 OH	7	243
$C_{15}H_{22}O$	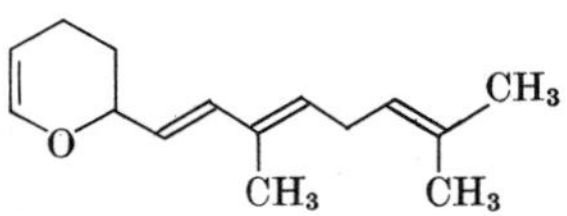 190°, 2.5 hr, sealed tube, Ar atm	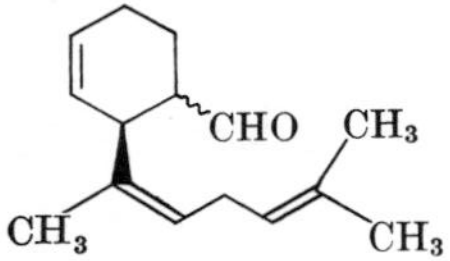 CHO CH_3 CH_3 CH_3	67	80
$C_{16}H_{26}O$	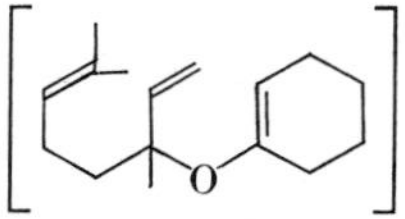 180°, 35 min	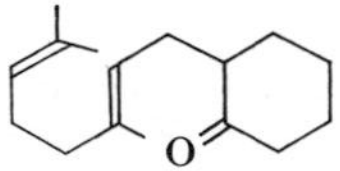	31	130

Note: References 344–439 are on pp. 251–252.

TABLE II. ALIPHATIC CLAISEN REARRANGEMENTS (Continued)

C. Ethers in Which the Vinyl Double Bond Is Part of a Ring (Continued)

Molecular Formula	Ether	Conditions	Product(s) and Ratio ( )	Yield(s), %	Refs.
$C_{16}H_{26}O$ (Contd.)	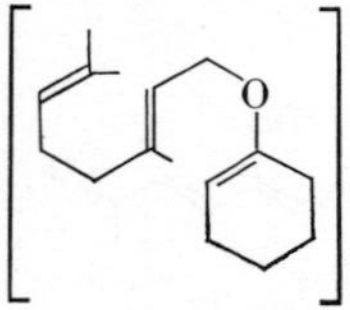	$185°$, 45 min	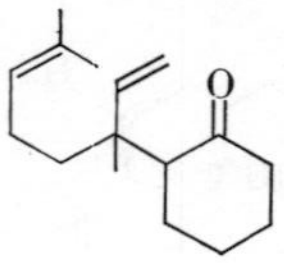	60	130
$C_{18}H_{23}NO_6$	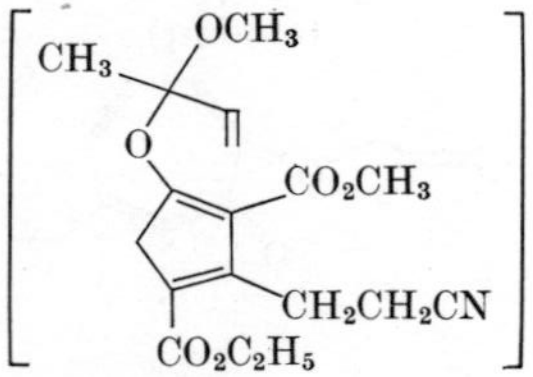	Reflux, 19 hr, C_6H_6, N_2 atm	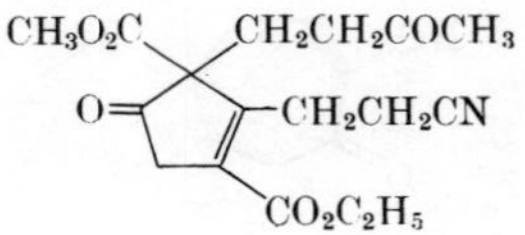	71	128b
$C_{22}H_{30}O_2$	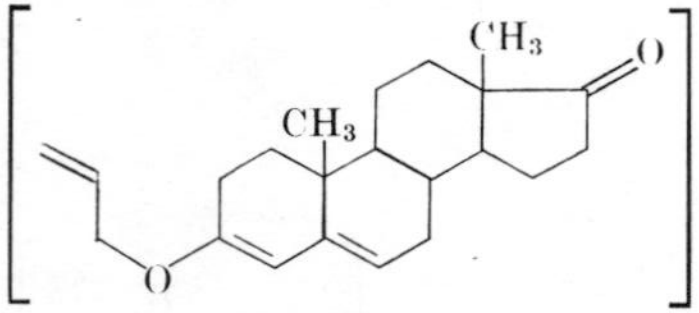	Reflux, 15 hr, $C_6H_5CH_3$	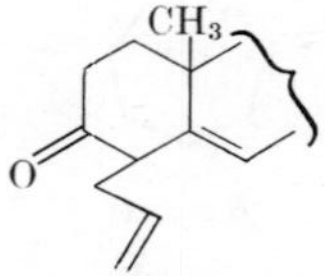	40	423
$C_{24}H_{36}O_3$	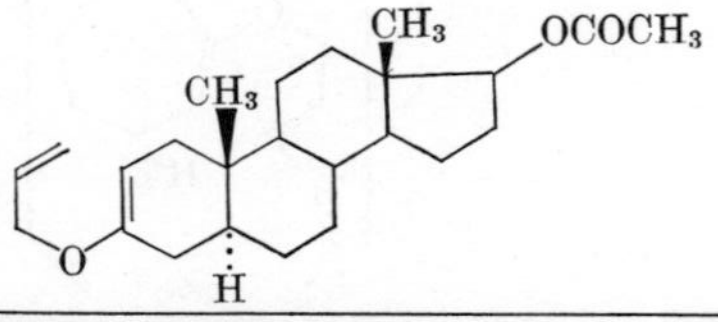	Reflux, 15 hr, pyridine	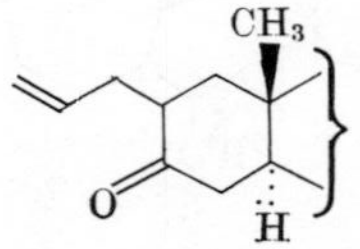	70	423

174

D. Propargyl Vinyl and Allenyl Vinyl Ethers

Molecular Formula	Ether	Conditions	Product(s) and Ratio ( )	Yield(s), %	Refs.
C_5H_6O	$HC{\equiv}CCH_2OCH{=}CH_2$	250°, heated tube, N_2 atm	$CH_2{=}C{=}CHCH_2CHO$	20–30	151
C_6H_8O	$HC{\equiv}CCH(CH_3)OCH{=}CH_2$	200°, heated tube, N_2 atm	$CH_3CH{=}C{=}CHCH_2CHO$	10–20	151
		200°	$CH_3CH{=}C{=}CHCH_2CHO$ (1.7), $CH_3CH{=}CHCH{=}CHCHO$ (1) (cis/trans 2/1)	—	120
$C_7H_3OF_4Cl$	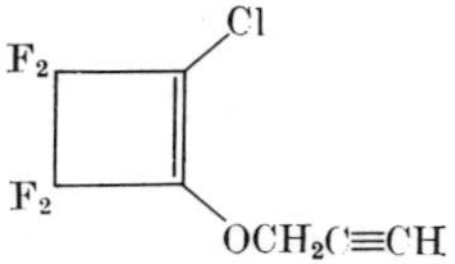 (see structure)	140°, 5 hr, reflux	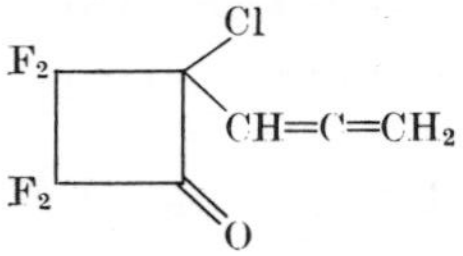 (see structure)	55	128a
$C_7H_{10}O$	$HC{\equiv}CC(CH_3)_2OCH{=}CH_2$	250°, heated tube, N_2 atm	$(CH_3)_2C{=}C{=}CHCH_2CHO$	10	151
	$HC{\equiv}CCH_2OCH{=}C(CH_3)_2$	140°, 4 hr, reflux	$CH_2{=}C{=}CHC(CH_3)_2CHO$	70	151
$C_7H_{10}O_2$	$[HC{\equiv}CCH_2OC{=}C(CH_3)OC_2H_5]$	140–150°, 4 hr, $C_2H_5CO_2H$	$CH_2{=}C{=}CHCH_2CO_2C_2H_5$	34	248
$C_8H_{12}O$	$[HC{\equiv}CC(CH_3)_2OC(CH_3){=}CH_2]$	60–80°, 15–24 hr, p-TsOH, hydroquinone	$(CH_3)_2C{=}C{=}CHCH_2COCH_3$	95	133
	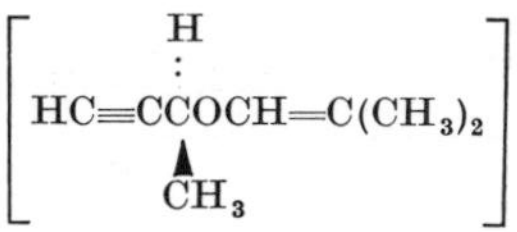 (see structure)[c]	210°, over silica, flow system	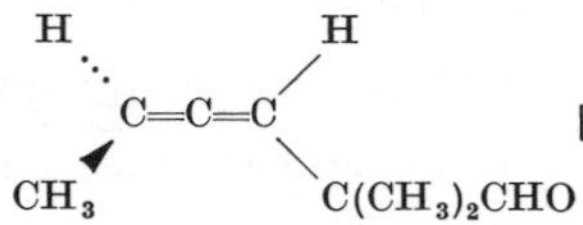 (see structure) $[\alpha]_D^{26} -15.9°$	—	116

Note: References 344–439 are on pp. 251–252.

[c] Prepared from alcohol of $[\alpha]_D^{17} +11.2°$

175

TABLE II. ALIPHATIC CLAISEN REARRANGEMENTS (Continued)

D. Propargyl Vinyl and Allenyl Vinyl Ethers (Continued)

Molecular Formula	Ether	Conditions	Product(s) and Ratio ( )	Yield(s), %	Refs.
$C_8H_{12}O$	$[HC{\equiv}CCH(CH_3)OCH{=}C(CH_3)_2]$	150°, 90 min, reflux	$CH_3CH{=}C{=}CHC(CH_3)_2CHO$	60	151
$C_8H_{12}O_2$	$\left[HC{\equiv}CCH(CH_3)OC{=}CH_2 \atop OC_2H_5\right]$	140–150°, 1.5 hr, $C_2H_5CO_2H$	$CH_3CH{=}C{=}CHCH_2CO_2C_2H_5$	63	248
$C_8H_{13}NO$	$\left[CH_3C{\equiv}CCH_2OC{=}CH_2 \atop N(CH_3)_2\right]$	200°, 4 hr	$CH_2{=}C{=}C(CH_3)CH_2CON(CH_3)_2$	—	247
$C_9H_{12}O$	$[HC{\equiv}CCHOC(CH_3){=}CH_2$, cyclopropyl]	80–100°, 20 hr, reflux, pet ether, p-TsOH, N_2 atm; $NaOCH_3$	$\triangle{-}CH{=}CH{\cdot}CH{=}CH(COCH_3)$	73	220
$C_9H_{14}O$	$[HC{\equiv}CCHOC(CH_3){=}CH_2$, $C_3H_7{-}i]$	Reflux, 15 hr, pet ether, p-TsOH	$(CH_3)_2CHCH{=}C{=}CHCH_2COCH_3$	60	220
	$HC{\equiv}CC(CH_3)_2OCH{=}C(CH_3)_2$	140°, 15 min, reflux	$(CH_3)_2C{=}C{=}CHC(CH_3)_2CHO$	76	151
$C_9H_{14}O_2$	$\left[HC{\equiv}CC(CH_3)_2OC{=}CH_2 \atop OC_2H_5\right]$	140–150°, 1.5 hr, $C_2H_5CO_2H$	$(CH_3)_2C{=}C{=}CHCH_2CO_2C_2H_5$	54	248
$C_{10}H_{16}O$	$[HC{\equiv}CC(CH_3)OC(CH_3){=}CH_2$, $C_3H_7{-}i]$	Reflux, 15 hr, pet ether, p-TsOH; $NaOCH_3$	$i{-}C_3H_7C(CH_3){=}CHCH{=}CHCOCH_3$	16	220
	$[HC{\equiv}CCH(C_3H_7{-}n)OCH{=}C(CH_3)_2]$	Reflux, 20 min	$n{-}C_3H_7CH{=}C{=}CHC(CH_3)_2CHO$	93	151
	$\left[HC{\equiv}CC(CH_3)_2OC{=}CHCH_3 \atop OC_2H_5\right]$	140–150°, 5 hr, $C_2H_5CO_2H$	$(CH_3)_2C{=}C{=}CHCH(CH_3)CO_2C_2H_5$	59	248
$C_{10}H_{16}O_2$	$\left[CH_3C{\equiv}CC(CH_3)_2OC{=}CH_2 \atop OC_2H_5\right]$	140–150°, 2 hr, $C_2H_5CO_2H$	$(CH_3)_2C{=}C{=}C(CH_3)CH_2CO_2C_2H_5$	61	248

Molecular formula	Propargyl ester / starting material	Conditions	Product	Yield (%)	Refs.
$C_{10}H_{17}NO$	$\left[\begin{array}{l} HC\equiv CCH(C_3H_7\text{-}n)OC{-}CH_2 \\ \qquad\qquad\quad OC_2H_5 \end{array}\right]$	$140\text{--}150°$, 3 hr, $C_2H_5CO_2H$	$n\text{-}C_3H_7CH{=}C{=}CHCH_2CO_2C_2H_5$	60	248
	$\left[\begin{array}{l} HC\equiv CCH_2OC{-}CHCH_3 \\ \qquad\qquad\quad N(C_2H_5)_2 \end{array}\right]$	Reflux, 5 hr, C_6H_6, $BF_3{\cdot}(C_2H_5)_2O$	$CH_2{=}C{=}CHCH(CH_3)CON(C_2H_5)_2$	60	126
$C_{11}H_{17}NO$	$\left[\begin{array}{l} CH_3C\equiv CCH_2OC{=}CH_2 \\ \qquad\quad N(C_2H_5)CH_2CH{=}CH_2 \end{array}\right]$	Reflux, 5 hr, C_6H_6	,,	70	126
	$\left[\begin{array}{l} CH_3C\equiv CCH_2OC{-}CHCH_3 \\ \qquad\qquad\qquad N(C_2H_5)_2 \end{array}\right]$	Heat	$CH_2{=}C{=}C(CH_3)CH_2CON(C_2H_5)CH_2CH{=}CH_2$	—	247
$C_{11}H_{19}NO$	$\left[\begin{array}{l} HC\equiv CCH(CH_3)OC{-}CHCH_3 \\ \qquad\qquad\qquad N(C_2H_5)_2 \end{array}\right]$	Reflux, 5 hr, C_6H_6	$CH_2{=}C{=}C(CH_3)CH(CH_3)CON(C_2H_5)_2$	85	126
		$120°$, 15 hr	$CH_3CH{=}C{=}CHCH(CH_3)CON(C_2H_5)_2$	60	126
$C_{12}H_{12}O$	$\left[\begin{array}{l} C_6H_5C\equiv CCH(CH_3)OCH{-}CH_2 \\ \qquad\qquad\qquad N(C_2H_5)_2 \end{array}\right]$	$200°$	$CH_3CH{=}C{=}C(C_6H_5)CH_2CHO$ (4), $CH_3CH{=}CHC(C_6H_5){=}CHCHO$ (1)	—	120
$C_{12}H_{18}O_2$	cyclohexane $\left[\begin{array}{l} C\equiv CH \\ OC{-}CH_2 \\ \quad OC_2H_5 \end{array}\right]$	$140\text{--}150°$, 1 hr, $C_2H_5CO_2H$	(cyclohexylidene)${=}C{=}CHCH_2CO_2C_2H_5$	51	248
$C_{12}H_{19}NO$	$H(C\equiv CCH_2OC{=}C(CH_3)_2$, piperidine-$N)$	Heat	$CH_2{=}C{=}CHC(CH_3)_2CON$(piperidine)	Quant	247
$C_{12}H_{21}NO$	$\left[\begin{array}{l} CH_2{=}C{=}C(CH_3)CH_2OC{=}CH(CH_3) \\ \qquad\qquad\qquad N(C_2H_5)_2 \end{array}\right]$	Heat, $C_6H_5CH_3$	$CH_2{=}C{=}C(CH_3)CCH(CH_3)CON(C_2H_5)_2$, $\overset{\|}{\ }CH_2$	50	127

Note: References 344–439 are on pp. 251–252.

177

TABLE II. ALIPHATIC CLAISEN REARRANGEMENTS (Continued)

D. Propargyl Vinyl and Allenyl Vinyl Ethers (Continued)

Molecular Formula	Ether	Conditions	Product(s) and Ratio ( )	Yield(s), %	Refs.
$C_{13}H_{14}O_2$	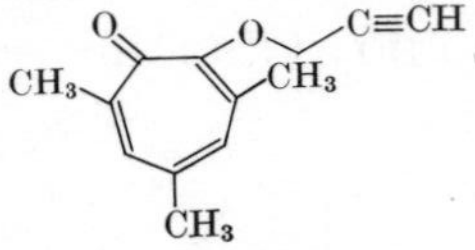	Reflux, 1 hr, *p*-cymene, N_2 atm	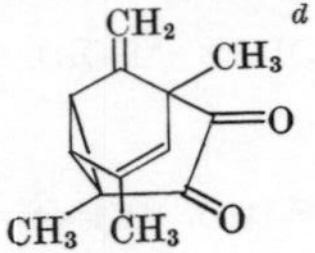 [d]	80	436
$C_{13}H_{20}O$	$[HC{\equiv}CC(CH_3)OC(CH_3){=}CH_2$ $(CH_2)_2CH{=}C(CH_3)_2]$	92°, 17 hr, ligroin, autoclave, N_2 atm	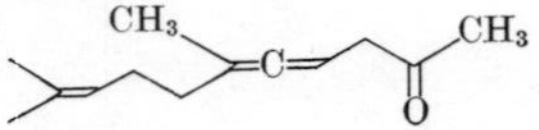	83	133
		60–80°, 15–24 hr, pet ether, *p*-TsOH	''	83	133
$C_{13}H_{22}O$	$[HC{\equiv}CC(CH_3)OC(CH_3){=}CH_2$ $C_6H_{13}\text{-}i$ $(CH_2)_2C(CH_3){=}CH_2]$	92°, 17 hr, ligroin, autoclave, N_2 atm	$i\text{-}C_6H_{13}C(CH_3)_2{-}C{=}CHCH_2COCH_3$	—	133
$C_{14}H_{22}O$	$[HC{\equiv}CC(CH_3)OC(CH_3){=}CH_2$ $(CH_2)_2C(CH_3){=}C(CH_3)_2]$	60–80°, 15–24 hr, pet ether, *p*-TsOH, hydroquinone	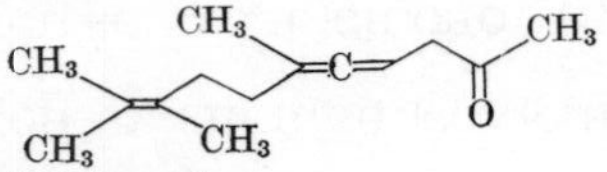	—	133
	Mixture of		Mixture of (2),		
		92°, 17 hr, ligroin, autoclave, N_2 atm	(1)	—	133

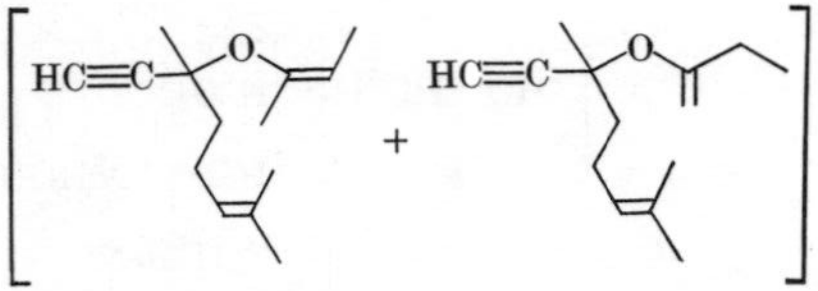

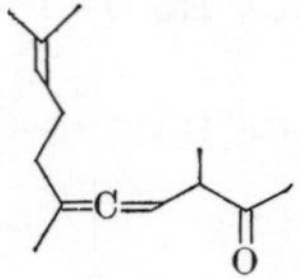

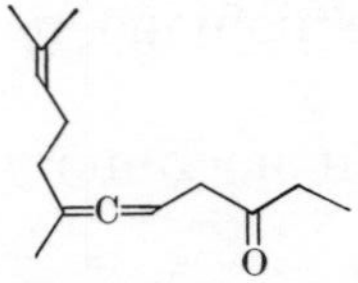

178

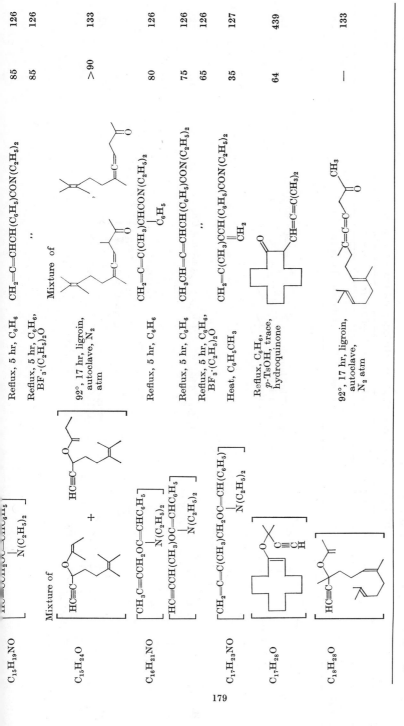

Formula	Conditions	Yield (%)	Refs.
$C_{15}H_{19}NO$	Reflux, 5 hr, C_6H_6	85	126
	Reflux, 5 hr, C_6H_6, $BF_3\cdot(C_2H_5)_2O$	85	126
$C_{15}H_{24}O$	92°, 17 hr, ligroin, autoclave, N_2 atm	>90	133
$C_{16}H_{21}NO$	Reflux, 5 hr, C_6H_6	80	126
	Reflux, 5 hr, C_6H_6	75	126
	Reflux, 5 hr, C_6H_6, $BF_3\cdot(C_2H_5)_2O$	65	126
$C_{17}H_{23}NO$	Heat, $C_6H_5CH_3$	35	127
$C_{17}H_{28}O$	Reflux, C_6H_6, p-TsOH, trace, hydroquinone	64	439
$C_{18}H_{28}O$	92°, 17 hr, ligroin, autoclave, N_2 atm	—	133

Note: References 344–439 are on pp. 251–252.

[a] Presumably, this is formed by an internal Diels-Alder reaction of the Claisen rearrangement product,

179

TABLE II. ALIPHATIC CLAISEN REARRANGEMENTS (*Continued*)

D. Propargyl Vinyl and Allenyl Vinyl Ethers (Continued)

Molecular Formula	Ether	Conditions	Product(s) and Ratio ( )	Yield(s), %	Refs.
C$_{18}$H$_{32}$O	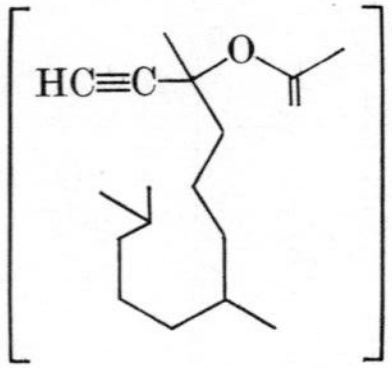	92°, 17 hr, ligroin, autoclave, N$_2$ atm	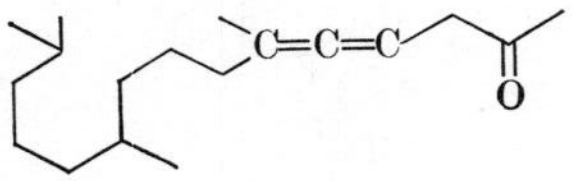	—	133
C$_{22}$H$_{30}$O$_2$	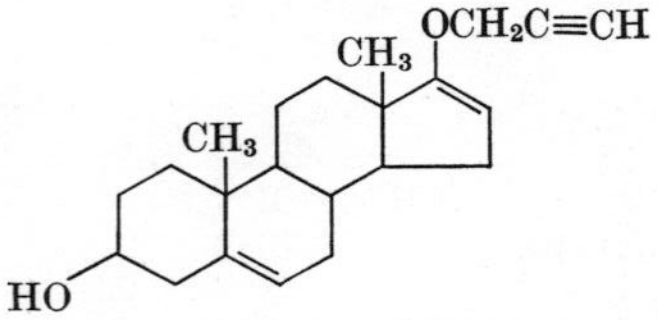	Reflux, 5 hr, C$_6$H$_5$CH$_3$	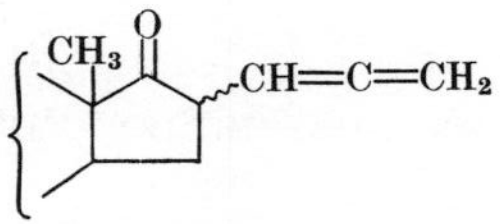 (epimers)	65	221
C$_{23}$H$_{32}$O$_3$	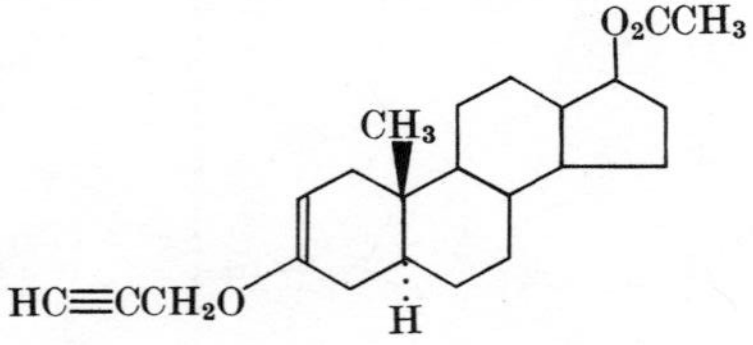	Reflux, 5 hr, C$_6$H$_5$CH$_3$	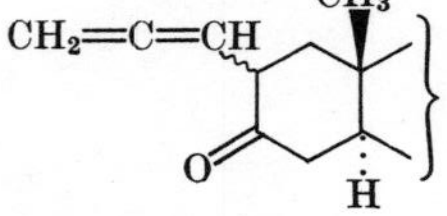 (epimers)	50	221

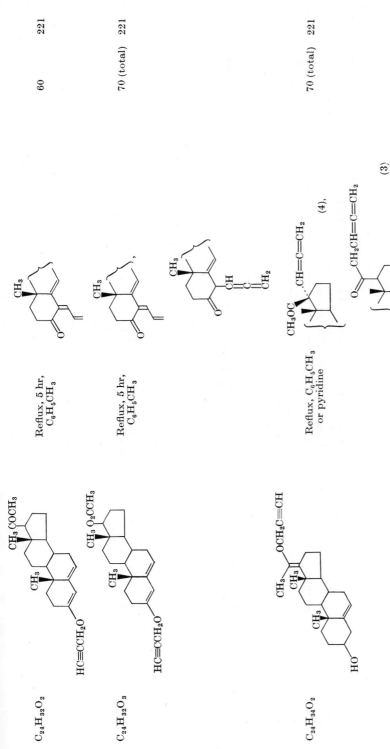

$C_{24}H_{32}O_2$ Reflux, 5 hr, $C_6H_5CH_3$ 60 221

$C_{24}H_{32}O_3$ Reflux, 5 hr, $C_6H_5CH_3$ 70 (total) 221

$C_{24}H_{34}O_2$ Reflux, $C_6H_5CH_3$ or pyridine (4), (3) 70 (total) 221

Note: References 344–439 are on pp. 251–252.

181

TABLE II. ALIPHATIC CLAISEN REARRANGEMENTS *(Continued)*

D. Propargyl Vinyl and Allenyl Vinyl Ethers (Continued)

Molecular Formula	Ether	Conditions	Product(s) and Ratio ( )	Yield(s), %	Refs.
$C_{26}H_{36}O_3$	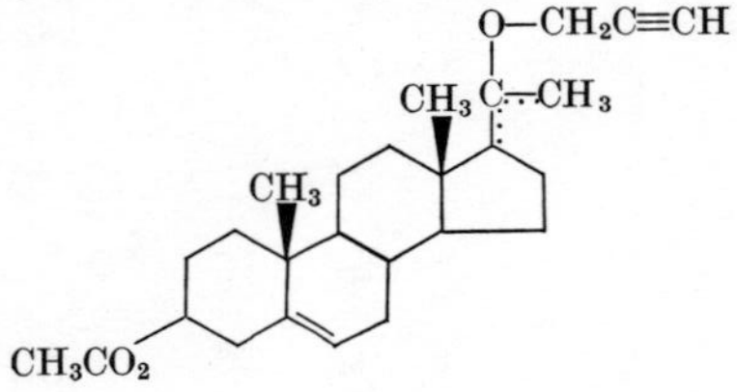	Reflux, 5 hr, $C_6H_5CH_3$, N_2 atm	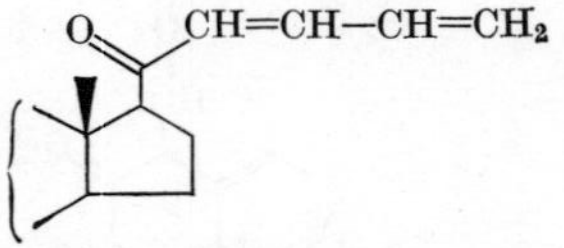 (1,2) 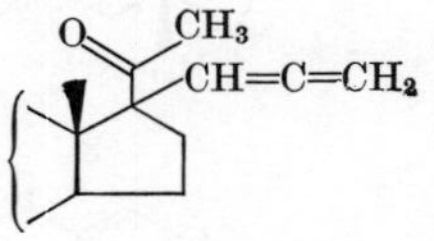 (1)	48 (total)	222

Note: References 344–439 are on pp. 251–252.

TABLE III. AMINO-CLAISEN REARRANGEMENTS

A. Aromatic and Heterocyclic Compounds

Molecular Formula	Amine	Conditions	Product(s) and Ratio ( )	Yield(s), %	Refs.
$C_9H_{13}N_3O_2$	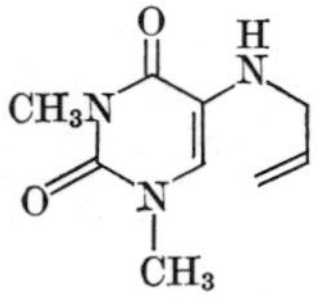	207°, 12 hr, tetralin	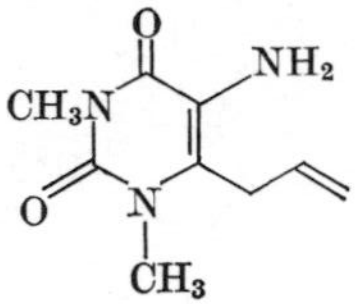	24	256
$C_{11}H_{12}NBr$	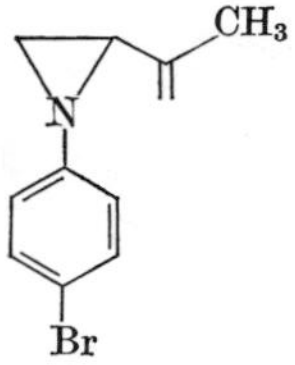	140°, 17 hr, xylene, N_2 atm	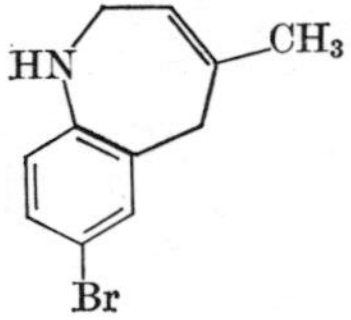	73	250
$C_{13}H_{13}N$	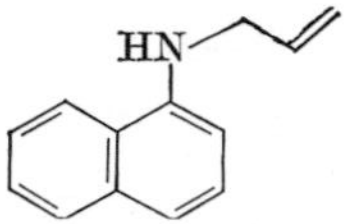	260°, 3 hr	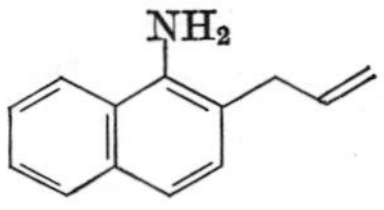	70	29
	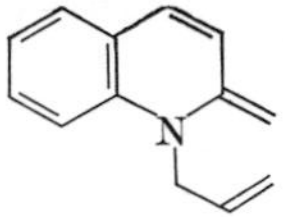	80°, 12–24 hr, C_6H_6, N_2 atm	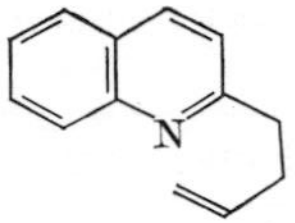	52.5	253

Note: References 344–439 are on pp. 251–252.

TABLE III. AMINO-CLAISEN REARRANGEMENTS (Continued)

A. Aromatic and Heterocyclic Compounds (Continued)

Molecular Formula	Amine	Conditions	Product(s) and Ratio ( )	Yield(s), %	Refs.
$C_{13}H_{13}NO_2$	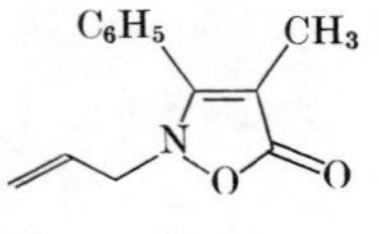	180°, 1 hr	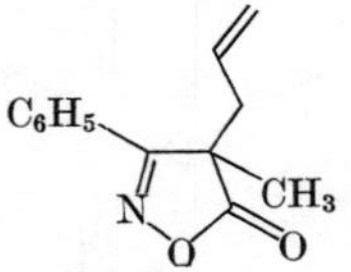 (99)[a]	—	255
$C_{14}H_{15}N$	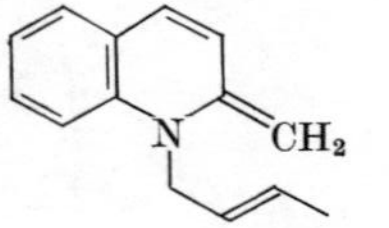	80°, 12–24 hr, C_6H_6, N_2 atm	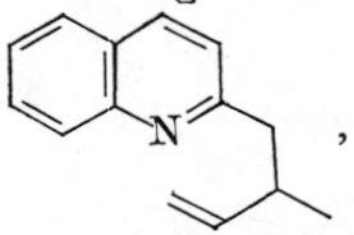 (6.8) , 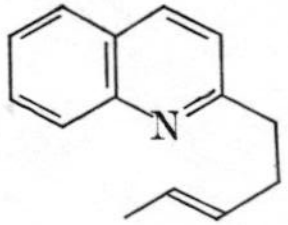 (1)	65 (total)	253
$C_{14}H_{15}NO_2$	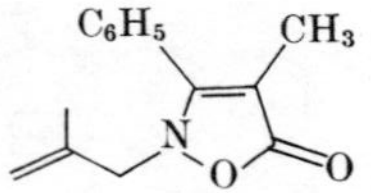	180°, 1 hr	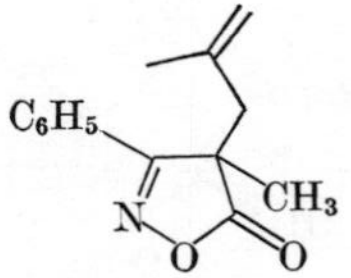 (99)[a]	—	255
$C_{14}H_{15}NO_2$	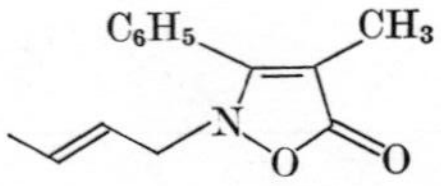	180°, 1 hr	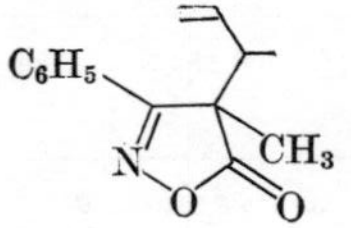 (46)[a]	—	255

Starting material	Conditions	Product	Yield (%)	Ref.
$C_{14}H_{16}N_2O$	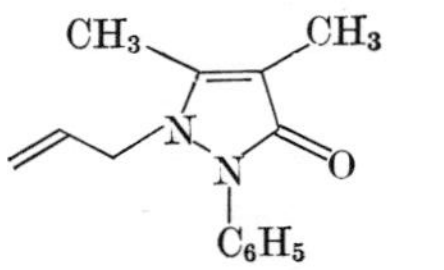 180°, 30 min	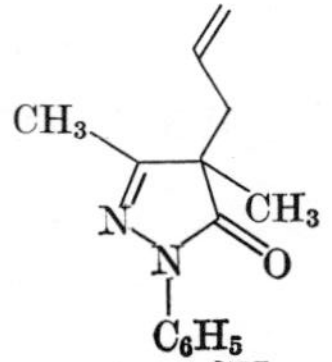	Quant	254
$C_{14}H_{17}N$	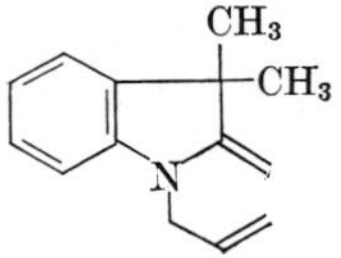 200°, 24 hr	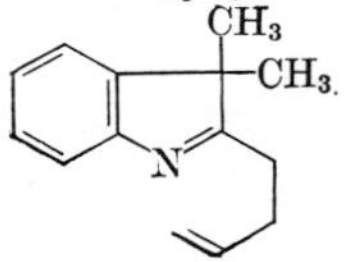	95	253
$C_{15}H_{18}N_2O$	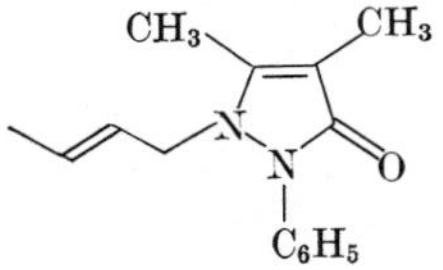 180°, 30 min	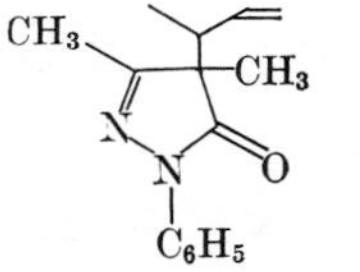	Quant	254
$C_{16}H_{18}N_2O$	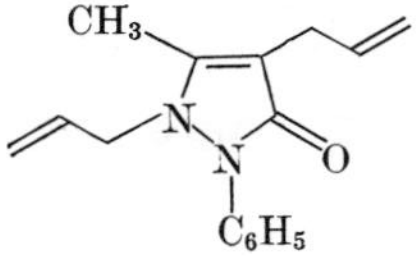 180°, 30 min	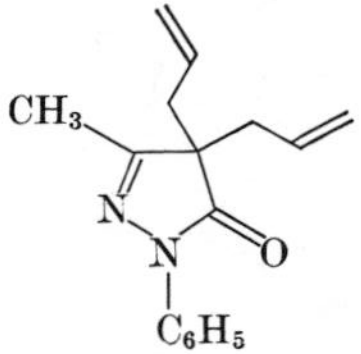	Quant	254
$C_{19}H_{17}N$	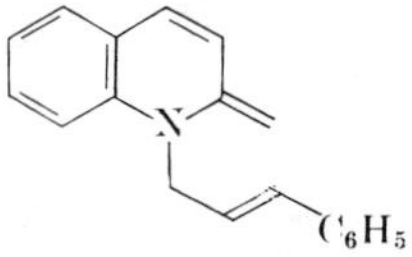 80°, 12–24 hr, C_6H_6, N_2 atm	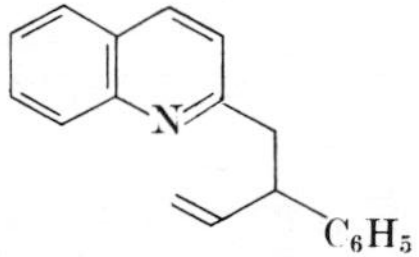	38	253

Note: References 344–439 are on pp. 251–252.

[a] The product is in equilibrium with the starting material. See p. 44.

TABLE III. AMINO-CLAISEN REARRANGEMENTS (Continued)

B. Aliphatic Compounds

Molecular Formula	Amine or Ammonium Ion	Conditions	Product(s) and Ratio ( ) After Hydrolysis[b]	Yield(s), %	Refs.
$C_8H_{13}N$	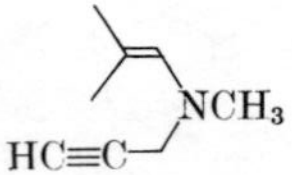	260°	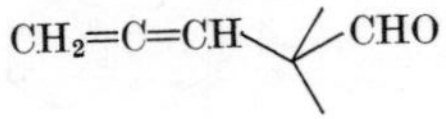 $CH_2=C=CH$ CHO	Quant	252
$C_8H_{15}N$	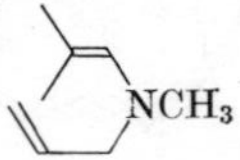	250°, 1 hr, sealed tube	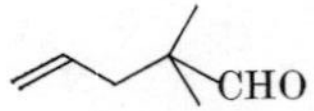 CHO	Quant	53
$C_{10}H_{19}NO$	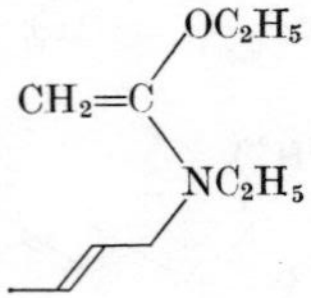	180°, overnight, sealed tube	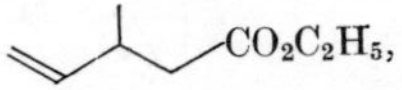 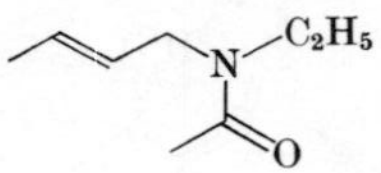 $CO_2C_2H_5$, NC_2H_5	Quant	247
$C_{10}H_{20}N$	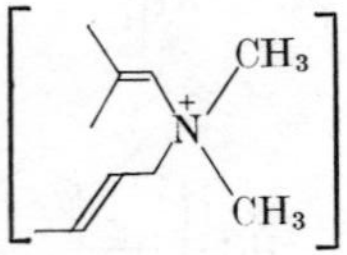	Reflux, 6.5 hr, CH_3CN	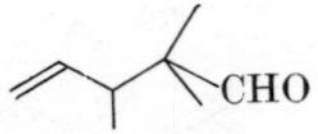 CHO	67	257
$C_{11}H_{18}N$	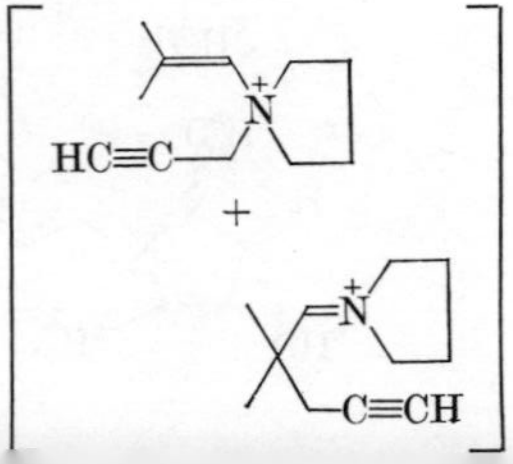	Reflux, 1–2 hr, CH_3CN	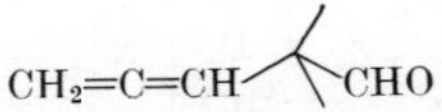 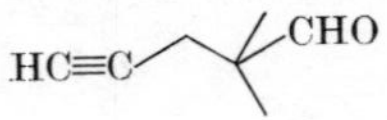 $CH_2=C=CH$ CHO (6), $HC\equiv C$ CHO (1)	30 (total)	261, 259

186

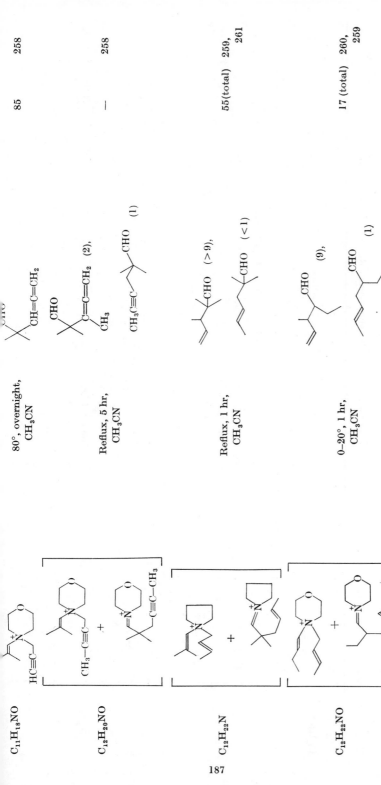

		Yield	Refs.
$C_{11}H_{18}NO$	80°, overnight, CH₃CN	85	258
$C_{12}H_{20}NO$	Reflux, 5 hr, CH₃CN	—	258
$C_{12}H_{22}N$	Reflux, 1 hr, CH₃CN	55(total)	259, 261
$C_{12}H_{22}NO$	0–20°, 1 hr, CH₃CN	17 (total)	260, 259

Note: References 344–439 are on pp. 251–252.

[b] In most cases the rearranged imines or immonium ions were directly hydrolyzed to carbonyl compounds. The stable amidines were isolated and are so reported.

187

TABLE III. AMINO-CLAISEN REARRANGEMENTS (Continued)

B. Aliphatic Compounds (Continued)

Molecular Formula	Amine or Ammonium Ion	Conditions	Product(s) and Ratio ( ) After Hydrolysis[b]	Yield(s), %	Refs.
$C_{13}H_{17}N$	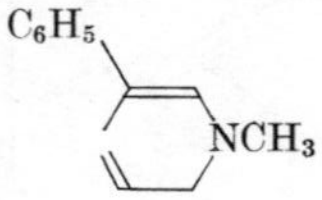	205°, 3 hr	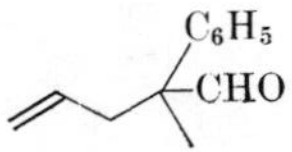	—	53
$C_{13}H_{22}N$	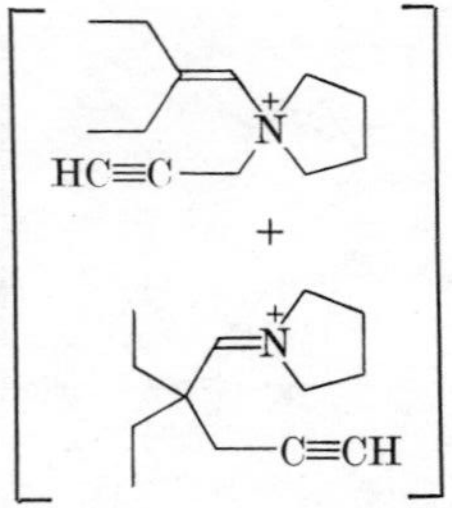	Reflux, 1–2 hr, CH_3CN	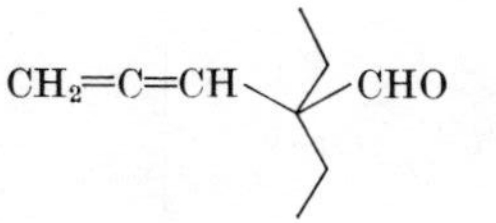 (3.3), 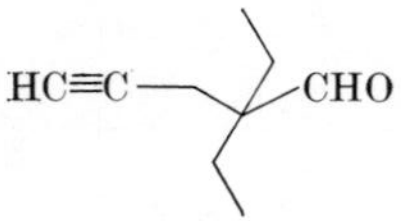 (1)	23 (total)	261, 259
$C_{14}H_{26}N$	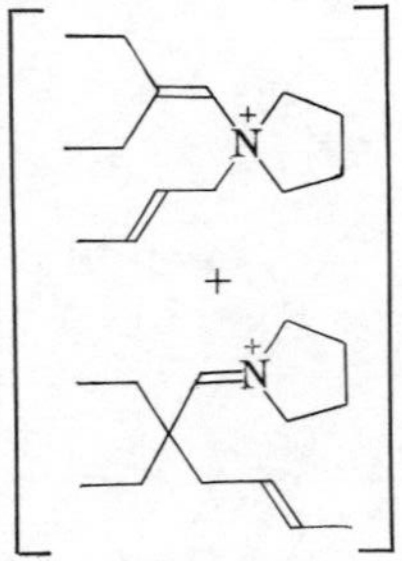	Reflux, 1 hr, CH_3CN	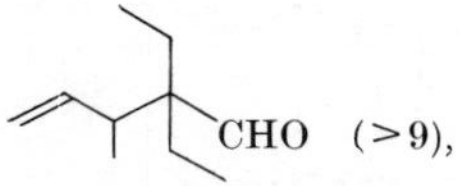 (>9), 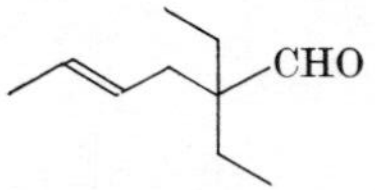 (<1)	56 (total)	259, 261

$C_{14}H_{26}N_2$	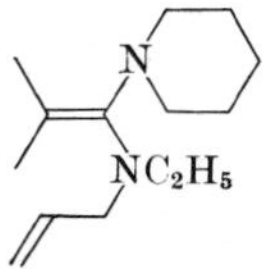 (piperidine–$N C_2H_5$ amidine structure, allyl)	200°, 4 hr	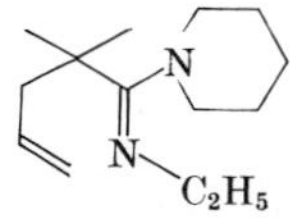 (piperidine N, $=N-C_2H_5$, allyl)	Quant	247
$C_{15}H_{20}N$	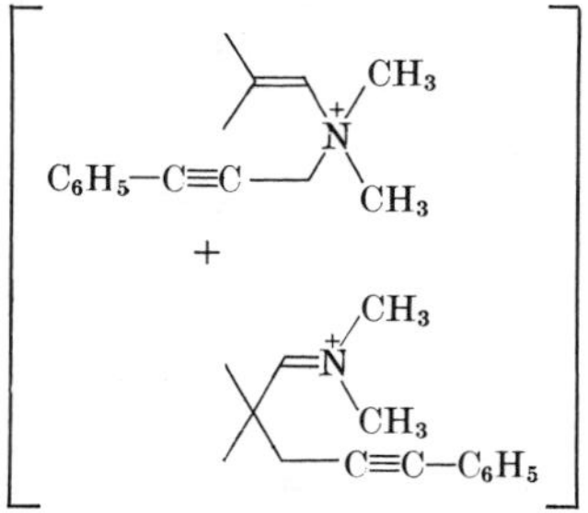 $C_6H_5-C\equiv C-$ $\overset{+}{N}(CH_3)_2$ + $\overset{+}{N}(CH_3)_2-C\equiv C-C_6H_5$	250°	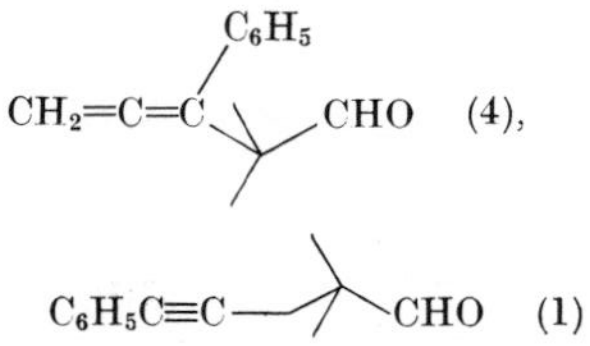 $CH_2=C=C\overset{C_6H_5}{-}CHO$ (4), $C_6H_5C\equiv C-C(CH_3)_2CHO$ (1)	—	252
$C_{15}H_{26}N$	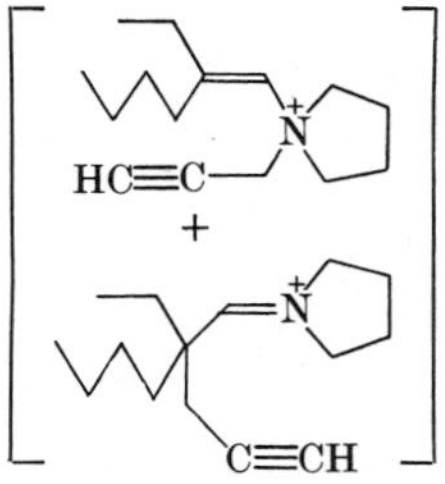 (pyrrolidinium) $HC\equiv C-CH_2$ + (pyrrolidinium) $-C\equiv CH$	Reflux, 1 hr, CH_3CN	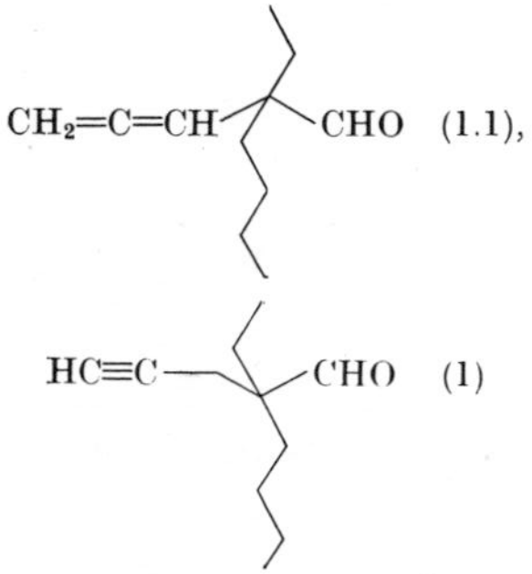 $CH_2=C=CH-CH(C_2H_5)CHO$ (1.1), $HC\equiv C-C(C_2H_5)CHO$ (1)	20 (total)	261, 259

Note: References 344–439 are on pp. 251–252.

[b] In most cases the rearranged imines or immonium ions were directly hydrolyzed to carbonyl compounds. The stable amidines were isolated and are so reported.

189

TABLE III. Amino-Claisen Rearrangements (Continued)

B. Aliphatic Compounds (Continued)

Molecular Formula	Amine or Ammonium Ion	Conditions	Product(s) and Ratio ( ) After Hydrolysis[b]	Yield(s), %	Refs.
$C_{16}H_{30}N$	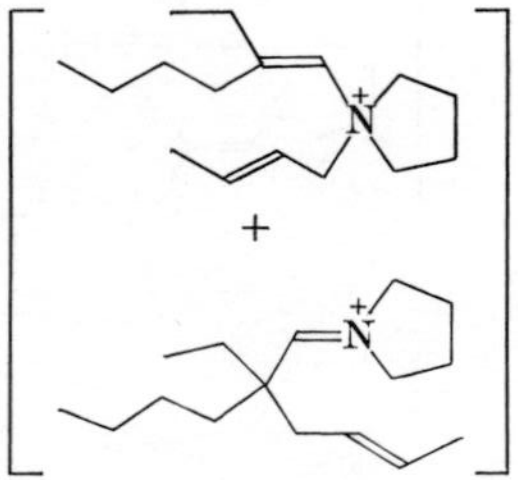	Reflux, 1–2 hr, CH_3CN	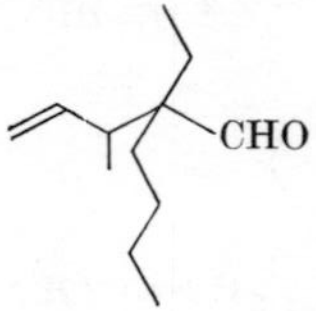 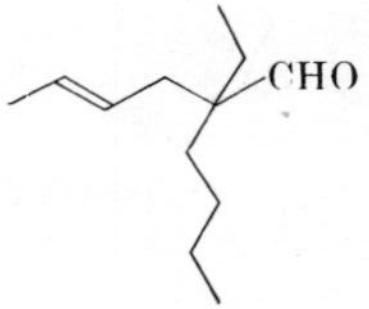	66 (total)	261, 259
$C_{17}H_{24}N$	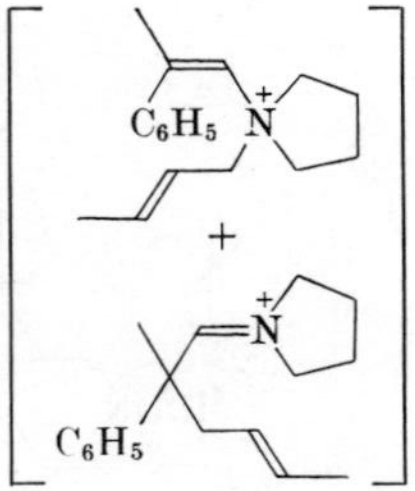	Reflux, 2.5 hr, CH_3CN	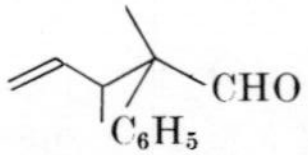 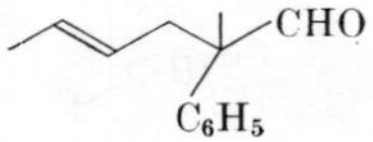	85 (total)	261, 259

$C_{19}H_{21}N$	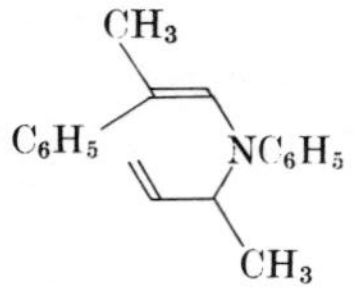	170–175°, distil	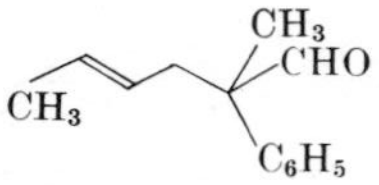 (*cis/trans* 1/7)	Quant	53
$C_{21}H_{26}N_2$	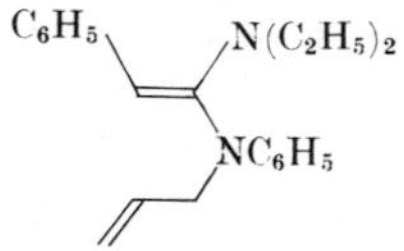	280°, 4 hr	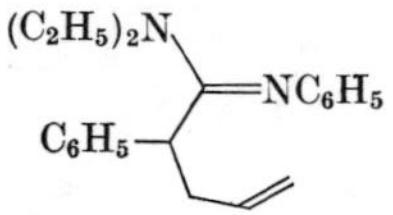	—	125

Note: References **344–439** are on pp. 251–252.

[b] In most cases the rearranged imines or immonium ions were directly hydrolyzed to carbonyl compounds. The stable amidines were isolated and are so reported.

TABLE IV. THIO-CLAISEN REARRANGEMENTS

A. Aromatic and Heterocyclic Compounds

Molecular Formula	Sulfide or Sulfonium Ion	Conditions	Product(s) and Ratio ( )	Yield(s), %	Refs.
$C_7H_6S_2$	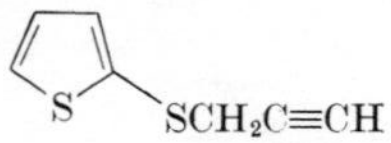 (2-thienyl)–$SCH_2C{\equiv}CH$	170–180°, 10 min, $[(CH_3)_2N]_3PO$, N_2 atm	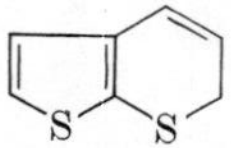 I (thieno-fused pyran structure)	92	274
		140–142°, 50 min, $(CH_3)_2SO$, $(i\text{-}C_3H_7)_2NH$	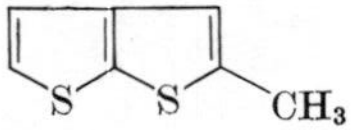 II (CH_3-substituted thienothiophene)	53	275
		150–170°, $(CH_3)_2SO$, $(i\text{-}C_3H_7)_2NH$	I (1), II (3)	—	275
	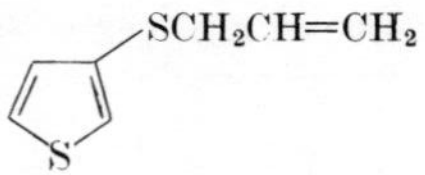 (3-thienyl)–$SCH_2CH{=}CH_2$	170–180°, 10 min, $[(CH_3)_2N]_3PO$, N_2 atm	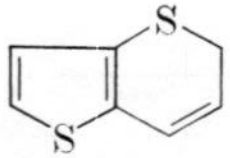 (thienothiophene structure)	87	274
		140–142°, 50 min, $(CH_3)_2SO$, $(i\text{-}C_3H_7)_2NH$	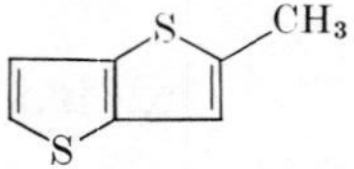 (CH_3-substituted thienothiophene)	65	275
$C_7H_8S_2$	 (2-thienyl)–$SCH_2CH{=}CH_2$	170°, 15 min, quinoline	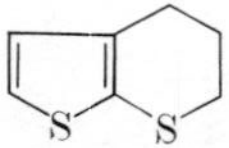 (1) 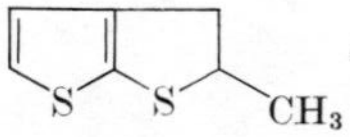 , (2), (3) CH_3 , dimer	83 (total) crude	424

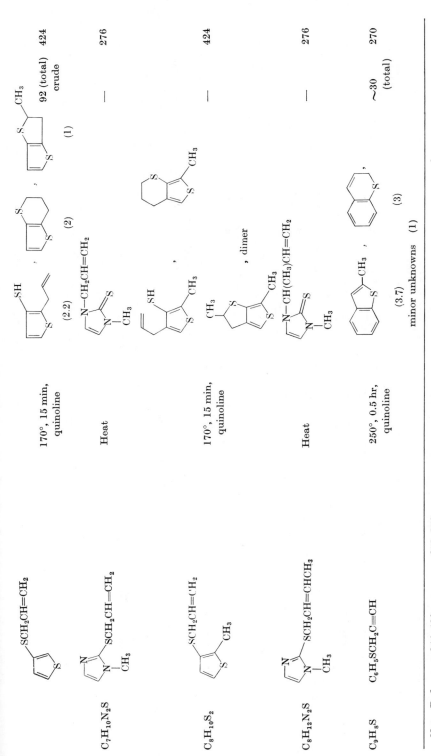

Formula	Conditions	Products (yield %)	Refs.
$C_7H_{10}N_2S$ (thiophene–$SCH_2CH{=}CH_2$)	170°, 15 min, quinoline	SH-thiophene–allyl (2.2), tetrahydrobenzothiophene (2), CH_3-fused thienothiophene (1), 92 (total) crude	424
$C_8H_{10}S_2$ (N-methylimidazole–$SCH_2CH{=}CH_2$)	Heat	N–$CH_2CH{=}CH_2$ imidazole-2-thione, N–CH_3 —	276
$C_8H_{12}N_2S$ (CH_3-thiophene–$SCH_2CH{=}CH_2$)	170°, 15 min, quinoline	SH–CH_3-thiophene–allyl, CH_3-fused thienothiophene, dimer —	424
C_9H_8S (N-methylimidazole–$SCH_2CH{=}CHCH_3$)	Heat	N–$CH(CH_3)CH{=}CH_2$ imidazole-2-thione, N–CH_3 —	276
C_9H_8S ($C_6H_5SCH_2C{\equiv}CH$)	250°, 0.5 hr, quinoline	2-methylbenzothiophene (3.7), 2H-thiochromene (3), minor unknowns (1), ~30 (total)	270

Note: References 344–439 are on pp. 251–252.

TABLE IV. THIO-CLAISEN REARRANGEMENTS (Continued)

A. Aromatic and Heterocyclic Compounds (Continued)

Molecular Formula	Sulfide or Sulfonium Ion	Conditions	Product(s) and Ratio ( )	Yield(s), %	Refs.
C_9H_9SCl	$C_6H_5SCH_2CCl{=}CH_2$	300°, 1.5 hr, quinoline	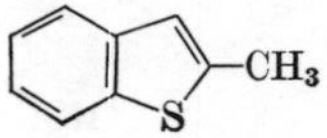 [2-methylbenzothiophene, CH_3]	22	264
		300°, 1.5 hr, octanoic acid	''	41	264
$C_9H_{10}S$	$C_6H_5SCH_2CH{=}CH_2$	230–240°, 2–4 hr, quinoline	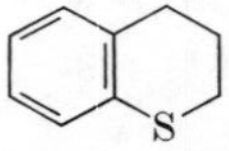 III [2,3-dihydrobenzothiophene] (1.2), 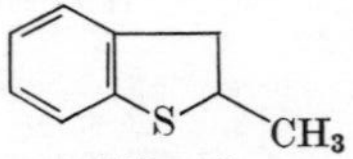 IV [2-methyl-2,3-dihydrobenzothiophene, CH_3] (1)	55	268
		300°, 1.5 hr, quinoline	III (1), IV (1.6)	75	264
		300°, 1.5 hr, octanoic acid	III (1), IV (1.7), $C_6H_5CH{=}CHCH_3$ (3.2)	60 (total)	264
		217–241°, 6 hr, quinoline	III (2), IV (2), polymer (1)	90 (total)	263
$C_{10}H_{10}S$	$C_6H_5SCH_2C{\equiv}CCH_3$	270°, 4 hr, quinoline	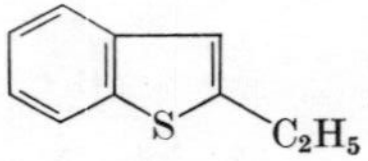 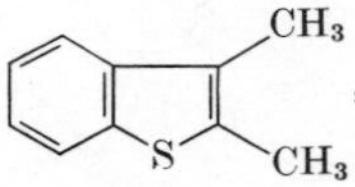 [2-ethylbenzothiophene, C_2H_5] (9.2), [2,3-dimethylbenzothiophene, CH_3, CH_3] (3), [4-methyl-2H-thiochromene, CH_3] 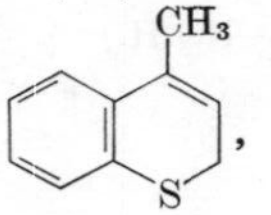 (1), minor unknowns and starting material (6.8)	~30 (total)	270

194

$C_{10}H_{12}S$	$C_6H_5SCH_2C(CH_3)=CH_2$	300°, 1.5 hr, quinoline	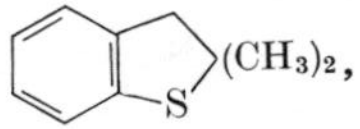 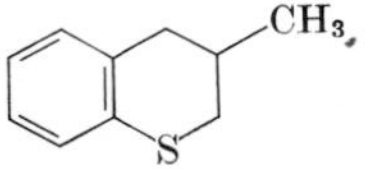 (structure) $(CH_3)_2$, V (6), (structure) CH_3, VI (10)	72 (total)	262, 264
		300°, 1.5 hr, octanoic acid	$C_6H_5SCH=C(CH_3)_2$ VII (1), V (1), VI (1), VII (2)	66 (total)	264
	$C_6H_5SCH_2CH=CHCH_3$	250°, 2 hr, quinoline	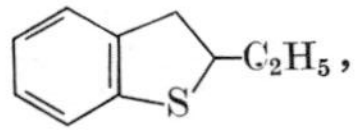 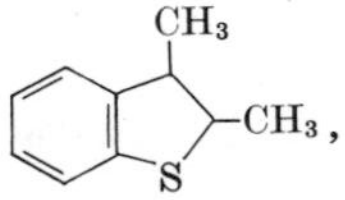 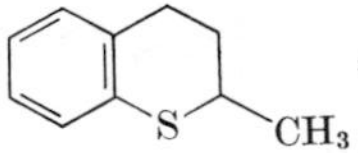 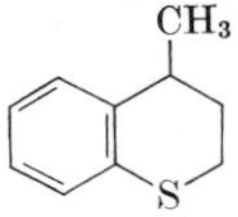 (structure) C_2H_5, VIII (10), (structure) CH_3 CH_3, IX (5), (structure) CH_3 X (2), (structure) CH_3 XI (1)	88 (total)	264
		250°, 2 hr, octanoic acid	Propenyl isomers (31), X (1), VIII (7)	60 (total)	264
$C_{11}H_{12}O_2S$	$o\text{-}HO_2CC_6H_4SCH_2CH=CHCH_3$	250°, 2 hr, neat	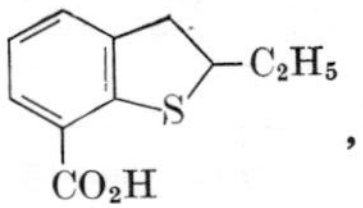 (structure) C_2H_5 CO_2H (36), cleavage products (24), X (1), VIII (6)	40 (total)	264

Note: References 344–439 are on pp. 251–252.

195

TABLE IV. Thio-Claisen Rearrangements (Continued)

A. Aromatic and Heterocyclic Compounds (Continued)

Molecular Formula	Sulfide or Sulfonium Ion	Conditions	Product(s) and Ratio ( )	Yield(s), %	Refs.
$C_{11}H_{14}S$	$m\text{-}CH_3C_6H_4SCH_2CH{=}CHCH_3$	Heat, quinoline	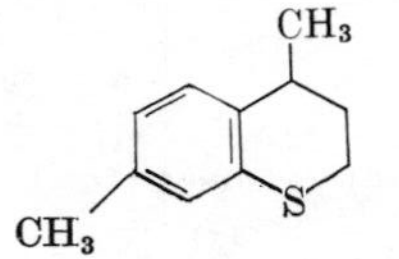	80	263
$C_{12}H_9NS$	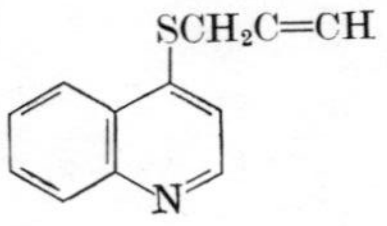 SCH$_2$C≡CH	200°, 2 hr, $(CH_3)_2NC_6H_5$	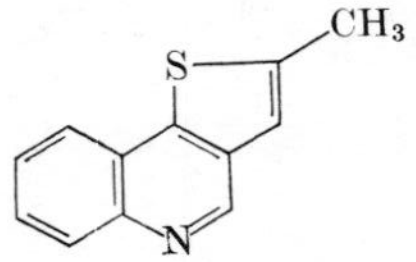 (36),	80	272
$C_{12}H_{11}NS$	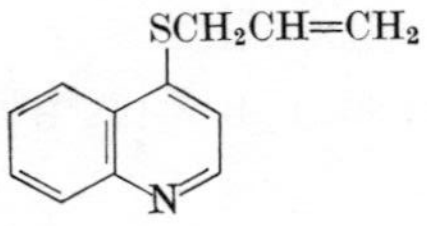 SCH$_2$CH=CH$_2$	200°, 1 hr, neat	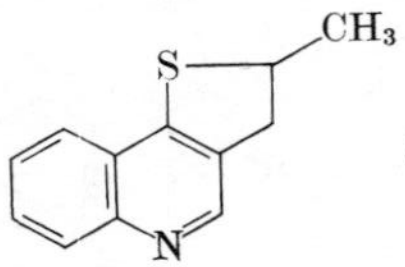 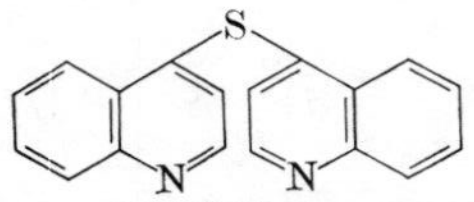 75 (total)	75 (total)	271
		200°, 2 hr, $(n\text{-}C_3H_7CO)_2O$	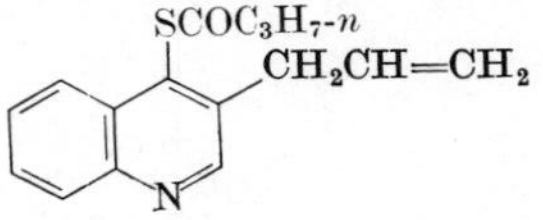 (1)	87	272

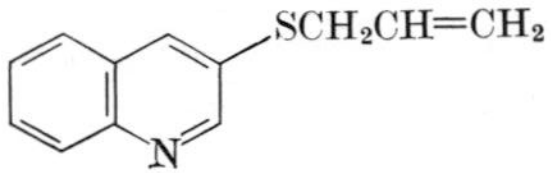

 SCH₂CH=CH₂ on quinoline (position 3)	200°, 2 hr, (CH₃)₂NC₆H₅, Ar atm	 XII (4.5), XIII (1)	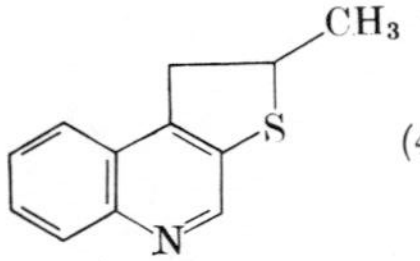 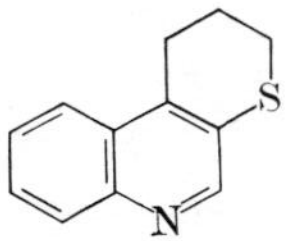 81.6 (total) 273
	200°, 1 hr, neat, Ar atm	 XII (11.5), XIII (1), 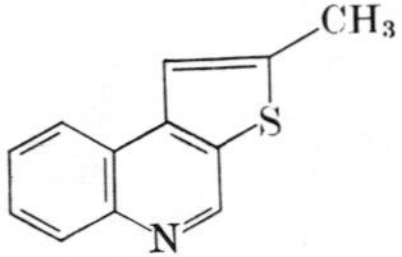 CH₃ (1.5)	89 (total) 273
C₁₂H₁₃NS SCH₂CH=CH₂, CH₃ (indole)	Reflux, C₆H₅CH₃	 CH₂CH=CH₂, S, CH₃	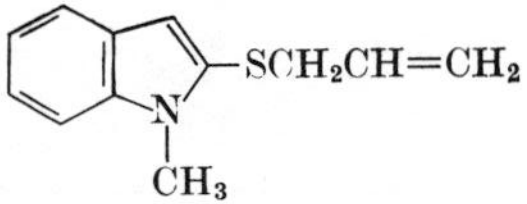 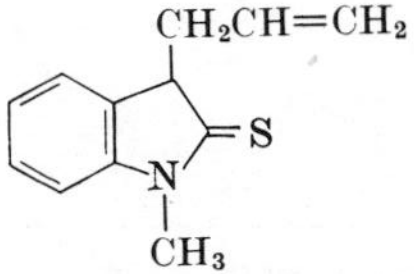 Quant 34

Note: References 344–439 are on pp. 251–252.

TABLE IV. Thio-Claisen Rearrangements (*Continued*)

A. *Aromatic and Heterocyclic Compounds* (*Continued*)

Molecular Formula	Sulfide or Sulfonium Ion	Conditions	Product(s) and Ratio ( )	Yield(s), %	Refs.
$C_{13}H_{13}NS$	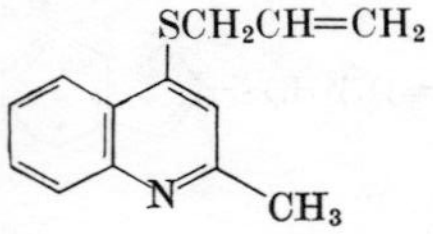 SCH$_2$CH=CH$_2$	200°, 1 hr, neat	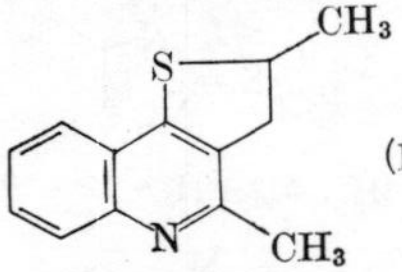 (15) 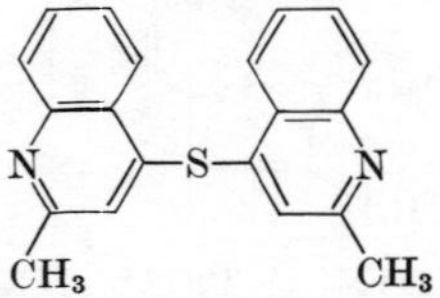 ,	73.5 (total)	271
	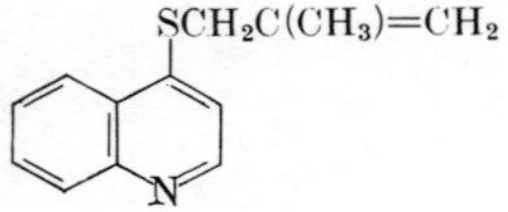 SCH$_2$C(CH$_3$)=CH$_2$	200°, 2 hr, neat or in quinoline	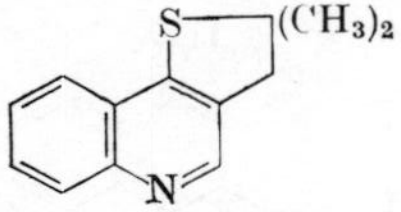 (1)	85–90	272
	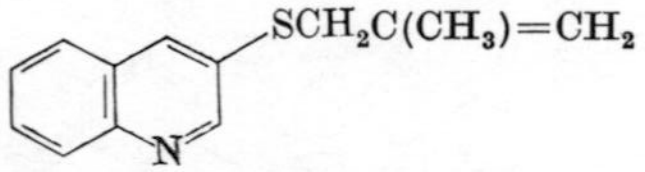 SCH$_2$C(CH$_3$)=CH$_2$	200°, 2 hr, (CH$_3$)$_2$NC$_6$H$_5$	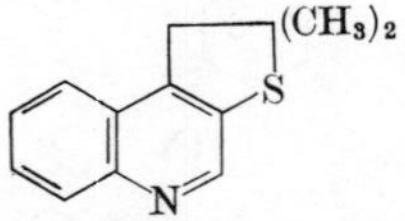 XIV (1), 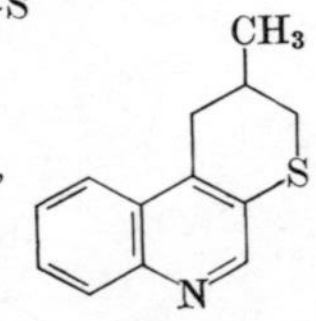	85 (total)	273
		200°, 1 hr, neat Ar atm	XIV (1), XV (2.5)	85	273

Molecular formula	Starting material	Conditions	Product	Yield	Ref.
$C_{13}H_{15}NS$	(indole with CH_3, N–CH_3, $SCH_2C\equiv CH$)	Heat, $(n\text{-}C_3H_7CO)_2O$	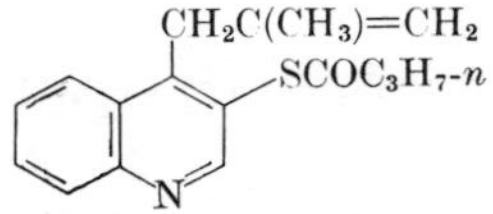 (quinoline with $CH_2C(CH_3)=CH_2$, $SCOC_3H_7\text{-}n$)	~Quant	273
	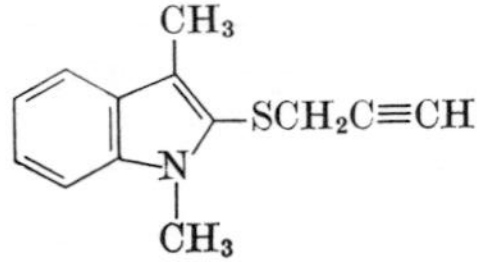 (indole with CH_3, N–CH_3, $SCH_2CH=CH_2$)	Reflux, $C_6H_5CH_3$	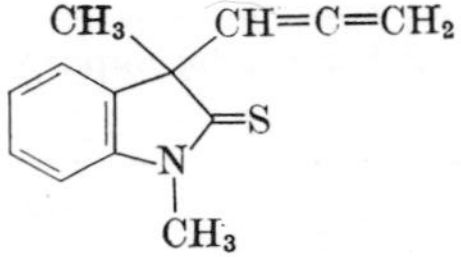 (CH_3, $CH=C=CH_2$, N–CH_3 thione)	Quant	34
	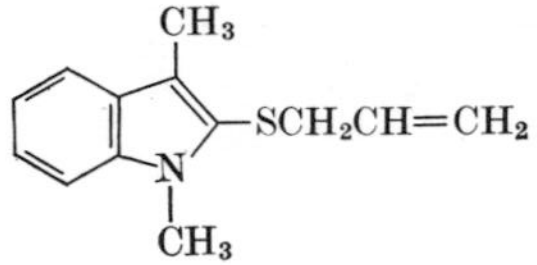 ($(CH_3)_2$ indole, $SCH_2CH=CH_2$)	Reflux, $C_6H_5CH_3$	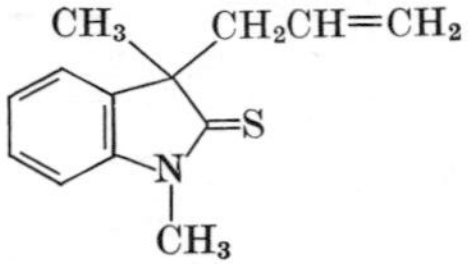 (CH_3, $CH_2CH=CH_2$, N–CH_3 thione)	Quant	34
	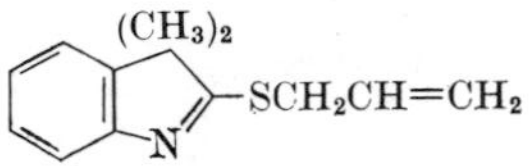	Reflux, tetralin	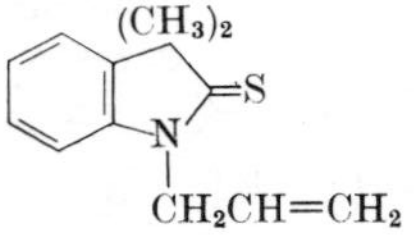 ($(CH_3)_2$, N–$CH_2CH=CH_2$ thione)	Quant	34
$C_{14}H_{17}NS$	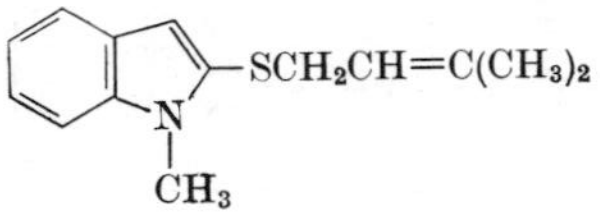 (indole, N–CH_3, $SCH_2CH=C(CH_3)_2$)	Standing	In equilibrium with 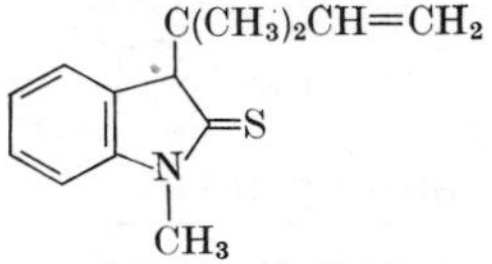 ($C(CH_3)_2CH=CH_2$, N–CH_3 thione)	—	34
$C_{15}H_{20}NS$	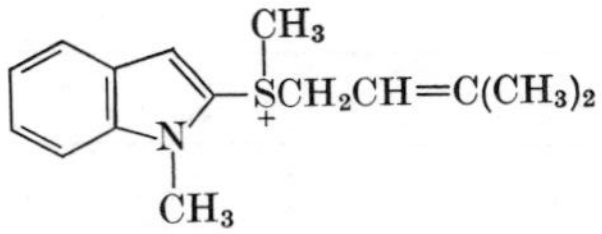 (indolium, N–CH_3, $\overset{+}{S}CH_2CH=C(CH_3)_2$, CH_3)	—	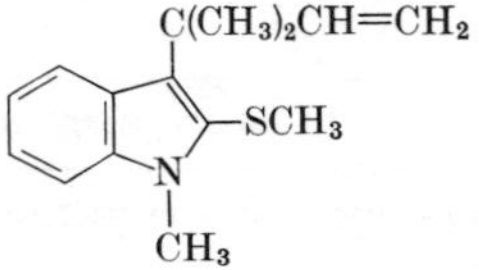 ($C(CH_3)_2CH=CH_2$, SCH_3, N–CH_3)	—	277

Note: References 344–439 are on pp. 251–252.

199

TABLE IV. THIO-CLAISEN REARRANGEMENTS (*Continued*)

A. *Aromatic and Heterocyclic Compounds* (*Continued*)

Molecular Formula	Sulfide or Sulfonium Ion	Conditions	Product(s) and Ratio ( )	Yield(s), %	Refs.
$C_{15}H_{20}NS$ (*Contd.*)	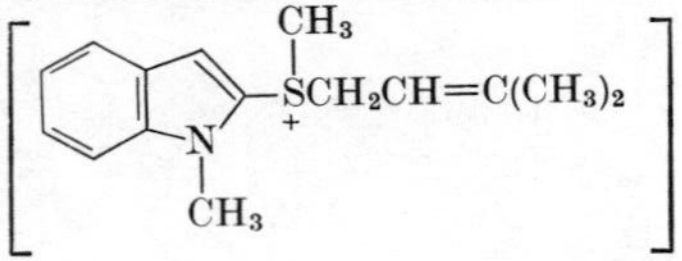	$HCON(CH_3)_2$, K_2CO_3	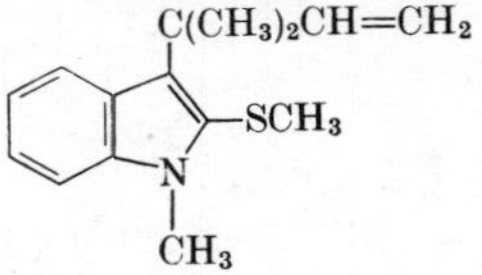	Good	277
	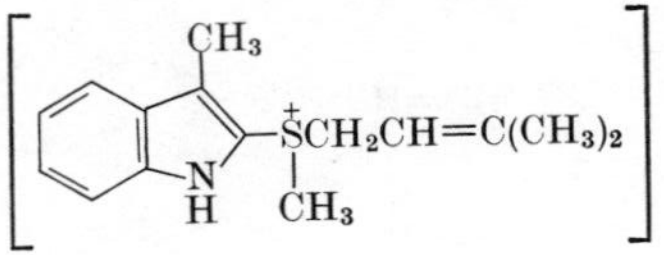	,,	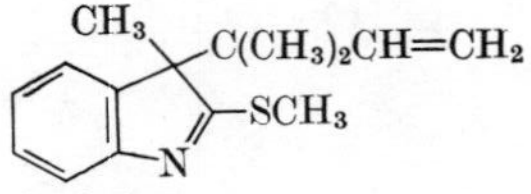 (major) 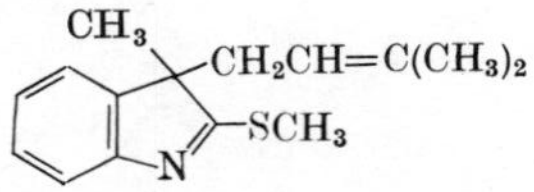	Good	277

B. *Aliphatic Compounds*

Molecular Formula	Sulfide	Conditions	Product(s) and Ratio ( )	Yield(s), %	Refs.
C_5H_6S	$HC{\equiv}CCH_2SCH{=}CH_2$	$115°$, $[(CH_3)_2N]_3PO$, pyridine, N_2 atm	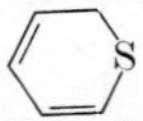	~80	278
$C_7H_{10}S_2$	$CH_2{=}C{-}CHSC(SC_2H_5){=}CH_2$	$80{-}100°$	$HC{\equiv}CCH_2CH_2CS_2C_2H_5$	40	281

Formula	Reactant	Conditions	Product	Yield (%)	Ref.
C7H12S2	[CH2=CHCH2SC(SC2H5)=CH2]	15°, <5 min	CH2=CHCH2CH2CS2C2H5	77	279
C8H12S2	CH3C=CCH2SC(SC2H5)=CH2	100–120°, 15 min	CH2=C(CH3)CH2CS2C2H5	60	280
C8H14OS	[CH2=CHCH2SC(OC2H5)=CHCH3]	15°, <5 min	CH2=CHCH2CH(CH3)CSOC2H5	74	279
C8H14S2	[CH2=C=C(C2H5)SC(SC2H5)=CHCH3]	15°, <5 min	CH2=CHCH2CH(CH3)CS2C2H5	91	279
C9H14S2	HC=CCH2SC(SC2H5)=C(CH3)2	80–100°	C2H5C=CCH2CH2CS2C2H5	36	281
	CH3C=CCH2SC(SC2H5)=CHCH3	100–120°, 15 min	CH2=C=CHC(CH3)2CS2C2H5	75	280
C9H16OS	(CH3)2C=CHCH2SC(OC2H5)=CH2	100–120°, 15 min	CH2=C=C(CH3)CH(CH3)CS2C2H5	80	280
	CH3CH=CHCH2SC(OC2H5)=CHCH3	15°, <5 min	CH2=CHC(CH3)2CH2CSOC2H5	65	279
C9H16S2	(CH3)2C=CHCH2SC(SC2H5)=CH2	15°, <5 min	CH2=CHCH(CH3)CH(CH3)CSOC2H5	58	279
	CH3CH=CHCH2SC(SC2H5)=CH2	15°, 5 min	CH2=CHC(CH3)2CH2CS2C2H5	45	279
		15°, 5 min	CH2=CHCH(CH3)CH2CS2C2H5	70	279
C9H17NS	[CH2=CHCH2SC=CHCH2CH3, N(CH3)2]	15°, 5 min	CH2=CHCH2CH(C2H5)CSN(CH3)2	89	279
C10H16N2S	[thiopyran ring; (C2H5)2N, CN, CH2=CHCH2–][a]	overnight, ether, NaOCH3	(C2H5)2N, CN, S-ring (allyl)	26	434
C10H16S	cyclohexylidene=CHSCH2CH=CH2	190°, 10 min, HgO, Ar atm	1-(CHO)cyclohexyl-CH2CH=CH2	82	282
C10H16S2	*CH3C=CCH2SC(SC2H5)=C(CH3)2	100–120°, 15 min	CH2=C(CH3)C(CH3)2CS2C2H5	74	280
	HC=CCH2SC(SC2H5)=CH(C3H7-i)	100–120°, 15 min	CH2=C=CHCH(C3H7-i)CS2C2H5	75	280
	CH3C=CCH2SC(SC2H5)=CHC2H5	100–120°, 15 min	CH2=C=C(C2H5)CH(C2H5)CS2C2H5	60	280
C10H18OS	(CH3)2C=CHCH2SC(OC2H5)=CHCH3	15°, 5 min	CH2=CHC(CH3)2CH(CH3)CSOC2H5	67	279
C10H18S2	[CH2=CHCH2SC(SC2H5)=CHC3H7-i]	15°, 5 min	CH2=CHCH2CH(C3H7-i)CS2C2H5	60	279
C11H12S	C6H5CH=CHSCH2CH=CH2	190°, 10 min, HgO, Ar atm	C6H5CHCHO, CH2CH=CH2	39	282

Note: References 344–439 are on pp. 251–252.

a This was obtained from the reaction of CH2=CHCH2SH and (C2H5)2NC=CCN.

TABLE IV. THIO-CLAISEN REARRANGEMENTS (Continued)

B. Aliphatic Compounds (Continued)

Molecular Formula	Sulfide	Conditions	Product(s) and Ratio ( )	Yield(s), %	Refs.
$C_{11}H_{16}S$	[bicyclic structure]$={=}CHSCH_2CH{=}CH_2$	190°, 10 min, HgO, Ar atm	[bicyclic structure with CHO, $CH_2CH{=}CH_2$]	83	282
$C_{11}H_{18}S_2$	$HC{\equiv}CCH_2SC(SC_2H_5){=}CH(C_4H_9\text{-}t)$	100–120°, 15 min	(3:2 mixture of epimers) $CH_2{=}C{=}CHCH(C_4H_9\text{-}t)CS_2C_2H_5$	75	280
	$CH_3C{\equiv}CCH_2SC(SC_2H_5){=}CH(C_3H_7\text{-}i)$	100–120°, 15 min	$CH_2{=}C{=}C(CH_3)CH(C_3H_7\text{-}i)CS_2C_2H_5$	71	280
$C_{11}H_{20}S_2$	$[CH_2{=}CHCH_2SC(SC_2H_5){=}CH(C_4H_9\text{-}n)]$	15°, 5 min	$CH_2{=}CHCH_2CH(C_4H_9\text{-}n)CS_2C_2H_5$	65	279
	$[CH_2{=}CHCH_2SC(SC_2H_5){=}CH(C_4H_9\text{-}t)]$	15°, 5 min	$CH_2{=}CHCH_2CH(C_4H_9\text{-}t)CS_2C_2H_5$	72	279
$C_{12}H_{20}S$	[cyclooctane structure]$={=}CHSCH_2CH{=}CH_2$	190°, 2 hr, HgO, Ar atm	[cyclooctane structure with CHO, $CH_2CH{=}CH_2$]	72	282
$C_{12}H_{20}S_2$	$CH_3C{\equiv}CCH_2SC(SC_2H_5){=}CH(C_4H_9\text{-}t)$	100–120°, 15 min	$CH_2{=}C{=}C(CH_3)CH(C_4H_9\text{-}t)CS_2C_2H_5$	60	280
$C_{12}H_{23}NS$	$CH_3CH{=}CHCH_2SC{=}CH(C_2H_5)$, $-N(C_2H_5)_2$	15°, 5 min	$CH_2{=}CHCH(CH_3)CHCSN(C_2H_5)_2$, $-C_2H_5$	65	279
$C_{13}H_{25}NS$	$[CH_2{=}CHCH_2SC{=}CH(C_4H_9\text{-}n)$, $-N(C_2H_5)_2$ / $N(C_2H_5)_2]$	15°, 5 min	$CH_2{=}CHCH_2CH(C_4H_9\text{-}n)CSN(C_2H_5)_2$	83	279

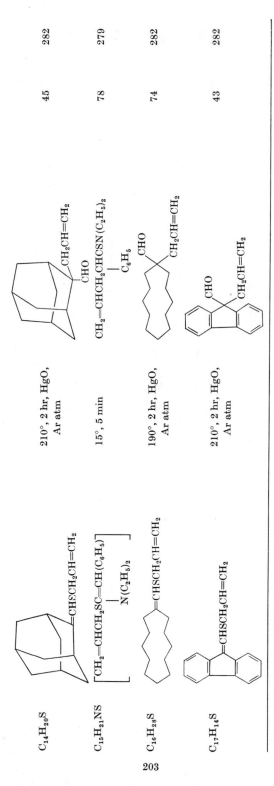

	Reactant	Conditions	Product	Yield (%)	Reference
$C_{14}H_{20}S$	(adamantylidene)=CHSCH$_2$CH=CH$_2$	210°, 2 hr, HgO, Ar atm	(adamantane)—CHO, CH$_2$CH=CH$_2$	45	282
$C_{15}H_{21}NS$	$[CH_2=CHCH_2SC=CH(C_6H_5)]$ N$(C_2H_5)_2$	15°, 5 min	CH$_2$=CHCH$_2$CHCH$_2$CHCSN$(C_2H_5)_2$, C$_6$H$_5$, CHO	78	279
$C_{16}H_{28}S$	(cyclododecylidene)=CHSCH$_2$CH=CH$_2$	190°, 2 hr, HgO, Ar atm	(cyclododecane)—CHO, CH$_2$CH=CH$_2$	74	282
$C_{17}H_{14}S$	(fluorenylidene)=CHSCH$_2$CH=CH$_2$	210°, 2 hr, HgO, Ar atm	(fluorene)—CHO, CH$_2$CH=CH$_2$	43	282

Note: References 344–439 are on pp. 251–252.

TABLE V. COPE REARRANGEMENTS

A. 1,5-Hexadiene Systems—Acyclic and Cyclic

Molecular Formula	Starting Material	Conditions	Product(s) and Ratio ( )	Yield(s), %	Refs.
C_6H_8		80°, 13 hr, $n\text{-}C_5H_{12}$		90	425
C_6H_8O	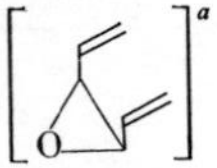 [a] (93% pure)		 I	—	178
		Heat (<100°), liquid phase	I	Quant.	296, 295
	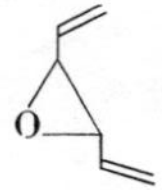	230°, 17 hr, sealed tube	I	43	295
		170–200°	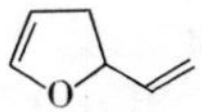 (2.3) I (1)	Quant.	300
$C_6H_8O_2S$	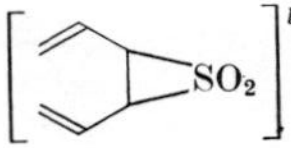 [b]	(Spontaneously)	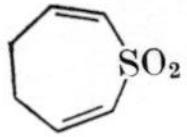	—	299
		−20°; then 3 hr, r.t.		29	437
C_7H_{10}	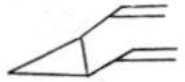 [c]	−20–35°	 (1.5)	67[b]	67[b]
	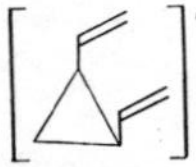	80°, Ag_2O	  (1)	75 (total)	62

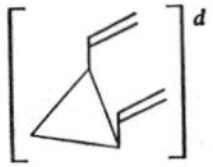

C_7H_{12} [d]	170–180°	(cycloheptadiene)	46	62
(CH₃ diene)	210–260°, gas phase	At equilibrium: (5.6/1 with starting material)	—	39
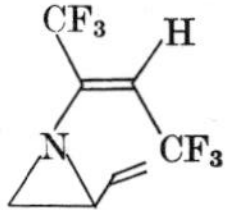 $C_7H_7NF_6$	Standing overnight	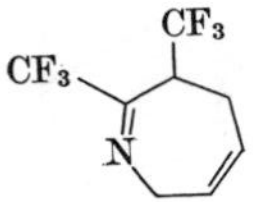 (CF₃ azepine)	—	298
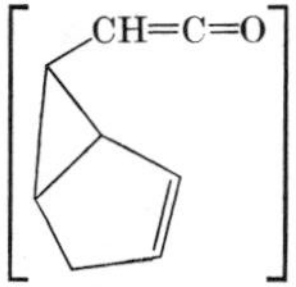 C_8H_8O [e]	THF	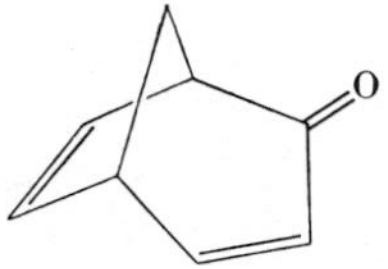 (major)	~50	290

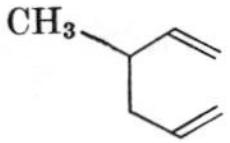

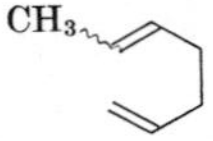

Note: References 344–439 are on pp. 251–252.

[a] This starting material, which was not isolated, was obtained from (structure) and lithium chloride at 200–210°.

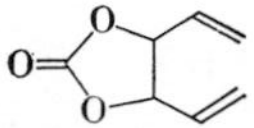

[b] This material was obtained from vinyl diazomethane and sulfur dioxide in ether solution.

[c] The compound was prepared from the *cis* quaternary ammonium salt and silver oxide at 80°.

[d] This material was obtained from the *cis* amine oxide.

[e] This compound was prepared by irradiation of the product from the reaction of the acid chloride and diazomethane in ether.

TABLE V. COPE REARRANGEMENTS (Continued)

A. 1,5-Hexadiene Systems—Acyclic and Cyclic (Continued)

Molecular Formula	Starting Material	Conditions	Product(s) and Ratio ( )	Yield(s), %	Refs.
$C_8H_8O_2$	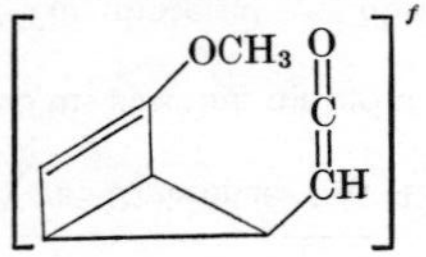	$< -70°$	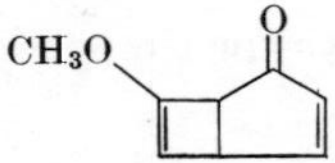 (1)	—	291
	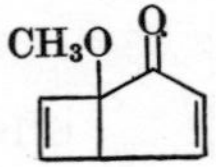	189°, 60 min, CHCl₃, sealed tube	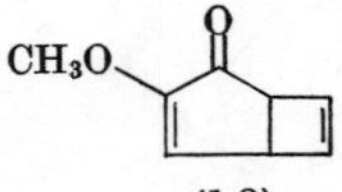 (1.8) 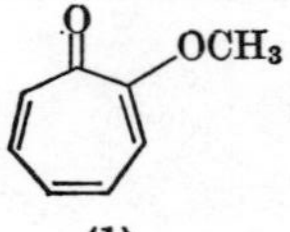	87.2 (total)	81
C_8H_{10}	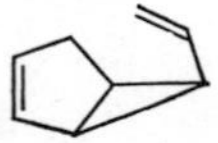	33°, 36 hr	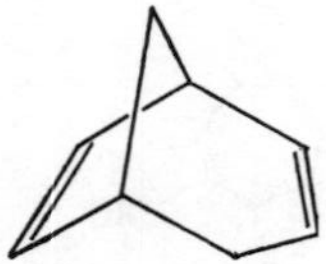	Quant.	64
C_8H_{12}	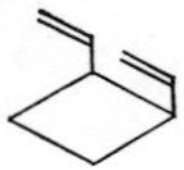	120°, 10 min, neat	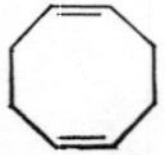 (II)	91	68, 38
$C_8H_{12}O$	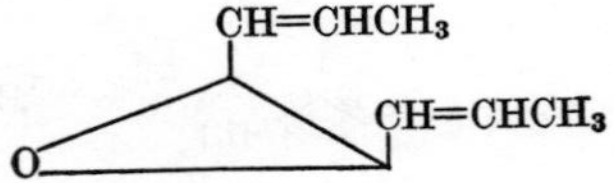	100°, liquid phase	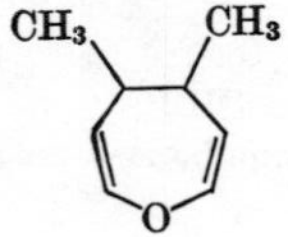 —	—	296
	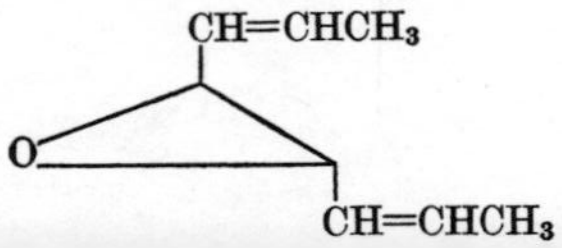	350°, gas phase	II, 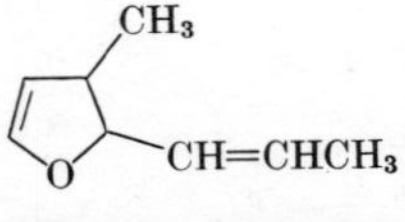	—	296

206

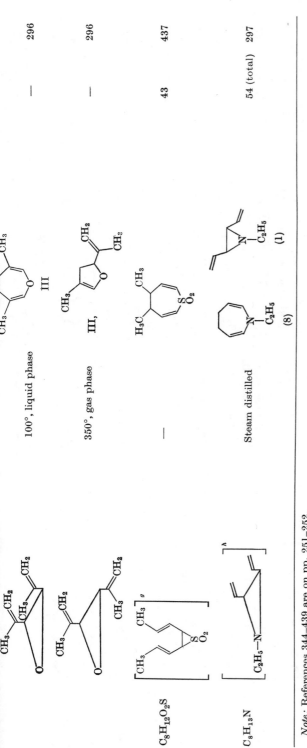

100°, liquid phase	III	—	296
350°, gas phase	III,	—	296
—		43	437
Steam distilled	(8) (1)	54 (total)	297

Note: References 344–439 are on pp. 251–252.

f This is the intermediate in the irradiation of at −190°.

g This was obtained from 1-diazobut-2-ene and gaseous SO_2.

h This was prepared from , chlorosulfonic acid, and sodium hydroxide.

$C_8H_{12}O_2S$

$C_8H_{13}N$

TABLE V. COPE REARRANGEMENTS (Continued)

A. 1,5-Hexadiene Systems—Acyclic and Cyclic (Continued)

Molecular Formula	Starting Material	Conditions	Product(s) and Ratio ( )	Yield(s), %	Refs.
C_8H_{14}		360°, gas phase	At equilibrium:	—	288
	(meso)	280°, 24 hr, neat, sealed tube, H_2 atm	(49/1 with starting material) CH_3 IV, (332) CH_3	97	57
	(racemic)	180°, 18 hr, neat, sealed tube, H_2 atm	CH_3 V (1) V (9), IV (<1), CH_3 (1) CH_3		57
C_9H_{12}		60°, 1 hr	At equilibrium:	—	292
		220°		—	301
C_9H_{14}		130°, gas phase	(19/1 with starting material)	—	301

$C_{10}H_6OCl_4$	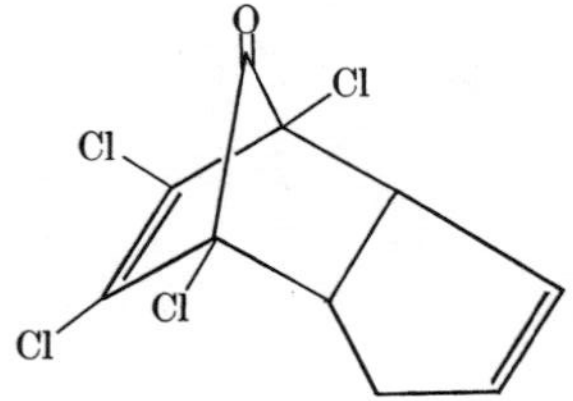	135°, 1 hr	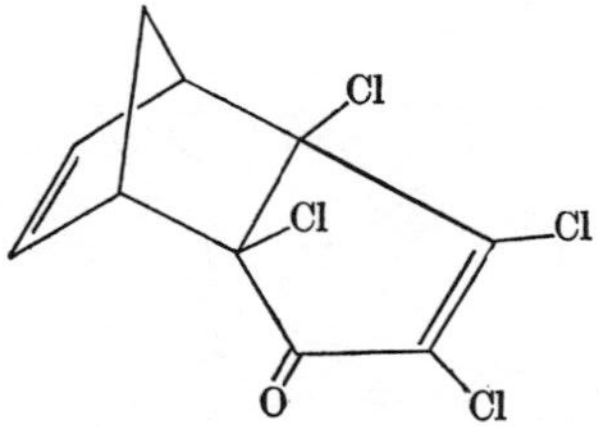 Quant	75
$C_{10}H_9NF_6$	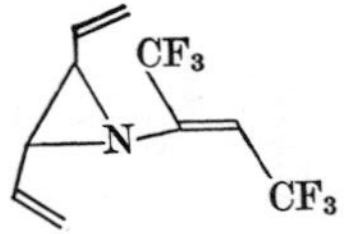	50°, 1–2 hr	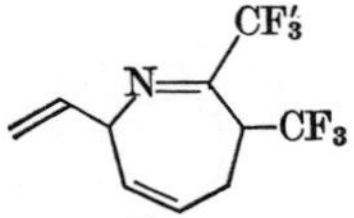 —	298
$C_{10}H_{10}$	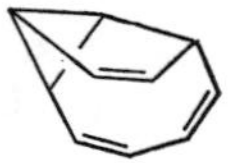	30°, $CDCl_3$	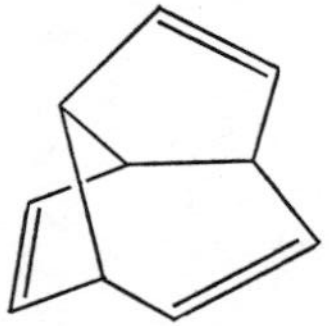 Quant	293
		20–35°, CCl_4	Quant	294
	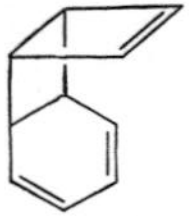	70°, in dilute solution	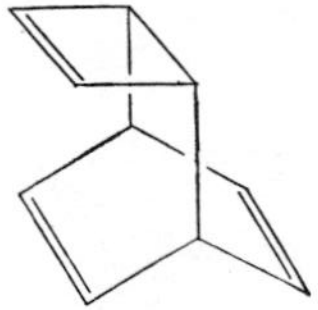 —	438

Note: References 344–439 are on pp. 251–252.

TABLE V. COPE REARRANGEMENTS (Continued)

A. 1,5-Hexadiene Systems—Acyclic and Cyclic (Continued)

Molecular Formula	Starting Material	Conditions	Product(s) and Ratio ( )	Yield(s), %	Refs.
$C_{10}H_{10}O$	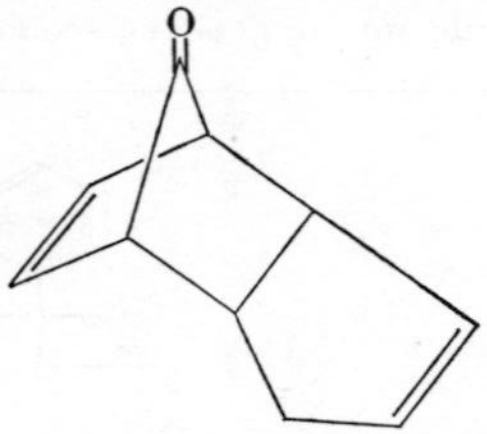	20°, C_2H_5OH	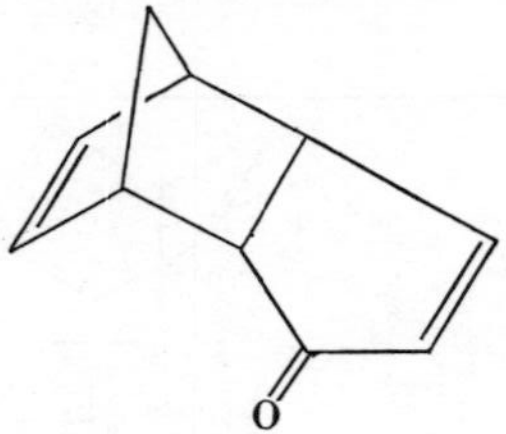	—	320
	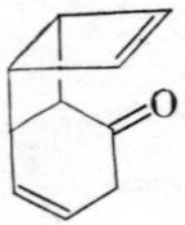	70°	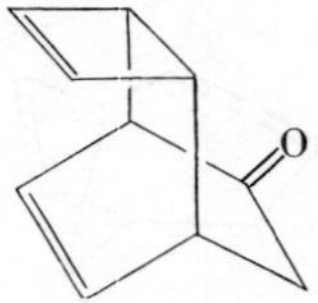	—	438
$C_{10}H_{10}O_2S$	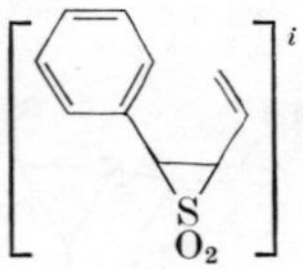	—	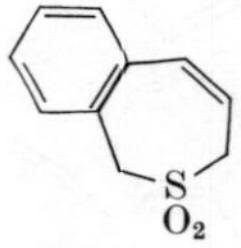	—	437
$C_{10}H_{12}O$	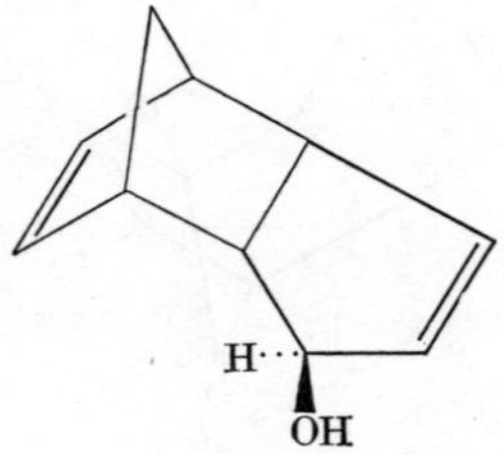	125–150°, decalin	At equilibrium: 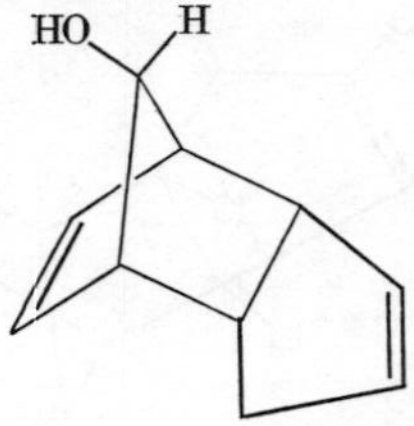	(2/1 with starting material)	48
		140°, 6 hr, neat		(1/1 with starting material)	73

210

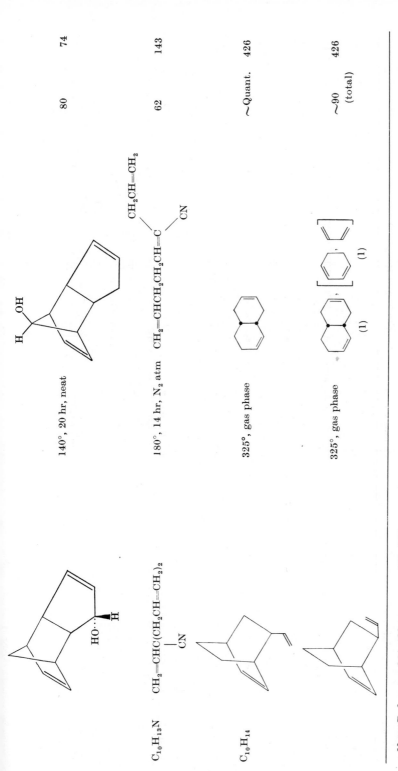

| | | 80 | 74 |

140°, 20 hr, neat

C$_{10}$H$_{13}$N   CH$_2$=CHC(CH$_2$CH=CH$_2$)$_2$ CH$_2$=CHCH$_2$CH$_2$CH=C⟨CH$_2$CH=CH$_2$ / CN⟩

| | | 62 | 143 |

180°, 14 hr, N$_2$ atm

C$_{10}$H$_{14}$

325°, gas phase   ~Quant.   426

325°, gas phase   ~90 (total)   426

(1)   (1)

Note: References 344–439 are on pp. 251–252.

^{i} This was obtained from vinyldiazomethane and C$_6$H$_5$CH=SO$_2$.

TABLE V. COPE REARRANGEMENTS (Continued)

A. 1,5-Hexadiene Systems—Acyclic and Cyclic (Continued)

Molecular Formula	Starting Material	Conditions	Product(s) and Ratio ( )	Yield(s), %	Refs.
$C_{10}H_{14}$ (contd.)	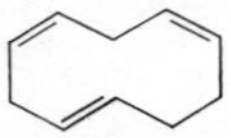	Heat	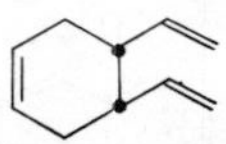	—	302
$C_{10}H_{15}N$	$CH_3CH=C(C_2H_5)CHCNCH_2CH=CH_2$	195°, 11 hr, N_2 atm	$CH_2=CHCH_2CH(CH_3)C(C_2H_5)=CHCN$	70	143
$C_{10}H_{16}$	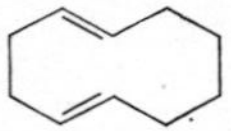	70°, 3 days	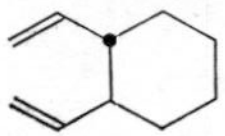	Quant.	303
	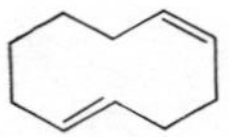	150°, 5.25 hr	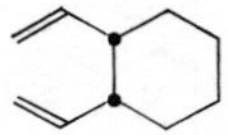	99	302
	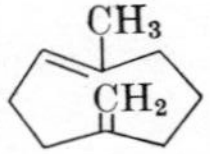	Heat	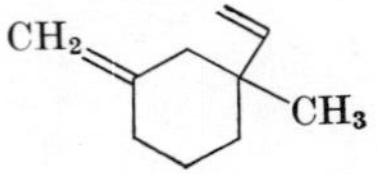	—	427
$C_{10}H_{16}N_2$	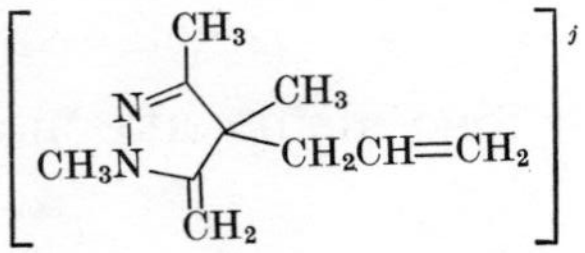	Reflux, 3.5 hr, C_2H_5OH, aq CH_3CO_2H	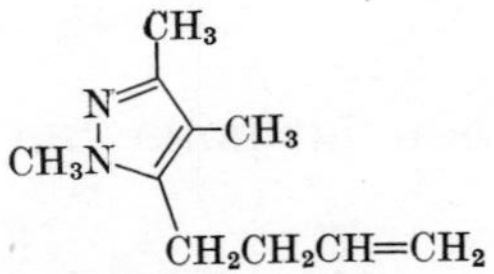 At equilibrium:	71	333
$C_{10}H_{16}O$	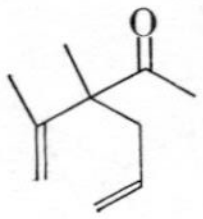	230°, 30 min	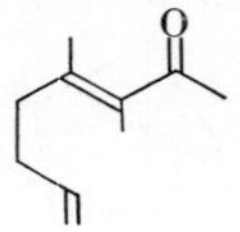 (1.3), 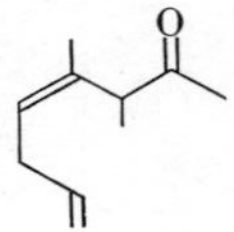 (1), 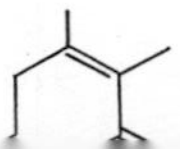 (1.2)	—	146

Formula	Conditions	Product(s)	Yield	Ref.
$C_{11}H_9O_2Cl$	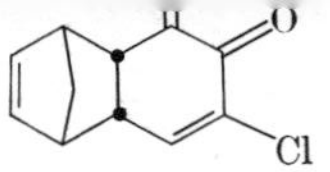 Reflux, C_6H_6	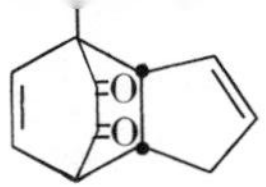 VI (R = Cl)	—	321
$C_{11}H_{10}O_2$	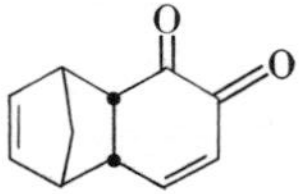 Reflux, C_6H_6	VI (R = H)	—	321
$C_{11}H_{14}N_2$	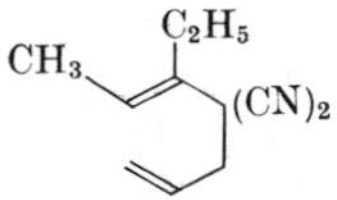 150°, 4 hr	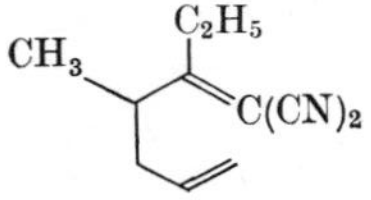	Quant.	138
$C_{11}H_{14}O$	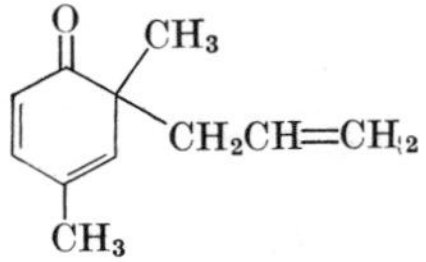 105°, 2 hr, neat	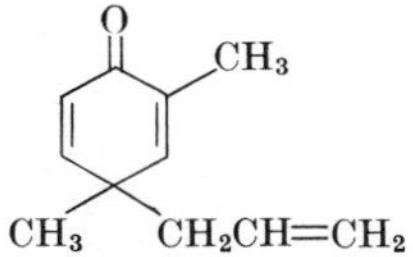	68	166
$C_{11}H_{14}O_2$	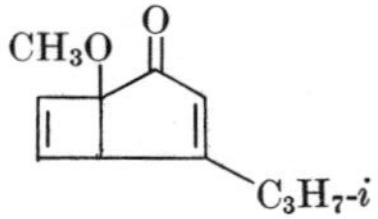 191°, 60 min, CHCl$_3$, sealed tube	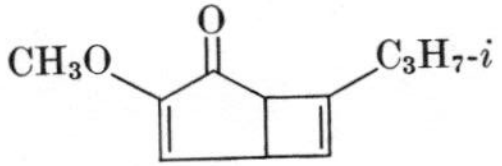 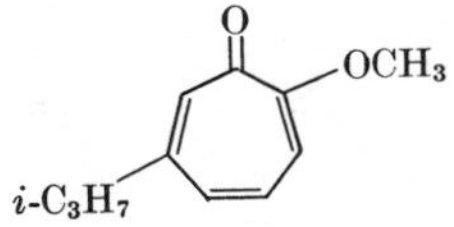 $C_3H_{7}\text{-}i$ (4.5),	83(total)	81
$C_{11}H_{15}N$	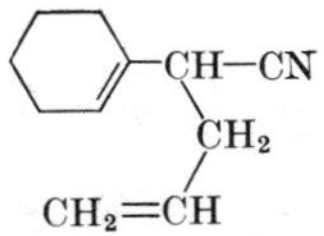 185°, 12 hr, N_2 atm	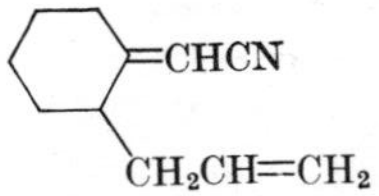 (1)	47	143

Note: References 344–439 are on pp. 251–252.

ʲ This material was generated *in situ* from the 3,3-disubstituted 2,4-pentanedione and a hydrazine.

213

TABLE V. COPE REARRANGEMENTS (Continued)

A. 1,5-Hexadiene Systems—Acyclic and Cyclic (Continued)

Molecular Formula	Starting Material	Conditions	Product(s) and Ratio ( )	Yield(s), %	Refs.
$C_{11}H_{16}$	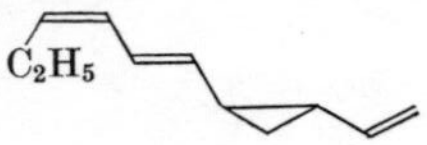	Heat	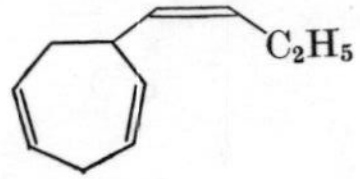	—	428
$C_{11}H_{16}N_2$	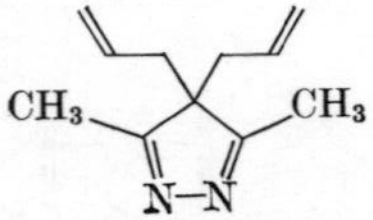	Reflux, 5.5 hr, xylene	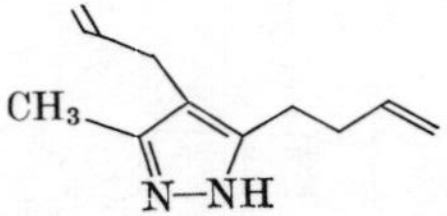	88	287
$C_{11}H_{16}O$	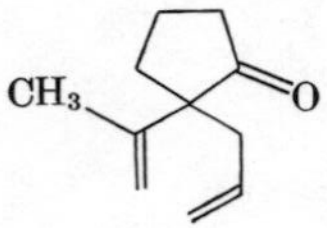	230°, 30 min	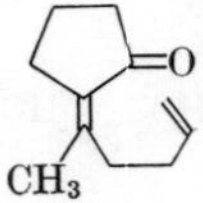 (1), 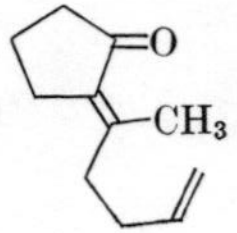 (1.4)	Quant.	148
	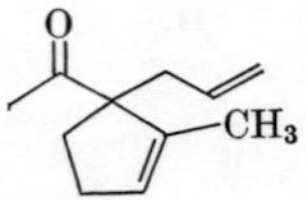	240°, 40 min	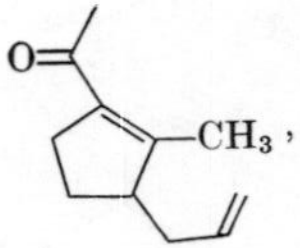 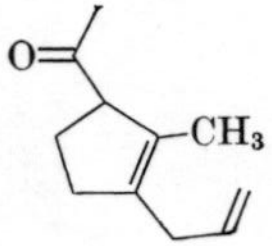	Quant.	149
$C_{11}H_{18}$	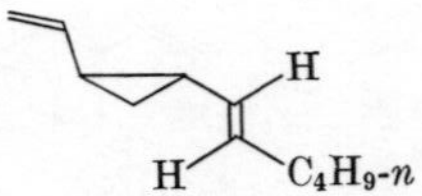	175°, 3 hr, sealed tube	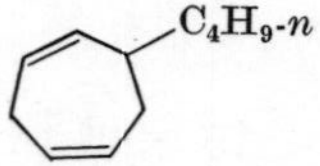 VII	—	65
	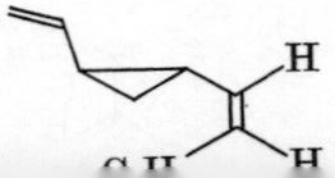	175°, 3 hr, sealed tube	VII	—	65

Reactant	Conditions	Product	Yield (%)	Ref.
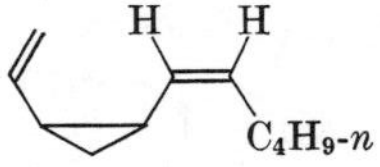 $C_{11}H_{18}N_2$ [k]	$75°$, 5 hr, sealed tube	VII	Quant.	65
		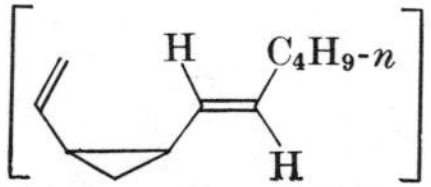 VII	Quant.	65
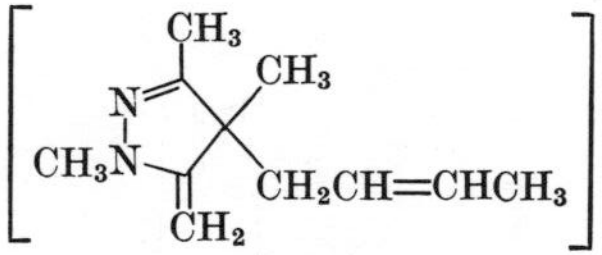 [j]	Reflux, 3.5 hr, $n\text{-}C_4H_9OH$, aq CH_3CO_2H	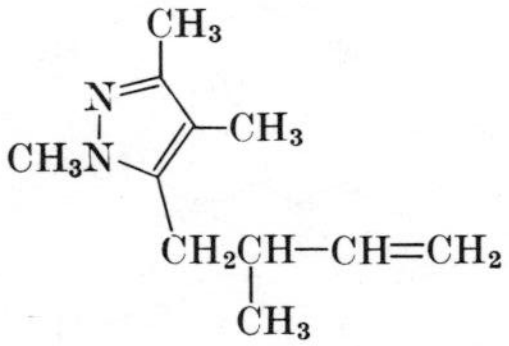	82	333
$C_{12}H_{12}O_2$	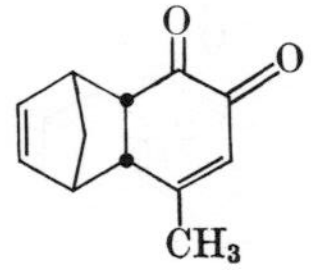 Reflux, C_6H_6	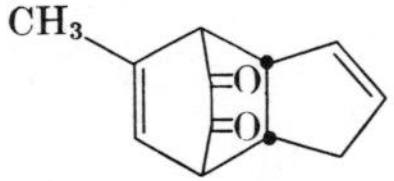	—	321
$C_{12}H_{12}O_3$	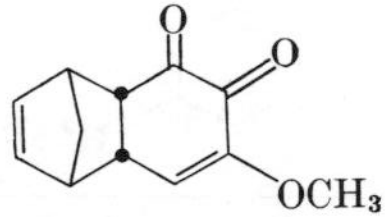 Reflux, C_6H_6	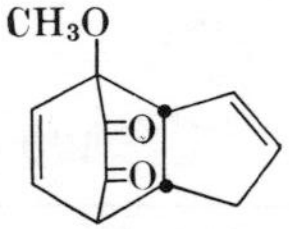	—	321

Note: References 344–439 are on pp. 251–252.

[j] This material was generated *in situ* from the 3,3-disubstituted 2,4-pentanedione and a hydrazine.

[k] The compound was not isolated; rearrangement occurs during preparation and workup.

215

TABLE V. COPE REARRANGEMENTS (*Continued*)

A. 1,5-Hexadiene Systems—Acyclic and Cyclic (*Continued*)

Molecular Formula	Starting Material	Conditions	Product(s) and Ratio ( )	Yield(s), %	Refs.
$C_{12}H_{14}$	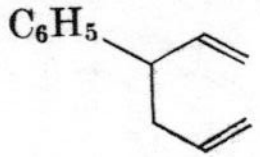 C_6H_5	176–178°, 26 hr, N_2 atm	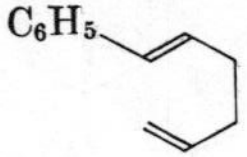 C_6H_5	72	155
$C_{12}H_{14}N_2$	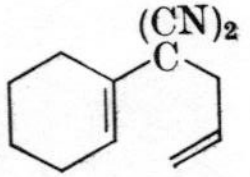 $C(CN)_2$	175°, 1.5 hr	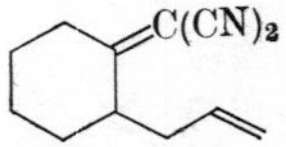 $C(CN)_2$	96	138
$C_{12}H_{16}O$	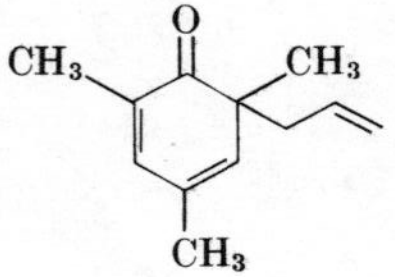 CH_3, CH_3, CH_3	108°, 2.5 hr, neat	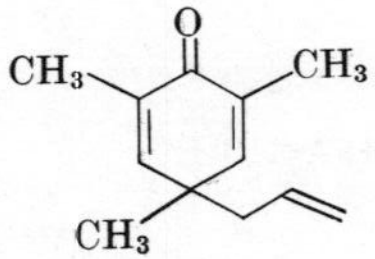 CH_3, CH_3, CH_3	71	166
	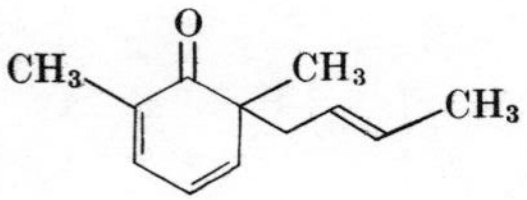 CH_3, CH_3, CH_3	105°, 1 hr, neat	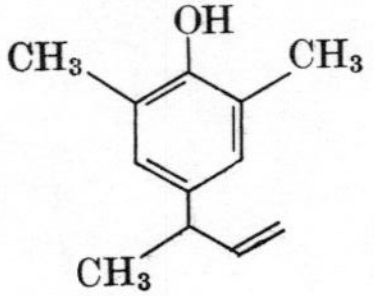 CH_3, CH_3, CH_3	83	166
$C_{12}H_{17}NO_2$	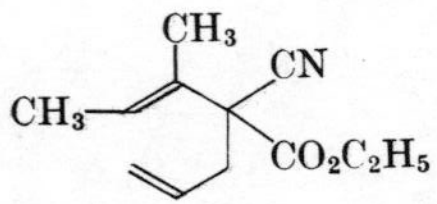 CH_3, CN, $CO_2C_2H_5$, CH_3	260°, 20 min	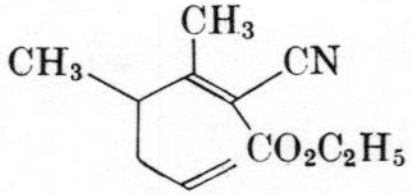 CH_3, CN, $CO_2C_2H_5$, CH_3	67	2

216

$C_{12}H_{18}$	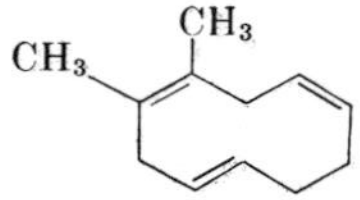 Heat	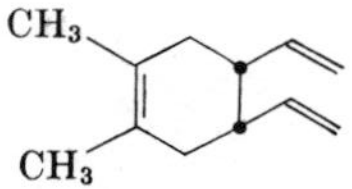	—	302
$C_{12}H_{18}O$	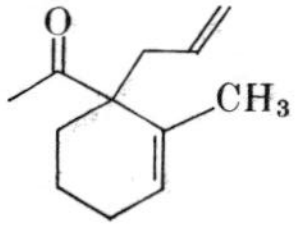 200°, 30 min, sealed tube	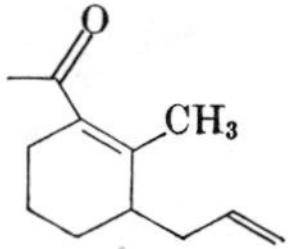	Quant	149
$C_{13}H_{12}O_5$	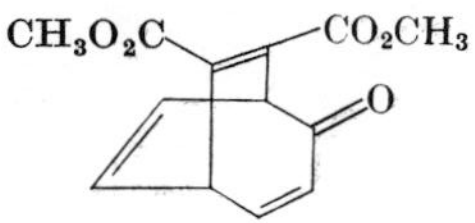 150°, xylene, CH_3OH	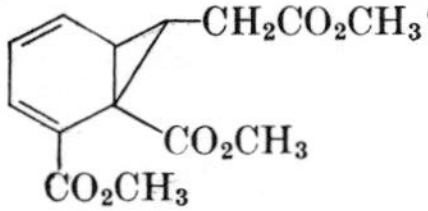	—	322
$C_{13}H_{14}O_2$	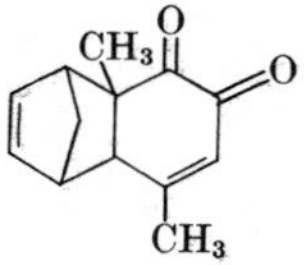 Reflux, C_6H_6	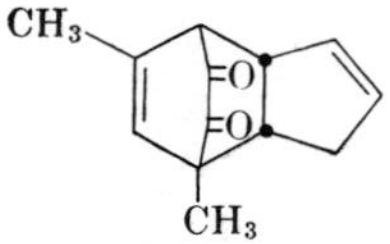	—	321
$C_{13}H_{14}O_2$	Reflux, C_6H_6		—	321

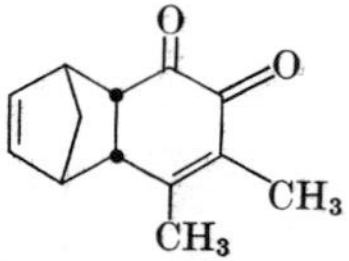

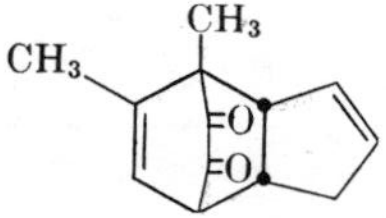

Note: References 344–439 are on pp. 251–252.

[l] The initially formed Cope product, the ketene, is trapped as the ester by reaction with methanol.

217

TABLE V. Cope Rearrangements (Continued)

A. 1,5-Hexadiene Systems—Acyclic and Cyclic (Continued)

Molecular Formula	Starting Material	Conditions	Product(s) and Ratio ( )	Yield(s), %	Refs.
$C_{13}H_{15}N$	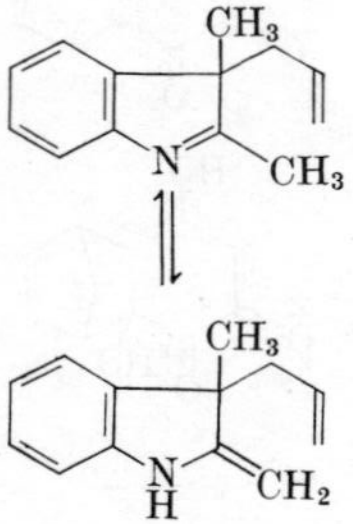	Reflux, 6.25 hr, tetralin	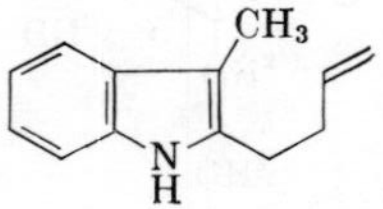	89	287
$C_{13}H_{16}$	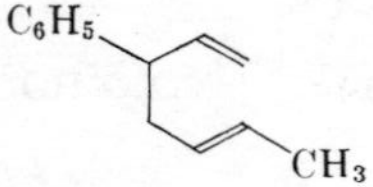	165–185°, 65 hr, N₂ atm	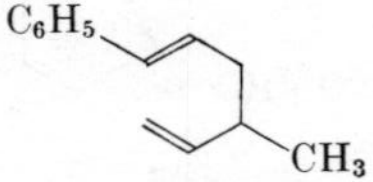	72	155
	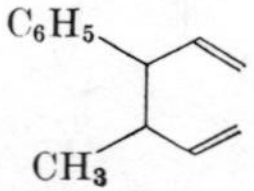	170–185°, 31 hr, N₂ atm	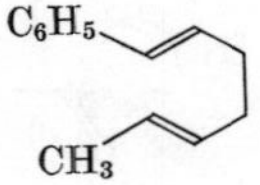	90	155
$C_{13}H_{18}O$	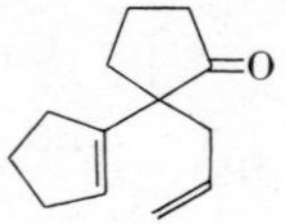	300°, 60 min	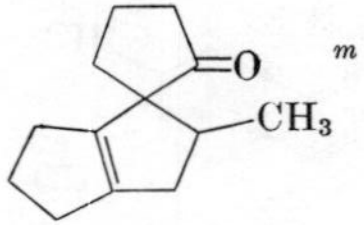	30	148

	Conditions		Yield (%)	Refs.
$C_{13}H_{19}NO_2$	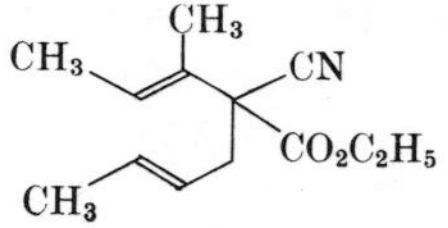 $180°$, 5.5 hr	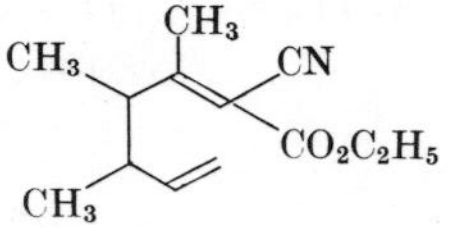	~80	139
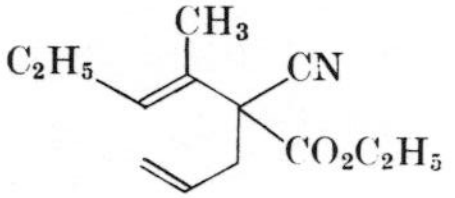	$170°$, 8 hr	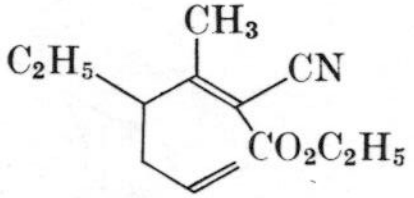	82	138
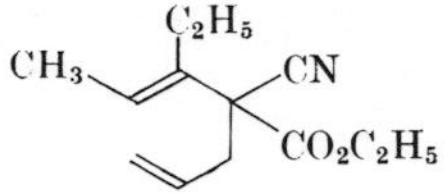	$200°$, 7 hr	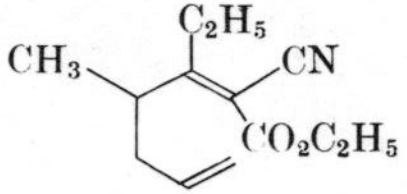	70	138
$C_{13}H_{20}O$	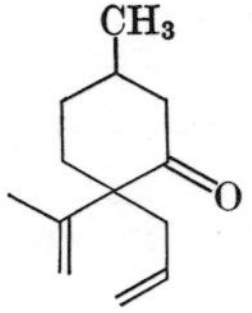 $220°$, 20 min, sealed tube	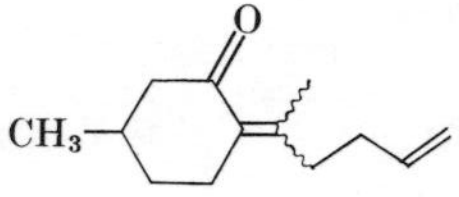 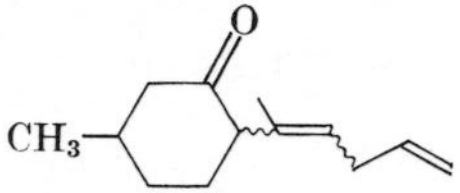	—	147
2R,5R + 2S,5R				

Note: References 344–439 are on pp. 251–252.

[m] Glpc evidence for the intermediate Cope products, ⟨structure⟩, is available. See pp. 52,–53.

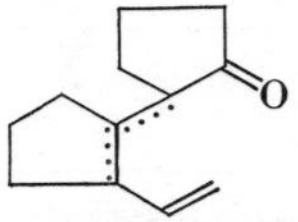

[n] Stereoisomers are formed from (+)-pulegone, allyl bromide, and sodium *t*-pentoxide.

[o] Stereoisomeric products are readily converted to spirans.

TABLE V. Cope Rearrangements (Continued)

A. 1,5-Hexadiene Systems—Acyclic and Cyclic (Continued)

Molecular Formula	Starting Material	Conditions	Product(s) and Ratio ( )	Yield(s), %	Refs.
$C_{13}H_{20}O_4$	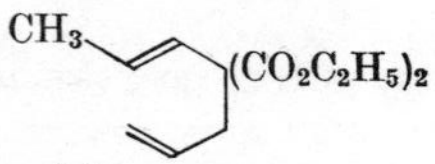 CH_3— ...(CO$_2$C$_2$H$_5$)$_2$	200°, 8 hr	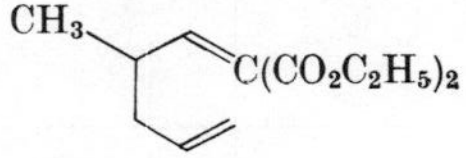 CH_3— ...C(CO$_2$C$_2$H$_5$)$_2$	90	138
$C_{14}H_{12}O_2$	CH_3 ...(CO$_2$C$_2$H$_5$)$_2$	150°, 48 hr, N$_2$ atm	CH_3 ...C(CO$_2$C$_2$H$_5$)$_2$	—	85b
$C_{14}H_{16}O_2$	CH_3O ...C_6H_5	189°, 150 min, CH$_3$OH, sealed tube	CH_3O ...C_6H_5 (5), OCH_3 ...C_6H_5 ...C_6H_5 (1)	86.5 (total)	81
$C_{14}H_{16}O_2$	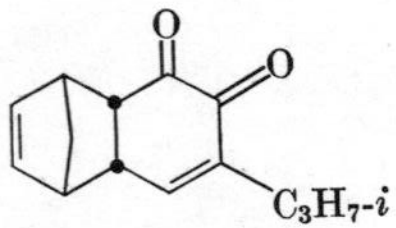 ...C_3H_{7}-i	Reflux, C$_6$H$_6$	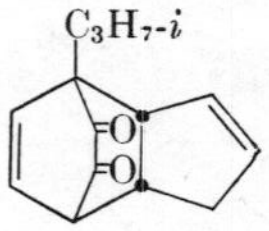 ...C_3H_{7}-i	—	321
$C_{14}H_{18}$	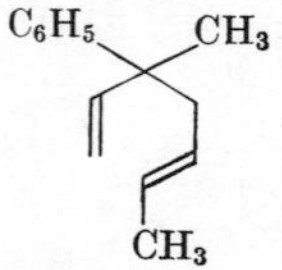 C_6H_5 ...CH_3 ...CH_3 $R(+)$ 95% optical purity	250°	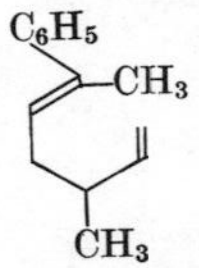 CH_3 ...C_6H_5 ...CH_3 $R(+)$ 89% opt purity (1); C_6H_5 ...CH_3 ...CH_3 $S(+)$ 91% opt purity (6.7)	Quant	52

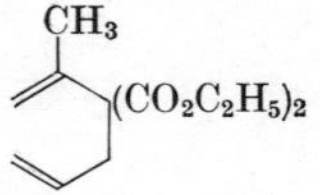

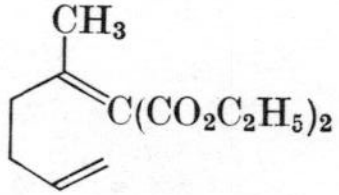

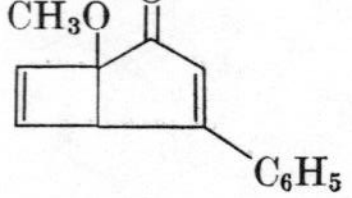

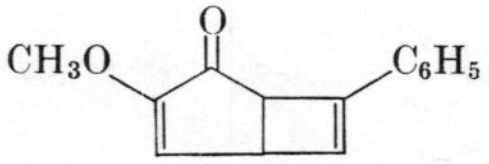

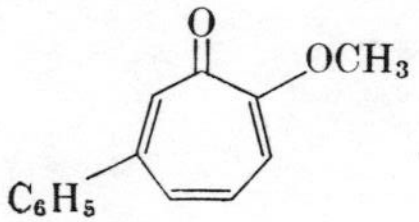

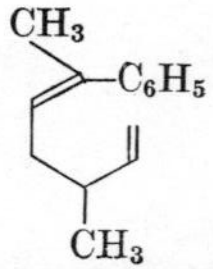

Molecular formula	Reactant	Conditions	Product	Yield (%)	Ref.
$C_{14}H_{19}N$	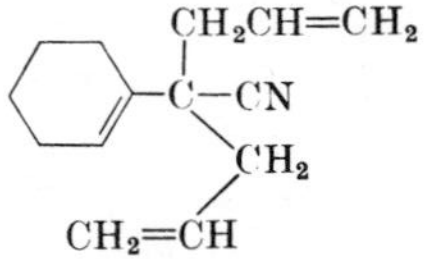 (1-cyclohexenyl, $CH_2CH=CH_2$, $CH_2CH=CH_2$, CN)	175°, 11 hr	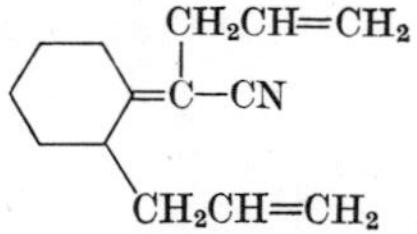	78	143
$C_{14}H_{19}NO_2$	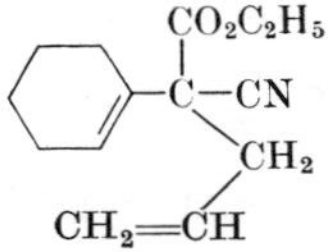 ($CO_2C_2H_5$, CN, $CH_2CH=CH_2$, 1-cyclohexenyl)	170°, 10 hr	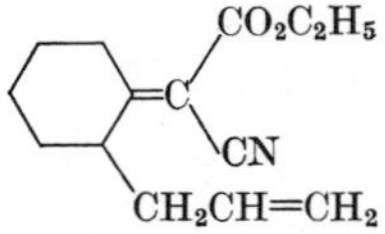	88	138
$C_{14}H_{20}O$	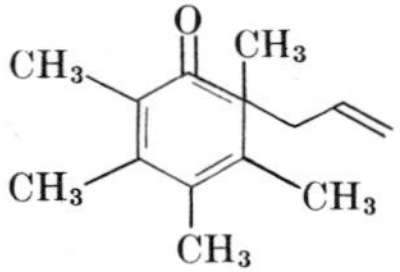 (allyl, CH_3 substituted cyclohexadienone) (1)	110°, 2.5 hr, neat	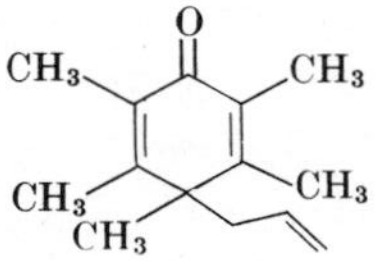 (2) 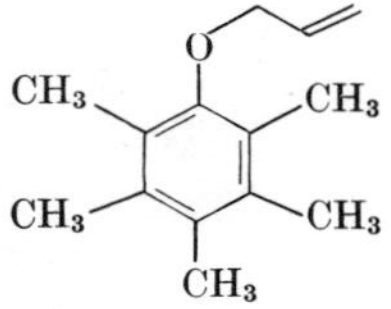	—	165
$C_{14}H_{21}NO_2$	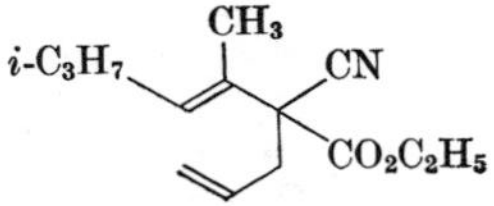 (CH_3, CN, $i\text{-}C_3H_7$, $CO_2C_2H_5$)	170°, 9 hr	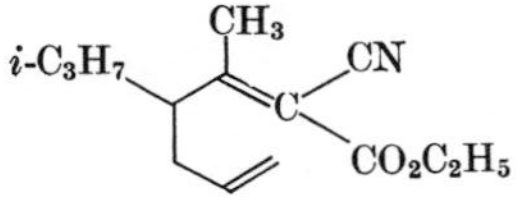	82	138
$C_{14}H_{22}O$	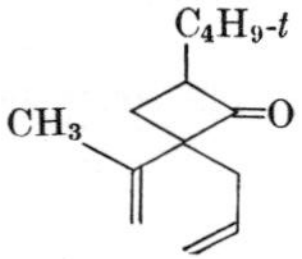 ($C_4H_9\text{-}t$, cyclobutanone, CH_3, allyl)	220°, 30 min	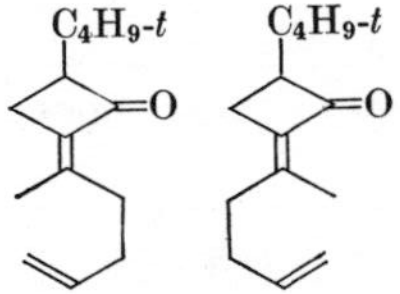 ($C_4H_9\text{-}t$ cyclobutanone derivatives)	Quant	148

Note: References 344–439 are on pp. 251–252.

221

TABLE V. Cope Rearrangements (Continued)

A. 1,5-Hexadiene Systems—Acyclic and Cyclic (Continued)

Molecular Formula	Starting Material	Conditions	Product(s) and Ratio ( )	Yield(s), %	Refs.
$C_{14}H_{22}O_4$	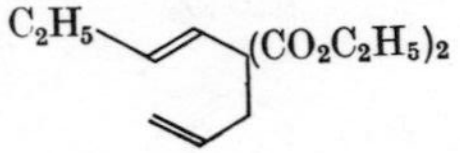	200°, 10 hr	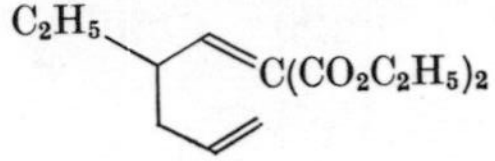	68	138
	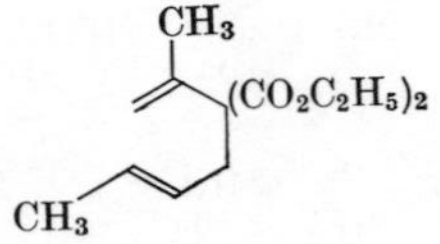	185°, 2 hr, reduced pres.	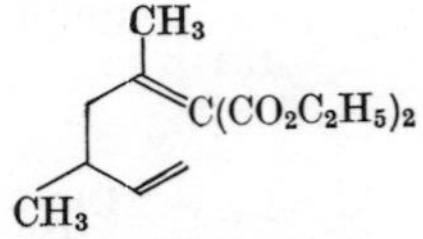	Quant	139
$C_{15}H_{16}O_3$	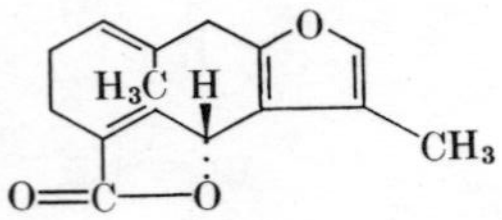	160–170°, 15 min	At equilibrium: 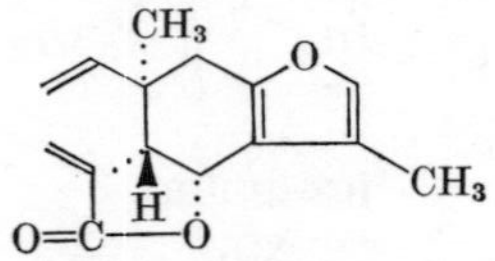 (1/1 with starting material)	—	317, 315
	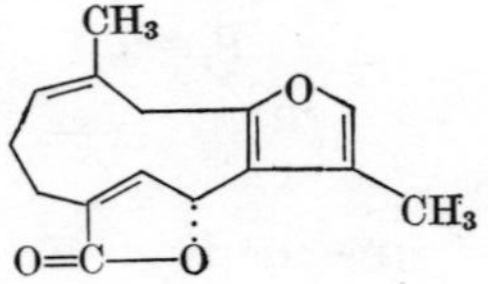	300°, 1–2 min	At equilibrium: 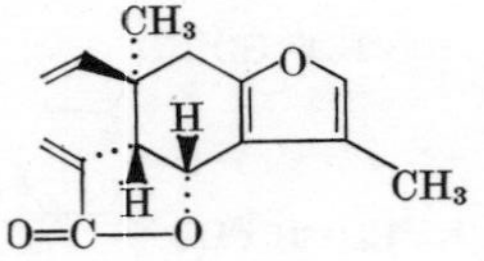 	3^p	312, 313

222

$C_{15}H_{18}O_2$	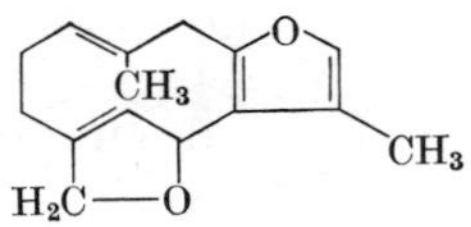 150°, 4 hr p-cymene	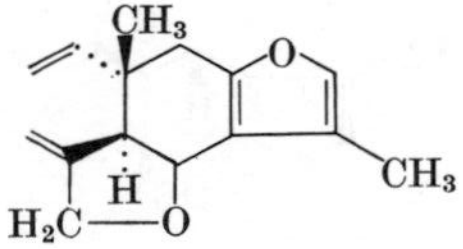 ~76	310
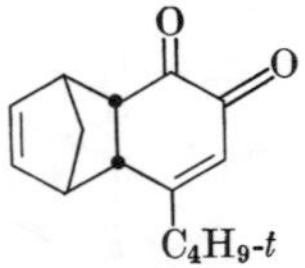	Reflux, C_6H_6	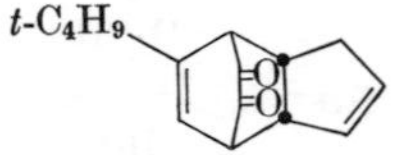 —	321
$C_{15}H_{19}N$	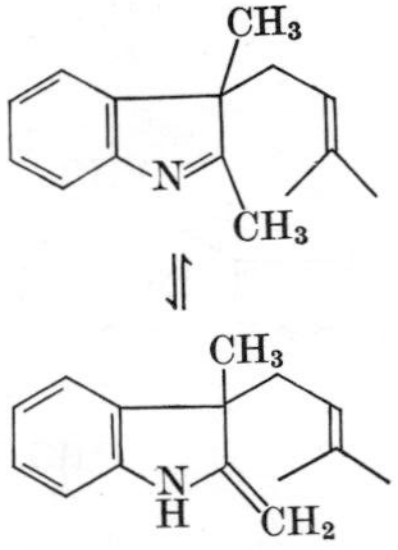 Reflux, 3.5 hr, tetralin	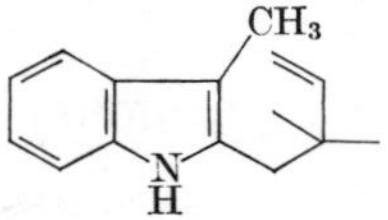 —	287
$C_{15}H_{20}O$	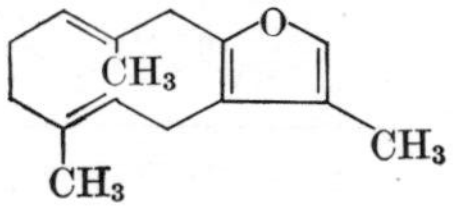 Heat	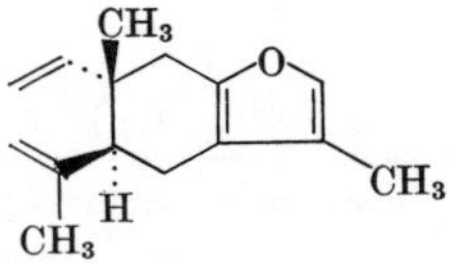 —	306

Note: References 344–439 are on pp. 251–252.

p Largely recovered starting material is observed.

223

TABLE V. COPE REARRANGEMENTS (Continued)

A. 1,5-Hexadiene Systems—Acyclic and Cyclic (Continued)

Molecular Formula	Starting Material	Conditions	Product(s) and Ratio ( )	Yield(s), %	Refs.
$C_{15}H_{20}O$ (Contd.)		200°, 23 hr sealed tube		~30	312, 313
$C_{15}H_{20}O_2$		Reflux, 23 hr, n-C_3H_7OH, N_2 atm		88	312
		230 ± 10°, 3 min, N_2 atm	At equilibrium:	—	307
$C_{15}H_{20}O_3$		Reflux, 17 hr, n-C_3H_7OH, N_2 atm	(1/2 with starting material)	83	310

224

$C_{15}H_{20}O_4$	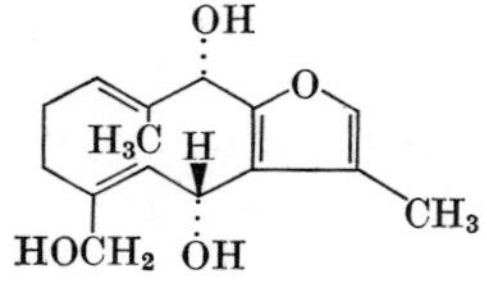	Reflux, 17 hr, n-C_3H_7OH, N_2 atm	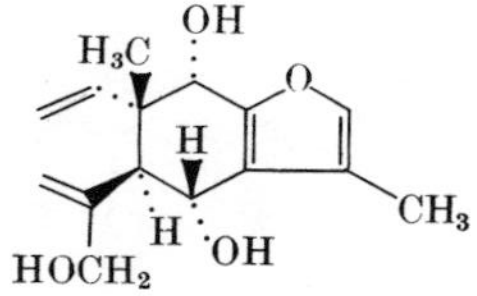	77	310
$C_{15}H_{22}O$	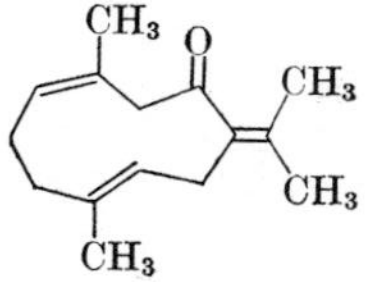	Heat	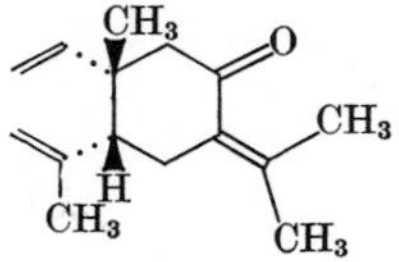	—	313
	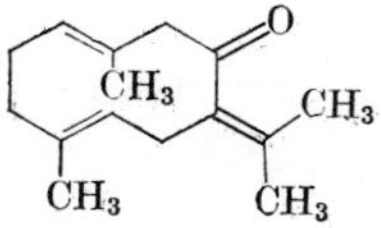	165°, distil, 10 mm	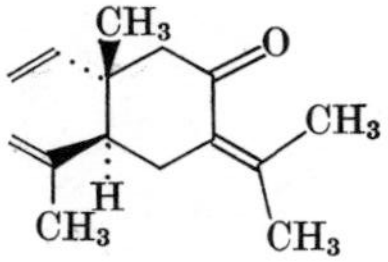	Quant.	308
	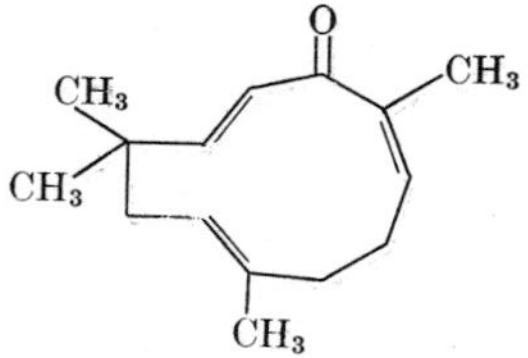	160° ± 10°	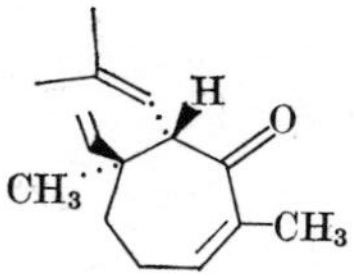	—	319
	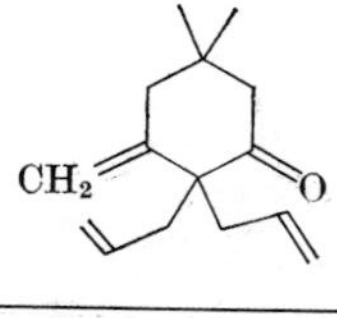	180–200°	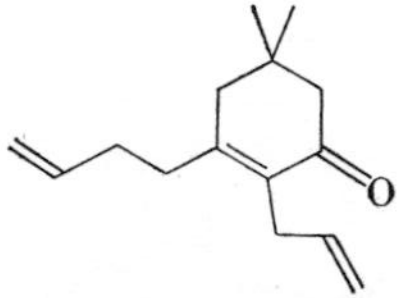	—	150

Note: References 344–439 are on pp. 251–252.

TABLE V. COPE REARRANGEMENTS (Continued)

A. 1,5-Hexadiene Systems—Acyclic and Cyclic (Continued)

Molecular Formula	Starting Material	Conditions	Product(s) and Ratio ( )	Yield(s), %	Refs.
$C_{15}H_{22}O$ (contd.)	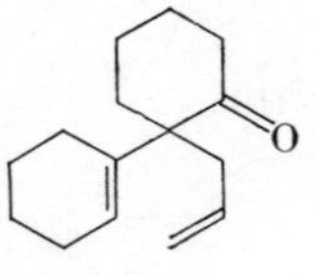	250°, 15 min	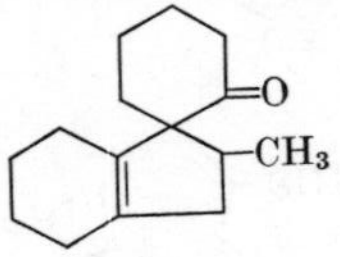 [a]	Quant.	148
$C_{15}H_{22}O_2$	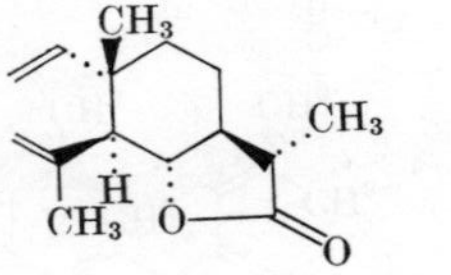	230 ± 10°, 3 min, N₂ atm	At equilibrium: 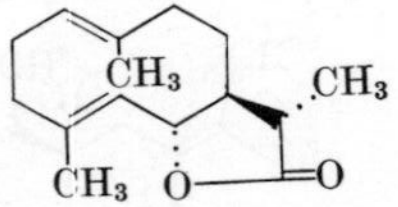	—	307
$C_{15}H_{23}NO_2$	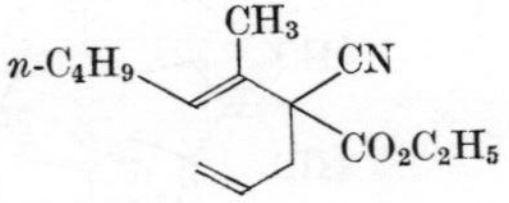	200°, 9 hr	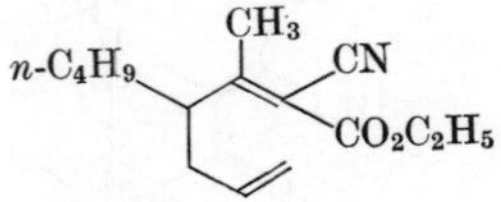 (1/2 with starting material)	77	138
$C_{15}H_{24}$	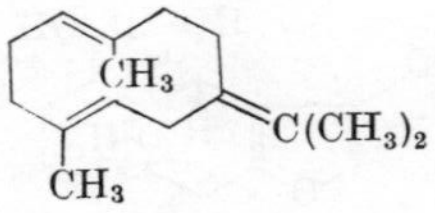	Heat	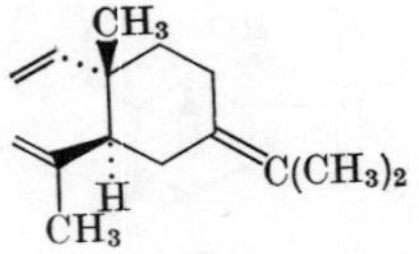	—	304

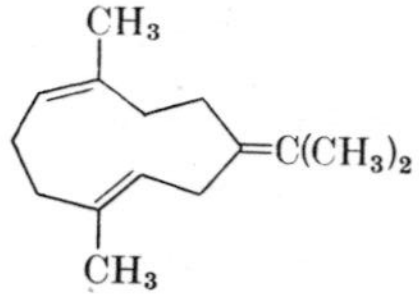

$C_{16}H_{17}NO_2$ 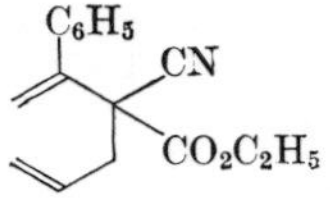 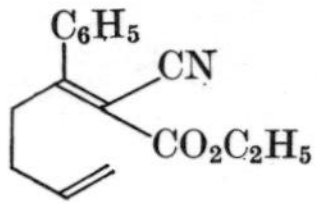	Heat		313
		—	
$C_{16}H_{20}$	170°, 7 hr	(13)	138
			79
	Heat	(1) (1)	302
		—	
$C_{16}H_{20}O$	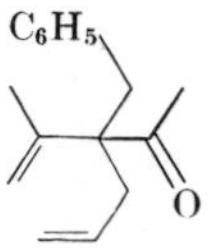 180–200°	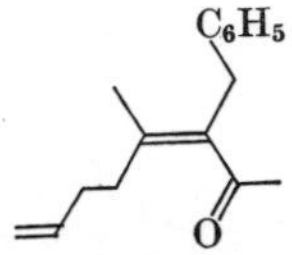	150
		—	

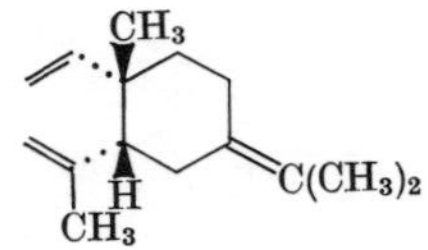

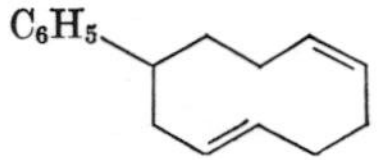

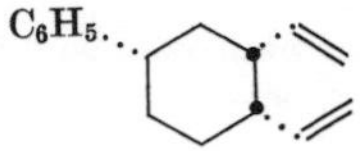

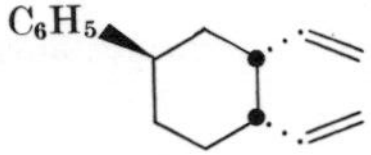

Note: References 344–439 are on pp. 251–252.

[a] The Cope product was not observed. The reaction was assumed to proceed via ⟨structure⟩ . See pp. 52–53.

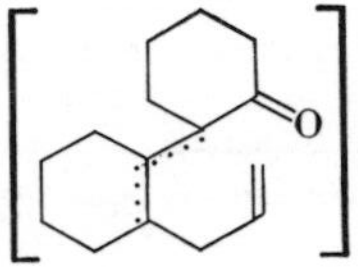

227

TABLE V. COPE REARRANGEMENTS (Continued)

A. 1,5-Hexadiene Systems—Acyclic and Cyclic (Continued)

Molecular Formula	Starting Material	Conditions	Product(s) and Ratio ( )	Yield(s), %	Refs.
$C_{16}H_{20}O_3$	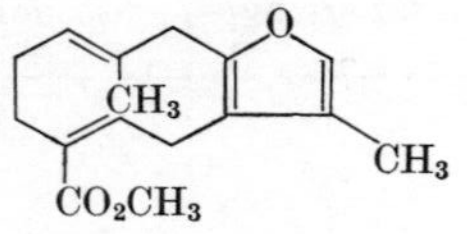	160–170°, 15 min, N₂ atm	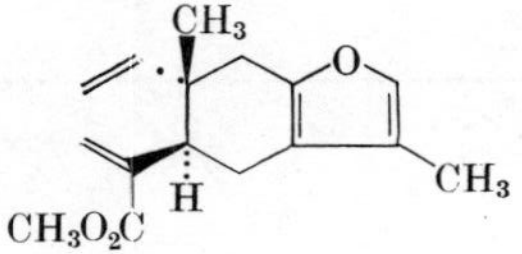	Good	305, 311
		Reflux, 3 hr, n-C₃H₇OH, N₂ atm	″	Quant	311
	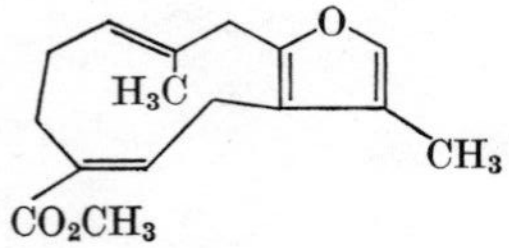	140°, 2.5 hr, p-cymene, N₂ atm	plus recovered starting material	5	311
$C_{17}H_{14}O_2$	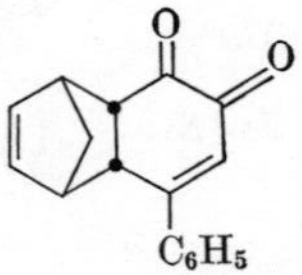	Reflux, C₆H₆	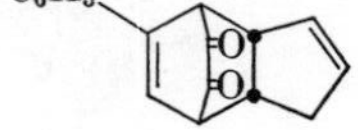	—	321
$C_{17}H_{17}NO_2$	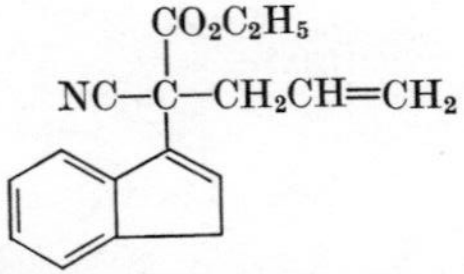	124–128°, 3 hr, sealed tube, N₂ atm	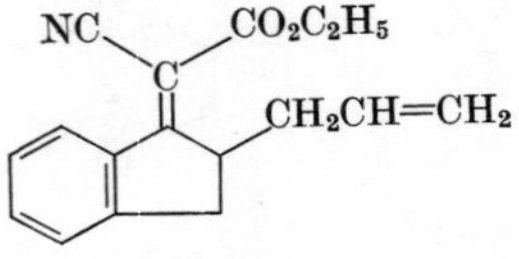	63	140

C$_{17}$H$_{18}$O$_5$

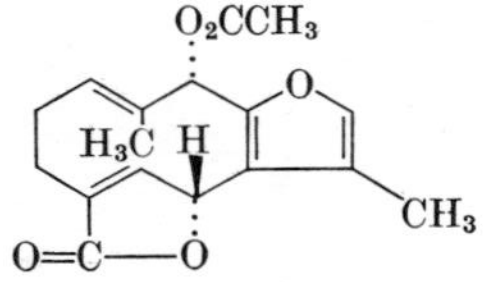

180°, 5 min, N$_2$ atm

At equilibrium:

— 316

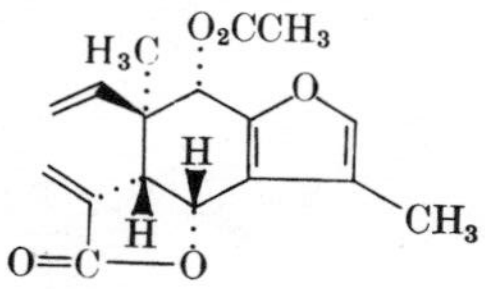

(3/2 with starting material)

C$_{17}$H$_{19}$N

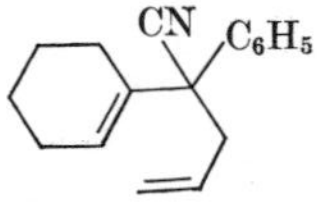

220°, 3.5 hr, N$_2$ atm

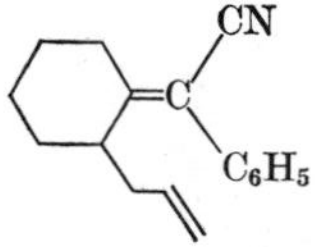

85 143

C$_{17}$H$_{24}$O$_3$

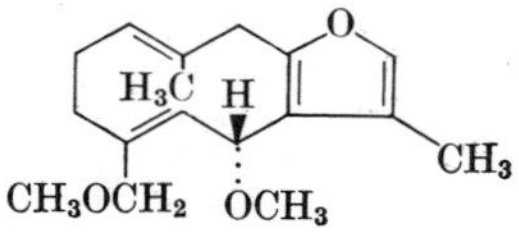

Reflux, 2.5 hr, pyridine, N$_2$ atm

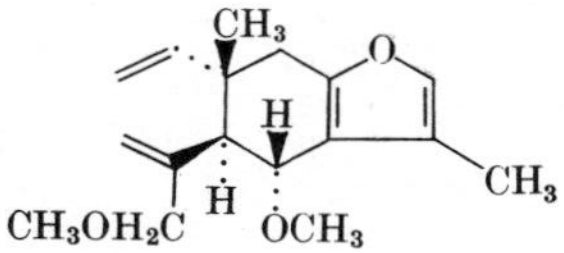

~90 310

C$_{17}$H$_{26}$O

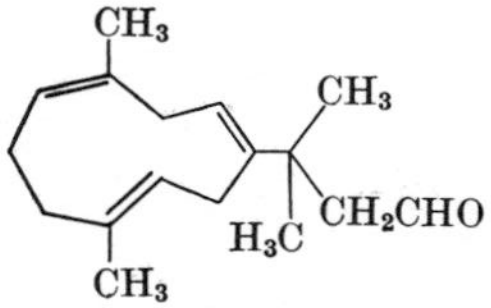

Heat

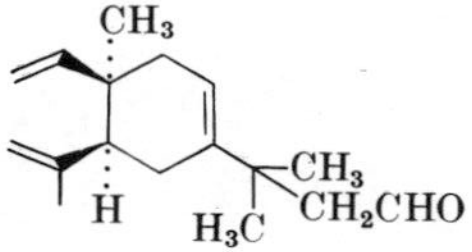

— 313

C$_{17}$H$_{27}$NO$_2$

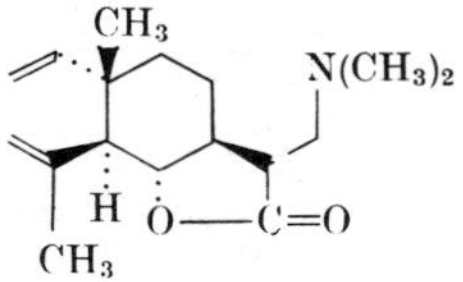

205–210°, 5 min, N$_2$ atm

At equilibrium:

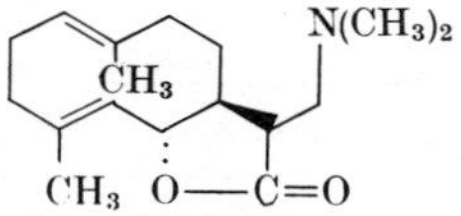

(~1/2 with starting material)

— 307

Note: References 344–439 are on pp. 251–252.

229

TABLE V. Cope Rearrangements (*Continued*)

A. 1,5-Hexadiene Systems—Acyclic and Cyclic (Continued)

Molecular Formula	Starting Material	Conditions	Product(s) and Ratio ( )	Yield(s), %	Refs.
$C_{18}H_{18}$	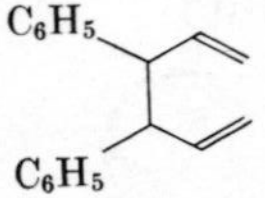 (meso)	120°, 93 hr, evacuated tube	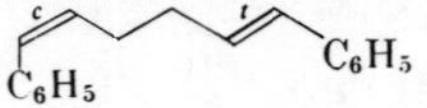 C_6H_5 VIII (1.7), 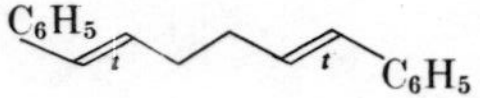 C_6H_5 IX (1)	98 (total)	289
	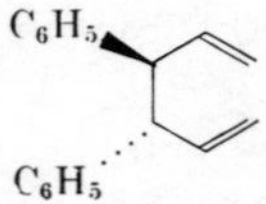 (racemic)	80°, 47 hr, evacuated tube	IX	98	289
$C_{18}H_{20}O_2Cl_4$	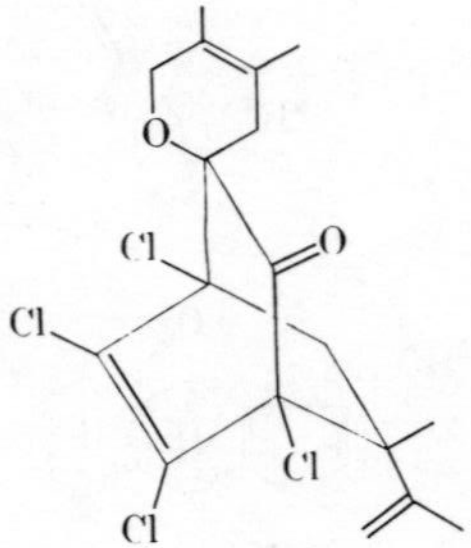	Heat, xylene	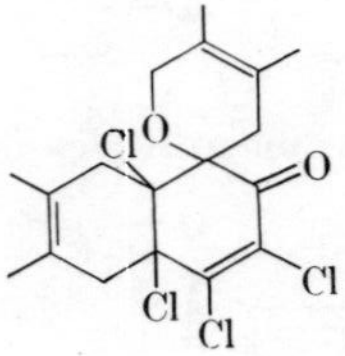	80	213b

$C_{18}H_{24}O$

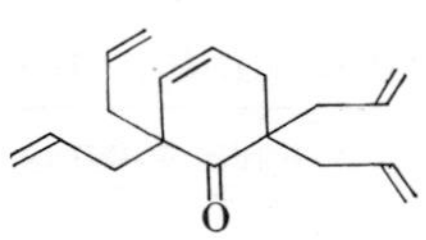

180–200°

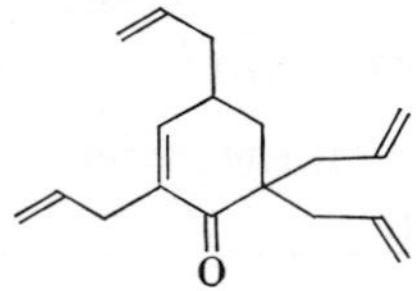

—

150

$C_{19}H_{24}O$

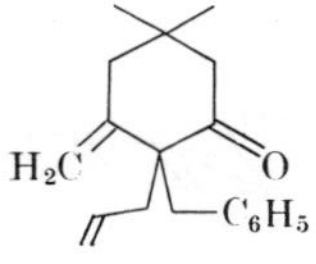

180–200°

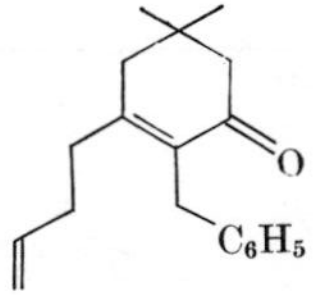

—

150

$C_{22}H_{30}O_2$

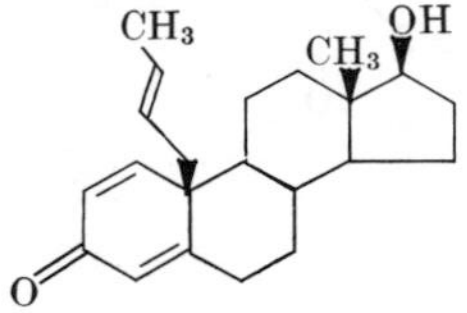

120°, decane

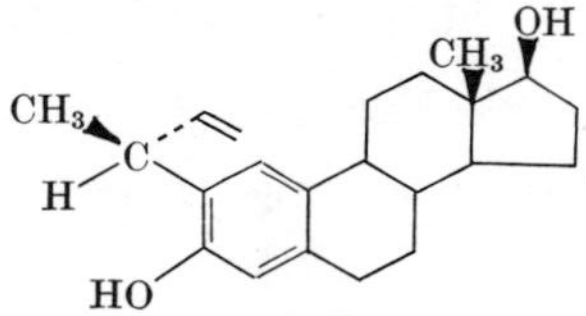

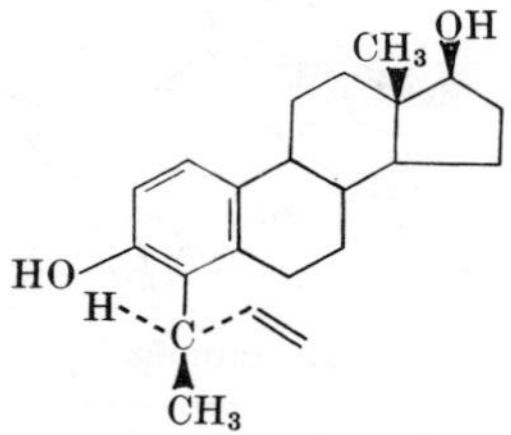

—

58

Note: References 344–439 are on pp. 251–252.

231

TABLE V. COPE REARRANGEMENTS (Continued)

A. 1,5-Hexadiene Systems—Acyclic and Cyclic (Continued)

Molecular Formula	Starting Material	Conditions	Product(s) and Ratio ( )	Yield(s). %	Refs.
$C_{24}H_{22}O$	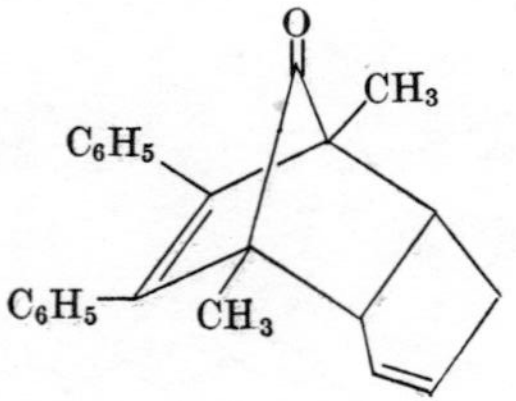	90°, $CCl_2{=}CCl_2$	At equilibrium: 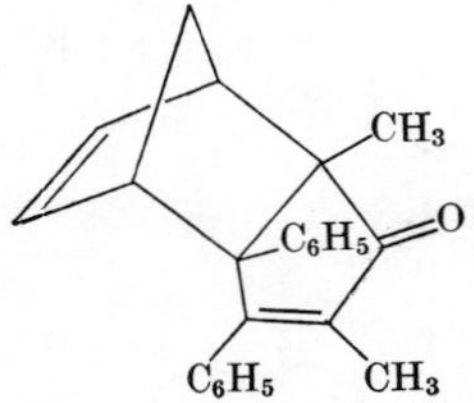	—	76
$C_{26}H_{24}O$	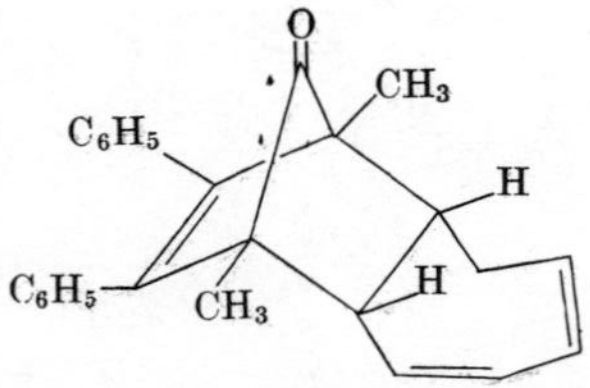	120°, <30 hr	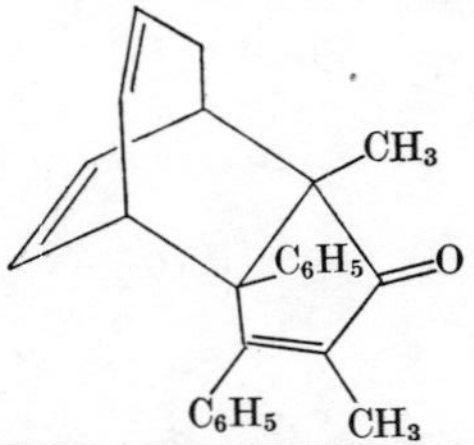 ($\sim$1/1/1 with starting material)	Quant	77

B. Oxy-Cope Rearrangements

Molecular Formula	Starting Material	Conditions	Product(s) and Ratio ( )	Yield(s), %	Refs.
C_6H_8O	$CH_2{=}CHCHOHCH_2C{\equiv}CH$	370–430°, through hot column with N_2, 25–30 mm	$CH_2{=}C{=}CHCH_2CH_2CHO$ X  XI	—	329

Formula	Structure	Conditions	Products (yield)	Ref.
	CH$_2$=CHCH$_2$CHOHC≡CH	390°, vapor through column, 15 mm	X (1), XI (4), fragmentation products (4) —	169
C$_6$H$_{10}$O	CH$_2$=CHCH$_2$CHOHCH=CH$_2$	370°, vapor through column, 20 mm	XI 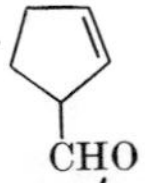 + CH$_2$=CHCH$_2$CH=CHCHO (6), —; CH$_3$CH=CHCH=CHCHO (3); fragmentation products (1); (7)	50
	CH$_2$=CH(CH$_2$)$_3$CHO	380°, distilled through packed column		168
C$_6$H$_{10}$O$_2$	CH$_2$=CHCHOHCHOHCH=CH$_2$	240–260°, reduced pressure, distil product 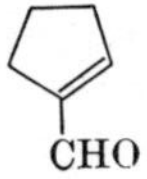	57	180
			40	
C$_7$H$_{10}$O	CH$_2$=CHC(OH)CH$_2$C≡CH |\nCH$_3$	375°, through packed tube with N$_2$	CH$_2$=C=CHCH$_2$CH$_2$COCH$_3$ XII (1), 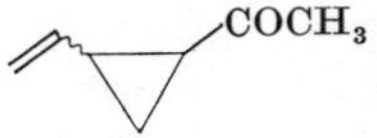 XIII (1.4), fragmentation products (3.6)	330
		330°, flow system, 1 mm	XII (1.6), XIII (1) 69 (total)	429
		370–430°, through hot column with N$_2$, 25–30 mm	XII  + —	329

Note: References 344–439 are on pp. 251–252.

233

TABLE V. COPE REARRANGEMENTS (Continued)

B. Oxy-Cope Rearrangements (Continued)

Molecular Formula	Starting Material	Conditions	Product(s) and Ratio ( )	Yield(s), %	Refs.
$C_7H_{10}O$ (contd.)	$CH_2=C(CH_3)CHOHCH_2C\equiv CH$	370–430°, through hot column with N_2, 25–30 mm	$CH_2=C=CHCH_2CH(CH_3)CHO$, $CH_3CH=CHCH_2CCHO$ ‖ CH_2	—	329
$C_7H_{12}O$	$CH_3CH=CHCHOHCH_2CH=CH_2$	340–380°, flow system, packed column	$CH_2=CHCH_2CH(CH_3)CH_2CHO$ (1), fragmentation products and starting material (2.5)	22[r]	167
	$CH_2=CHCH_2CHOHC(CH_3)=CH_2$	340–380°, flow system, packed column	$CH_2=CHCH_2CH_2CH(CH_3)CHO$ (3), fragmentation products (1)	64[r]	167
	$CH_2=CHC(CH_3)OHCH_2CH=CH_2$	340–380°, flow system, packed column	$CH_2=CHCH_2CH_2CH_2COCH_3$ (2), fragmentation products and starting material (1)	58[r]	167
	$CH_2=CHCHOHCH(CH_3)CH=CH_2$	340–380°, flow system, packed column	$CH_3CH=CHCH_2CH_2CH_2CHO$ (4), fragmentation products and starting material (1)	64[r]	167
	$CH_2=CHCHOHCH_2C(CH_3)=CH_2$	370–380°, flow system, packed column	$CH_2=C(CH_3)(CH_2)_3CHO$ (1), fragmentation products (1)	35[r]	167
$C_8H_{10}O_2$	$HC\equiv CC(CH_3)(OH)C(CH_3)(OH)C\equiv CH$ meso/racemic 1/1	350°, 0.01 mm	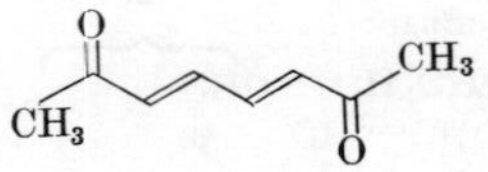	25	329
$C_8H_{12}O$	$CH_2=C(CH_3)C(OH)CH_2C\equiv CH$ | CH_3	370–430°, through hot tube, with N_2, 25–30 mm	$CH_2=C=CHCH_2CH(CH_3)COCH_3$, $CH_3CH=CHCH_2CCOCH_3$	—	329

234

Reactant	Conditions	Product(s) (ratio)	Yield (%)	Refs.
(continued)	system, reduced pressure	fragmentation products and starting material (1)	9.5ʳ	167
$CH_3CH=CHCH(OH)CH_2C(CH_3)=CH_2$	360–375°, flow system, reduced pressure	$CH_2=C(CH_3)CH_2CH_2CH(CH_3)CH_2CHO$ (1), fragmentation products and starting material (6.7)	22ʳ	167
$CH_2=C(CH_3)CHOHCH_2C(CH_3)=CH_2$	360–370°, flow system, reduced pressure	$CH_2=C(CH_3)CH_2CH_2CH_2CH(CH_3)CHO$ (1), fragmentation products and starting material (2)	42ʳ	167
$CH_2=CHC(OH)CH_2C(CH_3)=CH_2$ with CH_3 (i.e. $CH_2=CHC(OH)(CH_3)CH_2C(CH_3)=CH_2$)	360–370°, flow system, reduced pressure	$CH_2=C(CH_3)CH_2CH_2CH_2CH_2COCH_3$ (1), fragmentation products (1)		
$C_8H_{14}O_2$ $CH_2=CHC(OH)C(OH)CH=CH_2$ with CH_3 CH_3	145–190°, 1 hr	$CH_3CO(CH_2)_4COCH_3$ XIV	~90	181, 49
	240–260°, reduced pressure, distil product	XIV	—	180
	240–260°	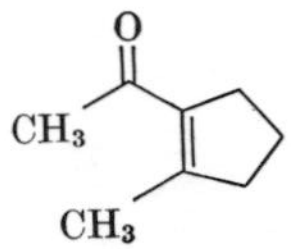 [2-methylcyclopent-1-en-1-yl methyl ketone] $O=C(CH_3)$ – ring – CH_3 XV	—	173
	300–320°, neat, Al_2O_3 column	XIV (1), XV (2), $(CH_2=CH)_2C(CH_3)COCH_3$ (1)	66 (total)	183
$CH_3CH=CHCH(OH)CH(OH)CH=CHCH_3$	240–260°, reduced pressure, distil product	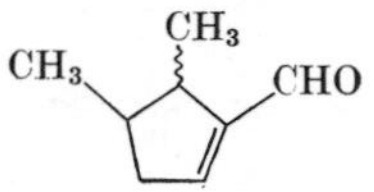 [cyclopentene with CH_3, CH_3, CHO]	40	180
meso/racemic 1/1	240–260°, reduced pressure, under N_2, until product distils (6–7 hr)	(*trans/cis* 3/1)	—	181

Note: References 344–439 are on pp. 251–252.

ʳ The percentage yield of the oxy-Cope product is based on a material balance of 70–80%.

235

TABLE V. Cope Rearrangements (Continued)

B. Oxy-Cope Rearrangements (Continued)

Molecular Formula	Starting Material	Conditions	Product(s) and Ratio ( )	Yield(s), %	Refs.
$C_9H_{14}O$	[structure]	370–375°, gas phase, 5 mm	[structure]	63	430
$C_9H_{14}O_2$	[structure]	170°, 15 min	[structure]	55	173
	[structure]	160°, 10 min, sealed tube, Pyrex	"	35	49
	[structure]	160°, 10 min, sealed tube, ordinary glass	[structure] (1), [structure] (1)	70 (total)	49
$C_9H_{16}O$	$(CH_3)_2C{=}CHC(OH)CH_2CH{=}CH_2$ $-CH_3$	370–380°, 15 min flow system	$CH_2{=}CHCH_2C(CH_3)_2CH_2COCH_3$ (1), fragmentation products and starting material (19)	3ʳ	167
$C_{10}H_{14}O$	[structure]	250–320°, gas phase	[structure] (9), unidentified material (1)	50 (total)	170

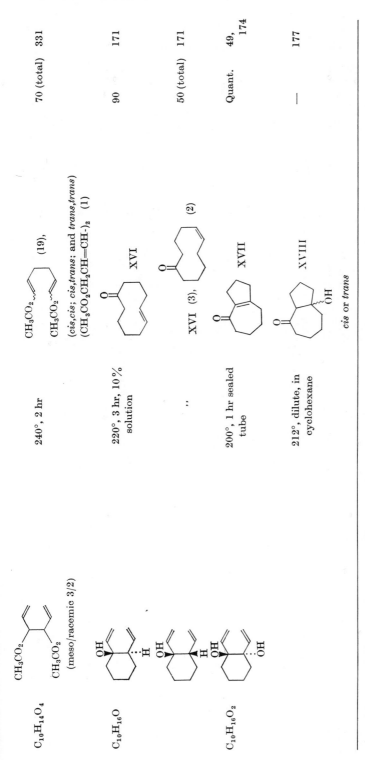

Reactant	Conditions	Products	Yield (%)	Ref.
C₁₀H₁₄O₄ (meso/racemic 3/2)	240°, 2 hr	(19), (cis,cis; cis,trans; and trans,trans) (1)	70 (total)	331
C₁₀H₁₆O	220°, 3 hr, 10% solution	XVI (3),	90	171
	″	XVI (2)	50 (total)	171
C₁₀H₁₆O₂	200°, 1 hr sealed tube	XVII	Quant.	49, 174
	212°, dilute, in cyclohexane	XVIII cis or trans	—	177

Note: References 344–439 are on pp. 251–252.

ʳ The percentage yield of the oxy-Cope product is based on a material balance of 70–80%.

TABLE V. COPE REARRANGEMENTS (*Continued*)

B. *Oxy-Cope Rearrangements* (*Continued*)

Molecular Formula	Starting Material	Conditions	Product(s) and Ratio ( )	Yield(s), %	Refs.
C₁₀H₁₆O₂ (*contd.*)	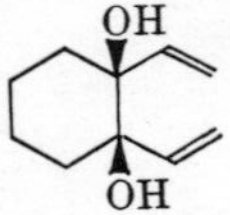	212°, dilute, in cyclohexane	XVII, XVIII *cis* and *trans* (p. 237)	—	177
C₁₁H₁₆O	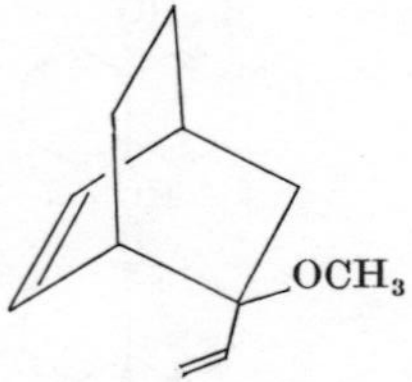	325°, gas phase	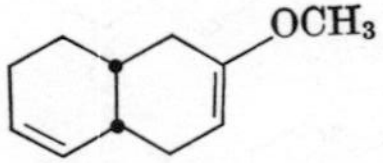 (6.7), 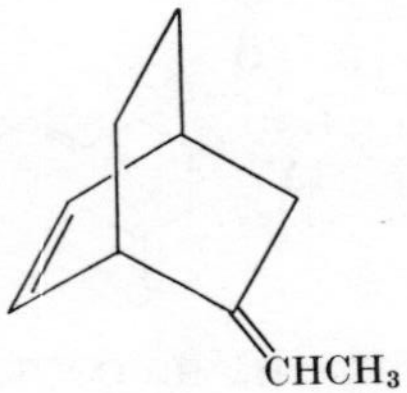 (1)	~90 (total)	324
C₁₁H₁₈O₂	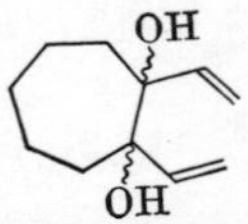	220°, 1 hr sealed tube, Pyrex	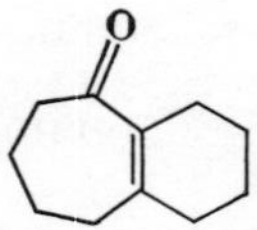 XIX	Quant	174, 49

	Conditions		Yield	Reference
C$_{12}$H$_{20}$O$_2$	220°, 1 hr, sealed tube, ordinary glass	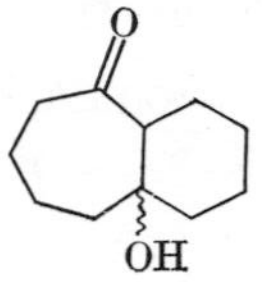 (1), XIX (1.5)	Quant.	174, 49
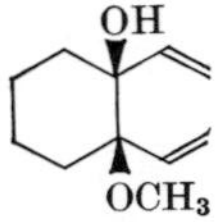	230°, 3 hr	XVII (p. 237)	Good	177
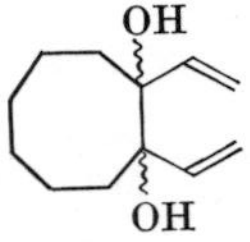 C$_{13}$H$_{14}$O$_2$	220°, 1 hr	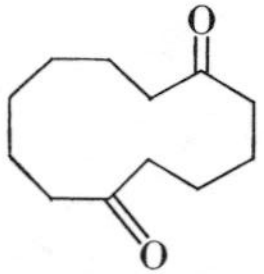	Quant.	174, 49
	225°, 12 hr, in cyclohexane, sealed tube	,,	40–50	176
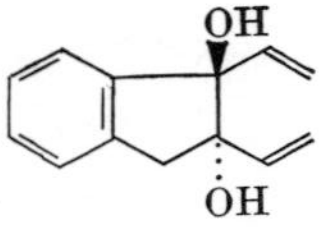	163°, 0.5 hr, sealed tube	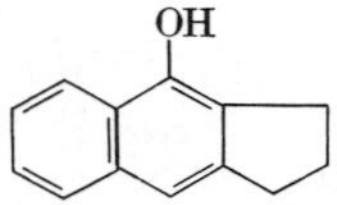	50	172
C$_{13}$H$_{18}$O$_2$	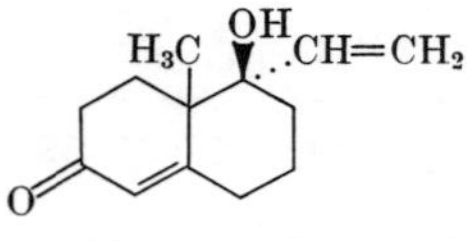 Reflux, 6 hr, diethylene glycol	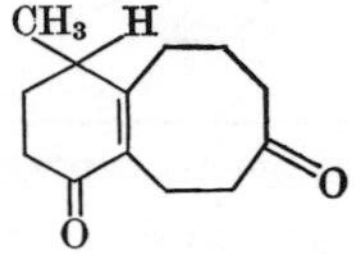	60	431

Note: References 344–439 are on pp. 251–252.

TABLE V. COPE REARRANGEMENTS (Continued)

B. Oxy-Cope Rearrangements (Continued)

Molecular Formula	Starting Material	Conditions	Product(s) and Ratio ( )	Yield(s), %	Refs.
$C_{14}H_{20}O_2$		Reflux, 3 hr, ethylene glycol		5 (pure)	431
$C_{14}H_{22}O_2$		220°, 1 hr	(1) (2)	Quant.	175
$C_{14}H_{22}O_4$	$i\text{-}C_3H_7CO_2$ $i\text{-}C_3H_7CO_2$	300°, 100 min, N₂ atm, steel reactor	$i\text{-}C_3H_7CO_2$ $i\text{-}C_3H_7CO_2$ (20)	51(total)	331

(cis,cis; cis,trans; and trans,trans),
($i\text{-}C_3H_7CO_2CH_2CH=CH\cdot)_2$ (1)

C₁₅H₂₂O₃ ... 185°, 5 min ... Quant. ... 327

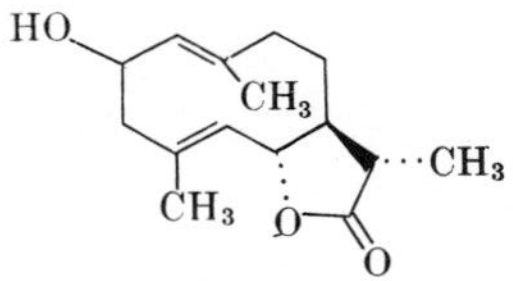

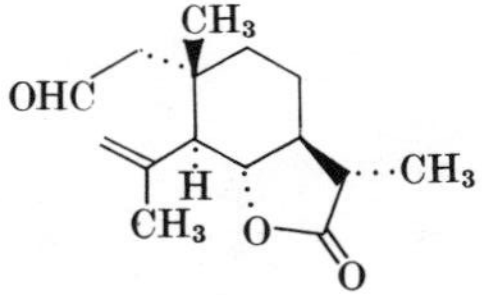

C₁₅H₂₄O₂ ... 220°, 1 hr ... (2), ... 82(total) ... 175

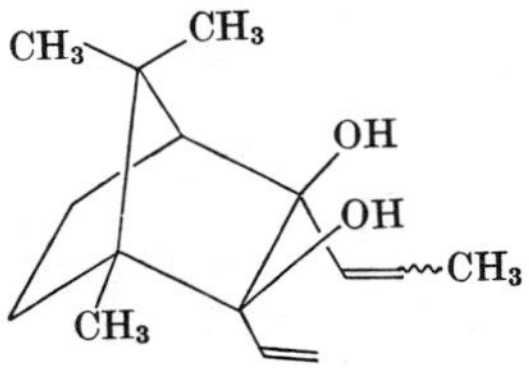

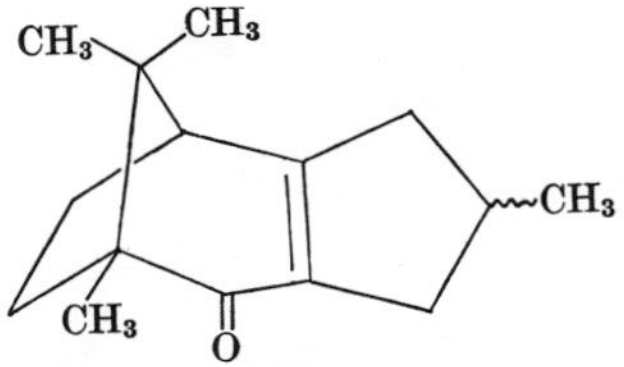

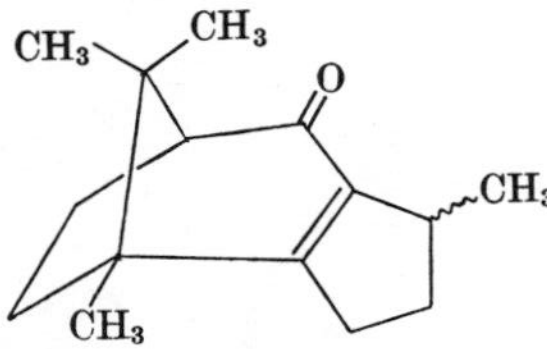

... (1)

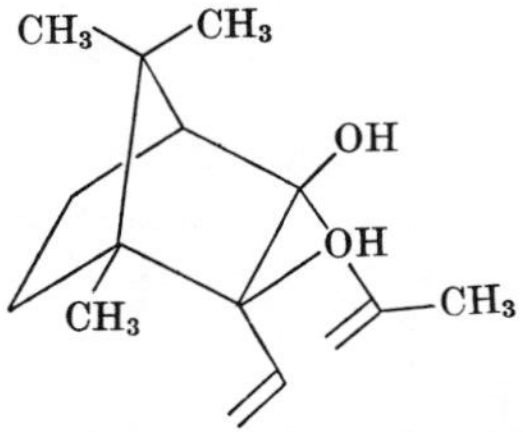

... 220°, 1 hr ... — ... 175

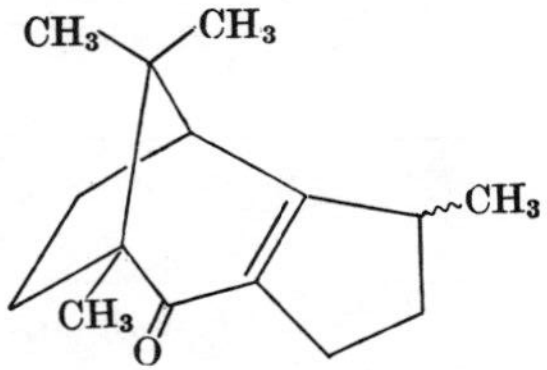

Note: References 344–439 are on pp. 251–252.

TABLE V. COPE REARRANGEMENTS (*Continued*)

B. Oxy-Cope Rearrangements (Continued)

Molecular Formula	Starting Material	Conditions	Product(s) and Ratio ( )	Yield(s), %	Refs.
$C_{16}H_{28}O_2$	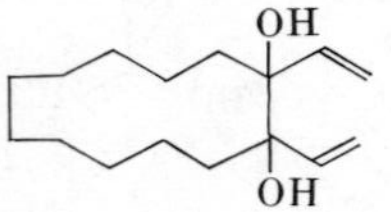	240°, 2 hr	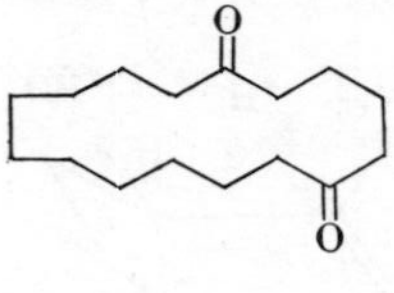	80	174, 49
$C_{17}H_{16}O_2$	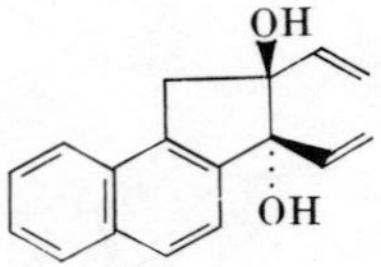	160°, 1.25 hr, sealed tube	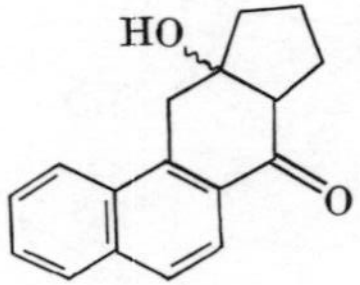	19	172
$C_{17}H_{24}O_4$	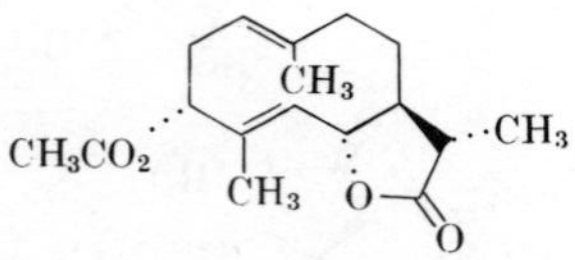	220°, 3 min, N₂ atm	At equilibrium: 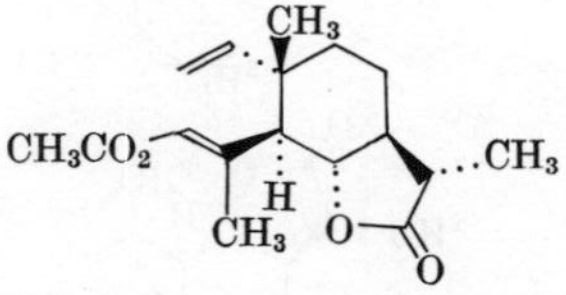 (3/2 with starting material)	—	326

242

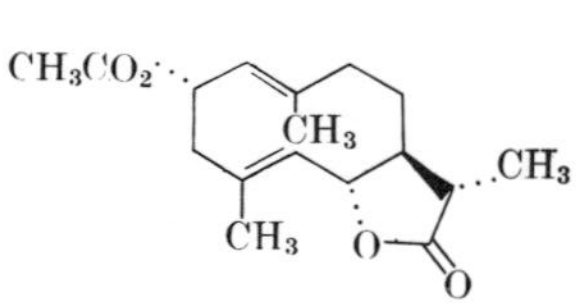

At equilibrium:

Substrate	Product	Conditions	Yield	Ref.
$C_{17}H_{24}O_5$ 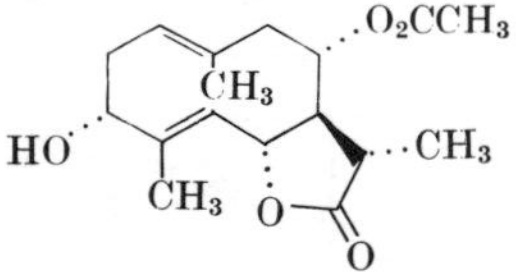 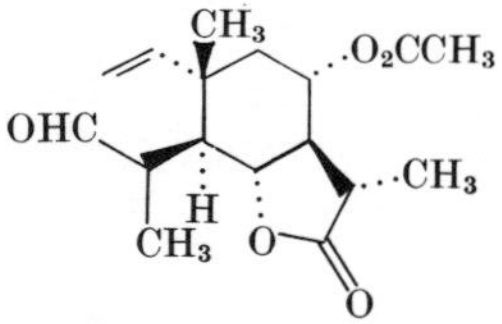		$220°$, 4 min	—	327
$C_{18}H_{18}O_2$	(1/1 with starting material)	$205°$, 5 min	~Quant.	328
$C_{19}H_{26}O_6$		$240–260°$, reduced pres., product distils	~Quant.	180
	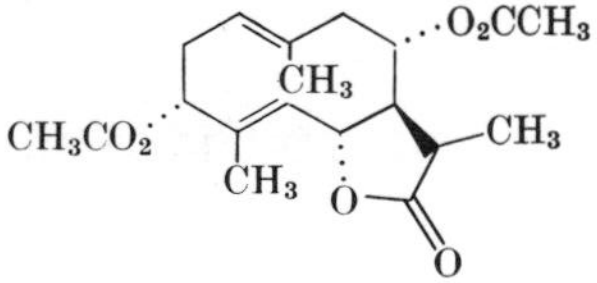	$205°$, 5 min	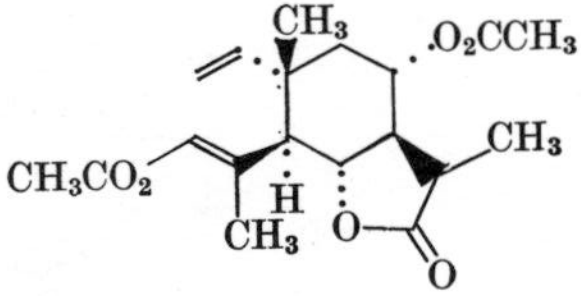 ~Quant.	328
$C_{19}H_{32}O$	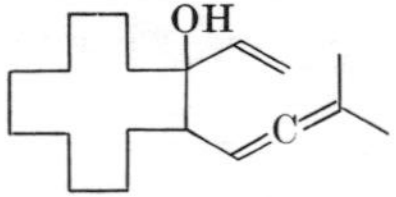	$350°$, through packed column	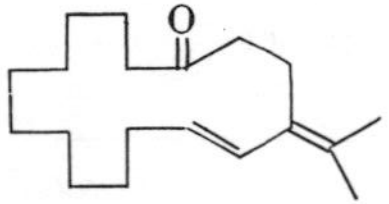 80	439

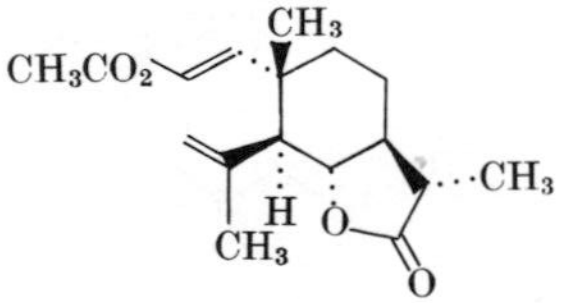

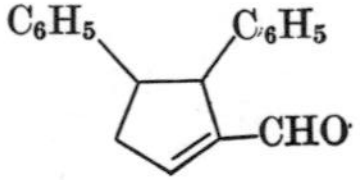

Note: References 344–439 are on pp. 251–252.

TABLE V. Cope Rearrangements (Continued)

C. Miscellaneous Cope Rearrangements

Molecular Formula	Starting Material	Conditions	Product(s) and Ratio ( )	Yield(s), %	Refs.
C_6H_6	$HC{\equiv}CCH_2CH_2C{\equiv}CH$	350°, flow system	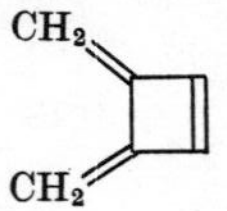	85	335
C_6H_7NO	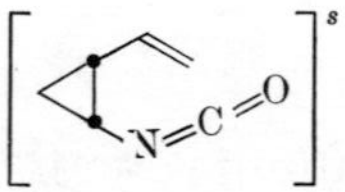 [s]	80°, C_6H_6	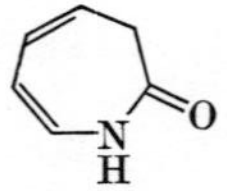	32[t]	70
	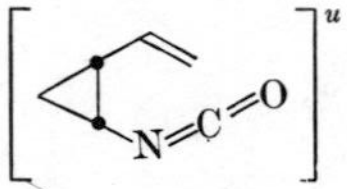 [u]	Room temp., overnight, $n\text{-}C_5H_{12}$	"	16[t]	71
	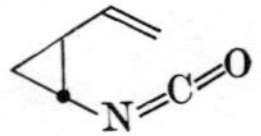	400°, >2 hr, 12 mm	"	86	70
C_6H_8	$HC{\equiv}CCH_2CH_2CH{=}CH_2$	340°, 62 sec contact time	$CH_2{=}C{=}CHCH_2CH{=}CH_2$ (32), 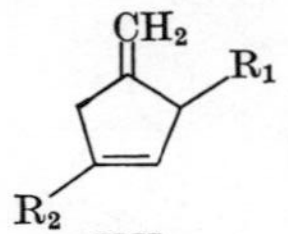 (1), XX ($R_1, R_2 = H$), 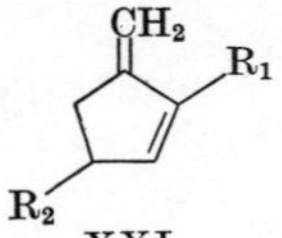 (1), XXI ($R_1, R_2 = H$)	70 (total)	152

C_7H_8	$CH_3C{\equiv}CCH_2CH_2C{\equiv}CH$	377°, flow system	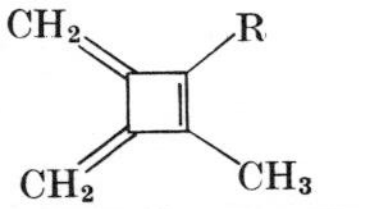 **XXII** (R=H)	Quant	153
C_7H_{10}	$CH_3C{\equiv}CCH_2CH_2CH{=}CH_2$	385°, 63 sec contact time	$CH_2{=}C{=}C(CH_3)CH_2CH{=}CH_2$ (1), **XX** ($R_1 = CH_3, R_2 = H$) (3), **XXI** ($R_1 = CH_3, R_2 = H$) (4)	68 (total)	152
	$HC{\equiv}CCH_2CH_2C(CH_3){=}CH_2$	340°, 36 sec contact time	$CH_2{=}C{=}CHCH_2C(CH_3){=}CH_2$ (12), **XX** ($R_1 = H, R_2 = CH_3$) (1), **XXI** ($R_1 = H, R_2 = CH_3$) (1)	79 (total)	152
	$CH_2{=}C{=}CHCH_2CH_2CH{=}CH_2$	300°, gas phase, hot tube, 10 mm	$CH_2{=}CHCH_2CH{=}CH_2$ with CH_2	98	337
C'_8H_8	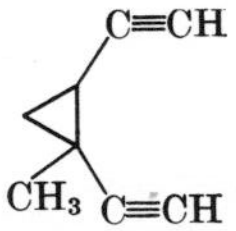 (cyclopropane–acetylene structure)	350°, flow pyrolysis in N_2 stream	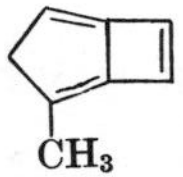 (bicyclic CH_3 structure)	30–40	334
C_8H_{10}	$CH_3C{\equiv}CCH_2CH_2C{\equiv}CCH_3$	410°, flow system	**XXII** (R = CH_3)	—	153
	$HC{\equiv}CCH(CH_3)CH(CH_3)C{\equiv}CH$ (meso)	350°, 15 min, static system	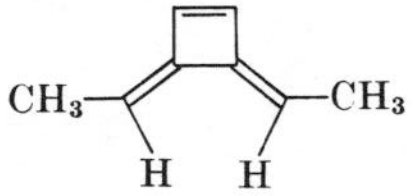 (dimethyl structure, H, H)	—	153
	racemic	,,		—	153

Note: References 344–439 are on pp. 251–252.

[s] This material was obtained by a Curtius rearrangement of the corresponding azide.

[t] The yield is based on acid chloride.

[u] The isocyanate was obtained by irradiation of the corresponding azide at −78°.

245

TABLE V. COPE REARRANGEMENTS (*Continued*)

C. Miscellaneous Cope Rearrangements (Continued)

Molecular Formula	Starting Material	Conditions	Product(s) and Ratio ( )	Yield(s), %	Refs.
C_8H_{10} (*contd.*)	$CH_2=C=CHCH_2CH_2CH=C=CH_2$	310°, gas phase, 0.2 mm, hot tube	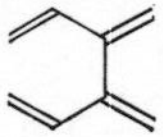	~50	337
C_9H_{12}	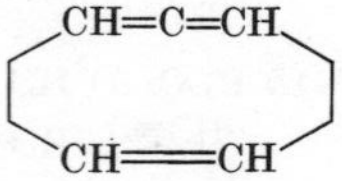 CH≡C=CH / CH=CH	140°, gas phase	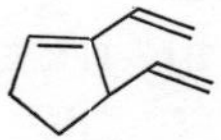 XXIII	—	301
		140°, 40 min, sealed tube, N_2 atm	XXIII, dimers	55	338
		235°, gas phase, 0.5 mm, hot tube	XXIII	92	337
		120–130°, 17 hr, 2 M solution in hexane, sealed tube	XXIII, dimers	42	337
$C_{10}H_{12}$	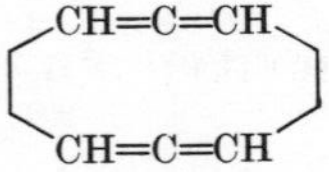 CH≡C=CH / CH=C=CH	300°, 0.5 mm, hot tube	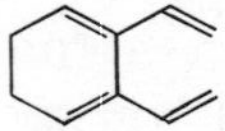	Quant.	337, 336
$C_{10}H_{14}$	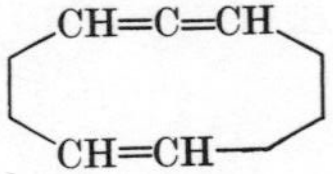 CH≡C=CH / CH=CH	180°, static system	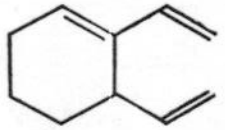	Quant.	336

246

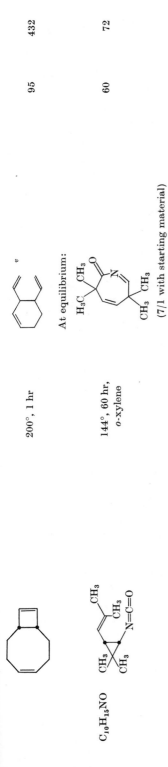

$C_{10}H_{15}NO$ — 200°, 1 hr — 95 — 432

At equilibrium: (7/1 with starting material) — 95 — 432

$C_{10}H_{15}NO$ — 144°, 60 hr, o-xylene — 60 — 72

$C_{11}H_{16}N_2$ — Reflux, 7.5 hr, C_2H_5OH, aq CH_3CO_2H — 98 — 333

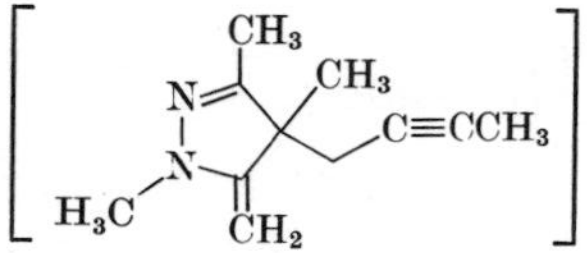

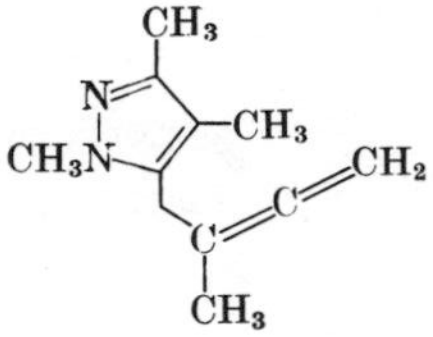

$C_{12}H_9NO$ — Reflux, 24 hr, C_6H_6, $C_6H_5CH_2OH$ — 69 — 69

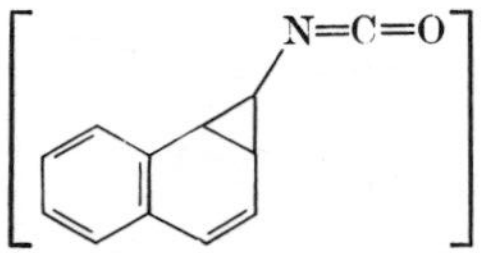

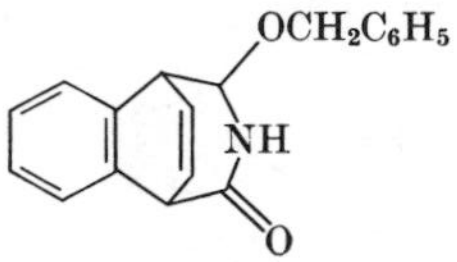

$OCH_2C_6H_5$

Note: References 344–439 are on pp. 251–252.

j This material was generated *in situ* from the 3,3-disubstituted 2,4-pentanedione and a hydrazine.

s This material was obtained by a Curtius rearrangement of the corresponding azide.

v This is thought to have been formed via [ ] produced by electrocyclic ring opening of the starting material.

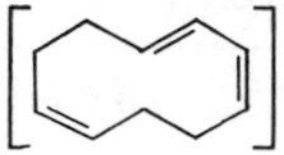

247

TABLE V. Cope Rearrangements (Continued)

C. Miscellaneous Cope Rearrangements (Continued)

Molecular Formula	Starting Material	Conditions	Product(s) and Ratio ( )	Yield(s), %	Refs.
$C_{12}H_{14}N_2$	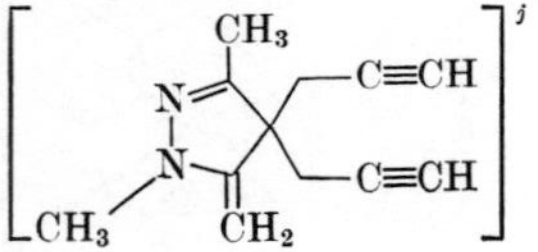	Reflux, 3.5 hr, C_2H_5OH, aq CH_3CO_2H	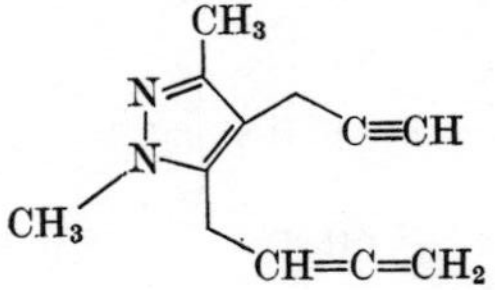	72	333
$C_{12}H_{16}N_2$	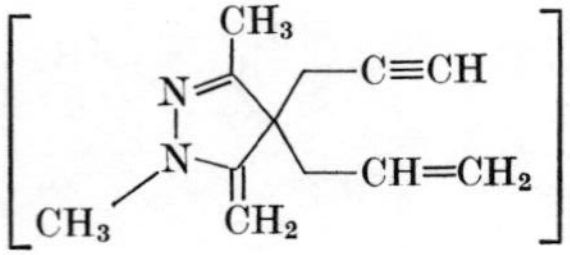	Reflux, C_2H_5OH, aq CH_3CO_2H	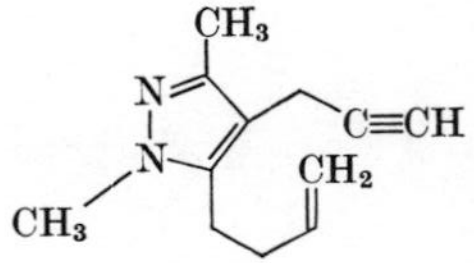 (1.6), 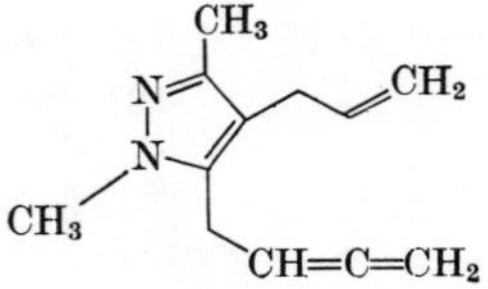 (1)	—	333
$C_{13}H_{13}NO_2S$	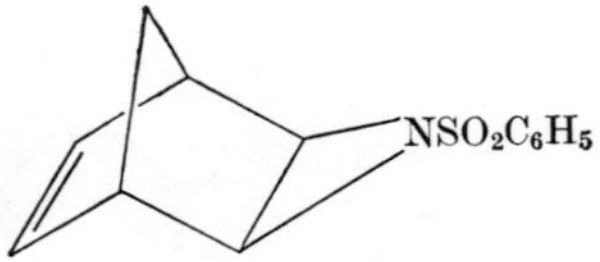	Standing	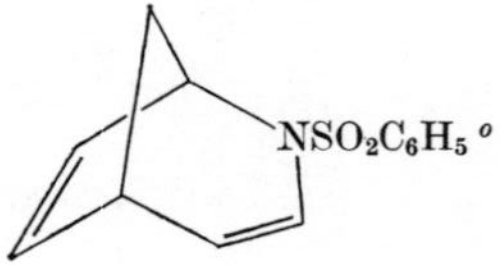	~Quant.	339

248

$C_{14}H_{20}O_4$	$(CH_3)_2C=CHCHCH_2C\equiv CH$ $(CO_2C_2H_5)_2$	270–280°, 30 min, N_2 atm	$CH_2=C=CHCH(CH_3)CH=C(CO_2C_2H_5)_2$,	20	151
			$(CH_3)_2CHCH=C(CO_2C_2H_5)_2$, polymer		
$C_{17}H_{16}N_2$	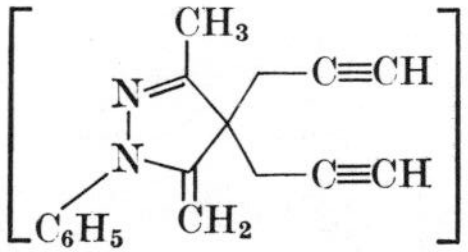	Reflux, 13 hr, n-C_4H_9OH, aq CH_3CO_2H	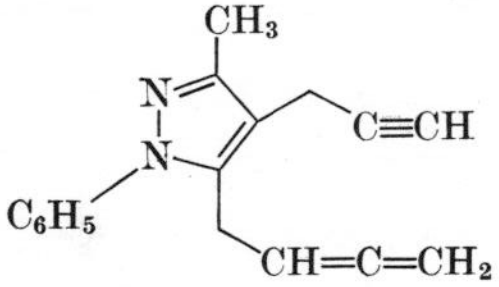	45	333
	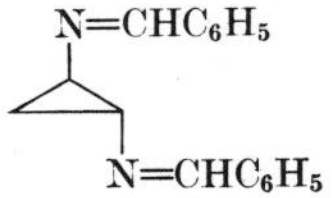	120–130° (melt), 30 min	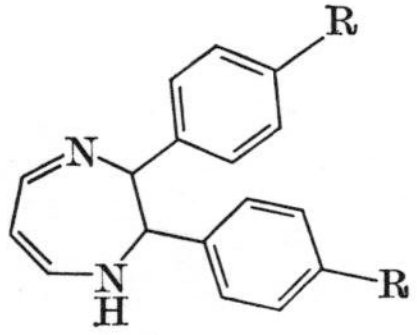 XXIV (R = H)	Quant.	73
$C_{19}H_{20}N_2O_2$	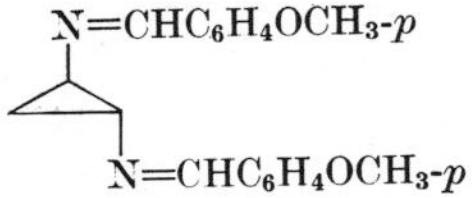	125°, 1 hr	XXIV (R = OCH$_3$)	Quant.	73

Note: References 344–439 are on pp. 251–252.

^{j} This material was generated *in situ* from the 3,3-disubstituted 2,4-pentanedione and a hydrazine.

^{o} This compound is thought to have been formed *via* the sulfonyl imine, 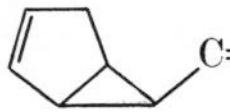 $C=NSO_2C_6H_5$; see p. 67.

TABLE V. Cope Rearrangements (Continued)

C. Miscellaneous Cope Rearrangements (Continued)

Molecular Formula	Starting Material	Conditions	Product(s) and Ratio ( )	Yield(s), %	Refs.
$C_{21}H_{26}N_4$	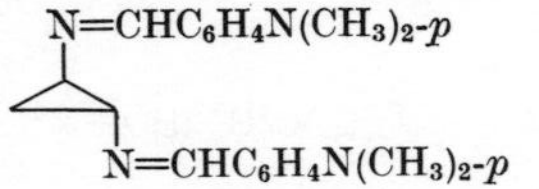	130°, 5 min	XXIV [R = N(CH$_3$)$_2$]	Quant.	73
$C_{24}H_{21}N_3$	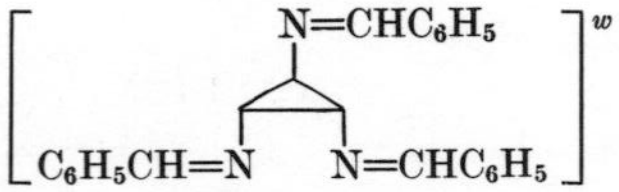 [w]	70–80°; then standing several days cold	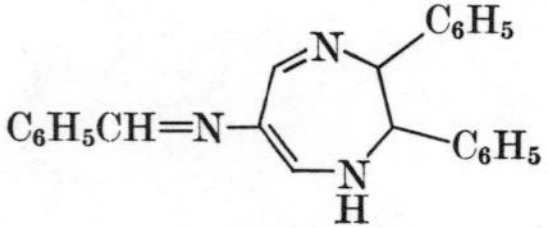	—	73
$C_{25}H_{20}N_2$	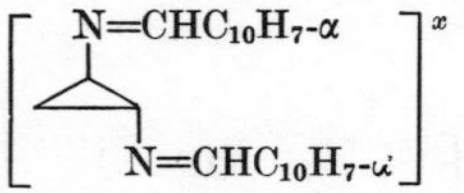 [x]	140°, 30 min, 10^{-2} mm	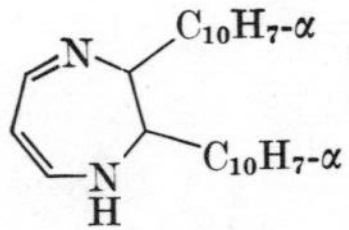	16[v]	73

Note: References 344–439 are on pp. 251–252.

[w] This is generated in situ from the reaction of the corresponding triaminocyclopropane and benzaldehyde.

[x] The compound was obtained from trans-1,2-diaminocyclopropane and α-naphthaldehyde.

[v] The yield is based on the starting material, the dihydrochloride of 1,2-diaminocyclopropane.

REFERENCES TO TABLES

344 W. N. White, D. Gwynn, R. Schlitt, C. Girard, and W. Fife, *J. Amer. Chem. Soc.*, **80**, 3271 (1958).

345 Z. Budĕsínsky and E. Ročková, *Chem. Listy*, **48**, 427 (1954) [*C.A.*, **49**, 3880b (1955)].

346 L. P. Sinyavskaya and A. A. Shamshurin, *Zh. Org. Khim.*, **3**, 2195 (1967) [*C.A.*, **68**, 68604j (1968)].

347 E. N. Marvell and R. Teranishi, *J. Amer. Chem. Soc.*, **76**, 6165 (1954).

348 A. B. Sen and R. P. Rastogi, *J. Indian Chem. Soc.*, **30**, 556 (1953) [*C.A.*, **49**, 921c (1955)].

349 W. N. White and C. D. Slater, *J. Org. Chem.*, **26**, 3631 (1961).

350 L. P. Sinyavskaya and A. A. Shamshurin, *Zh. Org. Khim.*, **4**, 1267 (1968) [*C.A.*, **69**, 76829x (1968)].

351 A. B. Sen and R. P. Rastogi, *J. Indian Chem. Soc.*, **30**, 355 (1953) [*C.A.*, **48**, 10649e (1954)].

352 A. A. Shamshurin and L. P. Sinyavskaya, *Zh. Org. Khim.*, **6**, 1682 (1970) [*C.A.*, **73**, 98547p (1970)].

353 D. Y. Curtin and H. W. Johnson, Jr., *J. Amer. Chem. Soc.*, **78**, 2611 (1956).

354 E. Piers and R. K. Brown, *Can. J. Chem.*, **41**, 329 (1963).

355 D. S. Tarbell and S. S. Stradling, *J. Org. Chem.*, **27**, 2724 (1962).

356 Chem. Fabrik Schweizerhall, Belg. Pat. 652198 (1964) [*C.A.*, **64**, 12608g (1966)].

357 C. S. Marvel and N. A. Higgins, *J. Amer. Chem. Soc.*, **70**, 2218 (1948).

358 E. D. Laskina, V. N. Belov, T. A. Rudol'fi, and M. M. Shchedrina, *Zh. Obshch. Khim.*, **34**, 4015 (1964) [*C.A.*, **62**, 9045g (1965)].

359 K. D. Kaufman and W. E. Russey, *J. Org. Chem.*, **30**, 1320 (1965).

360 K. D. Kaufman, *J. Org. Chem.*, **26**, 117 (1961).

361 E. T. McBee and E. Rapkin, *J. Amer. Chem. Soc.*, **73**, 2375 (1951).

362 T. Padmanathan and M. U. S. Sultanbawa, *J. Chem. Soc.*, **1963**, 4210.

363 J. Hlubucek, E. Ritchie, and W. C. Taylor, *Chem. Ind.* (London), **1969**, 1780.

364 E. N. Marvell, J. L. Stephenson, and J. Ong, *J. Amer. Chem. Soc.*, **87**, 1267 (1965).

365 A. W. Burgstahler, *J. Amer. Chem. Soc.*, **82**, 4681 (1960).

366 W. N. White and B. E. Norcross, *J. Amer. Chem. Soc.*, **83**, 3265 (1961).

367 D. McHale, S. Marcinkiewicz, and J. Green, *J. Chem. Soc.*, C, **1966**, 1427.

368 S. J. Rhoads and R. L. Crecelius, *J. Amer. Chem. Soc.*, **77**, 5060 (1955).

369 L. D. Huestis and L. J. Andrews, *J. Amer. Chem. Soc.*, **83**, 1963 (1961).

370 W. N. White and W. K. Fife, *J. Amer. Chem. Soc.*, **83**, 3846 (1961).

371 E. A. Vdovtsova, *Zh. Org. Khim.*, **5**, 498 (1969) [*C.A.*, **71**, 12719v (1969)].

372 V. G. Yagodin, L. I. Bunina-Krivorukova, and Kh. V. Bal'yan, *Zh. Org. Khim.*, **6**, 2513 (1970) [*C.A.*, **74**, 46009v (1971)].

373 B. S. Thyagarajan, K. K. Balasubramanian, and R. B. Rao, *Tetrahedron Lett.*, **1963**, 1393.

374 F. Weiss and A. Isard, *Bull. Soc. Chim. Fr.*, **1967**, 2033.

375 B. S. Thyagarajan, K. K. Balasubramanian, and R. B. Rao, *Tetrahedron*, **23**, 3205 (1967).

376 B. S. Thyagarajan, K. K. Balasubramanian, and R. B. Rao, *Tetrahedron*, **23**, 3533 (1967).

377 B. S. Thyagarajan, K. K. Balasubramanian, and R. B. Rao, *Chem. Ind.* (London), **1967**, 401.

378 E. Kaiser and E. Gunther, *J. Org. Chem.*, **25**, 1765 (1960).

379 R. G. Cooke, *Aust. J. Sci. Res.*, **3**, 481 (1950).

380 P. G. Holton, *J. Org. Chem.*, **27**, 357 (1962).

381 T. L. Patton, *J. Org. Chem.*, **27**, 910 (1962).

382 H. J. Minnemeyer, J. A. Egger, J. F. Holland, and H. Tieckelmann, *J. Org. Chem.*, **26**, 4425 (1961).

383 R. B. Moffett, *J. Org. Chem.*, **28**, 2885 (1963).

384 J. E. Litster and H. Tieckelmann, *J. Amer. Chem. Soc.*, **90**, 4361 (1968).

385 H. J. Minnemeyer, P. B. Clarke, and H. Tieckelmann, *J. Org. Chem.*, **31**, 406 (1966).

386 F. J. Dinan, H. J. Minnemeyer, and H. Tieckelmann, *J. Org. Chem.*, **28**, 1015 (1963).

387 Y. Makisumi, *Chem. Pharm. Bull.* (Tokyo), **11**, 851 (1963) [*C.A.*, **59**, 10048c (1963)].
388 J. K. Elwood and J. W. Gates, Jr., *J. Org. Chem.*, **32**, 2956 (1967).
389 H. Win and H. Tieckelmann, *J. Org. Chem.*, **32**, 59 (1967).
390 Y. Makisumi, *Tetrahedron Lett.*, **1964**, 2833.
391 Y. Makisumi, *Chem. Pharm. Bull.* (Tokyo), **12**, 1424 (1964) [*C.A.*, **62**, 10407b (1965)].
392 D. F. O'Brien and J. W. Gates, Jr., *J. Org. Chem.*, **31**, 1538 (1966).
393 E. Bergmann and H. Heimbold, *J. Chem. Soc.*, **1935**, 1365.
394 Y. Makisumi, *Chem. Pharm. Bull.* (Tokyo), **12**, 789 (1964) [*C.A.*, **61**, 9461d (1964)].
395 T. R. Seshadri and M. S. Sood, *Indian J. Chem.*, **3**, 354 (1965) [*C.A.*, **63**, 18009d (1965)].
396 H. Plieninger, H. Sirowej, and D. Rau, *Chem. Ber.*, **104**, 1863 (1971).
397 Y. Makisumi, *J. Org. Chem.*, **30**, 1986 (1965).
398 T. R. Chamberlain and M. F. Grundon, *J. Chem. Soc.*, *C*, **1971**, 910.
399 F. Scheinmann and H. Suschitzky, *Tetrahedron*, **7**, 31 (1959).
400 A. Mustafa, M. M. Sidky, S. M. A. D. Zayed, and F. M. Soliman, *Tetrahedron*, **19**, 1335 (1963).
401 Y. Makisumi, *Tetrahedron Lett.*, **1964**, 1635.
402 A. J. Quillinan and F. Scheinmann, *Chem. Commun.*, **1971**, 966.
403 H. D. Locksley, I. Moore, and F. Scheinmann, *J. Chem. Soc.*, *C*, **1966**, 2265.
404 W. Heimann and H. Baer, *Chem. Ber.*, **98**, 114 (1965).
405 S. Marcinkiewicz, D. McHale, and J. Green, *Proc. Chem. Soc.*, **1964**, 228.
406 Y. Makisumi, *Chem. Pharm. Bull.* (Tokyo), **11**, 859 (1963) [*C.A.*, **59**, 10048d (1963)].
407 J. A. Findlay and W. D. MacKay, *Chem. Commun.*, **1969**, 733.
408 P. Cresson, *C. R. Acad. Sci.*, *Ser. C*, **273**, 1382 (1971).
409 M. Bertrand and J. Le Gras, *C. R. Acad. Sci.*, *Ser. C*, **260**, 6926 (1965).
410 W. Sucrow and W. Richter, *Chem. Ber.*, **104**, 3679 (1971).
411 W. Sucrow, *Angew. Chem.*, **80**, 626 (1968).
412 W. Sucrow, *Tetrahedron Lett.*, **1970**, 4725.
413 R. P. Barnes and F. E. Chigbo, *J. Org. Chem.*, **28**, 1644 (1963).
414 J. P. Marion and Y. Bessière-Chrétien, *Chimia*, **24**, 72 (1970).
415 W. Sucrow and B. Girgensohn, *Chem. Ber.*, **103**, 750 (1970).
416 W. Sucrow, B. Schubert, W. Richter, and M. Slopianka, *Chem. Ber.*, **104**, 3689 (1971).
417 A. F. Thomas, *Helv. Chim. Acta*, **53**, 605 (1970).
418 A. F. Thomas and G. Ohloff, *Helv. Chim. Acta*, **53**, 1145 (1970).
419 R. J. Ferrier and N. Vethaviyasar, *Chem. Commun.*, **1970**, 1385.
420 S. Kukolja and V. Hahn, *Croat. Chem. Acta*, **33**, 229 (1961) [*C.A.*, **58**, 498d (1963)].
421 V. M. Dashunin and M. S. Tovbina, *Zh. Obshch. Khim.*, **34**, 1438 (1964) [*C.A.*, **61**, 5598h (1964)].
422 W. M. McLamore, E. Gelblum, and A. Bavley, *J. Amer. Chem. Soc.*, **78**, 2816 (1956).
423 R. Gardi and P. P. Castelli, *Gazz. Chim. Ital.*, **93**, 1681 (1963) [*C.A.*, **60**, 13290f (1964)].
424 J. Z. Mortensen, B. Hedegaard, and S. O. Lawesson, *Tetrahedron*, **27**, 3831 (1971).
425 T. C. Shields, W. E. Billups, and A. R. Lepley, *J. Amer. Chem. Soc.*, **90**, 4749 (1968).
426 J. A. Berson and E. J. Walsh, Jr., *J. Amer. Chem. Soc.*, **90**, 4730 (1968).
427 G. L. Buchanan, A. McKillop, and R. A. Raphael, *J. Chem. Soc.*, **1965**, 833.
428 A. Ali, D. Sarantakis, and B. Weinstein, *Chem. Commun.*, **1971**, 940.
429 A. Viola and J. H. MacMillan, *Chem. Commun.*, **1970**, 301.
430 A. Viola and E. J. Iorio, *J. Org. Chem.*, **35**, 856 (1970).
431 S. Swaninathan, K. G. Srinivasan, and P. S. Venkataramani, *Tetrahedron*, **26**, 1453 (1970).
432 P. Radlick and W. Fenical, *Tetrahedron Lett.*, **1967**, 4901.
433 G. M. Brooks, *Tetrahedron Lett.*, **1971**, 2377.
434 T. Sasaki, A. Kojima and M. Ohta, *J. Chem. Soc.* *C*, **1971**, 196.
435 Y. Tamura, Y. Kita, M. Shimagaki and M. Terashima, *Chem. Pharm. Bull.* (*Tokyo*), **19**, 571 (1971) [*C. A.*, **74**, 141069u (1971)].
436 R. M. Harrison, J. D. Hobson and M. M. Al Holly, *J. Chem. Soc. C*, **1971**, 3084.
437 L. A. Paquette and S. Maiorana, *Chem. Commun.* **1971**, 313.
438 E. Vedejs, *Chem. Commun.*, **1971**, 536.
439 R. C. Cookson and P. Singh, *J. Chem. Soc. C*, **1971**, 1477.

CHAPTER 2

SUBSTITUTION REACTIONS USING ORGANOCOPPER
REAGENTS

Gary H. Posner

The Johns Hopkins University, Baltimore, Maryland

CONTENTS

253

ACKNOWLEDGMENT

The help of Miss Susan A. Vladuchick, of E. I. du Pont de Nemours and Company, in searching the chemical literature for examples of organocopper substitution reactions is gratefully noted. Thanks are made also to Mrs. Ginny Selby for her meticulous typing of this manuscript.

INTRODUCTION

Carbon–carbon σ-bond formation, one of the most fundamental operations in organic chemistry, is often accomplished by interaction of an organometallic reagent with an organic substrate having a suitable leaving group. Organoalkali metal reagents derived from stabilized carbanions (e.g., lithium enolates) generally bring about effective substitution of X by R (Eq. 1),[1] whereas reagents derived from weak hydrocarbon acids[2] (e.g.,

$$RM + R'X \rightarrow R\text{--}R' + MX \qquad \text{(Eq. 1)}$$
$$(X = I, Br, Cl, OSO_2R, OCOR)$$

lithium alkyls) frequently undergo competing side reactions such as metal-halogen exchange (X = halogen),[3–6] α-metalation,[7, 8] α- and β-eliminations,[9–11] and coupling reactions that form the symmetrical dimers R–R

[1] H. O. House, Modern Synthetic Reactions, 2nd ed., W. A. Benjamin, New York, 1972.
[2] D. J. Cram, Fundamentals of Carbanion Chemistry, Academic Press, New York, 1965, Ch. 1.
[3] R. G. Jones and H. Gilman, Org. Reactions, 6, 339 (1951).
[4] H. Gilman, Org. Reactions, 8, 258 (1954).
[5] D. E. Applequist and D. F. O'Brien, J. Amer. Chem. Soc., 85, 743 (1963).
[6] H. J. S. Winkler and H. Winkler, J. Amer. Chem. Soc., 88, 964, 969 (1966).
[7] G. Köbrich, Angew. Chem., Int. Ed. Engl., 6, 41 (1967).
[8] W. Kirmse, Carbene Chemistry, Academic Press, New York, 1964.
[9] G. E. Coates, M. L. H. Green, and K. Wade, Organometallic Compounds, Vol. I, Methuen & Co. Ltd., London, 1967, pp. 298 ff.
[10] M. Tamura and J. Kochi, J. Organometal. Chem., 29, 111 (1971).
[11] G. M. Whitesides, E. R. Stedronsky, C. P. Casey, and J. San Filippo, Jr., J. Amer. Chem. Soc., 92, 1426 (1970).

and R'—R'. Organomagnesium reagents, although able to add across carbon-hetero atom multiple bonds, are not sufficiently reactive to effect substitution of X by R in most organic substrates unactivated toward displacement (e.g., nonallylic[12]); their reaction with activated substrates often leads to mixtures of products.[13, 14] Organoaluminum,[15] organocadmium,[16] and organozinc[16] reagents replace halogen by hydrocarbon groups in carboxylic acid halides but not in aliphatic halides; π-allylnickel reagents replace halogen by allylic groups in aliphatic halides.[17] Selective substitution of halogens and of alcohol derivatives (X = halide, sulfonyloxy, acyloxy) by various hydrocarbon groups in many different types of organic substrates has been achieved most successfully using organocopper reagents. The wide scope and effectiveness of these reagents in coupling with halides and with alcohol derivatives have made formation of the unsymmetrical coupling product R—R' a useful reaction in organic synthesis, allowing efficient and specific substitution of X by alkyl, alkenyl, alkynyl, or aryl groups. Several reviews are available.[18-24]

That copper metal and copper salts catalyze many organic reactions has been known for a long time. Two examples of such catalysis are the high-temperature coupling of aryl halides in the presence of finely powdered copper (Ullmann biaryl synthesis)[25, 26] and the conjugate addition of Grignard reagents to α,β-unsaturated carbonyl compounds in the presence

[12] A. Streitwieser, Jr., Solvolytic Displacement Reactions, McGraw-Hill Book Co., New York, 1962, Ch. 3.

[13] M. S. Kharasch and O. Reinmuth, Grignard Reactions of Nonmetallic Substances, Prentice-Hall, Englewood Cliffs, N.J., 1954.

[14] Hexamethylphosphoramide and tetrahydrofuran have been used as solvents with some success in promoting organomagnesium coupling with organic halides; the scope of this solvent effect, however, has yet to be determined; cf. H. Normant, Angew. Chem., Int. Ed. Engl., 6, 1046 (1967), and J. F. Normant, Bull. Soc. Chim. Fr. 1963, 1888.

[15] H. Reinheckel, K. Haage, and D. Jahnke, Organometal. Chem. Rev., A, 4, 55 (1969).

[16] D. A. Shirley, Org. Reactions, 8, 28 (1958).

[17] M. F. Semmelhack, Org. Reactions, 19, 115 (1972).

[18] J. F. Normant, Synthesis, 1972, 63.

[19] M. Nilsson, "Organocopper in Organic Synthesis," Carbocycl. Chem., manuscript submitted.

[20] G. Bähr and P. Burba in Methoden der Organischen Chemie, 4th ed., Vol. 13, Part 4, E. Müller, ed., Georg Theime Verlag, Stuttgart, 1970, p. 735.

[21] A. E. Jukes, "Organocopper Compounds," Advan. Organometal. Chem., manuscript in preparation.

[22] M. Nilsson, Svensk Kem. Tidskr., 80, 192 (1968) [C.A., 69, 87034k (1968)].

[23] K. Wada and H. Hashimoto, Senryo to Yakuhin, 12, 431 (1967) [C.A., 69, 87036n (1968)].

[24] (a) I. Kuwajima, Yuki Gosei Kagaku Kyokai Shi, 29, 616 (1971) [C.A., 75, 110354z (1971)]; J. Syn. Org. Chem. Jap., 29, 616 (1971); (b) W. Carruthers, Chem. Ind., 1973, 931; (c) T. Kaufmann, Angew Chem., Int. Ed. Engl., 13, 291 (1974).

[25] R. G. R. Bacon and H. A. O. Hill, Proc. Chem. Soc., 1962, 113.

[26] P. E. Fanta, Chem. Rev., 64, 613 (1964); Synthesis, 1974, 9.

of copper salts.[27] Both of these reactions were long thought to proceed through the intermediacy of σ-organocopper species.[28, 29] In 1966 it was shown conclusively that stoichiometric organocopper reagents prepared *in situ* prior to substrate introduction are indeed the reactive species in copper-catalyzed Grignard conjugate additions.[30] Experimentation since 1966 has led to a highly effective method for formation of carbon–carbon σ bonds by conjugate addition of organocopper reagents, a reaction that has been reviewed recently.[27] In the middle and late 1960s, work was begun to examine the mechanism, the scope, and the limitations of organocopper reagent coupling with organic halides.

Three types of organocopper reagents were prepared (Eqs. 2a–2c).* The insolubility and low reactivity of the mono-organocopper reagents (Eq. 2a) and the presence of the ligands (usually phosphines or sulfides

$$RLi + CuX \rightarrow RCu + LiX \qquad \text{(Eq. 2a)}$$

$$RLi + Lig\text{-}CuX \rightarrow Lig\text{-}CuR + LiX \qquad \text{(Eq. 2b)}$$

$$RLi + RCu \rightarrow R_2CuLi \qquad \text{(Eq. 2c)}$$

or their oxidation products) during workup of reactions using complexed organocopper reagents (Eq. 2b; Lig = ligand) limited the utility of these two types of reagents. The third group (Eq. 2c) involves lithium diorgano-cuprate(I) (or diorganocopperlithium) reagents and has proven extraordinarily useful for replacement of X by R in a wide variety of organic substrates.

Because selective coupling between organic substrate and organocopper reagent is usually achieved more effectively by stoichiometric than by catalytic organocopper reagents,[33] and more effectively still by organo-cuprates(I) than by mono-organocopper or by complexed organocopper reagents,† the emphasis in this chapter is on organocuprates(I). Where appropriate, the various types of organocopper reagents are compared.

---

* Although Gilman and his students had prepared methylcopper in 1936[31] and lithium dimethylcuprate(I) in 1952,[32] it was not until the late 1960s that the effectiveness of organocuprates(I) in coupling with organic halides was clearly demonstrated.

† Cuprous acetylides are an exception to this generalization.

[27] G. H. Posner, *Org. Reactions*, **19**, 1 (1972).

[28] J. Munch-Petersen, C. Bretting, P. Moller Jorgensen, S. Refn, and V. K. Andersen, *Acta Chem. Scand.*, **15**, 277 (1961).

[29] A. H. Lewin and T. Cohen, *Tetrahedron Lett.*, **1965**, 4531.

[30] H. O. House, W. L. Respess, and G. M. Whitesides, *J. Org. Chem.*, **31**, 3128 (1966).

[31] H. Gilman and J. M. Straley, *Rec. Trav. Chim. Pays-Bas*, **55**, 821 (1936).

[32] H. Gilman, R. G. Jones, and L. A. Woods, *J. Org. Chem.*, **17**, 1630 (1952).

[33] G. M. Whitesides, W. F. Fischer, Jr., J. San Filippo, Jr., R. W. Bashe, and H. O. House, *J. Amer. Chem. Soc.*, **91**, 4871 (1969). [33a] G. Fouquet and M. Schlosser, *Angew Chem., Int. Ed. Engl.*, **13**, 82 (1974).

Consideration is given to possible mechanisms of substitution reactions using organocopper reagents, to the scope, limitations, and synthetic utility of these reactions, and to optimal experimental conditions for their application.

MECHANISM

Despite a rapidly growing body of information, the available data are not sufficient to allow formulation of a detailed mechanism for substitution reactions using organocopper reagents. Indeed, depending on the type of reaction, different mechanisms may operate: thermal dimerization, oxidative dimerization, organocopper coupling with alkyl halides, organocopper coupling with alkenyl halides, or organocopper coupling with aryl halides.

Thermal dimerization of isomerically pure alkenylcopper species with essentially complete retention of configuration has been taken to indicate the absence of typical free radicals.[34] Thermolysis of alkylcopper compounds having a beta hydrogen atom produces a roughly equal mixture of alkene and alkane, but no dimers;[35] on the basis of these results and a deuterium-labeling experiment[36] a copper hydride mechanism has been proposed (Eq. 3).[36] Study of the electron spin resonance of the thermal de-

$$RCH_2CH_2Cu \rightarrow RCH=CH_2 + CuH \qquad \text{(Eq. 3a)}$$

$$RCH_2CH_2Cu + CuH \rightarrow RCH_2CH_3 + 2\ Cu \qquad \text{(Eq. 3b)}$$

$$2\ RCH_2CH_2Cu \rightarrow RCH=CH_2 + RCH_2CH_3 + 2\ Cu \qquad \text{(Eq. 4)}$$

composition of such alkylcopper species has produced evidence for the intermediacy of a binuclear copper cluster in which the two copper atoms are in different oxidation states;[37] based on this and related results, a direct dismutation mechanism occurring within a copper aggregate has been proposed (Eq. 4).[38] Choice between these two mechanisms is impossible with the evidence now available. Pyrolysis of an octameric arylcopper cluster compound was studied by nmr and by cryoscopic molecular weight determinations and was shown to involve pairwise formation of a biaryl in a unimolecular reaction.[39] Mechanistic discussion of organocopper thermolysis is increasingly centered on the role of polynuclear copper aggregates and the properties of the organic groups within these species.

[34] G. M. Whitesides and C. P. Casey, Jr., *J. Amer. Chem. Soc.*, **88**, 4541 (1966).

[35] G. M. Whitesides, C. P. Casey, J. San Filippo, Jr., and E. J. Panek, *Trans. N.Y. Acad. Sci.*, **29**, 572 (1967).

[36] G. M. Whitesides, E. R. Stedronsky, C. P. Casey, and J. San Filippo, Jr., *J. Amer. Chem. Soc.*, **92**, 1426 (1970).

[37] K. Wada, M. Tamura, and J. Kochi, *J. Amer. Chem. Soc.*, **92**, 6656 (1970).

[38] M. Tamura and J. K. Kochi, *J. Organometal. Chem.*, **42**, 205 (1972).

[39] A. Cairncross and W. A. Sheppard, *J. Amer. Chem. Soc.*, **93**, 247 (1971).

Oxidative dimerization of isomerically pure lithium diorganocuprate(I) species with essentially complete retention of configuration and of lithium dineophylcuprate(I) without rearrangement has been interpreted to indicate the absence of typical free radicals.[40] A mechanism has been suggested which involves oxidation of the dialkylcuprate(I) to a neutral, transient dialkylcopper(II) species that rapidly disproportionates to give alkylcopper(I) and alkyl-alkyl dimer.[40]

Organocopper coupling with alkyl and allylic halides and with oxygen derivatives is best considered a bimolecular nucleophilic substitution or an oxidative addition-reductive elimination.[41, 42] Substrate reactivity is typical for S_N2 reactions: primary > secondary > tertiary. The kinetics are roughly first order in organocopper reagent and first order in substrate.[38] Substitution with allylic rearrangement occurs in propargylic acetates[43] and in many allylic acetates.[44] Here, the stereochemical consequence is usually inversion.[33, 45, 45a] Distinction between an S_N2 mechanism (e.g., Eq. 5) and an oxidative addition mechanism involving a formal

$$R—\overset{\overset{R}{|}}{Cu}(I)^- + \overset{/}{\underset{/}{C}}\overset{\frown}{-X} \xrightarrow{-X^-} R—\overset{/}{C}\overset{\diagdown}{—} + RCu(I) \qquad \text{(Eq. 5)}$$

copper(III)[46-51] organometallic [or a copper(II)-radical complex[52]] (Eq. 6)

$$R_2\overset{\frown}{Cu}(I)^- + \overset{/}{\underset{/}{C}}\overset{\frown}{-X} \xrightarrow{-X^-} \left[R_2Cu(III)—\overset{/}{C}\overset{\diagdown}{—} \right] \longrightarrow R—\overset{/}{C}\overset{\diagdown}{—} + RCu(I)$$

$$\text{(Eq. 6)}$$

is not now possible.[33] Arguments for[53] and against[38] the oxidative addition pathway have recently appeared.

[40] G. M. Whitesides, J. San Filippo, Jr., C. P. Casey, Jr., and E. J. Panek, J. Amer. Chem. Soc., 89, 5302 (1967).

[41] J. P. Collman, Accts. Chem. Res., 1, 136 (1968).

[42] J. P. Collman, S. R. Winter and D. R. Clark, J. Amer. Chem. Soc., 94, 1788 (1972), inter alia.

[43] P. Rona and P. Crabbé, J. Amer. Chem. Soc., 91, 3289 (1969).

[44] R. J. Anderson, C. A. Henrick, and J. B. Siddall, J. Amer. Chem. Soc., 92, 735 (1970).

[45] C. R. Johnson and G. A. Dutra, J. Amer. Chem. Soc., 95, 7783 (1973).

[45a] P. H. Anderson, B. Stephenson, and H. S. Mosher, J. Amer. Chem. Soc., 96, 3171 (1974).

[46] G. H. Posner, Ph.D. Thesis, Harvard University, 1968 [Diss. Abstr., 29, 1613-B (1968)].

[47] J. K. Kochi, A. Bemis, and C. J. Jenkins, J. Amer. Chem. Soc., 90, 4616 (1968).

[48] H. O. House and M. J. Umen, J. Amer. Chem. Soc., 94, 5495 (1972).

[49] E. J. Corey and I. Kuwajima, J. Amer. Chem. Soc., 92, 395 (1970).

[50] R. W. Herr, D. M. Wieland, and C. R. Johnson, J. Amer. Chem. Soc., 92, 3813 (1970).

[51] For discussion of an oxidative addition mechanism involving Au(III) in reaction of alkylgold(I) with organic halides, see A. Tamiaki and J. K. Kochi, J. Organometal. Chem., 40, C81 (1972) and A. Tamiaki, S. A. Magennis, and J. K. Kochi, J. Amer. Chem. Soc., 95, 6487 (1973).

[52] J. Kochi and F. Rust, J. Amer. Chem. Soc., 83, 2017 (1961).

[53] A. H. Lewin and N. L. Goldberg, Tetrahedron Lett., 1972, 491.

Unlike organocopper coupling with alkyl halides, organocopper coupling with alkenyl halides proceeds stereospecifically with *retention* of configuration in the substrate.[54, 33] Typical free radicals can therefore be excluded. An S_N2 mechanism with retention of configuration has been proposed,[55] but an oxidative addition pathway is also possible.[56, 57]

Several mechanisms have been proposed for organocopper coupling with aryl halides. On the basis of isolation of several Meisenheimer complexes from interaction of arylcopper compounds and 1,3,5-trinitrobenzene, an aromatic nucleophilic substitution is suggested for arylcopper coupling with aryl halides.[58] Study of aryl substituent effects in coupling of lithium dimethylcuprate with aryl halides has also led to proposal of an aromatic nucleophilic substitution mechanism.[59] Interaction of aryl halides with some lithium dialkylcuprates and lithium diarylcuprates has been shown to cause initial transmetalation, forming a mixed homocuprate (ArRCuLi or ArAr'CuLi, Eq. 7a), which is in rapid equilibrium with the two related

$$\text{ArX} + (\text{Ar}')_2\text{CuLi} \rightleftharpoons \text{ArAr}'\text{CuLi} + \text{Ar}'\text{X} \qquad \text{(Eq. 7a)}$$

$$2\ \text{ArAr}'\text{CuLi} \rightleftharpoons (\text{Ar})_2\text{CuLi} + (\text{Ar}')_2\text{CuLi} \qquad \text{(Eq. 7b)}$$

$$\text{ArAr}'\text{CuLi} + \text{O}_2 \rightarrow \text{Ar-Ar}' \qquad \text{(Eq. 7c)}$$

homocuprates (Eq. 7b); subsequent addition of an oxidant then causes formation of the coupled products Ar—R (usually along with Ar—Ar and R—R) or Ar—Ar' (and Ar—Ar plus Ar'—Ar', Eq. 7c).[33, 60] The composition of the mixture of coupled products is usually predicted statistically from the amount of Ar and R or Ar and Ar' groups present before oxidation. Cuprous acetylide coupling with aryl halides has been suggested to involve a four-center transition state.[61]

As the pieces of the mechanistic jigsaw puzzle are slowly put together and a clear picture of the mechanism of substitution reactions using organocopper reagents is formed, there will emerge a more thorough understanding of why copper among the transition metals promotes

[54] E. J. Corey and G. H. Posner, *J. Amer. Chem. Soc.*, **89**, 3911 (1967).

[55] J. Klein and R. Levene, *J. Amer. Chem. Soc.*, **94**, 2520 (1972).

[56] Iron-catalyzed Grignard substitution of halogen in alkenyl halides proceeds with retention of configuration and is proposed to involve double bond-iron coordination: M. Tamura and J. Kochi, *J. Amer. Chem. Soc.*, **93**, 1487 (1971).

[57] R. G. Pearson and W. R. Muir, *J. Amer. Chem. Soc.*, **92**, 5519 (1970).

[58] O. Wennerström, *Acta Chem. Scand.*, **25**, 2341 (1971).

[59] V. N. Drozd and O. I. Trifonova, *Zh. Org. Khim.*, **6**, 2493 (1970); *J. Org. Chem. USSR*, **6**, 2504 (1970).

[60] H. O. House, D. G. Koespell, and W. J. Campbell, *J. Org. Chem.*, **37**, 1003 (1972).

[61] C. E. Castro, R. Havlin, V. K. Honwad, A. Malte, and S. Mojé, *J. Amer. Chem. Soc.*, **91**, 6464 (1969).

coupling so effectively. Undoubtedly the uniqueness of copper is attributable in large part to the relatively low ionic character of a copper-carbon bond,* to the low oxidation potential (0.15 V) separating cuprous from cupric ions, and to the tendency of copper to form polynuclear copper clusters[39, 62] and mixed-valence copper compounds.[33, 37, 63]

SCOPE AND LIMITATIONS

The Organocopper Reagent

The scope and limitations of substitution reactions using organocopper reagents are discussed in this section with emphasis on the role of the organocopper reagent. Of the by-products formed in these substitution reactions, two types are typical: reduced substrate arising via metal-halogen exchange, and symmetrical dimers (R—R) arising from the organic group in the organocopper reagent itself. The limitations imposed by these side products can usually be minimized by proper selection of organocopper reagent and experimental conditions (see p. 296).

Nature and Preparation of Stoichiometric Organocopper Reagents

The nature and preparation of organocopper reagents have recently been reviewed in *Organic Reactions*;[27] discussion in this section is therefore limited to a short summary of the appropriate material in that review and to new developments.

RCu Reagents. The most general preparation of RCu species involves metathesis between an organometallic reagent and a copper(I) salt (Eq. 8)

$$R—Met + CuX \rightarrow RCu + Met—X \qquad \text{(Eq. 8)}$$

at low temperature (0° or below) in an inert atmosphere. Metals (Met) used for this metathesis include lead,[64] zinc,[65] mercury,[66] magnesium,[67] lithium, boron,[68] and apparently aluminum.[69] Magnesium and lithium are

---

* On the Pauling electronegativity scale, copper is 1.9 and carbon is 2.5.

[62] L. E. McCandlish, E. C. Bissell, D. Coucouvanis, J. P. Fackler, and K. Knox, *J. Amer. Chem. Soc.*, **90**, 7357 (1968), and references cited therein.

[63] G. H. Posner, *Methodicum Chimicum Houben-Weyl*, Vol. VIII, "Synthesis of Derivatives of Copper," F. Korte, ed., Georg Thieme Verlag, Stuttgart, 1973.

[64] H. Gilman and L. A. Woods, *J. Amer. Chem. Soc.*, **65**, 435 (1941).

[65] K. H. Thiele and J. Köhler, *J. Organometal. Chem.*, **12**, 225 (1968).

[66] G. M. Whitesides and D. E. Bergbreiter, *J. Amer. Chem. Soc.*, in press.

[67] N. T. Luong-Thi, H. Rivière, J. Bégué, and C. Forrestier, *Tetrahedron Lett.*, **1971**, 2113.

[68] A. N. Nesmeyanov, V. A. Sazonova, and N. N. Sedova, *Dokl. Akad. Nauk SSSR*, **202**, 362 (1972) [*C.A.*, **76**, 140984t (1972)].

[69] G. Zweifel and R. L. Miller, *J. Amer. Chem. Soc.*, **92**, 6678 (1970).

used most often, with cuprous iodide[70] in diethyl ether as solvent. 2,5-Dicuprio-3,4-dichlorothiophene has been prepared from the corresponding dilithium compound and cuprous iodide.[71] The effect of trace metal impurities in decreasing the stability (i.e., catalyzing the decomposition) of RCu organocopper species has been noted.[37, 72] Several perhaloarylcopper species have been shown by mass spectroscopic and cryoscopic data to exist as tetrameric and octameric copper cluster compounds;[39] the insolubility of alkylcopper species in most organic solvents suggests a polymeric structure for them as well.[31] Infrared[73, 74] and nmr studies of a limited number of organocopper compounds have been done, and the effect of the magnetic anisotropy between carbon and copper has been discussed.[75] The chemical reactivity of RCu species appears to be lower than that of analogous R_2Cd compounds.[76]

Several methods of limited generality have also been used to prepare stoichiometric RCu organocopper reagents. Thermal decarboxylation of cuprous carboxylates has produced various arylcopper species[77-79] and, in one instance, a benzylic copper compound (Eq. 9).[80] Treating perfluoro-

$$RCO_2Cu \xrightarrow{\text{Heat}} RCu + CO_2 \quad \text{(Eq. 9)}$$

$$R_FI + Cu \xrightarrow[\text{Heat}]{\text{DMSO}} R_FCu \quad \text{(Eq. 10)}$$

alkyl and perhaloaryl iodides with activated copper bronze in polar aprotic solvents (dimethylformamide, dimethyl sulfoxide) causes generation in situ of the corresponding organocopper reagents (Eq. 10); evidence for the presence of these species rests in some cases on their protonolysis to form perfluoroalkanes and perhaloarenes and in some cases on their isolation.[81] Several dicopper compounds have been prepared in this way from the corresponding α,ω-diiodides.[81-83] It should be noted that the Ullmann

[70] A. E. Jukes, S. S. Dua, and H. Gilman, J. Organometal. Chem., 21, 241 (1970).
[71] M. R. Smith, Jr., and H. Gilman, J. Organometal. Chem., 42, 1 (1972).
[72] M. Boussu and J. E. Dubois, C.R. Acad. Sci., Ser. C, 273, 1270 (1971).
[73] A. Camus and N. Marsich, J. Organometal. Chem., 14, 441 (1968).
[74] G. Costa, A. Camus, L. Gatti, and N. Marsich, J. Organometal. Chem., 5, 568 (1966).
[75] A. Baici, A. Camus, and G. Pellizer, J. Organometal. Chem., 26, 431 (1971), and references cited therein.
[76] S. S. Dua, A. E. Jukes, and H. Gilman, Organometal. Chem. Syn., 1, 87 (1970).
[77] A. Cairncross, J. R. Roland, R. M. Henderson, and W. A. Sheppard, J. Amer. Chem. Soc., 92, 3187 (1970).
[78] T. Cohen and R. A. Shambach, J. Amer. Chem. Soc., 92, 3189 (1970).
[79] J. Chodowska-Palicka and M. Nilsson, Acta Chem. Scand., 25, 3451 (1971).
[80] B. M. Trost and P. L. Kinson, J. Org. Chem., 37, 1273 (1972).
[81] V. C. R. McLoughlin and J. Thrower, Tetrahedron, 25, 5921 (1969).
[82] J. Burdon, P. L. Coe, C. R. Marsh, and J. C. Tatlow, Chem. Commun., 1967, 1259.
[83] J. Burdon, P. L. Coe, C. R. Marsh, and J. C. Tatlow, J. Chem. Soc., Perkin I, 1972, 639.

biaryl synthesis,[25, 26] a highly useful method for coupling of aryl halides, is thought to involve generation *in situ* of arylcopper compounds. Cuprous acetylides are routinely prepared from the corresponding acetylene and a cuprous halide in ammoniacal solution;[84] recently cuprous acetylides have been prepared in high yield from acetylenes and cuprous chloride in hexamethylphosphoramide.[85] Monoalkenylcopper reagents have been generated stereospecifically by *cis* addition of a monoalkylcopper species to an acetylene[86, 87] (Eq. 11).

$$RCu + R'C{\equiv}CH \xrightarrow{\text{Et}_2\text{O}} \underset{R}{\overset{R'}{\diagdown}}C=C\underset{Cu}{\overset{H}{\diagup}} \qquad \text{(Eq. 11)}$$

RCu· Ligand Reagents.   Metathesis between an organometallic compound and a cuprous halide coordinated with a ligand or cupric acetyl-acetonate[88] in presence of a ligand leads to the corresponding RCu·Ligand species (Eq. 12). Organolithium and, in one case, organothallium[89] compounds have been used with cuprous iodide-phosphine,[33, 90, 91] -sulfide,[33] or -amine[91] complexes. Infrared[89] and X-ray crystallographic data[92, 39] show the triphenylphosphine(cyclopentadienyl)copper(I) species to have pentahaptocyclopentadienyl rings.

$$R\text{—Met} + \text{Ligand· }CuX \rightarrow RCu\text{· Ligand} + \text{Met—X} \qquad \text{(Eq. 12)}$$

R$_2$CuLi Reagents.   Lithium diorganocuprate(I) reagents have been prepared in three ways at low temperature (0° or below) and in an inert atmosphere (Eqs. 2c, 13a, 13b).* The reagent prepared via Eq. 13a differs from that prepared by Eq. 2c only by the presence of LiX; the effect of such inorganic salts on the course of organocopper substitution reactions is deleterious in only a few cases (*e.g.*, in coupling with aryl iodides),[33] and

---

* The nature of the diorganocopper reagent represented as R$_2$CuLi is unclear; the R$_2$CuLi notation itself was introduced merely to indicate the stoichiometry of the reaction represented in Eq. 13a (p. 264) and was not intended to convey structural information.

[84] R. D. Stephens and C. E. Castro, *J. Org. Chem.*, **28**, 3313 (1963).

[85] M. Bourgain and J. F. Normant, *Bull. Soc. Chim. Fr.*, **1969**, 2477.

[86] J. F. Normant and M. Bourgain, *Tetrahedron Lett.*, **1971**, 2583.

[87] J. F. Normant, G. Cahiez, C. Chuit, A. Alexakis, and J. Villieras, *J. Organometal. Chem.*, **40**, C49 (1972).

[88] A. Yamamoto, A. Miyashita, T. Yamamoto, and S. Ikeda, *Bull. Chem. Soc. Jap.*, **45**, 1583 (1972).

[89] F. A. Cotton and T. J. Marks, *J. Amer. Chem. Soc.*, **91**, 7281 (1969).

[90] G. Costa, G. Pellizer, and F. Rubessa, *J. Inorg. Nucl. Chem.*, **26**, 961 (1964).

[91] G. Costa, A. Camus, N. Marsich, and L. Gatti, *J. Organometal. Chem.*, **8**, 339 (1967).

[92] F. A. Cotton and J. Takats, *J. Amer. Chem. Soc.*, **92**, 2353 (1970).

[93] L. T. J. Delfaere and D. W. McBride, *Acta Crystallogr.*, **B26**, 515 (1970).

therefore Eq. 13a offers the most convenient and practical way for preparing and using primary lithium diorganocuprates(I). Cuprous iodide is

$$RCu + RLi \rightarrow R_2CuLi \qquad \text{(Eq. 2c)}$$

$$2\,RLi + CuX \rightarrow R_2CuLi + LiX \qquad \text{(Eq. 13a)}$$

$$2\,RLi + Ligand \cdot CuX \rightarrow R_2CuLi \cdot Ligand + LiX \qquad \text{(Eq. 13b)}$$

most often used; cuprous bromide, however, must be used to prepare di*aryl*cuprates(I).[33] It should be noted that Grignard reagents cannot in general be substituted for organolithium reagents in this procedure.[33, 94] The preparation and use of R_3CuLi_2 in a substitution reaction has just recently been reported.[60] Secondary and tertiary organocuprates(I) are less stable thermally than primary organocuprates(I). Hence they are best prepared in the presence of a stabilizing ligand such as a phosphine or sulfide, according to Eq. 13b, even though such ligands usually make product isolation difficult.[33]

Data concerning the structure of organocuprate(I) species are limited. A recent nmr study indicates a tetranuclear metal cluster structure with bridging organic groups.[95]

It is now well established that organocuprates(I) are more reactive than mono-organocopper species in substitution reactions. Thus, methylcopper does not react with iodocyclohexane,[46] whereas methylcyclohexane is formed in 75% yield with lithium dimethylcuprate(I).[54] Likewise, 1,2-epoxycyclohexane is inert to methylcopper but is converted to *trans*-2-methylcyclohexanol by lithium dimethylcuprate(I).[50] Allylic and propargylic acetates behave similarly; they are inert to methylcopper and n-butylcopper but undergo substitution with the corresponding lithium dialkylcuprates(I).[43, 50] Transition metal impurities have been shown to increase the reactivity of organocuprates(I) in substitution reactions.[33]

The nucleophilicity of organocuprates(I) toward carbonyl compounds is substantially lower than that of analogous Grignard or lithium reagents. Thus organocuprates(I) have been used at $-78°$ for selective reaction with the acid chloride portion of molecules containing remote halo-, cyano-, acyl-, and alkoxycarbonyl groups.[96]

Organocuprates(I) appear to be more basic than RCu organocopper species. Lithium dimethylcuprate(I) in ether, for example, abstracts the acidic proton of terminal acetylenes, whereas methylcopper in ether adds

[94] L. M. Seitz and R. Madl, *J. Organometal. Chem.*, **34**, 415 (1972).

[95] G. van Koten and J. G. Noltes, *Chem. Commun.*, **1972**, 940.

[96] G. H. Posner, C. E. Whitten, and P. E. McFarland, *J. Amer. Chem. Soc.*, **94**, 5106 (1972).

to terminal acetylenes to give an alkenylcopper species (Eq. 14).[86] It is

$$RH + (R)(R'C{\equiv}C)CuLi \xleftarrow[Et_2O]{R_2CuLi} R'C{\equiv}CH \xrightarrow[Et_2O]{RCu(MgBr_2)} \quad \begin{array}{c} R' \\ R \end{array}{>}{=}{<} \begin{array}{c} H \\ Cu \end{array}$$

$$(Eq.\ 14)$$

noteworthy that, when the solvent is tetrahydrofuran or hexamethyl-phosphoramide, even methylcopper is sufficiently basic to abstract the acidic proton of the acetylene.[86]

RR'CuLi Reagents. Mixed cuprates(I) have been prepared recently in three ways (Eqs. 15–17). Use of isomerically pure organomercurials to

$$RCu + R'Li \to RR'CuLi \qquad (Eq.\ 15)$$

R	R'
CH_3	Ar^{70}
CH_3	$(C_6H_5)_3Si^{21}$
$n\text{-}C_5H_{11}C{\equiv}C$	$CH_3^{18,\ 98}$
$n\text{-}C_3H_7C{\equiv}C$	$CH_2{=}CH^{97}$

$$RC{\equiv}CH + R'_2CuLi \to (RC{\equiv}C)(R')CuLi \qquad (Eq.\ 16)^{86}$$

$$t\text{-}C_4H_9OCu + RLi \to (t\text{-}C_4H_9O)(R)CuLi \qquad (Eq.\ 17)^{100}$$
$$(R = sec\text{-}C_4H_9, t\text{-}C_4H_9)$$

prepare mixed homocuprates has been demonstrated with the 2-norbornyl system.[66] In at least several cases the mixed homocuprates RR'CuLi have been shown to be in rapid equilibrium with the corresponding homocuprates R_2CuLi and R'_2CuLi (see Eq. 7b).[33, 60] Several substitution reactions have been achieved with these mixed homocuprates in which the R (or R') group is transferred selectively to the substrate;[97a] the selectivity of this transfer may be due to the greater reactivity of the R group within the RR'CuLi species or to the greater reactivity of R_2CuLi over R'_2CuLi (or to both). Furthermore, with (alkyl)(acetylenic) mixed homocuprates, the alkyl group is transferred selectively in some conjugate addition reactions,[97b] but the acetylenic group is transferred selectively in some substitution reactions.[98] Unfortunately the data now available do not allow any generalizations about which factors (steric bulk, organic group nucleophilicity, or aggregate structure) determine the facility of group transfer from mixed homocuprates. The selective transfer of the alkyl group in substitutions with alkoxy-alkylcuprates (Eq. 17)[99, 100] presumably occurs

---

[97a] W. H. Mandeville and G. M. Whitesides, *J. Org. Chem.*, **39**, 400 (1974).
[97b] E. J. Corey and D. J. Beames, *J. Amer. Chem. Soc.*, **94**, 7210 (1972).
[98] (a) J. F. Normant and M. Bourgain, *Tetrahedron Lett.*, **1970**, 2659; (b) M. Bourgain, J. Villieras, and J. F. Normant, *C.R. Acad. Sci. Paris, Ser. C*, **276**, 1477 (1973).
[99] G. H. Posner and J. J. Sterling, *J. Amer. Chem. Soc.*, **95**, 3076 (1973).
[100] G. H. Posner and C. E. Whitten, *Tetrahedron Lett.*, 1815 (1973).

266 ORGANIC REACTIONS

directly from the heterocuprate rather than from any homocuprate, R_2CuLi, that might be present.[101]

Thermal and Oxidative Dimerization of R

Most organocopper reagents are extremely sensitive to heat and to oxidants. Ethylcopper, for example, decomposes at $-18°$ and phenylcopper at $80°$;[30] thermal stability decreases as follows: neopentyl > methyl > n-propyl > ethyl > isopropyl.[38] Perhaloalkylcopper and perhaloarylcopper species are much more thermally stable than the corresponding nonhalogenated species.[39, 81, 102, 103a] Complete oxidative decomposition of lithium diorganocuprate(I) reagents can be achieved rapidly even at $-78°$ simply by bubbling oxygen into the dilute (≤ 0.1 M)[33] reaction mixture.[40] Other oxidants (e.g., nitroaromatics and copper(II) salts) have also been used effectively.[40]

A large number of RCu and RCu· Ligand species have been thermally or oxidatively dimerized to symmetrical products R—R (see Table IIA). Examples include dimerization of alkyl, alkenyl, alkynyl, aryl,[103b] heteroaryl, benzylic, and functionalized alkyl groups. There are several outstanding features of these dimerizations. First, thermolysis or oxidation of alkenylcopper species produces dimers (butadiene derivatives) stereospecifically with retention of configuration.[104, 105] Second, dimerization of functionalized alkylcopper reagents has led to preparation of various unusual types of compounds (e.g., 1,2-bis-sulfones, 1,2-bisphosphine oxides)[106] and to fundamental classes of compounds having substantial synthetic utility (e.g., 1,4-diketones, Eq. 18).[107, 108] Finally, dimerization of organocopper species may be the preferred method to prepare

$$(C_6H_5)_2C=NN \qquad (C_6H_5)_2C=NN \qquad NN=C(C_6H_5)_2$$
$$\overset{\|}{R\overset{\|}{C}CH_2Cu} \xrightarrow{0-20°} R\overset{\|}{C}CH_2CH_2\overset{\|}{C}R$$
$$\xrightarrow{H_2O} \quad \overset{O}{R\overset{\|}{C}CH_2CH_2\overset{\|}{C}R} \quad \text{(Eq. 18)}$$

[101] G. H. Posner, C. E. Whitten, and J. J. Sterling, J. Amer. Chem. Soc., 95, 7778 (1973), and C. E. Whitten, Ph.D. Thesis, The Johns Hopkins University, February, 1974.

[102] A. Cairncross, H. Omura, and W. A. Sheppard, J. Amer. Chem. Soc., 93, 248 (1971).

[103] (a) A. Cairncross and W. A. Sheppard, J. Amer. Chem. Soc., 90, 2186 (1968); (b) For a recent study of arylcopper(I) autoxidation, see A. Camus and N. Marsich, J. Organometal. Chem., 46, 385 (1972).

[104] G. M. Whitesides, C. P. Casey, and J. K. Krieger, J. Amer. Chem. Soc., 93, 1379 (1971).

[105] T. Kauffmann and W. Sahm, Angew. Chem., 79, 101 (1967); Angew. Chem., Int. Ed. Engl., 6, 85 (1967).

[106] (a) T. Kauffmann and D. Berger, Chem. Ber., 101, 3022 (1968); (b) C. A. Maryanoff, B. E. Maryanoff, R. Tang, and K. Mislow, J. Amer. Chem. Soc., 95, 5839 (1973).

[107] T. Kauffmann, M. Schönfelder, and J. Legler, Ann., 731, 37 (1970).

[108] M. W. Rathke and A. Lindert, J. Amer. Chem. Soc., 93, 4605 (1971).

natural products accessible only with difficulty by other means (e.g., kotanin, Eq. 19).[109]

$$2,4,6\text{-}(CH_3O)_3C_6H_2Cu \xrightarrow[\text{(25\%)}]{CuCl_2}$$

$$2,4,6\text{-}(CH_3O)_3C_6H_2C_6H_2(OCH_3)_3\text{-}2,4,6 \quad \text{(Eq. 19)}$$

Relatively few lithium diorganocuprate(I) compounds have been intentionally thermolyzed or oxidized (see Table IIB). There are available several examples of alkyl, alkenyl, alkynyl, and aryl group dimerization. The most useful feature of these dimerizations is that typical radicals are not involved; thus lithium dineophylcuprate(I) (Eq. 20) and lithium

$$[C_6H_5C(CH_3)_2CH_2]_2CuLi \xrightarrow[O_2, -78°]{THF}$$

$$C_6H_5C(CH_3)_2CH_2CH_2C(CH_3)_2C_6H_5 \quad \text{(Eq. 20)}$$
$$\text{(88\%)}$$

dialkenylcuprates(I) are thermally and oxidatively dimerized without structural or geometric rearrangement.[40]

Thermal or oxidative dimerization of mixed homocuprates(I) has not been widely used (see Table IIC). This procedure for preparing unsymmetrical dimers R—R′ may, however, have distinct advantages over direct substitution of R′X by R_2CuLi, especially when such substitution is difficult, (for example with aryl halides).[33] A severe limitation on this method for forming dimers R—R′ is concomitant formation of undesired symmetrical dimers R—R and R′—R′, presumably due to the presence of some R_2CuLi and $R_2'CuLi$ species before oxidation.[33] Useful amounts of unsymmetrical coupling products R—R′ have nevertheless been formed in several instances from RR′CuLi reagents: (alkyl)(alkyl), (alkyl)(alkenyl), (alkyl)(aryl), and (aryl)(aryl).

Structural Variation of R

The aim in this section is to enumerate in one place all of the kinds of R groups in organocopper reagents which have undergone substitution reactions. The structural types of R are considered in terms of the formal hybridization of the carbon bound to copper: sp^3, sp^2, sp, and in terms of special cases (e.g., allylic, haloalkyl, haloaryl, and functionalized alkyl R). Only references to the more unusual examples are given. In later sections both the effect of the structure of R on the course of the substitution reaction and on the choice of an organocopper reagent (e.g., RCu·Ligand or R_2CuLi) for substitution with a particular type of R are discussed.

sp^3-Hybridized primary R groups from methyl through butyl have

[109] G. Büchi, D. H. Klaubert, R. C. Shank, S. M. Weinreb, and G. N. Wogan, J. Org. Chem., 36, 1143 (1971).

undergone successful substitution reactions. Isopropyl,[100, 110, 111] sec-butyl,[33, 100, 112] and 2-norbornyl[66] are the only examples of secondary sp^3-hybridized R groups, and t-butyl[100] and 3-ethyl-3-pentyl[112a] the only examples of tertiary sp^3-hybridized R groups that have undergone organo-copper substitution reactions.

Vinyl[33] and propenyl[113, 114] cuprates(I) and various alkenyl RCu species[87] have been used successfully in substitution reactions. Likewise, there have been reports of organocopper substitutions using phenyl, p-tolyl,[115] p-anisyl,[67] and p-dimethylaminophenyl[67] organocuprates(I) as well as phenyl, p-tolyl,[116] p-anisyl,[31] o-dimethylaminomethylphenyl,[117] 2,6-dimethoxyphenyl,[118] 2,4,6-trimethoxyphenyl,[119] and cyclopentadienyl[120] RCu species. Heteroarylcopper species of several types undergo substitution reactions: 2-furylcopper,[121] 2-thienylcopper,[121] 2-(1-methyl)pyrrolyl-copper,[122] and 4-pyridylcopper[123] and also a 2,5-dicopperthienyl system.[71]

sp-Hybridized R groups (cuprous acetylides) have long been known to couple effectively with organic halides.* The large number of structural types of R'C≡CCu having undergone successful substitution is indicated by the size of Table IV. Structural variation in R' has included alkyl,[125, 126] alkenyl,[125, 127] alkynyl,[128] aryl,[129–132] heteroaryl,[127–129, 133]

* Although cuprous cyanides are similar to cuprous acetylides, cuprous cyanides are excluded from this review; for leading references on cuprous cyanide chemistry, see ref. 124.

110 A. T. Worm and J. H. Brewster, J. Org. Chem., 35, 1715 (1970).

111 J. E. Dubois, C. Lion, and C. Moulineau, Tetrahedron Lett., 1971, 177.

112 C. R. Johnson and G. A. Dutra, J. Amer. Chem. Soc., 95, 7777 (1973).

112a J. E. Dubois, M. Boussu, and C. Lion, Tetrahedron Lett., 1971, 829.

113 G. Büchi and J. A. Carlson, J. Amer. Chem. Soc., 90, 5336 (1968); 91, 6470 (1969).

114 O. P. Vig, J. C. Kapur, and S. D. Sharma, J. Indian Chem. Soc., 45, 1026 (1968).

115 O. P. Vig, S. D. Sharma, and J. C. Kapur, J. Indian Chem. Soc., 46, 167 (1969).

116 T. Sato and S. Watanabe, Chem. Commun., 1969, 515.

117 G. van Koten, A. J. Leusink, and J. G. Noltes, Chem. Commun., 1970, 1107.

118 C. Bjorklund, M. Nilsson, and O. Wennerström, Acta Chem. Scand., 24, 3599 (1970).

119 J. G. Noltes, Institute for Organic Chemistry, Utrecht, The Netherlands, unpublished results.

120 M. Nilsson, R. Wahren, and O. Wennerström, Tetrahedron Lett., 1970, 4583.

121 M. Nilsson and C. Ullenius, Acta Chem. Scand., 24, 2379 (1970).

122 N. Gjos and S. Gronowitz, Acta Chem. Scand., 25, 2596(1971).

123 E. J. Soloski, W. E. Ward, and C. Tamborski, J. Fluorine Chem., 2, 361 (1972/1973).

124 H. O. House and W. F. Fischer, Jr., J. Org. Chem., 34, 3626 (1969), and references therein.

125 F. Bohlmann, P. Bonnet, and H. Hofmeister, Chem. Ber., 100, 1200 (1967).

126 S. A. Mladenovic and C. E. Castro, J. Heterocycl. Chem., 5, 227 (1968).

127 R. E. Atkinson, R. F. Curtis, and G. T. Phillips, Chem. Ind., 1964, 2101.

128 S. P. Korshunov, R. I. Katkevich, and L. I. Vereshchagin, Zh. Org. Khim. 3, 1327 (1967); J. Org. Chem. USSR, 3, 1288 (1967).

129 R. F. Curtis and J. A. Taylor, Tetrahedron Lett., 1968, 2919.

130 M. D. Rausch and A. Siegel, J. Org. Chem., 34, 1974 (1969).

131 C. C. Bond and M. Hooper, J. Chem. Soc., C, 1969, 2453.

132 J. Ipaktschi and H. A. Staab, Tetrahedron Lett., 1967, 4403.

133 C. E. Castro, E. J. Gaughan, and D. C. Owsley, J. Org. Chem., 31, 4071 (1966).

and haloaryl,[134, 135] as well as variously functionalized alkyl systems (e.g., phenoxymethyl[83] and hydroxymethyl[126, 127, 136]).

Lithium diallylcuprates have been used to couple with only two types of organic substrates: alkyl halides[33] and epoxides.[137]

Haloalkylcopper species studied to date vary from trifluoromethylcopper[81] through n-$C_9F_{19}Cu$[81] and include perfluoroisopropylcopper[81] and several α,ω-perfluoroalkyldicopper species, $Cu(CF_2)_nCu$, with n equal to 3, 4 and 7.[81] Haloarylcopper compounds include pentafluorophenyl,[103, 138] pentachlorophenyl,[123] pentabromophenyl,[123] p-bromotetrafluorophenyl,[123] 2,3,5,6-tetrafluorophenyl,[123] and o-(pentafluorophenyl)tetrafluorophenyl.[139] Halogenated heteroarylcopper species include tetrachloro-4-pyridyl,[76, 123] tetrafluoro-4-pyridyl,[123] trichloro-2-thienyl,[140] and dichloro-2,5-thienyldicopper.[71]

Several functionalized alkylcopper compounds have been reported to undergo effective and highly useful organocopper substitution reactions: phenylthiomethylcopper,[141] cyanomethylcopper,[142] and ethoxycarbonylmethylcopper.[143] α-Alkoxycarbonyl-substituted alkenylcopper species (1)[144, 145] and α-(dimethylaminomethyl)vinylcuprate (2)[146] also couple effectively with organic halides.

The types of R groups transferred in substitution reactions using organocopper reagents are probably not limited to those just mentioned. It is anticipated that R groups of many different and novel structural types will ultimately be used; the main limitation appears to be the availability of RMet species from which RCu and organocuprate(I) reagents can be prepared.

[134] F. Waugh and D. R. M. Walton, J. Organometal. Chem., **39**, 275 (1972).

[135] M. D. Rausch, A. Siegel, and L. P. Klemann, J. Org. Chem., **34**, 468 (1969).

[136] M. Stefanovic, Lj. Krstic, and S. Mladenovic, Tetrahedron Lett., **1971**, 3311.

[137] J. Fried, C. H. Lin, J. C. Sih, P. Dalven, and G. F. Cooper, J. Amer. Chem. Soc., **94**, 4342 (1972).

[138] R. J. DePasquale and C. Tamborski, J. Org. Chem., **34**, 1736 (1969).

[139] R. Filler and A. E. Fiebig, Chem. Commun., **1970**, 546.

[140] M. R. Smith, Jr., M. T. Rahman, and H. Gilman, Organometal. Chem. Syn., **1**, 295 (1971).

[141] E. J. Corey and M. Jautelat, Tetrahedron Lett., **1968**, 5787.

[142] E. J. Corey and I. Kuwajima, Tetrahedron Lett., **1972**, 487.

[143] I. Kuwajima and Y. Doi, Tetrahedron Lett., **1972**, 1163.

[144] J. A. Katzenellenbogen, Ph.D. Thesis, Harvard University, 1969 [Diss. Abstr., **31**, 1826–B (1970)].

[145] C. A. Henrick and J. B. Siddall, Zoëcon Corporation, unpublished results.

[146] E. J. Corey, D. Cane, and L. Libit, J. Amer. Chem. Soc., **93**, 7016 (1971).

Stereochemical Stability of R

Only one type of sp^3-hybridized R group capable of existing as one or both of two epimers has been reported; *endo*-2-norbornylcopper-tri-*n*-butylphosphine complex **3** is configurationally stable at $-78°$ in ether and undergoes highly stereoselective coupling reactions.[33] Mixed (*t*-butyl)(2-norbornyl)homocuprate species as well undergo oxidative dimerization and cross-coupling reactions with greater than 95 % stereoselectivity (Eq. 21).[66]

$$\text{[CuLiHg]·(C}_4\text{H}_9\text{-}t)_3$$

$$\xrightarrow{\text{C}_6\text{H}_5\text{NO}_2} (45\%) \quad \text{C}_4\text{H}_9\text{-}t$$

$$\xrightarrow[\text{(70\%)}]{\text{CH}_3\text{I}} \quad \text{CH}_3$$

(Eq. 21)

Many examples of sp^2-hybridized R groups are known in which R is configurationally stable. Tri-*n*-butylphosphine complexes of Z- and E-1-propenylcopper and Z- and E-2-butenylcopper and the corresponding organocuprates(I) undergo stereospecific thermal dimerizations (Eq. 22)[104, 105] and stereospecific reduction with cuprous hydride,[147] as well as coupling with halides (Eq. 23).[86, 87]

$$(\text{Z—CH}_3\text{CH}=\text{CH})_2\text{CuLi} \rightarrow \text{Z, Z—CH}_3\text{CH}=\text{CHCH}=\text{CHCH}_3 \quad \text{(Eq. 22)}$$

$$\begin{array}{c} n\text{-C}_4\text{H}_9 \\ \diagdown \\ C=C \\ \diagup \quad \diagdown \\ \text{C}_2\text{H}_5 \quad \text{Cu} \end{array} \xrightarrow[\text{HMPA}]{\text{CH}_2=\text{CHCH}_2\text{Br}} \begin{array}{c} n\text{-C}_4\text{H}_9 \quad\quad H \\ \diagdown \quad\quad \diagup \\ C=C \\ \diagup \quad\quad \diagdown \\ \text{C}_2\text{H}_5 \quad \text{CH}_2\text{CH}=\text{CH}_2 \end{array} \quad \text{(Eq. 23)}$$

147 G. M. Whitesides, J. San Filippo, Jr., E. R. Stedronsky, and C. P. Casey, Jr., *J. Amer. Chem. Soc.*, **91**, 6542 (1969).

α-Alkoxycarbonyl and α-carboxyvinylic R groups have been prepared *in situ* from conjugate addition of organocopper reagents to acetylenic esters and acids.[27] The configurational stability of these presumably carbon-copper enolates as a function of time, temperature, solvent, and complexing additives has been reviewed in detail.[27] Oxidative decomposition of such (methyl)(alkenyl)cuprates(I) (Eq. 24)[144] and reaction with

$$
\left(
\begin{array}{c}
\substack{n\text{-}C_7H_{15} \\ \diagdown \\ \\ CH_3 \diagup} C=C \substack{\diagup CO_2CH_3 \\ \\ \diagdown}
\end{array}
\right)_{(CH_3)CuLi}
\xrightarrow[-78°]{O_2}
\substack{n\text{-}C_7H_{15} \\ \diagdown \\ \\ CH_3 \diagup} C=C \substack{\diagup CO_2CH_3 \\ \\ \diagdown CH_3}
\qquad \text{(Eq. 24)}
$$
$$
(32\text{–}46\%)
$$

$$
\substack{CH_3 \\ \diagdown \\ \\ C_2H_5 \diagup} C=C \substack{\diagup CO_2CH_3 \\ \\ \diagdown Cu}
\xrightarrow[\text{THF}]{CH_2=CHCH_2Br}
\substack{CH_3 \\ \diagdown \\ \\ C_2H_5 \diagup} C=C \substack{\diagup CO_2CH_3 \\ \\ \diagdown CH_2CH=CH_2}
\qquad \text{(Eq. 25)}
$$
$$
(>80\%)
$$

allylic halides (Eq. 25)[145] both proceed stereospecifically with retention of configuration.

No optically active organocopper compound having an asymmetric carbon atom directly attached to copper has been reported.

Effect of Structure on the Substitution Reaction

Structure of Reagent. Different types of organocopper reagents generally undergo substitution reactions in different yields. Of the three most common types — RCu, RCu · Ligand, and R_2CuLi — the organocuprates(I) have been used most often and generally provide the highest yield of substitution products. Cuprous acetylides are exceptions to this generalization; they are used only as RCu species. Several typical examples illustrate the superiority in yield of the organocuprates(I). 1-Iododecane is converted to n-undecane by methylcopper in 68% yield and by lithium dimethylcuprate(I) in 90% yield.[46] 1-Bromo-2-phenylethylene is transformed into 1-propenylbenzene by methylcopper in 70% yield and by lithium dimethylcuprate(I) in 81% yield.[46] Benzoyl chloride is methylated to acetophenone by methylcopper in 53% yield and by lithium dimethylcuprate(I) in 94% yield.[46, 101] Unfortunately such direct comparisons between n-alkylcopper, sec-alkylcopper, or arylcopper RCu species and the corresponding R_2CuLi organocuprates(I) have not been reported, but the greater effectiveness of the organocuprates(I) is to be anticipated.[101]

On the basis of limited data, it appears that organocuprates(I) are also more effective in substitution reactions than RCu · Ligand species. Benzoyl chloride, for example, is converted to acetophenone by methyl(tri-n-butylphosphine)copper(I) in 60% yield[33] and by lithium dimethylcuprate

TABLE I. COMPARISON OF EFFECT OF STRUCTURE OF R IN THE ORGANOCOPPER REAGENT ON THE SUBSTITUTION REACTION

Entry	R	R'X	% R–R'	Refs.
1	$(CH_3)_2CuLi$	$E\text{-}C_6H_5CH{=}CHBr$	81	46
2	$(C_2H_5)_2CuLi$	$E\text{-}C_6H_5CH{=}CHBr$	65	46
3	$(n\text{-}C_4H_9)_2CuLi$	$E\text{-}C_6H_5CH{=}CHBr$	65	46
4	$(CH_3)_2CuLi$	C_6H_5I	90	33
5	$(n\text{-}C_4H_9)_2CuLi$	C_6H_5I	75	33
6	$(CH_3)_2CuLi$	3-Bromocyclohexene	75	148
7	$(n\text{-}C_4H_9)_2CuLi$	3-Bromocyclohexene	60	148
8	$(n\text{-}C_4H_9)_2CuLi$	$n\text{-}C_5H_{11}I$	53	33
9	$(sec\text{-}C_4H_9)_2CuLi$	$n\text{-}C_5H_{11}I$	7	33
10	$(i\text{-}C_3H_7)_2CuLi$	$i\text{-}C_3H_7CH_2COCH(Br)C_3H_7\text{-}i$	45	111, 149
11	$(t\text{-}C_4H_9)_2CuLi$	$i\text{-}C_3H_7CH_2COCH(Br)C_3H_7\text{-}i$	16	111, 149
12	$n\text{-}C_4H_9CuP(C_4H_9\text{-}n)_3$	$n\text{-}C_5H_{11}Br$	93	33
13	$sec\text{-}C_4H_9CuP(C_4H_9\text{-}n)_3$	$n\text{-}C_5H_{11}Br$	94	33
14	$t\text{-}C_4H_9CuP(C_4H_9\text{-}n)_3$	$n\text{-}C_5H_{11}Br$	92	33
15	$(CH_3)_2CuLi$	Bromocyclooctatetraene	93	150
16	$(C_6H_5)_2CuLi$	Bromocyclooctatetraene	58	150
17	$(CH_3)_2CuLi$	$C_6H_5CH_2Cl$	80	115
18	$(C_6H_5)_2CuLi$	$C_6H_5CH_2Cl$	50	115
19	C_6F_5Cu	CH_3COCl	66	31
20	C_6F_5Cu	CH_3COCl	82	151
21	C_6H_5Cu	C_6H_5COCl	55	31
22	C_6Br_5Cu	C_6H_5COCl	90	152
23	C_6H_5Cu	$o\text{-}IC_6H_4NO_2$	18	153, 121
24	2-ThienylCu	$o\text{-}IC_6H_4NO_2$	56	153, 121
25	C_6H_5Cu	$o\text{-}IC_6H_4CO_2CH_3$	17–28	121
26	2-ThienylCu	$o\text{-}IC_6H_4CO_2CH_3$	50	121

in 94% yield.[101] Even when the yield of substitution product from both types of reagent is comparable, the organocuprates(I) are preferred because the workup is not complicated by phosphines and phosphine oxides. Nevertheless, it may be that for replacement of halogen in aryl halides, RCu·Ligand species are better than R_2CuLi compounds; 1-iodonaphthalene is converted to 1-methylnaphthalene by lithium dimethylcuprate(I) in 33% yield and by methyl(tri-n-butylphosphine)copper(I) in 75% yield.[33] Methylation of aryl iodides, however, is apparently best accomplished by methyllithium itself.[33]

Structure of R. Within a given class of organocopper reagents — RCu, RCu· Ligand, or R_2CuLi — the structure of R often substantially affects the success of a substitution reaction. Five comparisons can be made on the basis of available data:* methyl $vs.$ n-alkyl, n-alkyl $vs.$ sec-alkyl $vs.$ t-alkyl, methyl $vs.$ aryl, phenyl $vs.$ perhalophenyl, and phenyl $vs.$ 2-thienyl.

The data in Table I allow the following generalizations. Substitution of halogen is usually achieved more effectively by methyl than by n-alkyl R groups (entries 1–7). The yield in replacement of halogen by n-alkyl, sec-alkyl or t-alkyl R groups depends on the reagent used. For lithium diorganocuprates prepared normally ($i.e.$, from 2 mol of RLi per mol of CuI) n-alkylation is much more favorable than sec-alkylation, and sec-alkylation is more favorable than t-alkylation (entries 8–11); for alkyl(tri-n-butylphosphine)copper(I) species, n-alkylation, sec-alkylation, and t-alkylation all proceed well, at least for substitution of primary bromine (entries 12–14). Methylation is usually a higher-yield reaction than phenylation (entries 15–18), and phenylation is usually a lower-yield reaction than perhalophenylation and 2-thienylation (entries 19–26).

Selection of Reagent for the Substitution Reaction

Choosing a reagent for any chemical transformation requires evaluation of several basic factors about the product of the reaction and about the reagent. Ideally the desired product should be formed in high yield and with good stereochemical purity, and it should not react further with the reagent nor should it be accompanied by side products ($e.g.$, Ligand from RCu· Ligand) which make product isolation difficult. The reagent should be

---

* To maximize their validity, these comparisons are based, whenever possible, on reactions run under very similar if not identical conditions in one laboratory.

[148] E. J. Corey and G. H. Posner, $J. Amer. Chem. Soc.$, **90**, 5615 (1968).

[149] J. E. Dubois and C. Lion, $C.R. Acad. Sci.$, $Ser. C$, **272**, 1377 (1971).

[150] J. Gasteiger, G. E. Gream, R. Huisgen, W. E. Konz, and U. Schmegg, $Chem. Ber.$, **104**, 2412 (1971).

[151] A. F. Webb and G. Gilman. $J. Organometal. Chem.$, **20**, 281 (1969).

[152] C. F. Smith, G. J. Moore, and C. Tamborski, $J. Organometal. Chem.$, **42**, 257 (1972).

[153] M. Nilsson and O. Wennerström, $Acta Chem. Scand.$, **24**, 482 (1970).

easy to prepare and to handle, and it should give reproducible results. On the basis of these considerations, selection of an organocopper reagent for substitution with a particular type of R group can be discussed. Organocopper reagents will be treated in the following order: methyl and n-alkyl; sec-alkyl and t-alkyl; alkenyl; alkynyl; aryl, haloaryl, and heteroaryl; allylic; and functionalized alkyl.

Clearly, substitution reactions of methyl and n-alkyl groups are best achieved using lithium diorganocuprate(I) reagents, usually in ether solvents at 25° or below. sec-Alkyl and t-alkyl groups have been transferred using four types of organocopper reagents: R_2CuLi at low temperature,[110, 111, 154] $R_2CuLi \cdot Ligand$ species,[33] mixed homocuprates ($RR'CuLi$), which are oxidized to unsymmetrical dimers R—R' (accompanied in most cases by symmetrical dimers R—R and R'—R'),[33] plus alkoxy- and thioalkoxy(alkyl)cuprates [Het(R)CuLi].[99, 101] Choice among the first three of these four types of organocopper reagents for transfer of sec- or t-alkyl R groups is difficult and will probably depend on the peculiarities of the substrate and the desired product of substitution ($e.g.$, whether presence of the Ligand from $R_2CuLi \cdot Ligand$ is tolerable, or whether dimers R—R and R'—R' can be separated easily from desired product R—R'). The fourth type of organocopper reagent, Het(R)CuLi, appears very promising for selective, high-yield transfer of sec- and t-alkyl groups; the full scope and limitations of these hetero(alkyl)cuprate reagents have yet to be determined.

Alkenyl species of the form RCu and RCu·Ligand have been used most often for thermal or oxidative dimerization to form symmetrical dimers R—R, whereas for coupling with organic halides both RCu and R_2CuLi reagents have been used effectively. If RCu alkenylcopper reagents are used in substitution reactions, hexamethylphosphoramide is the best solvent.[87]

Substitution reactions using alkynyl R groups have been achieved only with RCu species — cuprous acetylides.

Transfer of aryl groups has been achieved successfully using both RCu and R_2CuLi reagents; choice between these two types of organocopper reagents is difficult. If reaction with substrate proceeds at a tolerable rate with the RCu arylcopper reagent, it is recommended that this reagent be used rather than the R_2CuLi species, which usually gives substantial quantities of biaryl (R—R) side product; if reaction is too slow with RCu arylcopper, then the R_2CuLi reagent should be tried. Both haloaryl and heteroaryl R groups have been transferred using only RCu arylcopper species.

[154] C. R. Johnson, R. W. Herr, and D. M. Wieland, unpublished results.

Allylic R groups have been used in substitution reactions involving only R_2CuLi organocopper species,[33, 137] and functionalized alkyl R groups have been used in substitution reactions involving only RCu organocopper species.[141-143]

The Organic Substrate

The scope and limitations of substitution reactions using organocopper reagents are discussed in this section, with emphasis on the role of the organic substrate.

Halides

Alkyl Halides. Two generalizations can be made about organocopper interaction with alkyl halides. First, the halide reactivity decreases in the order primary > secondary > tertiary.* Second, cuprous acetylides apparently do not couple effectively with alkyl halides.

Primary alkyl halides undergo effective substitution of halogen by R using most often R_2CuLi organocopper reagents. Replacement of iodine by R is usually a high-yield process. Even one *gem*-diiodide has been doubly substituted (Eq. 26), and several iodides bearing other remote functional

$$CH_2I_2 \xrightarrow[(70\%)]{C_6F_5Cu} CH_2(C_6F_5)_2 \qquad \text{(Eq. 26)}^{156}$$

$$CH_2{=}CHCH_2OCH_2CH_2CD_2I \xrightarrow{(CH_3)_2CuLi} CH_2{=}CHCH_2OCH_2CH_2CD_2CH_3$$
$$\text{(Eq. 27)}^{157}$$

$$RCO(CH_2)_{10}I \xrightarrow[\substack{R' = CH_3, n\text{-}C_4H_9,\\R = HO, CH_3O, C_6H_5(CH_3)N}]{R'_2CuLi} RCO(CH_2)_{10}R'$$
$$\text{(60-80\%)}$$
$$\text{(Eq. 28)}^{33, 54, 148}$$

$$n\text{-}C_5H_{11}Br \xrightarrow[(94\%)]{(sec\text{-}C_4H_9)_2CuLi\cdot P(C_4H_9\text{-}n)_3} \text{3-Methyloctane} \qquad \text{(Eq. 29)}^{33}$$

$$n\text{-}C_5H_{11}Cl \xrightarrow[(80\%)]{(n\text{-}C_4H_9)_2CuLi,\ THF} n\text{-Nonane} \qquad \text{(Eq. 30)}^{33}$$

groups (*e.g.*, olefin, carboxy, ethoxycarbonyl, amide) have been selectively substituted by methyl and *n*-alkyl groups (Eqs. 27, 28). Substitution of bromine also is usually a satisfactory, albeit somewhat slower, reaction (Eq. 29). Replacement of chlorine is sluggish, but nevertheless proceeds in reasonably good yield when carried out at high temperature (*e.g.*, 25°; Eq. 30); tetrahydrofuran has been recommended as a solvent to facilitate

---

* This order is reversed for coupling with perfluorophenylcopper.[155]

[155] A. Cairncross and W. A. Sheppard, Central Research Department, E. I. du Pont de Nemours, personal communication.

[156] A. E. Jukes, S. S. Dua, and H. Gilman, *J. Organometal. Chem.*, **24**, 791 (1970).

[157] J-P. Morizur and C. Djerassi, *Org. Mass Spectrom.*, **5**, 895 (1971).

this coupling reaction. There has been no report of organocopper reaction with an alkyl fluoride.

Secondary alkyl halides undergo somewhat erratic substitution reactions with organocopper reagents; the success of the reactions may depend on the organocopper reagent or on the nature of the substrate. Despite the small number of examples, some tentative generalizations appear valid. Methylation of cyclic secondary alkyl iodides and bromides with lithium dimethylcuprate(I) proceeds reasonably well, although some competing α,β elimination and reduction also occur (Eq. 31). Butylation of acyclic

45% *trans*,   (5%)
10% *cis*
(Eq. 31)[46.54]

(35%)

2-Bromopentane $\xrightarrow[\text{(10\%)}]{(n\text{-}C_4H_9)_2CuLi}$ 4-Methyloctane   (Eq. 32)[33]

2-Iodopropane $\xrightarrow[\text{15\%}]{Z-C_2H_5(CH_3)C=CHCu}$ $Z-(CH_3)_2CHCH=C(CH_3)C_2H_5$
(Eq. 35)[87]

$\xrightarrow[\text{(80\%)}]{[CH_2=C(CH_3)]_2CuLi}$ (Eq. 34)[114.15]

$(-)\text{-}(R)\text{-}2\text{-Bromobutane}$ $\xrightarrow[\text{(87\%)}]{(C_6H_5)_2CuLi}$ $(+)\text{-}(S)\text{-}2\text{-Phenylbutane}$ (Eq. 35)[33]

Bromocyclohexane $\xrightarrow[\text{(10\%)}]{(C_6H_5)_2CuLi}$ Phenylcyclohexane   (Eq. 36)[33]

(Eq. 37)[159a]

(36%)   (40%)

or cyclic secondary bromides, however, appears to be a low-yield reaction (Eq. 32). Vinylation of secondary halides proceeds poorly with RCu and well with R_2CuLi vinylcopper species (Eqs. 33, 34). Replacement of bromine

158 O. P. Vig, J. C. Kapur, and S. D. Sharma, *J. Indian Chem. Soc.*, **45**, 734 (1968).
159 (a) G. H. Posner and D. J. Brunelle, The Johns Hopkins University, unpublished results; (b) G. H. Posner and J.-S. Ting, *Synthetic Commun.*, **3**, 281 (1973).

by phenyl occurs satisfactorily in acyclic bromides, but unsatisfactorily in cyclic bromides (Eqs. 35, 36). Finally, only one secondary chloride has been treated with an organocopper reagent (Eq. 37). The stereochemistry of substitution is predominantly inversion of configuration (see Eq. 35), although in at least one case (Eq. 31) some retention has been observed.

Only three tertiary alkyl halides (all bromides) have been studied. Their reactivity is so low that organocopper substitution with methyl,[46] n-butyl,[33] or phenyl[33] groups does not occur. The only successful substitution of a tertiary bromide has been achieved using pentafluorophenylcopper (Eq. 38). An alternative approach to the unsymmetrical coupling

$$1\text{-Bromoadamantane} \xrightarrow[\;(93\%)\;]{\text{C}_6\text{F}_5\text{Cu}} 1\text{-Pentafluorophenyladamantane}$$

$$\text{(Eq. 38)}^{103}$$

$$t\text{-C}_4\text{H}_9\text{Li} + 4\,\text{C}_6\text{H}_5\text{Li} \xrightarrow[\;\;2.\ \text{O}_2,\ -78°\;\;]{1.\ 2.5\ \text{CuBr, THF, } -78°} t\text{-C}_4\text{H}_9\text{C}_6\text{H}_5 \quad \text{(Eq. 39)}$$
$$\text{(73\%)}$$

product R–R′ in which R is tertiary alkyl involves conversion of RX to RLi and then to RR′CuLi, which can be oxidatively dimerized to R–R′ (Eq. 39);[33] the main product in this reaction (Eq. 39), however, is R′–R′ (biphenyl).

1,2-Dibromoalkanes react with lithium dialkylcopper reagents to produce olefins.[159b]

Alkenyl Halides.   Two features of organocopper reaction with alkenyl halides are outstanding. First, the relatively high reactivity of haloolefins stands in contrast to their inertness in typical nucleophilic substitution reactions. Second, the replacement of halogen by organic groups proceeds with essentially complete retention of configuration in the substrate.[33, 54, 110] Both of these features have led to wide use of organocopper reagents for substitution of alkenyl halogen; several total syntheses of natural products have used such a substitution reaction with an organocopper compound as a key step: farnesol,[160] fulvoplumierin,[113] and insect juvenile hormone.[144, 161]

Primary alkenyl iodides (C=CHI) react with halogenated RCu reagents to replace iodide with R; double replacements also have been achieved on 1,2-diiodoethylenes,[83] and one quadruple replacement has been reported using a cuprous acetylide (Eq. 40). Halogen reactivity (I > Br > Cl) is

[160] E. J. Corey, J. A. Katzenellenbogen, and G. H. Posner, *J. Amer. Chem. Soc.*, **89**, 4245 (1967).
[161] E. J. Corey, J. A. Katzenellenbogen, S. A. Roman, and N. W. Gilman, *Tetrahedron Lett.*, **1971**, 1821.

illustrated by Eq. 41 in which iodine is replaced by perfluoroalkyl while chlorine remains unperturbed. Lithium dialkylcuprates(I) transform E-1-iodononene into the corresponding E-olefins stereospecifically and in

$$I_2C{=}CI_2 \xrightarrow[\text{(40\%)}]{C_6H_5C{\equiv}CCu} (C_6H_5C{\equiv}C)_2C{=}C(C{\equiv}CC_6H_5)_2 \qquad \text{(Eq. 40)}^{82}$$

$$ClCH{=}CHI \xrightarrow[\text{(96\%)}]{Cu(CF_2)_3Cu} ClCH{=}CH(CF_2)_3CH{=}CHCl \qquad \text{(Eq. 41)}^{82}$$

$$C_6H_5CH{=}CHBr \xrightarrow{(t\text{-}C_4H_9)_2CuLi}
\begin{array}{l}
C_6H_5CH{=}CHC_4H_9\text{-}t \\
\text{(50\%)} \\[2em]
C_6H_5CH_2CH_2C_4H_9\text{-}t \\
\text{(50\%)}
\end{array}
\qquad \text{(Eq. 42)}^{110}$$

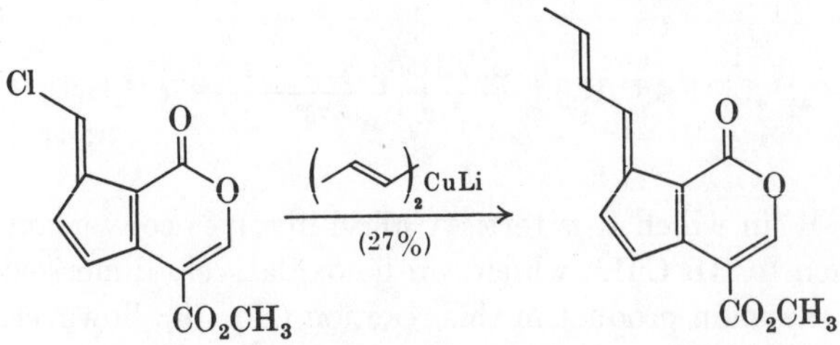

(Eq. 43)^{113}

high yield.$^{46, 148}$ Primary alkenyl bromides also undergo effective replacement of bromine by R using RCu,^{46} halogenated RCu,^{83} and R$_2$CuLi reagents.^{33} With lithium diisopropyl- or di-t-butylcuprate(I), alkylation and reduction appear to compete with substitution (Eq. 42). Only one primary vinylic chloride has been studied; replacement of chlorine by the propenyl group occurs stereospecifically and in moderate yield even in the presence of ester and lactone functions (Eq. 43). No reactions have been reported between vinylic fluorides and organocopper reagents.

Reaction of secondary alkenyl halides [C=C(R)Hal] with only lithium diorganocuprate(I) reagents (R$_2$CuLi) has been studied. A large number of secondary vinylic iodides have been alkylated with retention of configuration, as illustrated by Eq. 44 in which double ethylation occurs even with

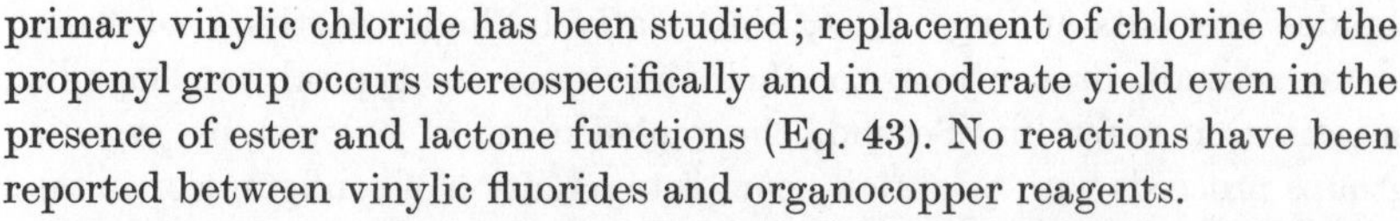

(Eq. 44)^{161}

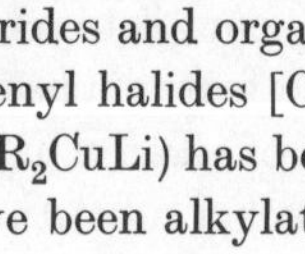

4

a free hydroxyl group in the diiodide substrate.* Likewise, bromoolefin **4**, also bearing an allylic hydroxyl, is methylated by excess lithium dimethylcuprate(I) in 95% yield.[162] 1,4-Dibromocyclooctatetraene undergoes replacement of both bromine atoms by methyl,[150, 163] and 1-bromocyclohexene reacts sluggishly with lithium dimethylcuprate(I) (starting material was recovered) and reasonably well with lithium di-n-butylcuprate(I), an indication that dialkylcuprates(I) are more reactive than dimethylcuprate(I).[46, 148] Reaction of organocuprates(I) with 2-bromoacrylates has been reported to give a mixture of alkylated and reduced (*i.e.*, acrylate) products.[55] Chloroolefins are rather unreactive; 1-chlorocyclohexene, for example, cannot be methylated with lithium dimethylcuprate(I), but it can be butylated with the more reactive lithium di-n-butylcuprate(I).[46, 148] Chloroolefin **5** reacts with lithium dimethylcuprate(I) to replace chlorine

5

$$C_6H_5CH{=}CHSCH_3 \xrightarrow[\text{(50\%)}]{(n\text{-}C_4H_9)_2CuLi} C_6H_5CH{=}CHC_4H_9\text{-}n \quad \text{(Eq. 45)}^{165}$$

by methyl in 39% yield.[164] Replacement of vinyl substituents other than halogen has not been examined in detail; however, a methylthio group attached to a vinylic carbon has been replaced (Eq. 45).

Alkynyl Halides. Substitution of alkynyl halogen by R has been reported in a few cases where R is acetylide, in a few cases where R is

$$(C_2H_5)_3SiC{\equiv}CBr \xrightarrow[\text{(85\%)}]{C_6F_5Cu} (C_2H_5)_3SiC{\equiv}CC_6F_5 \quad \text{(Eq. 46)}^{134}$$

$$\xrightarrow[\text{(71\%)}]{BrC{\equiv}CCH_2OH}$$

$$\quad \text{(Eq. 47)}^{129}$$

---

* Presumably the hydroxyl group exists as a metal alkoxide during the reaction and is regenerated on workup.

162 D. Whalen, University of Maryland, Baltimore Campus, unpublished results.
163 W. E. Konz, W. Hechtl, and R. Huisgen, *J. Amer. Chem. Soc.*, **92**, 4104 (1970).
164 L. A. Paquette and J. C. Stowell, *J. Amer. Chem. Soc.*, **93**, 5735 (1971).
165 G. H. Posner and D. J. Brunelle, *J. Org. Chem.*, **38**, 2747 (1973).

aryl,[166] and in one case where R is pentafluorophenyl (Eq. 46). Replacement of alkynyl halogen by an acetylide produces a 1,3-butadiyne (e.g., Eq. 47).

Aryl Halides. A large number of aryl and heteroaryl iodides and bromides have undergone useful substitution reactions with organocopper reagents; substitution of chlorine is successful only for unusually reactive aromatic chlorides (e.g., nitrochlorobenzenes). Most of the organocopper reagents used for aryl halide replacement have been of the RCu type, including many perfluoroalkyl- and perfluoroaryl-copper species; lithium diorganocuprate(I) reagents have been used sporadically, presumably because with these reagents metal-halogen exchange[4] is often an undesirable competing reaction.[33]

Replacement of iodine in iodobenzene by methyl has been accomplished in good yield using lithium dimethylcuprate(I) or methyllithium itself; it has been suggested that methylation of aryl iodides generally may best be achieved using methyllithium.[33] Treating iodobenzene with lithium di-n-butylcuprate(I) causes substantial metal-halogen exchange, as shown by deuterium oxide quenching which produces mainly monodeutero-benzene.[33, 46] Two methods have been used to overcome this undesirable side reaction: addition of excess n-butyl iodide at the end of the reaction produces n-butylbenzene in 75% yield,[148] or oxidation of the presumed intermediate, mixed (phenyl)(n-butyl)cuprate(I), produces n-butylbenzene in 50% yield.[33] As noted previously (p. 267), oxidative dimerization of mixed homocuprates usually leads to a statistical distribution of symmetrical (Ar—Ar and R—R) and unsymmetrical (Ar—R) dimers; therefore, to maximize the yield of unsymmetrical coupling product Ar—R from an aryl halide and an organocopper reagent, a large excess of the organocopper reagent should be used.

Substitution of aromatic iodine by cyclopentadienyl (Eq. 48), by various perfluoroalkyl groups (e.g., Eq. 49), and by heteroaryl groups (Eq. 50) has

$$p\text{-CH}_3\text{C}_6\text{H}_4\text{I} \xrightarrow[\text{(50\%)}]{\text{C}_5\text{H}_5\text{Cu}\cdot\text{P(C}_4\text{H}_9\text{-}n)_3} p\text{-CH}_3\text{C}_6\text{H}_4\!-\!\langle\!\!\langle\rangle\!\!\rangle \quad \text{(Eq. 48)}^{153}$$

$$p\text{-BrC}_6\text{F}_4\text{I} \xrightarrow[\text{(50\%)}]{\text{H(CF}_2)_6\text{Cu}} p\text{-BrC}_6\text{H}_4(\text{CF}_2)_6\text{H} \quad \text{(Eq. 49)}^{81}$$

$$\text{(Eq. 50)}^{122}$$

166 R. Oliver and D. R. M. Walton, *Tetrahedron Lett.*, **1972**, 5209.

been reported. Note that the reactivity of aromatic bromine is sufficiently lower than that of aromatic iodine to allow selective replacement of iodine (Eq. 49). This selectivity also appears in cuprous acetylide coupling with aromatic halides.[167a]

Aromatic bromine has been replaced by a variety of groups, among which the following are typical: alkyl (Eq. 51), fluoroalkyl (Eq. 52),

$$\text{Bromo[18]annulene} \xrightarrow[\text{(52\%)}]{\text{(CH}_3)_2\text{CuLi}} \text{Methyl[18]annulene} \quad \text{(Eq. 51)}^{168}$$

$$p\text{-HOCH(CH}_3)\text{C}_6\text{H}_4\text{Br} \xrightarrow[\text{(65\%)}]{n\text{-C}_7\text{F}_{15}\text{Cu}} p\text{-HOCH(CH}_3)\text{C}_6\text{H}_4\text{C}_7\text{F}_{15}\text{-}n$$
$$\text{(Eq. 52)}^{81}$$

$$p\text{-O}_2\text{NC}_6\text{H}_4\text{Br} \xrightarrow[\text{(85\%)}]{\text{C}_6\text{F}_5\text{Cu}} p\text{-O}_2\text{NC}_6\text{H}_4\text{C}_6\text{F}_5 \quad \text{(Eq. 53)}^{169}$$

$$2,4,6\text{-}(\text{O}_2\text{N})_3\text{C}_6\text{H}_2\text{Cl} \xrightarrow[\text{(36\%)}]{\text{CuC}_6\text{H}_3(\text{OCH}_3)_2\text{-}2,6} 2,4,6\text{-}(\text{O}_2\text{N})_3\text{C}_6\text{H}_2\text{C}_6\text{H}_3(\text{OCH}_3)_2\text{-}2,6$$
$$\text{(Eq. 54)}^{118}$$

$$\text{(Eq. 55)}^{170}$$

fluoroaryl (Eq. 53), vinyl,[158] and acetylide.[167b] Only a limited number of reports are available concerning substitution of aromatic chlorine. Two illustrative examples (Eqs. 54 and 55) show that activated chlorine is usually required for effective substitution.

Cuprous acetylides have been widely used for replacement of aromatic halogen by the acetylide group. An excellent short review is available.[61] One of the most useful aspects of these replacement reactions is the synthesis of a broad spectrum of heterocyclic compounds starting from *ortho*-substituted aryl halides and cuprous acetylides (Eq. 56, X = I, Br, or Cl).

[167a] A. N. Nesmeyanov, V. A. Sazonova, and V. N. Drozd, *Dokl. Akad. Nauk SSSR*, **154**, 158 (1964) [*C.A.*, **60**, 9309g (1964)].
[167b] G. Martelli, P. Spagnolo, and M. Tiecco, *J. Chem. Soc., B*, **1970**, 1413.
[168] E. P. Woo and F. Sondheimer, *Tetrahedron*, **26**, 3933 (1970).
[169] W. A. Sheppard, *J. Amer. Chem. Soc.*, **92**, 5419 (1970).
[170] M. S. Manhas and S. D. Sharma, *J. Heterocycl. Chem.*, **8**, 1051 (1971).

Thus a facile route is available to indoles $(Q = NR)$,[133] benzofurans $(Q = O)$,[133] benzothiophenes $(Q = S)$,[171] phthalides (Eq. 57),[133] and other "polynuclear multiheterocyclic arrays."[126] (See Table IVC).

$$\underset{QH}{\overset{X}{\bigcirc\!\!\!\!\bigcirc}} \xrightarrow[\text{(Q = NR,O,S)}]{R'C\!\equiv\!CCu} \underset{Q}{\overset{}{\bigcirc\!\!\!\!\bigcirc}}\!\!-R' \qquad \text{(Eq. 56)}^{61}$$

$$\underset{CO_2H}{\overset{X}{\bigcirc\!\!\!\!\bigcirc}} \xrightarrow{RC\!\equiv\!CCu} \underset{O}{\overset{CHR}{\bigcirc\!\!\!\!\bigcirc}}\!\!O \qquad \text{(Eq. 57)}^{61}$$

Benzylic, Allylic, and Propargylic Halides. These halides are highly reactive toward organocopper reagents. Benzylic bromides and chlorides (there are no reports on iodides or fluorides) undergo facile replacement of halogen by R with either RCu or R_2CuLi organocopper reagents. A bibenzyl is typically a minor side product. Primary benzylic halides are more reactive than the corresponding secondary halides.[172] Benzal chloride undergoes double substitution with lithium dimethylcuprate(I), and it has been shown that "loss of the second chlorine atom . . . must precede or occur simultaneously with introduction of the first methyl group" even though no carbenes were formed (Eq. 58).[171] Typical sub-

$$C_6H_5CHCl_2 + (CH_3)_2CuLi \nearrow^{\displaystyle \underset{(40)\%}{C_6H_5CH(CH_3)_2}}_{\searrow \displaystyle C_6H_5CH(CH_3)Cl} \qquad \text{(Eq. 58)}^{171}$$

$$C_6H_5CH_2Cl \xrightarrow[\text{(50\%)}]{(p\text{-}CH_3C_6H_4)_2CuLi} C_6H_5CH_2C_6H_4CH_3\text{-}p \qquad \text{(Eq. 59)}^{115}$$

$$o\text{-}C_6H_4(CH_2Br)_2 \xrightarrow[\text{(87\%)}]{(CH_3)_2\,CuLi} o\text{-}C_6H_4(CH_2CH_3)_2 \qquad \text{(Eq. 60)}^{172}$$

stitution reactions of benzylic halides and organocopper reagents are illustrated by Eqs. 59 and 60. A limited number of cuprous acetylide couplings with benzylic halides have been reported.[173]

There has been no report of an organocopper reaction with allylic iodides or fluorides; allylic bromides and chlorides react with alkyl (Eq. 61),

[171] A. M. Malte and C. E. Castro, *J. Amer. Chem. Soc.*, **89**, 6770 (1967).

[172] G. H. Posner and D.J. Brunelle, *Tetrahedron Lett.*, **1972**, 293.

[173] K. Gump, S. W. Moje, and C. E. Castro, *J. Amer. Chem. Soc.*, **89**, 6770 (1967).

alkenyl,[86] aryl,[71] functionalized alkyl (Eq. 62), and acetylide-copper reagents (Eq. 63) to replace halogen by R in good yield. Note that a vinylic bromine atom survives at least one cuprous acetylide substitution reaction

$$\text{Cl} \xrightarrow[\text{(65\%)}]{(CH_3)_2CuLi} \text{CH}_3 \qquad \text{(Eq. 61)}^{46}$$

$$\xrightarrow[\text{(94\%)}]{C_2H_5O_2CCH_2Cu} \qquad \text{(Eq. 62)}^{143}$$

$$\text{CO}_2\text{C}_2\text{H}_5$$

$$CH_2{=}C(Br)CH_2Br \xrightarrow[\text{(50\%)}]{(CH_3)_2C(OH)C{\equiv}CCu} CH_2{=}C(Br)CH_2C{\equiv}CC(OH)(CH_3)_2$$

$$\text{(Eq. 63)}^{175a}$$

(Eq. 63). It should be emphasized that organomagnesium and organolithium reagents themselves often couple effectively with allylic halides.[13, 14, 174] For a discussion of substitution with allylic rearrangement, see p. 286.

Propargyl chloride reacts with cuprous acetylide 6 to replace chlorine by acetylide in unspecified yield,[175a] and several prepargylic chlorides react with lithium dialkylcuprate reagents to form allenes in good yields.[175b]

$$(CH_3)_2C(OH)C{\equiv}CCu$$
6

α-Halocarbonyl Substrates. Replacement of halogen alpha to a carbonyl group may be tricky. Competing reduction of substrate upon aqueous workup (Eq. 64) and furan formation in cuprous acetylide coupling (Eq. 65) are often serious side reactions. Overall reduction suggests the intermediacy of a copper (or lithium) enolate;[99] such species have been trapped as enol acetates,[27] and recently they have been found effective for coupling with some organic halides.[124,175d-f] Direct organocopper substitution of α-bromocarbonyl and α-chlorocarbonyl systems has been achieved in moderate yields using tetrahydrofuran as solvent (Eq. 66).[115] (See p. 284.)

α,α'-Dibromoketones have been found to react with lithium dialkyl-cuprates(I) and alkoxy-alkylcuprates(I) under rigorous exclusion of oxygen

[174] B. Rickborn, University of California, Santa Barbara, personal communication.

[175] (a) J. Colonge and R. Falcotet, *Bull. Soc. Chim. Fr.*, **1957**, 1166; (b) M. Kalli, P. D. Landor, and S. D. Landor, *J. Chem. Soc.*, *Perkin* I, 1347 (1973); (c) J. R. Bull and A. Tuinman, *Tetrahedron Lett.*, **1973**, 4349 and J. R. Bull and H. H. Lachman, *ibid.*, **1973**, 3055; (d) R. K. Boeckman, Jr., *J. Org. Chem.*, **38**, 4450 (1973); (e) R. M. Coates and L. O. Sandepur, *J. Org. Chem.*, **39**, 275 (1974); (f) G. H. Posner, J. J. Sterling, C. E. Whitten, and D. J. Brunelle, manuscript submitted.

$$C_6H_5COCH_2Br \xrightarrow[\text{(95\%)}]{(CH_3)_2CuLi} C_6H_5COCH_3 \qquad \text{(Eq. 64)}^{46,175c}$$

(Eq. 65)[173]

(Eq. 66)[114,158]

to give α-alkylketones in synthetically useful yields (Eq. 67).[99, 101, 149] The reaction presumably proceeds through the intermediacy of cyclopropanones, one of which has been trapped.[99]

(Eq. 67)

Acyl Halides. Acyl fluorides, chlorides, and bromides undergo highly effective substitution reactions with a wide variety of organocopper reagents. For example, acetyl bromide reacts with pentafluorophenylcopper[70], and benzoyl fluoride[96] reacts with lithium di-n-butylcuprate(I) to give the corresponding ketones in 83–87 % yields. Diacyl chlorides from oxalyl virtually through adipyl undergo double substitution with arylcopper reagents.[71, 76] Many functional groups present in the substrate do not interfere with organocopper coupling with acid halides. Thus, acid chlorides bearing the following substituents have undergone dialkylcuprate(I) substitutions, usually at low temperature: α-chloro,[176] ω-iodo, α,β-alkenyl, ω-alkoxycarbonyl, ω-acyl, and ω-cyano (e.g., Eq. 68).

$$Z(CH_2)_nCOCl \xrightarrow[\text{(83–95\%)}]{(n\text{-}C_4H_9)_2CuLi} Z(CH_2)_nCOC_4H_9\text{-}n \qquad \text{(Eq. 68)}^{96}$$

$$(n = 4: Z = n\text{-}C_4H_9CO, CH_3O_2C)$$
$$(n = 10: Z = I, CN)$$

Furthermore, replacement of acyl halogen by alkyl or aryl groups proceeds without epimerization at the carbon atom alpha to the carbonyl; cis-4-t-butylcyclohexanecarbonyl chloride is converted with retention of configuration to the corresponding aryl and t-butyl ketones by organocuprate reagents.[67, 177]

176 C. Jallabert, N. T. Luong-Thi, and H. Rivière, Bull. Soc. Chim. Fr., 1970, 797.
177 K. Kojima and K. Sakai, Tetrahedron Lett., 1972, 3333.

Many different types of R groups have been used in organocopper substitution reactions with acyl halides. Some unusual examples (*t*-alkyl, acetylenic, and perhaloaryl) are provided by Eqs. 69–71.

$$n\text{-}C_4H_9CO(CH_2)_4COCl \xrightarrow[\text{(73\%)}]{(t\text{-}C_4H_9OCuC_4H_9\text{-}t)\text{Li}} n\text{-}C_4H_9CO(CH_2)_4COC_4H_9\text{-}t$$

(Eq. 69)[100]

$$E\text{-}CH_3CH{=}CHCOCl \xrightarrow[\text{(78\%)}]{(CH_3)(n\text{-}C_5H_{11}C{\equiv}C)CuLi} E\text{-}CH_3CH{=}CHCOC{\equiv}CC_5H_{11}\text{-}n$$

(Eq. 70)[98]

$$C_6H_5COCl \xrightarrow[\text{(90\%)}]{C_6Br_5Cu} C_6H_5COC_6Br_5$$ (Eq. 71)[152]

Miscellaneous Halides. Reaction of organocopper reagents with allenic bromides and iodides and with *gem*-dibromocyclopropanes has been reported.[178] Three different *gem*-dibromocyclopropyl systems **(7–9)** have

7^{54} 8^{179} 9^{180}

been doubly methylated or ethylated in poor to good yield using the appropriate organocuprate(I) reagents.

Alcohol Derivatives

p-**Toluenesulfonate Esters.** A wide variety of primary and secondary alkyl tosylates undergo substitution reactions with organocuprate(I) reagents. The replacement of tosyloxy by R proceeds with inversion of

$$(+)\text{-}(S)\text{-}sec\text{-}C_4H_9OTs \xrightarrow[\text{(45\%)}]{(C_6H_5)_2CuLi} (-)\text{-}(R)\text{-}sec\text{-}C_4H_9C_6H_5$$

(Eq. 72)[182]

(Eq. 73)[182]

configuration (Eqs. 72, 73). Elimination is a serious side reaction, especially in secondary alkyl tosylate systems. The list of organic groups that have

[178] M. Kalli, P. D. Landor, and S. R. Landor, *Chem. Commun.*, **1972**, 593.
[179] W. G. Dauben and W. M. Welch, *Tetrahedron Lett.*, **1971**, 4531; and personal communication.
[180] C. Descoins, M. Julia, and H. van Sang, *Bull. Soc. Chim. Fr.*, **1971**, 4087.

been used in organocuprates(I) to displace the tosyloxy group includes
methyl, ethyl, n-butyl, sec-butyl, norbornyl,[147] t-butyl, and phenyl.[181, 182]

Allylic Acetates.   There are only four reports of the acetoxyl group
in allylic acetates undergoing substitution with organocopper reagents.[44, 183–185] In many cases the reaction proceeds with allylic transposition and produces olefins stereoselectively (Eqs. 74 and 75). The

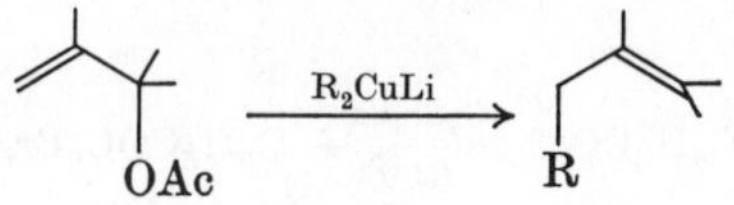

$$\text{(Eq. 74)}$$

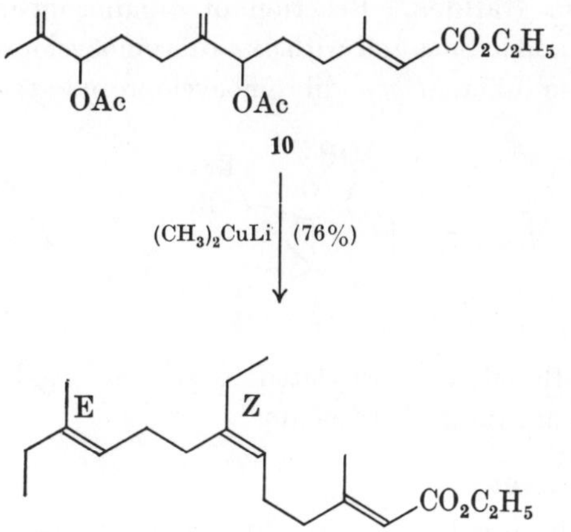

10

$$(CH_3)_2CuLi \quad (76\%)$$

$$\text{(Eq. 75)}^{44}$$

stereoselectivity of these reactions is higher for n-alkyl than for aryl
organocuprate(I) reagents.[44] Displacement of acetoxy on a cyclohexane
ring with lithium dimethylcuprate(I) occurs with predominant inversion of
configuration.[185] Equation 75 illustrates the functional group selectivity
of organocuprate(I) reagents; at $-10°$ for 0.5 hour lithium dimethyl-
cuprate(I) reacts with the allylic acetate functionality in **10**, but does not
undergo conjugate addition to the α,β-unsaturated ester system.[27] Only
three R groups have been examined in organocuprate(I) reactions with
allylic acetates: methyl, n-butyl, and phenyl.

[181] C. R. Johnson and G. A. Dutra, *J. Amer. Chem. Soc.*, **95**, 7777 (1973).

[182] C. R. Johnson and G. A. Dutra, *J. Amer. Chem. Soc.*, **95**, 7783 (1973).

[183] P. Rona, L. Tökes, J. Tremble, and P. Crabbé, *Chem. Commun.*, **1969**, 43.

[184] R. J. Anderson, C. A. Henrick, J. B. Siddall, and R. Zurflüh, *J. Amer. Chem. Soc.*, **94**, 5379 (1972).

[185] B. Rickborn and J. A. Staroscik, University of California, Santa Barbara, unpublished results.

Lithium dimethylcuprate(I) reacts with the 4-isopropenyl lactone 11, but not with the allylic ether 12.[44]

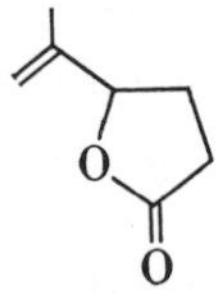

E-$(CH_3)_3CCH_2CH$=$CHC(CH_3)_2OCH_3$
12

11

Propargylic Acetates. Propargylic acetates react with lithium dialkylcuprate(I) reagents to produce secondary, tertiary, or quaternary allenes.[43, 186, 187] This replacement of acetoxy by alkyl is nonstereoselective.[43] It has been applied to cycloalkyl, steroidal (Eq. 76), and acyclic

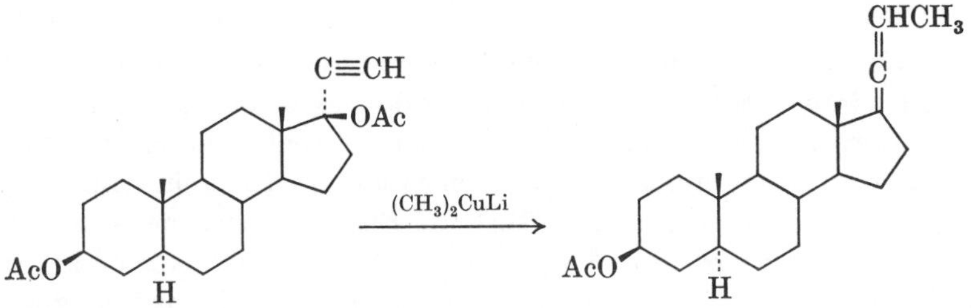

(Eq. 76)

ethynylcarbinol acetates.[187a] Isolated acetoxy and methoxy groups in the substrate survive the reaction.

Epoxides

Although nucleophilic opening of an epoxide by an organocuprate(I) may be considered an addition reaction, with respect to one carbon atom of the oxirane this process involves replacement of oxygen by carbon and therefore also legitimately may be considered a substitution reaction. Alkyl, aryl, and allyl organocuprates(I) generally attack epoxides at the sterically more accessible carbon atom of the oxirane ring to give the corresponding alcohols, but a carbalkoxy group tends to favor attack at the carbon atom to which it is attached (an arrow shows the site of predominant attack in epoxides 13–16). The tetrasubstituted epoxide 17

[186] P. Rona and P. Crabbé, *J. Amer. Chem. Soc.*, **90**, 4733 (1968).

[187] L. A. Van Dijck, B. J. Lankwerden, J. G. C. M. Vermeer, and A. J. M. Weber, *Rec. Trav. Chim.*, **90**, 801 (1971).

[187a] C. Descoins, C. A. Henrick, and J. B. Siddall, *Tetrahedron Lett.*, **1972**, 3777.

13 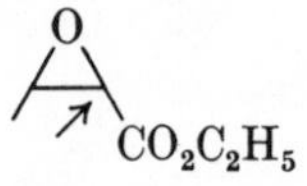   **14** 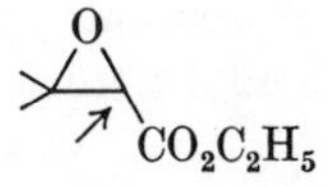   **15**

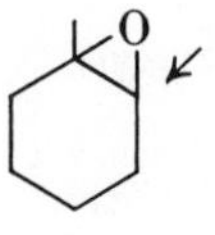

16 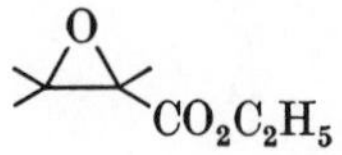   **17**

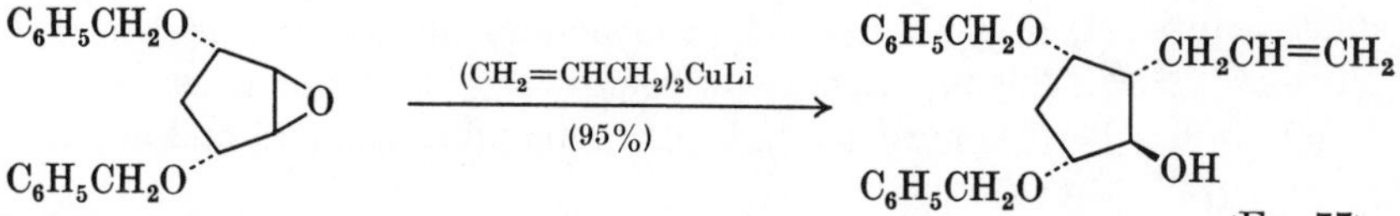

$$\xrightarrow[\text{(95\%)}]{(CH_2=CHCH_2)_2CuLi}$$

(Eq. 77)

does not react with lithium dimethylcuprate(I), probably because of steric hindrance.[188] Replacement of an oxirane carbon-oxygen bond by a carbon-carbon bond occurs with inversion of configuration; the high stereoselectivity of this reaction has been used to good avail in synthesis of a prostaglandin precursor (Eq. 77).[137] 1,4-Epoxycyclohexane is inert to lithium dimethylcuprate(I).[46] Epoxides bearing remote ester or ketone functionalities have undergone reaction selectively at the oxirane group with lithium dimethylcuprate(I) at low temperature.[50, 154]

Several 1,2-epoxycyclohexanes having an oxygen functionality in the 3 position have been treated with lithium dimethylcuprate(I) to study the possible directive influence of an adjacent oxygen function.[188] Although selective reactions occur in some cases (Eqs. 78–80), no rationale for the selectivity is available as yet.

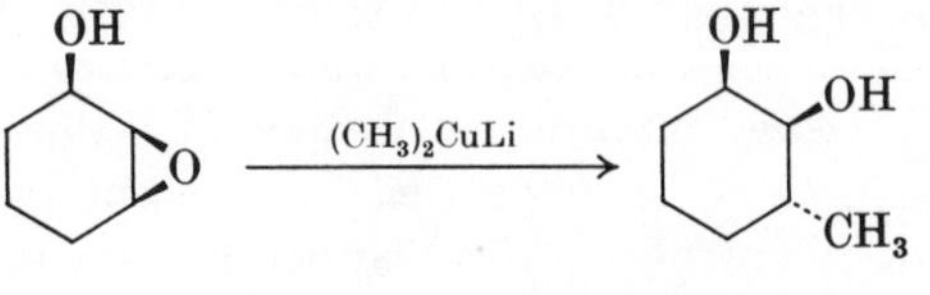

$$\xrightarrow{(CH_3)_2CuLi}$$

(Eq. 78)

(Major)

$$\xrightarrow{(CH_3)_2CuLi}$$

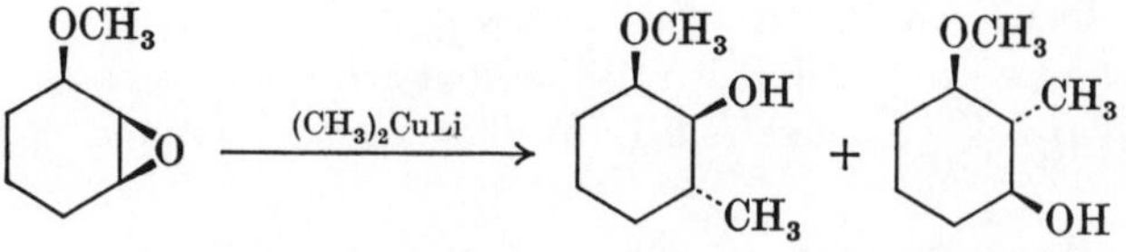

(Eq. 79)

(49%)   (45%)

[188] B. C. Hartman, T. Livinghouse, and B. Rickborn, *J. Org. Chem.*, **38**, 4346 (1973) and personal communication from B. Rickborn.

$$\text{(CH}_3)_2\text{CuLi(4.7eq)} \xrightarrow{\quad (76\%) \quad} \qquad \qquad \text{(Eq. 80)}$$

Acyclic vinylic epoxides undergo organocuprate(I) substitution with formal allylic rearrangement (*i.e.*, conjugate addition) to form *trans*-allylic alcohols stereoselectively (Eq. 81).[189-192] This reaction with organo-

$$\xrightarrow[\text{R = CH}_3, \, n\text{-C}_4\text{H}_9, \, \text{C}_6\text{H}_5]{\text{R}_2\text{CuLi}} \qquad \qquad \text{OH} \qquad \text{(Eq. 81)}$$

cuprate(I) reagents compares favorably in yield and mildness of conditions with analogous transformations using organolithium or organomagnesium[189, 191] and organoboron species.[193] Cyclic vinylic epoxides also react easily with organocuprates(I); the site of reagent attack on the epoxide varies with the epoxide structure and the organocuprate(I) used.[192] On the basis of recent studies, the relative reactivity of various types of substrates toward lithium diorganocuprate reagents appears to be as follows: acid chlorides > aldehydes > tosylates ≈ epoxides > iodides > ketones > esters > nitriles.[96, 181]

Miscellaneous Substrates

Organocopper reagents undergo some unusual reactions to achieve substitution of hydrogen, halogen, alkoxy, and nitrogen functions.

Replacement of aromatic hydrogen by aryl, heteroaryl, and acetylide[194] groups has been accomplished with polynitrobenzene substrates (Eq. 82).[118] *gem*-Difluorocyclopropenyl acetates react with lithium dimethylcuprate(I)

$$\text{O}_2\text{N} \overset{\text{NO}_2}{\underset{\text{NO}_2}{\diagdown}} \text{H} \xrightarrow[\text{(45\%)}]{} \text{O}_2\text{N} \overset{\text{NO}_2}{\underset{\text{NO}_2}{\diagdown}} \qquad \text{(Eq. 82)}[195]$$

[189] R. J. Anderson, *J. Amer. Chem. Soc.*, **92**, 4978 (1970).
[190] R. W. Herr and C. R. Johnson, *J. Amer. Chem. Soc.*, **92**, 4979 (1970).
[191] J. Staroscik and B. Rickborn, *J. Amer. Chem. Soc.*, **93**, 3046 (1971).
[192] D. M. Wieland and C. R. Johnson, *J. Amer. Chem. Soc.*, **93**, 3047 (1971).
[193] A. Suzuki, N. Miyaura, M. Itoh, H. C. Brown, G. W. Holland, and E. Negishi, *J. Amer. Chem. Soc.*, **93**, 2792 (1971).
[194] O. Wennerström, *Acta Chem. Scand.*, **25**, 789 (1971).

(Eq. 83)[183]

to form dienes (Eq. 83), and the arenesulfenyl chloride $2,4-(O_2N)_2C_6H_3SCl$ reacts with cuprous phenylacetylide to replace chlorine by phenylacetylene.[61]

Conversion of esters having no alpha protons to the corresponding methyl ketones has been achieved using lithium dimethylcuprate(I) with ethyl benzoate and with ester 18.[196] α-Diazoketones,[116] an α-diazoester,

$$C_2H_5C(OC_2H_5)_2CO_2C_2H_5$$
18

$$N_2CHCO_2C_2H_5 \xrightarrow[(43\%)]{C_6F_5Cu} C_6F_5CH_2CO_2C_2H_5 \qquad \text{(Eq. 84)}^{103}$$

and a diazoalkane[116] all react with arylcopper reagents to form the corresponding α-arylketones, α-aryl ester, and arylalkane on acidic workup (Eq. 84).

SUBSTITUTION REACTIONS USING OTHER ORGANOMETALLIC REAGENTS

Although some organic derivatives of magnesium and lithium, of Group II metals (Zn, Cd), of Group III metals (B, Al), and of transition metals [Ni, Ag (closest neighbors to Cu), Fe, Mn, Co, Rh] have undergone substitution reactions with organic substrates, none of these organometallic compounds approaches organocopper reagents in general utility for substitution reactions. Although Grignard reagents are able to add across carbon-hetero atom multiple bonds, they are not sufficiently reactive to effect substitution in most organic substrates not activated toward displacement (e.g., nonallylic[12]); their reaction with activated substrates often leads to mixtures of products.[13, 14] Furthermore, their basic properties make them less useful than organocuprate(I) reagents for nucleophilic opening of epoxides.[190] For substitution of bridgehead

[195] M. Nilsson, C. Ullenius, and O. Wennerström, *Tetrahedron Lett.*, **1971**, 2713.

[196] S. A. Humphrey, J. L. Herrmann, and R. H. Schlesinger, *Chem. Commun.*, **1971**, 1244.

bromine atoms by alkyl groups (especially methyl), Grignard reagents are preferred over organocuprates(I); heating a mixture of 1-bromoadamantane and methyl Grignard at 100° in an aerosol pressure bottle produces 1-methyladamantane in 83 % yield.[197]

Simple organolithium reagents themselves cannot effect most of the transformations discussed in this review. However, charge-delocalized or complexed organolithium reagents have been used in substitution reactions with a variety of organic substrates. Thus allylic and benzylic lithium compounds couple with secondary alkyl halides stereospecifically with inversion of configuration,[198] and the complexed organolithium species 19 reacts with organic halides to give coupled products RCH_2R' (Eq. 85).

$$RCH_2X \longrightarrow RCH_2S \overset{S}{\underset{N}{\diagup}} \longrightarrow$$

$$RCH\overset{S}{\underset{\underset{N}{Li}}{\diagup}}\overset{S}{\diagdown} \xrightarrow[\text{2. Raney Ni}]{\text{1. R'X}} RCH_2R' \qquad (Eq.\ 85)^{199}$$

19

Organocadmium and organozinc reagents displace chlorine in acid chlorides, but are unreactive toward most organic halides and oxygen derivatives.[16]

Vinylic boranes derived from hydroboration of acetylenes have been stereoselectively dimerized by treatment with iodine to 1,3-butadienes,[200] and alkyl boranes derived from hydroboration of olefins have caused oxirane opening in diene monoepoxides leading to predominantly *trans* allylic alcohols.[193] Silver-promoted dimerization of organoboranes has also been reported.[201] Organoaluminum compounds undergo substitution reactions with acid halides but not with other organic substrates.[15]

Bis-(1,5-cyclooctadiene)nickel(0) has been used to effect symmetrical dimerization of aryl halides,[202] and bis-π-allylnickel(I) complexes have

[197] E. Osawa, Z. Majerski, and P. von R. Schleyer, *J. Org. Chem.*, **36**, 205 (1971).

[198] (a) L. H. Sommer and W. D. Korte, *J. Org. Chem.*, **35**, 22 (1970); (b) S. Akiyama and J. Hooz, *Tetrahedron Lett.*, **1973**, 4115.

[199] (a) K. Hirai, H. Matsuda, and Y. Kishida, *Tetrahedron Lett.*, **1971**, 4359; (b) D. A. Evans and G. C. Andrews, *Accts. Chem. Res.*, **7**, 147 (1974); (c) W. D. Korte, K. Cripe, and R. Cooke, *J. Org. Chem.*, **39**, 1168 (1974).

[200] G. Zweifel, N. L. Polston, and C. C. Whitney, *J. Amer. Chem. Soc.*, **90**, 6243 (1968).

[201] H. C. Brown, C. Verbrugge, and C. H. Snyder, *J. Amer. Chem. Soc.*, **83**, 1002 (1961), and references therein.

[202] M. F. Semmelhack, P. M. Helquist, and L. D. Jones, *J. Amer. Chem. Soc.*, **93**, 5908 (1971).

been used to replace halogen by allylic groups in a wide variety of organic halides.[17] Potassium hexacyanodinickelate(I) reacts with organic halides to give reduced, coupled, or cyanated products, depending on the structure of the halide substrate; coupling to form symmetrical dimers predominates with some benzylic halides and with β-bromostyrene.[203] Nickel-phosphine complexes have been observed to catalyze Grignard coupling with allylic alcohols[204] and with vinylic and aryl chlorides;[205, 206] this nickel-catalyzed replacement of C_{sp^2}-chlorine by alkyl and aryl groups may be more effective than analogous reactions with organocopper reagents.

The stability and reactivity of organometallic compounds of Group I-B metals decrease in the same order: RCu > RAg > RAu.[31] Pentafluorophenylsilver[207] and one perfluoroalkenylsilver[208, 209] species have been shown to couple with some organic halides in reasonable yields, and thermal dimerization of lithium di(pentafluorophenyl)argentate(I) has been reported.[210] Phenylsilver reacts with acetyl chloride and allyl bromide to form the corresponding substitution products in 21% and 30% yields, respectively,[31] and the argentodiazoacetate 20 couples with alkyl halides

$$\underset{\textbf{20}}{Ag-\overset{\overset{N_2}{\|}}{C}-CO_2C_2H_5}$$

to form diazocarboxylic esters.[211] Thermal and oxidative dimerization of organosilver species has been studied from mechanistic[212] and synthetic[213] viewpoints, and silver-catalyzed coupling of Grignard reagents with primary alkyl bromides has been reported.[214]

Several halogenated organoiron reagents have been oxidatively dimerized,[215] and n-alkyliron species have been coupled most effectively with vinylic (and allylic[216]) halides to form new olefins stereospecifically with

[203] I. Hashimoto, N. Tsuruta, M. Ryang, and S. Tsutsumi, J. Org. Chem., 35, 3748 (1970).
[204] C. Chuit, H. Felkin, C. Frajerman, G. Roussi, and G. Swierczewski, Chem. Commun., 1968, 1604.
[205] K. Tamao, K. Sumitani, and M. Kumada, J. Amer. Chem. Soc., 94, 4374 (1972).
[206] R. J. P. Corriu and J. P. Masse, Chem. Commun., 1972, 144.
[207] K. K. Sun and W. T. Miller, J. Amer. Chem. Soc., 92, 6985 (1970).
[208] W. T. Miller, R. H. Snider, and R. J. Hummel, J. Amer. Chem. Soc., 91, 6532 (1969).
[209] R. E. Banks, R. N. Hazeldine, D. R. Taylor, and G. Webb, Tetrahedron Lett., 1970, 5215.
[210] V. B. Smith and A. G. Massey, J. Organometal. Chem., 23, C9 (1970).
[211] U. Schöllkopf and N. Rieber, Angew. Chem., 79, 238 (1967); Angew. Chem., Int. Ed. Engl., 6, 261 (1967).
[212] M. Tamura and J. Kochi, J. Amer. Chem. Soc., 93, 1483 (1971).
[213] G. Köbrich, H. Fröhlich, and W. Drishel, J. Organometal. Chem., 6, 194 (1966).
[214] M. Tamura and J. Kochi, Synthesis, 1971, 303.
[215] G. Köbrich and H. Buttner, Tetrahedron, 25, 883 (1969).
[216] E. J. Corey, H. Yamamoto, D. K. Herron, and K. Achiwa, J. Amer. Chem. Soc., 92, 6635 (1970).

retention of configuration.[56, 217] Disodium tetracarbonylferrate(II) has been used recently to convert acid chlorides to ketones.[218]

Although n-alkylmanganese reagents do not couple effectively with alkyl halides, they do undergo stereospecific substitution reactions with alkenyl halides.[10, 217] n-Alkylcobalt species react similarly but in somewhat lower yields.[46]

Alkylrhodium(I) complexes have been shown to transform acyl chlorides into ketones[219a] and to couple with various alkenyl and aryl halides.[219b]

SYNTHETIC UTILITY

Carbon-carbon bond formation between a nucleophile and an electrophile normally proceeds best when the nucleophile is strongly nucleophilic and weakly basic. Carbanions stabilized by adjacent substituents (*e.g.*, carbonyl group or heteroatom) usually satisfy these requirements and have frequently been used in substitution reactions.[1] Removal of the "adjacent substituent" after coupling is most often difficult, and therefore hydrocarbons are not easily produced in this way. An exception to this generalization is illustrated by Eq. 86 in which a phosphorus (or sulfur) ylide

$$\text{RCH}_2\text{X} \xrightarrow[\text{2. Base}]{\text{1. } \text{R}_3\text{P}} \text{RCHPR}_3^{-\;+} \xrightarrow[\text{2. Base}]{\text{1. } \text{R}'\text{X}} \text{R}\bar{\text{C}}(\text{R}')\text{PR}_3 \xrightarrow{\text{H}_2\text{O}} \text{RCH}_2\text{R}'$$

$$\text{(Eq. 86)}$$

couples with an organic halide to yield a hydrocarbon after reductive or hydrolytic workup;[1] this sequence, however, is limited for all practical purposes to formation of hydrocarbons of the type $\text{RCH}_2\text{R}'$.

Attachment of one hydrocarbon group to another has long been a serious challenge to synthetic organic chemists. Most organometallic reagents derived from unstabilized carbanions are too basic to act solely as nucleophiles in displacing a halogen or oxygen leaving group, and undesired α and β eliminations are often side reactions. Organocopper compounds derived from unstabilized carbanions, however, are unique in that the basicity of the organic group is sufficiently low to allow highly effective coupling with a wide variety of organic substrates, as described in this review.

Some general synthetic transformations which have been achieved using organocopper reagents include the following: carbonyl → allene (Eq.

[217] E. J. Corey and G. H. Posner, *Tetrahedron Lett.*, **1970**, 315.

[218] J. P. Collman and N. W. Hoffman, *J. Amer. Chem. Soc.*, **95**, 2689 (1973).

[219] (a) L. S. Hegedus, S. M. Lo, and D. E. Bloss, *J. Amer. Chem. Soc.*, **95**, 3042 (1973); (b) M. F. Semmelhack and L. Ryono, *Tetrahedron Lett.*, **1973**, 2967.

87),[186] carbonyl → *gem*-dialkyl (Eq. 88),[165, 172, 182, 220] allylic alcohol → olefin (Eq. 89),[184] cyclic epoxide → *trans*-2-alkylcycloalkanol (Eq. 90),[50, 137] and vinylic epoxide → *trans*-allylic alcohol (Eq. 81).[189] Especially noteworthy also is the selectivity of organocopper reagents, which makes

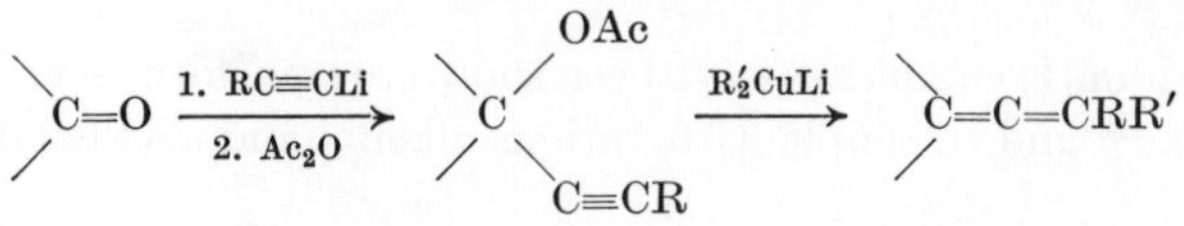

$$\underset{/}{\overset{\diagdown}{C}}=O \quad \xrightarrow[\text{2. Ac}_2\text{O}]{\text{1. RC}\equiv\text{CLi}} \quad \underset{\underset{C\equiv CR}{|}}{\overset{OAc}{\underset{|}{\overset{\diagup}{\underset{/}{\overset{\diagdown}{C}}}}}} \quad \xrightarrow{\text{R}_2'\text{CuLi}} \quad \underset{/}{\overset{\diagdown}{C}}=C=CRR' \qquad (\text{Eq. 87})$$

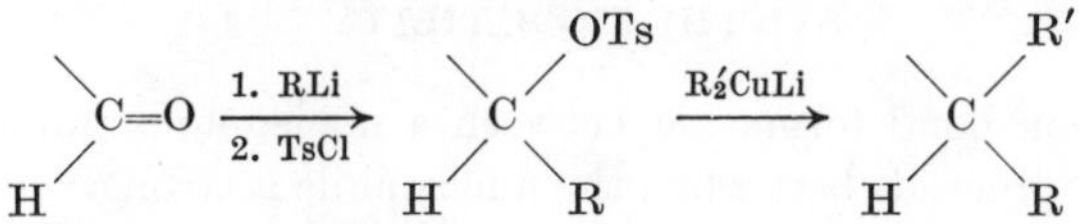

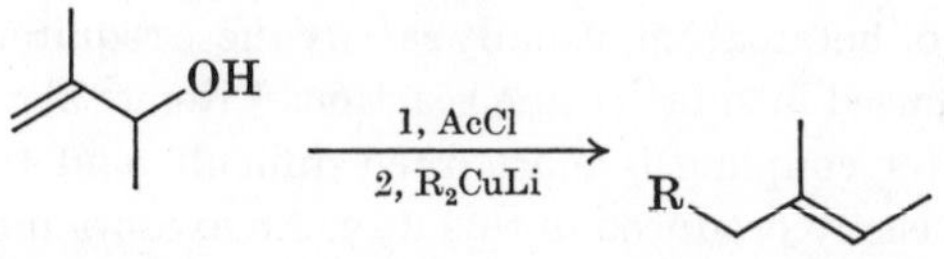

$$\underset{H}{\overset{\diagdown}{\underset{/}{\overset{\diagdown}{C}}}}=O \quad \xrightarrow[\text{2. TsCl}]{\text{1. RLi}} \quad \underset{H}{\overset{OTs}{\underset{R}{\overset{\diagup}{\underset{\diagdown}{C}}}}} \quad \xrightarrow{\text{R}_2'\text{CuLi}} \quad \underset{H}{\overset{R'}{\underset{R}{\overset{\diagup}{\underset{\diagdown}{C}}}}} \qquad (\text{Eq. 88})$$

$$\underset{}{\overset{}{}} \quad \xrightarrow[\text{2, R}_2\text{CuLi}]{\text{1, AcCl}} \quad \underset{}{\overset{}{}} \qquad (\text{Eq. 89})$$

$$\underset{}{\overset{}{}} \quad \xrightarrow{\text{R}_2\text{CuLi}} \quad \underset{}{\overset{}{}} \qquad (\text{Eq. 90})$$

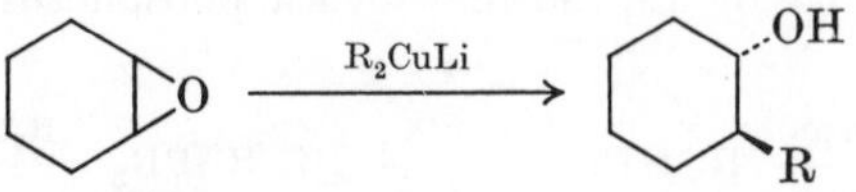

possible substitution without affecting remote, normally labile functional groups (*e.g.*, ketone, ester, halogen) in the substrate.[96, 100, 101]

In exploring the scope and limitations of organocopper reactions with organic substrates bearing suitable leaving groups, investigators have synthesized a large variety of substitution products that are often interesting and useful molecules.

A substantial number of these compounds, however, have been prepared via organocopper reagents *specifically as synthetic intermediates*, whose structural elaboration has led to a variety of natural products. Enumeration of these naturally occurring substances provides an indication of the broad synthetic utility of organocopper reagents.

Cuprous methylacetylide has been used to prepare heterocyclic junipal,*[127] and lithium diorganocuprate(I) reagents have been used in the synthesis of olivetol,[221] fulvoplumierin,[113] carvestrene,[114] isopulegone,[114] *trans,trans*-farnesol,[160] α-*trans*-bergamotene,[146] insect juvenile hormone,[161]

---

* The wavy lines in the accompanying structures indicate the site of organocopper carbon-carbon bond formation.

[220] G. H. Posner and D. J. Brunelle, *Tetrahedron Lett.*, **1973**, 935.

[221] T. Petrzilka, W. Haefliger, and C. Sikemeier, *Helv. Chim. Acta*, **52**, 1102 (1969).

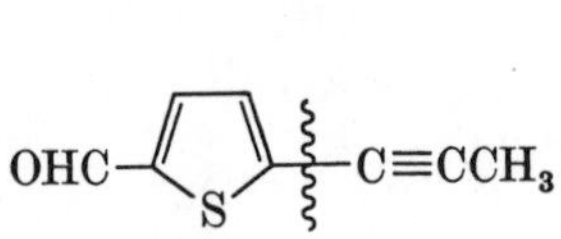

OHC—⟨S⟩—C≡CCH$_3$

Junipal

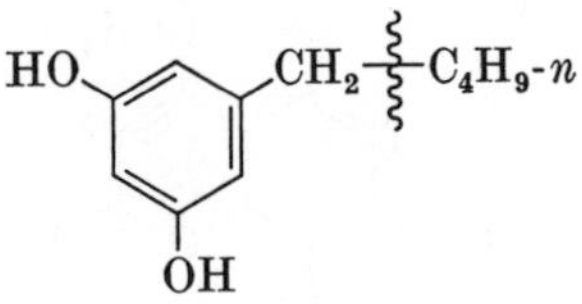

HO—⟨⟩—CH$_2$—C$_4$H$_9$-n

OH

Olivetol

CO$_2$CH$_3$

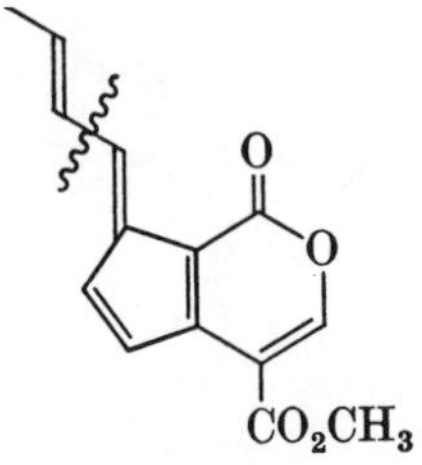

Fulvoplumierin

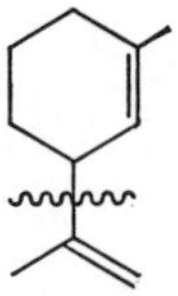

Carvestrene

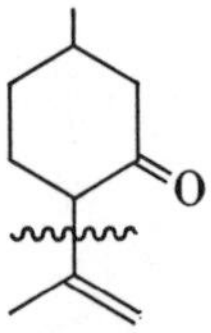

O

Isopulegone

OH

CH$_3$

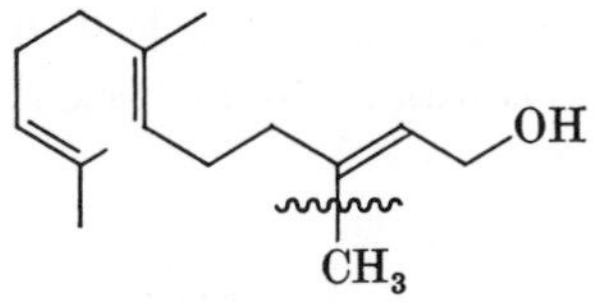

trans,trans-Farnesol

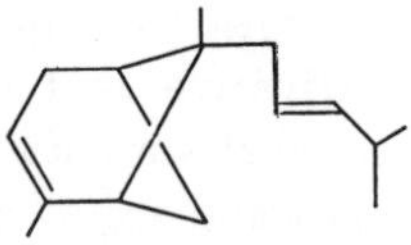

α-*trans*-Bergamotene

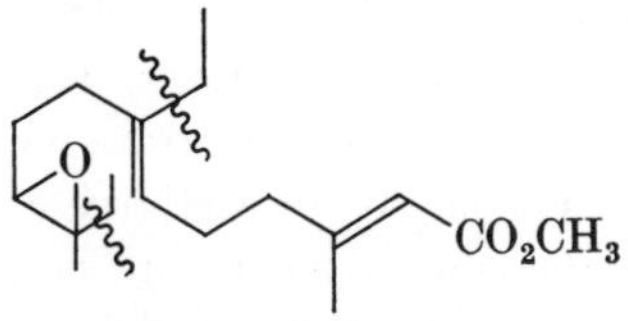

O

CO$_2$CH$_3$

Insect juvenile hormone

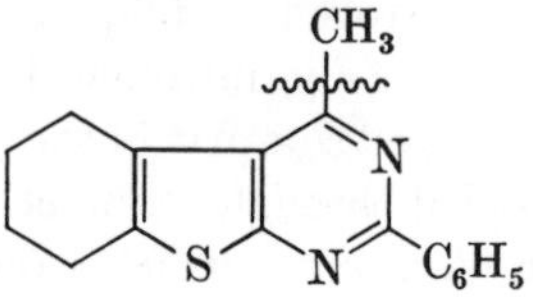

CH$_3$

S   N   C$_6$H$_5$

N

21

OH

(CH$_2$)$_3$CO$_2$H

OH   H OH

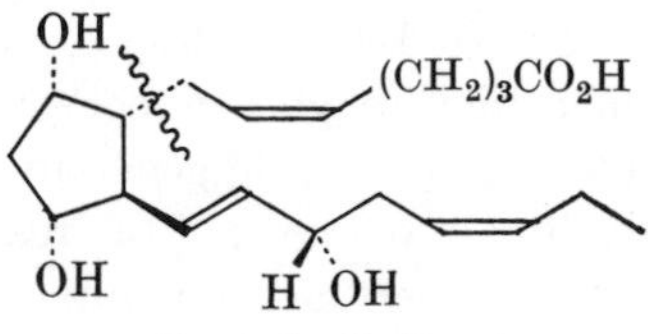

Prostaglandin F$_{3\alpha}$

the pyrimidine derivative **21**,[170] prostaglandin F$_{3\alpha}$,[137] kotanin,[109] and the corrin derivative **22**.[222] (Formulas on p. 296).

Clearly, many possibilities for synthetically useful organocopper substitution reactions remain to be explored. Within the next ten years, advances in experimental technique (*e.g.*, control of temperature, solvent,

[222] A. Hamilton and A. W. Johnson, *Chem. Commun.*, **1971**, 523; *J. Chem. Soc.*, *C*, **1971**, 3879.

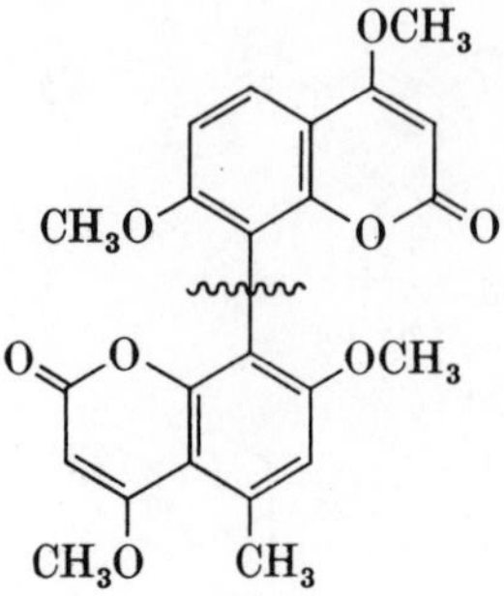

Kotanin

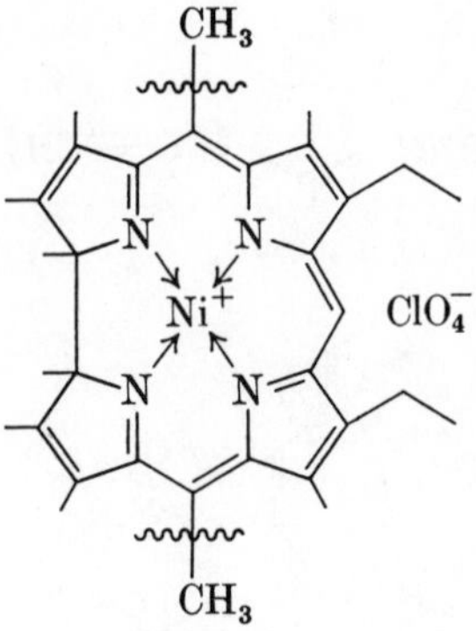

22

and complexing ligands), development of different organocopper reagents (e.g., XCH_2Cu, $RR'CuLi$, and $ROCu$), and use of diverse types of substrates should substantially increase the general synthetic utility of these novel reagents.

EXPERIMENTAL FACTORS

Preparation and Handling of Organocopper Reagents

For most coupling reactions, the necessary ratio of organocopper reagent to organic halide substrate (or alcohol derivative) is at least 2.5:1.

Because of their high reactivity and low thermal stability, stoichiometric organocopper reagents are prepared in situ and are used immediately. Air and moisture must be rigorously excluded; reactions are generally run in an atmosphere of argon or prepurified nitrogen, and liquid reagents are best transferred via dry hypodermic syringes and introduced into the reaction mixture through a rubber septum-capped side arm of the reaction flask. Solid reagents should generally be added through a funnel with its stem extending into a neck of the reaction flask out of which a rapid and constant flow of inert gas is maintained.

Complete formation of stoichiometric organocopper reagent can often be judged by a negative Gilman test with Michler's ketone[223] or, less accurately, by visually following the dissolution of copper salt. Alternatively, an aliquot quenched at low temperature with benzoyl chloride,[30] for example, might indicate whether organocopper formation is complete; if any alcoholic product is obtained organocopper formation is probably incomplete.[96, 100]

[223] H. Gilman and F. Schulze, J. Amer. Chem. Soc., 47, 2002 (1925).

Since most of the copper salts used to prepare the organocopper reagents are not hygroscopic, glove-bag or dry-box procedures are unnecessary; highly effective stoichiometric organocopper reagents have been prepared from carefully purified and dried cuprous iodide[30] as well as from commercial samples* (e.g., Fisher Chemical Co.).[96] Similarly, commercially available organolithium reagents are usually satisfactory; however, the solvent in which the reagent is prepared becomes part of the reaction mixture and occasionally influences the course of the substitution reaction.

Temperature

Temperature variation has been observed to affect both the rate of formation of organocopper reagents and their stability, once formed. Lithium di-n-butylcopper, for example, is formed in ether from n-butyllithium and cuprous iodide rapidly ($\ll 1$ minute) at $0°$, but slowly (>10 minutes) at $-40°$;[144] optimum conditions for generation of organocopper reagents, therefore, must involve careful control of reaction temperature.

The thermal stability of an organocopper reagent depends on the nature of the reagent and on the structure of the organic group. Phenylcopper, for example, is stable below $80°$, but different complexes of phenyl copper [e.g., $(C_6H_5Cu)_4C_6H_5Li\cdot(C_2H_5)_2O$, $(C_6H_5Cu)_2(C_6H_5)_2Mg(THF)$, and $C_6H_5CuP(C_6H_5)_3$] have different thermal stabilities.[74, 91, 94]

The structure of R strongly influences the stability and reactivity of organocopper reagents. Thus, whereas phenylcopper is stable below $80°$, methylcopper decomposes at room temperature and ethylcopper at $-18°$ in ether.[31, 32] This trend in organocopper thermal stability generally follows the thermal stability of the corresponding organolithium reagents (stability: $C_6H_5Li > CH_3Li > C_2H_5Li$).[224] Similarly, lithium dimethylcopper in ether under an inert atmosphere is stable for hours at $0°$, but the more reactive lithium di-n-alkylcopper and secondary and tertiary organocopper reagents rapidly decompose in ether above $-20°$. Thus, for most effective preparation and use of these less stable, more reactive organocopper reagents, the reaction temperature should be carefully controlled. Usually stoichiometric n-alkyl, secondary, and tertiary organocopper reagents are prepared either below $-20°$ for a sufficient amount of time (usually >5 minutes) to allow complete formation of reagent[33] or at $0°$ for several minutes (rapid reagent formation) followed immediately by cooling to below $-20°$.[144] (See experimental procedure for bis-homo-geraniol, p. 302). The exact reaction temperature used for a substitution

---

* Gentle flaming of the cuprous iodide under vacuum immediately before use generally produces a cleaner solution of organocopper reagent.

[224] T. L. Brown, Adv. Organometal. Chem., 3, 365 (1965).

reaction generally should be the lowest temperature that gives an acceptable reaction rate.

Solvent

Heteroaromatic solvents such as pyridine and quinoline have routinely been used for cuprous acetylide substitution reactions, and ethers such as diethyl ether and tetrahydrofuran have normally been used for substitution reactions with all other organocopper reagents. Cuprous acetylide coupling with *ortho*-substituted aryl halides is subject to a dramatic solvent effect; in pyridine, replacement of halogen by acetylide occurs to give an *ortho*-substituted arylacetylene, but in dimethylformamide the reaction proceeds further to give a heterocycle, as illustrated by Eq. 91 for synthesis of an indole.[61]

(Eq. 91)

Conflicting reports have appeared about whether diethyl ether or tetrahydrofuran is the better solvent for organocopper substitution reactions. Thus, lithium dialkylcuprates(I) react with 1-chloroalkanes[33] and with α-halocarbonyl substrates[115] to give desired products in higher yields in tetrahydrofuran than in diethyl ether. Copper catalysis of Grignard coupling with organic halides also proceeds better in tetrahydrofuran.[214] On the other hand, reaction of organocuprate(I) with toluenesulfonate esters,[182] with allylic acetates,[44] and with haloolefins[46] proceeds faster and more selectively in diethyl ether than in tetrahydrofuran. To optimize the yield of a desired substitution product, therefore, small-scale preliminary experiments should be done with tetrahydrofuran and with diethyl ether to determine which solvent is more effective for the particular transformation in question. The solvent of choice should normally be freshly distilled from a drying agent such as lithium aluminum hydride.

Organocopper reagents will undoubtedly continue to be used for effective substitution reactions. To systematize reporting of experimental results in this area, the following comments are made. Inorganic salts[30] and trace impurities[33] in the reagents used to prepare organocopper species have been found on occasion to alter the course of the substitution reaction.

Future authors are strongly urged, therefore, to indicate explicitly the source and/or method of purification of all reagents. Furthermore, since the reactivity of many organometallic reagents depends on their state of aggregation,[224, 225] which may change as concentration of reagent is varied, publications dealing with organocopper reagents should specify clearly the concentration of any organocopper reagent used.

EXPERIMENTAL PROCEDURES

The following examples of effective substitution reactions using organocopper reagents have been carefully chosen to illustrate useful and general experimental procedures, all performed under inert atmosphere. They are organized into four categories: (1) thermal and oxidative dimerization, (2) organocopper coupling with halide substrates, (3) cuprous acetylide coupling with halide substrates, and (4) organocopper coupling with epoxides.

Thermal and Oxidative Dimerization of Organocopper and Organocuprate Species

E,E-2,4-Hexadiene (Thermal Dimerization with Retention of Configuration of a Vinylcopper-Phosphine Complex).[104] To a flame-dried 12-ml centrifuge tube capped with a serum stopper was added 190 mg (0.483 meq) of tetrakis[iodo(tri-n-butylphosphine)copper(I)],[226] and the tube was flushed with nitrogen. Ether (*ca.* 2 ml) which had been distilled from lithium aluminum hydride under inert atmosphere immediately before use was transferred into the flask from a storage vessel using a stainless steel cannula. An ethereal solution (4 ml) containing n-decane as internal glpc standard and 1 eq E-1-propenyllithium[227] was transferred by cannula to the cold ($-78°$) solution of the phosphine complex. Mixing the reagents produced a yellow solution of the tri-n-butylphosphine complex of E-1-propenylcopper. Ether was added by cannula to give a 0.1 N solution. After 4–6 hours at room temperature a metallic mirror and E,E-2,4-hexadiene [89% yield, 97.1% isomeric purity by glpc using a 12-ft 25% 1,2,3-tris(2-cyanoethoxy)propane (TCEOP) on Chromosorb W column] were formed. Preparative glpc (same column) gave E,E-2,4-hexadiene, pure by ir comparison with an authentic sample.[228]

***t*-Butylbenzene** (Oxidative Dimerization of a mixed Homocuprate; Unsymmetrical Coupling of *t*-Butyl and Aryl Groups).[33] Copper(I)

[225] W. H. Glaze and C. H. Freeman, *J. Amer. Chem. Soc.*, **91**, 7198 (1969).
[226] F. G. Mann, D. Purdie, and A. F. Wells, *J. Chem. Soc.*, **1936**, 1503.
[227] D. Seyferth and L. G. Vaughan, *J. Amer. Chem. Soc.*, **86**, 883 (1964).
[228] L. K. Montgomery, K. Schueller, and P. D. Bartlett, *J. Amer. Chem. Soc.*, **86**, 622 (1964).

bromide[229] (1.426 g, 10.0 mmol) was placed in a 40-ml centrifuge tube. The tube was gently flamed dry under a stream of nitrogen and capped with a serum stopper. Tetrahydrofuran (5 ml) and an ether solution of phenyllithium[230] (9.0 ml, 18 mmol) were added to the tube at 25°. This mixture was shaken vigorously for 5 minutes. It was then cooled to −78° and 1.25 ml (2.0 mmol) of a pentane solution of t-butyllithium (Foote Mineral Co.) was added. The resulting mixture was shaken thoroughly for 10 seconds, then oxidized by addition of 60 ml of molecular oxygen by syringe at −78°. Hydrolysis and glpc analysis using a calibrated internal standard indicated the presence of t-butylbenzene (73%, based on t-butyllithium).

Organocopper Coupling with Halide Substrates

N-Methyldodecananilide (Substitution of Iodine by Methyl with Lithium Dimethylcuprate(I); Use of Low Temperature to Achieve Functional Group Selectivity).[46]

$$I(CH_2)_{10}CON(CH_3)C_6H_5 \xrightarrow{(CH_3)_2 CuLi} CH_3(CH_2)_{10}CON(CH_3)C_6H_5$$

To 464 mg (2.44 mmol) of cuprous iodide (Fisher Chemical Co) in 15 ml of anhydrous ether at −20° under nitrogen was added by hypodermic syringe 3.0 ml of 1.4 M (4.2 mmol) methyllithium in ether (Foote Mineral Co). After several minutes 108 mg (0.27 mmol) of N-methyl-11-iodoundec-ananilide (prepared from 11-iodoundecanoic acid via the acid chloride and N-methylaniline) in 1 ml of ether was added. After 21 hours at −20°, the reaction mixture was poured into a saturated aqueous ammonium chloride solution and extracted with ether. The combined ethereal extracts were dried over magnesium sulfate and evaporated under reduced pressure to give 50 mg (65%) of colorless, liquid N-methyldodecananilide; infrared (film): 6.0μ (C=O); nmr (CCl$_4$): 0.88δ (t, 3H, −CH$_2$C$\underline{H}_3$), 1.96 (t, 2H,

$$-CH_2C\underline{H}_2\overset{|}{C}ON), \text{ and } 3.20 \text{ (S, 3H, } -NCH_3).$$

2,2-Dimethyldecane (Substitution of Bromine by t-Butyl Using Lithium t-Butoxy-t-Butylcuprate).[100]   To 1.904 g (10.0 mmol) of cuprous iodide (Fisher Scientific Co.), which had been extracted for 6 hours with refluxing tetrahydrofuran and then dried under vacuum at 25°, was added via syringe 20.0 ml of dry tetrahydrofuran (distilled from lithium aluminum hydride). Into this suspension was injected 9.10 ml of a 1.10 M (10.0 mmol) tetrahydrofuran solution of t-butoxylithium, which had been pre-pared from dry t-butyl alcohol and n-butyllithium (Foote Mineral Co). This mixture was stirred at room temperature under an atmosphere of

[229] G. B. Kauffman and L. A. Teter, *Inorg. Syn.*, **7**, 9 (1963).
[230] M. Schlosser and V. Ladenberger, *J. Organometal. Chem.*, **8**, 193 (1967).

prepurified nitrogen until formation of cuprous t-butoxide was complete (as evidenced by the disappearance of gray, insoluble cuprous iodide, 10–15 minutes). Into the cooled ($-78°$) solution was then injected 8.90 ml of a 1.12 M solution of t-butyllithium (9.97 mmol) in pentane (Foote Mineral Co.), and the resulting cloudy brown mixture was stirred for 5 minutes at $-78°$. 1-Bromooctane (386 mg, 2.00 mmol, Aldrich Chemical Co.) in 2.0 ml of tetrahydrofuran was injected and stirring was continued at $-78°$ for 2 hours. The temperature was allowed to rise slowly (2 hours) to $-30°$. Absolute methanol (2.0 ml, 49 mmol) was added, and analytical glpc on a 10 ft $\times$ $\frac{1}{4}$ in. 10 % FFAP on 60/80 Chromosorb W column using cumene as internal standard indicated 83 % 2,2-dimethyldecane (retention time 2.0 minutes at 130°). Preparative glpc using a 20 ft $\times$ $\frac{3}{8}$ in. 20 % SE-30 on 45/60 Chromosorb W column at 200° gave 2,2-dimethyldecane (14 minutes retention time), n^{20}D 1.4203, bp 198° (760 mm).

Z-2-Undecene (Replacement of Iodine by an Alkenyl Group Using a Lithium Dialkenylcuprate; Substitution with Retention of Configuration).[231, 232] A solution of 2.42 g (20.0 mmol) of Z-1-propenyl bromide (99 % Z) in 6 ml of anhydrous ether was added to 0.28 g (40 mmol) of finely cut lithium (1.5 % Na) in 6 ml of ether at $-10°$. The resulting organolithium solution was added dropwise to a stirred suspension of 4.38 g (10 mmol) of $[(CH_3O)_3P]_2CuI$ in 15 ml of dry ether at -25 to $-30°$; the rate of addition was such as to obtain a deep-red solution. After all of the copper(I) complex had dissolved, 1.2 g (5 mmol) of 1-octyl iodide in 2 ml of dry ether was added during 5 minutes at $-30°$. The reaction mixture was stirred for an additional 15 minutes at $-30°$, and was then left for 90 minutes to reach room temperature. The mixture was poured into excess saturated aqueous ammonium chloride, filtered through Celite, and extracted with ether. The ethereal extract was washed with saturated aqueous solutions of ammonium chloride and sodium chloride, dried (MgSO$_4$), and concentrated. Filtration of a hexane solution of the oily crystalline residue through a column of 40 g of silica gel afforded 0.5 g (66 %) of Z-2-undecene, bp 90–100° (11 torr, bulb distillation), 97 % Z by glpc using a 200-ft $\times$ 0.02 in. UCON metal capillary column; infrared (neat): 3010 (C=CH), 1650 (C=C), 695 cm^{-1} (Z–CH=CH); nmr: 0.9 (3H, CH$_3$), 1.55 (d, 3H, $J = 5H_z$, C=CCH$_3$) 5.1–5.5 δ (m, 2H, CH=CH); mass spectrum: m/e 154 (m$^+$), 55 (base).

(+)-(S)-2-Phenylbutane (Replacement of Bromine by Phenyl Using Lithium Diphenylcuprate; Substitution with Inversion of Configuration).[33] A mixture of 120 mmol of lithium diphenylcuprate (prepared

[231] F. Näf and P. Degen, *Helv. Chim. Acta*, **54**, 1939 (1971).
[232] G. Linstrumelle, J. K. Krieger, and G. M. Whitesides, *Org. Syntheses*, **53**, Procedure No. 1842 (1973).

from 120 mmol of purified[229] cuprous bromide and 240 mmol of ethereal phenyllithium at 0°) and 5.00 g (36.5 mmol) of (−)-(R)-2-bromobutane ([α]^{25}D −26.02°, optical purity 73–78%) in a mixture of 129 ml of ether and 180 ml of tetrahydrofuran was refluxed (51–52°) for 72 hours. The mixture was then quenched in aqueous ammonium chloride and extracted with ether. After the ethereal solution had been washed with aqueous sodium chloride, dried, and concentrated, distillation of the residual liquid through a Teflon spinning-band column separated 4.33 g (87%) of (+)-(S)-2-phenylbutane, bp 170°, n^{25}D 1.4879, d_4^{25} 0.856, [α]^{27}D +18.20° (neat) (optical purity 67%). In a second run employing (−)-(R)-2-bromobutane ([α]^{25}D −27.01°, optical purity 76–81%) the (+)-(S)-2-phenylbutane was obtained in 67% yield; bp 169–170°, n^{25}D 1.4877, d_4^{25} 0.855 [α]^{26}D +18.70° (neat) (optical purity 68%). Each distilled product exhibited a single peak on glpc, using a column packed with a nitrile silicon gum, XE-60, suspended on Chromosorb P.

3-Ethyl-7-methyl-E,Z-2,6-nonadien-1-ol (Bis-homogeraniol) (Replacement of Alkenyl Iodine by Ethyl Using Lithium Diethylcuprate; Substitution with Retention of Configuration; Use of Ethyl Iodide To Minimize Deleterious Effect of Metal-Halogen Exchange).[144, 233] Ethyllithium (21.4 mmol; 33 ml of a 0.65 M benzene solution) was placed in a three-necked, round-bottomed flask, and the solvent was removed under aspirator pressure. Anhydrous ether (60 ml) was added; the vessel was cooled to −50°, and 2.03 g (10.7 mmol) of cuprous iodide was introduced while argon flowed through the flask. After stirring for 45 minutes at −50°, the copper salt had dissolved, giving a homogeneous green-brown reagent. A solution of 3-iodo-7-methyl-E,Z-2,6-nonadien-1-ol (604 mg, 2.16 mmol) in a few milliliters of ether was added at −70°, and the reaction was allowed to warm to −30° over 25 minutes and to stir at this temperature for an additional 1.8 hours. Glpc analysis indicated that starting material was absent and that 11% metal-halogen exchange had occurred. Ethyl iodide (4 ml, 50 mmol) was added, and the reaction was warmed to 0°. After 22 hours at this temperature, the reaction was quenched with water. Insoluble copper salts were removed by vacuum filtration through a pad of Hyflo Super Cel. The ethereal layer was separated, combined with three further extracts of the aqueous layer, dried (MgSO$_4$), and concentrated to 463 mg of an orange oil. Preparative tlc on silica gel (12:1 benzene:ethyl acetate, 2 developments) gave 309 mg (78%) of pure bishomogeraniol. The purified material showed a single spot of R$_f$ 0.41 upon tlc analysis (10:1 benzene:ethyl acetate, 2 developments). Glpc analysis showed that the material was >99% pure and that less than 1%

[233] E. J. Corey, J. A. Katzenellenbogen, N. W. Gilman, S. A. Roman, and B. W. Erickson, J. Amer. Chem. Soc., 90, 5618 (1968).

exchanged material was present: infrared (film): 2.96 (s, O—H stretch), 3.38 (s, C—H stretch), 6.02 (m, C=C stretch), 6.94 (s) and 7.32 (m, C—H bend), and 9.92 μ (s, C—O stretch); nmr (CCl$_4$): 0.96 δ (triplet, $J = 7$ Hz, 6H, C$\underline{H}_3$—CH$_2$— at C$_3$ and C$_7$), 1.66 (singlet, 3H, CH$_3$— at C$_7$), 1.8–2.3 (multiplet, 8H, methylenes except C$_1$), 3.58 (s, 1H, —OH), 4.04 (doublet, $J = 7$ Hz, 2H, —CH$_2$—O—), 5.05 (broad triplet, 1H, =CH—, C$_6$), and 5.30 (triplet, $J = 7$ Hz, 1H, =CH—, C$_2$). The molecular ion was found at m/e 182.1669 (calcd for C$_{12}$H$_{22}$O: 182.1671).

1,7-Dichloro-3,3,4,4,5,5-hexafluoro-E,E-1,6-heptadiene (Replacement of Alkenyl Iodine by a Perfluoroalkyl Group Using a Perfluoroalkylcopper Generated *in situ;* Substitution with Retention of Configuration.)[83] Hexafluoro-1,3-diiodopropane (4 g, 0.01 mol) was heated and stirred at 120° under nitrogen with E-1-chloro-2-iodoethylene (3.6 g, 0.02 mol) and activated copper bronze (3 g) in dimethylformamide (25 ml). The mixture was heated for 12 hours, after which time there was a copious white precipitate of copper iodide. The mixture was poured into water and extracted with ether (3 × 50 ml). The extracts were dried (MgSO$_4$), and distilled to leave an oily residue, which on distillation under reduced pressure afforded 2.6 g (66%) of 1,7-dichloro-3,3,4,4,5,5-hexafluoro-E,E-1,6-heptadiene, bp 42° (20 mm).

Pentafluorobiphenyl (Substitution of Aryl Iodine by a Perhaloaryl Group Using Pentafluorophenylcopper).[138] Ethylmagnesium bromide (0.101 mol in 80 ml of tetrahydrofuran) was added to a stirred solution containing bromopentafluorobenzene (25.0 g, 0.101 mol) dissolved in tetrahydrofuran (170 ml) containing *t*-butylbenzene (2.5 g) as an internal glpc standard. The rate of addition was such that the temperature of the reaction mixture did not exceed 30°. After the completion of addition, the reaction mixture was stirred for 0.5 hour at room temperature. On rapid addition of cuprous chloride (10.0 g, 0.101 mol), the reaction temperature rose to 35°. The resulting mixture was allowed to stir at room temperature for 18 hours. The precipitate was allowed to settle and an aliquot sample of the supernatant liquid was withdrawn, hydrolyzed, extracted with pentane and analyzed by glpc. The yields of pentafluorobenzene thus obtained ranged from 76 to 85%. Small amounts of decafluorobiphenyl were detected in solution. The supernatant liquid showed little change after storage for 2 weeks under nitrogen and was used for the following experiment. Iodobenzene (1.6 g, 7.9 mmol) was added to 25 ml of the tetrahydrofuran solution of pentafluorophenylcopper reagent (7.9 mmol) and the reaction mixture was heated at 66°. The reaction was monitored by glpc analysis which indicated that the reaction proceeded very slowly. After 7 days at 66°, glpc analysis indicated pentafluorobiphenyl (71%) as the only reaction product. There was no biphenyl or decafluorobiphenyl present. The reaction mixture was hydrolyzed and the crude product was recrystallized from a methanol/benzene mixture; mp 109–111°.

Ethyl 4-Bromo-4-pentenoate (Substitution of Allylic Bromine by Ethoxycarbonylmethyl Using a Functionalized Alkylcopper Reagent and Careful Temperature Control).[143] A solution of lithium diisopropylamide

$$CH_3CO_2C_2H_5 \xrightarrow{(i\text{-}C_3H_7)_2NLi} LiCH_2CO_2C_2H_5 \xrightarrow{CuI} LiCu(CH_2CO_2C_2H_5)_2$$

$$\xrightarrow{CH_2=CBrCH_2Br} CH_2=CBrCH_2CH_2CO_2H_5$$

(4 mmol) in tetrahydrofuran (5 ml) was slowly added to a solution of ethyl acetate (352 mg, 4 mmol) and cuprous iodide (1.52 g, 8 mmol) in tetrahydrofuran (15 ml) at −110° under nitrogen atmosphere. It was stirred until the cooling bath was warmed up to −30°; a solution of 2,3-dibromopropene (328 mg, 2 mmol) in tetrahydrofuran (5 ml) was added at that temperature and stirring was contined for 1 hour. The reaction mixture was hydrolyzed with aqueous ammonium chloride and the crude product was purified by preparative tlc giving ethyl 4-bromo-4-pentenoate in 83% yield. The infrared spectrum of the product (in CCl_4) manifested peaks due to ester at 1735 cm^{-1} and $CH_2=C$ at 1630 and 885 cm^{-1}; the nmr spectrum (CCl_4) showed peaks at 1.32 δ (triplet 3H, CH_3—C—O), 2.54–2.93 (multiplet, 4H, $-C=C-CH_2CH_2CO-$), 4.15 (quartet, 2H, $-CH_2-O-$), 5.42 (doublet, 1H, olefinic proton) and 5.64 (doublet, 1H, olefinic proton).

5,10-Tetradecanedione (Replacement of Chlorine in Acyl Chloride by n-Butyl Using Lithium Di-n-butylcuprate; Use of Low Temperature to Achieve Functional Group Selectivity).[96, 234] Into a dry 50-ml two-

$$n\text{-}C_4H_9CO(CH_2)_4COCl \xrightarrow{(n\text{-}C_4H_9)_2CuLi} n\text{-}C_4H_9CO(CH_2)_4COC_4H_9\text{-}n$$

necked, round-bottomed flask equipped with a rubber septum was placed a magnetic stirring bar and 571 mg (3.0 mmol) of cuprous iodide (Fisher Chemical Co.). A three-way stopcock bearing a nitrogen-filled balloon was used to evacuate and then fill the flask with nitrogen. The purging procedure was repeated two more times, and the flask was gently flamed during the third evacuation. Anhydrous diethyl ether (8 ml) was added and the system was cooled to −40°. n-Butyllithium (4.54 ml of a 1.32 M pentane solution, 6.0 mmol) was injected. After about 5 minutes at −40°, the temperature was lowered to −78°. A precooled ethereal solution (1 ml) of 6-oxodecanoyl chloride [bp 86–87.5° (0.23 mm), 213 mg, 1.04 mmol] was injected. After 15 minutes at −78°, absolute methanol (352 mg, 11.0 mmol) was injected and the reaction mixture was allowed to reach room

[234] G. H. Posner and C. E. Whitten, *Org. Syntheses*, **52**, Procedure No. 1775 (1972).

temperature. It was poured with stirring into an equal volume of saturated aqueous ammonium chloride; ether extraction followed by rotary evaporation gave 193 mg (83%) of 5,10-tetradecanedione, mp 59–62°. Recrystallization from n-pentane gave white needles (162 mg, 70%), mp 65–66°.

Cuprous Acetylide Coupling with Halide Substrates

2-Phenylfuro[3,2-b]pyridine (Replacement of Aryl Iodine by Phenylethinyl Followed by Cyclization of the Initial Product; Use of Refluxing Pyridine to Drive Reaction to Completion).[235] To a 300-ml

three-necked flask fitted with a nitrogen inlet stopcock, a magnetic stirring bar, and a condenser attached to a nitrogen outlet stopcock and a mercury trap is added 2.47 g (0.150 mol) of copper(I) phenylacetylide [prepared from copper(II) sulfate pentahydrate, concentrated aqueous ammonia, hydroxylamine hydrochloride, and phenylacetylene]. The system is purged with nitrogen for 20 minutes and then 80 ml of pyridine is added. The resulting mixture is stirred for 20 minutes under a nitrogen atmosphere and then 3.30 g (0.0149 mol) of 2-iodo-3-pyridinol is added. The mixture, which changes in color from yellow to dark green as the acetylide dissolves, is warmed in an oil bath at 110–120° for 9 hours with continuous stirring under a nitrogen atmosphere. The reaction solution is then transferred to a 500-ml round-bottomed flask and concentrated to 20 ml at 60–70° (30–80 mm) in a rotary evaporator. The pyridine solution is treated with 100 ml of concentrated aqueous ammonia.

The resulting deep-blue mixture is stirred for 10 minutes and then extracted with five 100-ml portions of ether. The combined ethereal extracts are washed with three 250-ml portions of water, dried ($MgSO_4$), and concentrated in a rotary evaporator. The crude product, 2.6–2.8 g of orange semisolid, is dissolved in 100 ml of boiling cyclohexane, and the solution is filtered, concentrated to about 30 ml, and cooled in an ice bath. The partially purified product crystallizes as 2.3–2.7 g of orange solid, mp 83–89°. Further purification is effected by sublimation at 110–120° and 0.01–0.2 mm. 2-Phenylfuro[3,2-b]pyridine, a yellow solid melting at 90–91°, amounts to 2.2–2.4 g (75–82%).

1-Decen-4-yne [Substitution of Allylic Bromine by an Acetylide Using a Mixed (Alkyl)(Alkynyl) Homocuprate].[18] A 1.0 N solution of n-butyllithium (0.05 mol) is added, with stirring at room temperature, to a

235 D. C. Owsley and C. E. Castro, *Org. Syntheses*, **52**, 128 (1972).

suspension of 1-heptynylcopper (7.92 g, 0.05 mol) in ether (30 ml). Within 30 minutes, a dark solution is obtained. To this is added with stirring at −10° a solution of allyl bromide (12.1 g, 0.1 mol) in ether (40 ml). The mixture is stirred at 20° for 1 hour. Then hexamethylphosphortriamide (30 ml) is added, whereupon the temperature rises to 30°. The mixture is stirred at room temperature for 15 hours. It is then treated with 5 N hydrochloric acid and filtered, and the aqueous phase is extracted with ether (2 × 50 ml). The ethereal layers are dried with magnesium sulfate. The solvent is evaporated and the residue is distilled to give two fractions: (1) bp 70–95° (760 mm): allyl bromide + heptene + 1,5-hexadiene; (2) bp 64–90° (13 mm), 6.9 g. Rectification gives 6.1 g (89 %) of 1-decen-4-yne; bp 65° (13 mm); $n^{25}D$ 1.4500. The residue is 5-undecyne.

Organocopper Coupling with an Epoxide

trans-2-Methylcyclohexanol (Replacement of Epoxide Oxygen by Methyl Using Lithium Dimethylcuprate).[154, 46] To a solution of 5 mmol of lithium dimethyl-cuprate [prepared from methyllithium (Foote Mineral Co) and purified[229] cuprous iodide] in 21 ml of ether, at 0°, there was added dropwise with stirring 0.263 g (2.5 mmol) of cyclohexene oxide in 20 ml of ether over a 10-minute period. No reaction was immediately discernible, but after a few minutes a yellow solid began to precipitate from solution. The mixture was stirred for 5 hours at 0°, then hydrolyzed by addition of 20 ml of a saturated ammonium chloride solution. The mixture was stirred for 2 hours at room temperature, then the aqueous layer was separated and extracted with two 10-ml portions of ether. The combined ether extracts were washed with 10 ml of saturated sodium chloride solution and dried over anhydrous sodium sulfate. The ether was removed by distillation, and the product mixture was analyzed by glpc (¼″ × 8′, 10% Carbowax 20-M, on Chromosorb W, 60–80 mesh column; column temperature 125°, helium flow rate 60 ml/min). Three peaks were obtained with retention times of 4.3, 5.5, and 8.1 minutes. Material was collected from the glpc column, and ir spectra were obtained for the three compounds. Comparison with the ir spectra of authentic samples confirmed the following assignments: 4.3 minutes (8%), cyclohexene oxide; 5.5 minutes (22%), cyclohexanone. The remaining peak, 8.1 minutes (70%), was proved to be *trans*-2-methylcyclohexanol by comparison of its ir spectrum with Sadtler IR spectrum 13371. Analysis of the *trans*-2-methylcyclohexanol on a 0.020″ ID × 50′ Perkin–Elmer diethyleneglycol succinate SCOT column (column temperature 85°, nitrogen flow rate ∼1.5 ml/min) showed conclusively that no *cis*-2-methylcyclohexanol was present. With the conditions used, a mixture of *cis*- and *trans*-2-methylcyclohexanol was separated cleanly into its components.

TABULAR SURVEY

An attempt has been made to include in the tables all substitution reactions using organocopper reagents reported in the literature through December 1973; there are some references through April 1974. Table II covers thermal and oxidative dimerization of organocopper and organocuprate species. Table III includes organocopper coupling with halide substrates, whereas Table IV is limited specifically to cuprous acetylide coupling with halide substrates. Tables V and VI refer to organocopper coupling with alcohol derivatives and with epoxides, respectively. Table VII covers organocopper coupling with miscellaneous substrates.

Within each table the substrates are listed in order of increasing number of carbon atoms, subdivided in order of increasing number of hydrogen atoms, and isomers are arranged according to increasing substituent number (e.g., 1-bromobutane before 2-bromobutane). Halogen compounds having the same number of carbon and hydrogen atoms are listed alphabetically (e.g., 1-bromobutane before 1-chlorobutane). Derivatives of alcohols, amines, ketones, and carboxylic acids are listed by the number of carbon atoms in the parent alcohol, amine, ketone, and acid, respectively (e.g., 4-iodoanisole is listed under C_6, and methyl 4-iodobenzoate is listed under C_7). When a number of different organocopper reagents or reaction conditions have been used for the same substrate, they are listed mainly in order of increasing complexity (e.g., alkyl-, alkenyl-, haloalkyl-, aryl-, haloaryl-, heteroaryl-, and mixed-copper reagents).

When there is more than one reference, the experimental data are taken from the first reference, and the remaining references are arranged in numerical order.

In all the tables, yields are based on the reactant present in lowest concentration; a dash means that no yield was given in the reference. Diethyl ether was used as solvent unless noted otherwise.

Abbreviations used in the tables are as follows:

DMAC   N,N-Dimethylacetamide
DME   1,2-Dimethoxyethane
DMF   N,N-Dimethylformamide
DMSO   Dimethyl sulfoxide
Ether Diethyl ether
HMPA   Hexamethylphosphoramide
OAc   Acetoxy
THF   Tetrahydrofuran
THP   Tetrahydropyranyl
Ts   Tosyl, $p\text{-}CH_3C_6H_4SO_2$

TABLE II. THERMAL AND OXIDATIVE DIMERIZATION OF ORGANOCOPPER AND ORGANOCUPRATE SPECIES

Organocopper Species	Reaction Conditions[a] (Solvent)	Product(s) and Yield(s) (%)	Refs.
A. Organocopper Compounds (RCu)			
C₁			
CH_3Cu	25° (THF)	Ethane (84)	38, 236
$(C_2H_5S)_3CCu$	25° (THF)	$(C_2H_5S)_2C=C(SC_2H_5)_2$ (54)	237
$(C_6H_5S)_2CHCu$	25° (THF)	$C_6H_5SCH=CHSC_6H_5$ (82)	237
$(C_6H_5S)_3CCu$	25° (THF)	$(C_6H_5S)_2C=C(SC_6H_5)_2$ (61)	237
C₂			
$CH_2=CHCu$	20° (THF)	1,3-Butadiene (61)	105
$C_2H_5O_2CCH_2Cu$	−50°, O_2 (THF)	$(C_2H_5O_2CCH_2)_2$ (73)	143
C_2H_5Cu	2° (THF)	n-Butane (74)	236
C_2H_5Cu	2°, NO_2 (THF)	n-Butane (80)	38
C₃			
$Z-CH_3CH=CHCu$ (0.1 M)	90° (THF)	2,4-Hexadiene (I), Z,Z-I (84; 98.2% isomeric purity)	104, 105
$Z-CH_3CH=CHCuP(C_4H_9\text{-}n)_3$ (0.1 M)	25°, 4 hr	Z,Z-I (99; 94.7% isomeric purity)	104
$E-CH_3CH=CHCu$ (0.1 M)	25°, 4 hr	E,E-I (100; 97.7% isomeric purity)	104
$E-CH_3CH=CHCuP(C_4H_9\text{-}n)_3$ (0.1 M)	25°, 4 hr	E,E-I (89; 97.1% isomeric purity)	104
$n\text{-}C_3H_7Cu$	Heat (THF)	Propylene (51), propane (49)	37
	−50°, O_2 (THF)	n-Hexane (95)	38
C₄			
$Z-CH_3CH=C(CH_3)Cu$	25°, 4 hr	$CH_3CH=C(CH_3)C(CH_3)=CHCH_3$ (I), Z,Z-I (72; 96.2% isomeric purity)	104
$Z-CH_3CH=C(CH_3)CuP(C_4H_9\text{-}n)_3$	25°, 4 hr	Z,Z-I (92; 93.5% isomeric purity)	104
$E-CH_3CH=C(CH_3)Cu$	25°, 4 hr	E,E-I (92; 95% isomeric purity)	104
$E-CH_3CH=C(CH_3)CuP(C_4H_9\text{-}n)_3$	25°, 4 hr	E,E-I (99; 93% isomeric purity)	104
$(CH_3)_2C=CHCu$	20° (THF)	$(CH_3)_2C=CHCH=C(CH_3)_2$ (97)	105
$n\text{-}C_4H_9CuP(C_4H_9\text{-}n)_3$	0°, 4 hr	n-Butane (49), 1-butene (51)	11, 35
$n\text{-}C_3H_7CD_2CuP(C_4H_9\text{-}n)_3$	0°	1-Butene-1-d₂ (—)	11
$C_2H_5CD_2CH_2CuP(C_4H_9\text{-}n)_3$	0°	1-Butene-2-d₂ (53)	11
C₅			
[chlorothienyl–Cu structure]	25°, 1 hr (pyridine)	[bis(chlorothienyl) structure] (75)	238a
$NN=C(C_6H_5)CH_3$, $C_2H_5CCH_2Cu$	60°, 3 hr (1:1 THF:ether)	$NN=C(C_6H_5)CH_3$ (54), $[C_2H_5CCH_2]_2$	107
[2-bromo-6-pyridyl–Cu structure]	−78°, $CuCl_2$, O_2	[bipyridyl-Br structure] (50)	238b

C₅	t-C₄H₉CH₂CH₂Cu	2° (THF)	t-C₄H₉CH₂CH₂CH₂C₄H₉-t (19)	236
C₆	C₆F₅Cu(MgBrCl)	25°, O₂ (THF)	C₆F₅C₆F₅ (75)	138, 239
	p-C₆F₅OC₆F₄Cu	25°, O₂ (THF)	(p-C₆F₅OC₆F₄)₂ (71)	239
	p-FC₆H₄Cu(MgBr₂)	Air	(p-FC₆H₄-)₂ (25)	155
	C₆H₅Cu	Air, 2 hr	Biphenyl (—)	31
		Reflux (benzene)	Biphenyl (—)	240, 241
		35°, 2 hr (ether) then 80°, 2 hr (xylene)	Biphenyl (65–80)	242, 243
	o-CH₃OC₆H₄Cu	150° (neat)	(o-CH₃OC₆H₄)₂ (—)	73
	p-CH₃OC₆H₄Cu	—	(p-CH₃OC₆H₄)₂ (62)	31
	2,4,6-(CH₃O)₃C₆H₂Cu	−78°, 1 hr; then 25°, 2 hr, CuCl₂	[2,4,6-(CH₃O)₃C₆H₂]₂ (25)	109
	o-O₂NC₆H₄CO₂Cu	145–165° (quinoline)	o-O₂NC₆H₄C₆H₄NO₂-o (27)	77–79

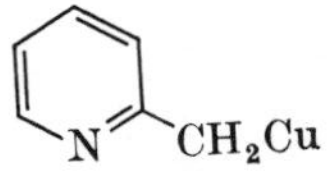

2-(CH₂Cu)pyridine ; 60° (1:1 THF:ether) ; product (78) ; 244

4-(CH₂Cu)pyridine ; 60° (1:1 THF:ether) ; product (54) ; 245

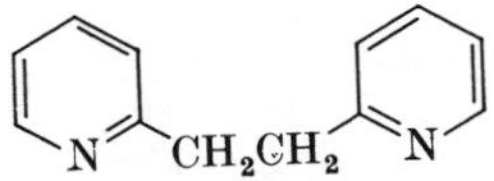

(pyridine dioxolane)Cu ; −78°, CuCl₂, O₂ ; product (46) ; 238b

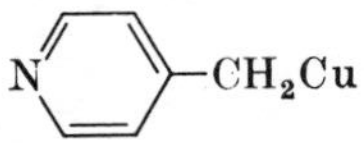

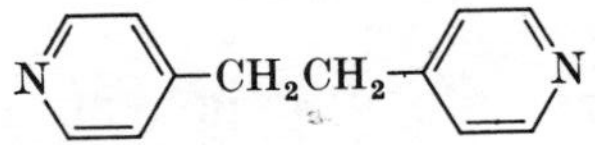

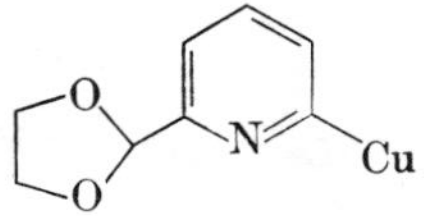

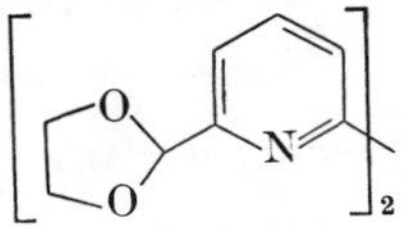

NN=C(C₆H₅)₂ cyclohexanone–Cu ; 0–10°, 4.5 hr (1:1 THF:ether) ; NN=C(C₆H₅)₂ product (25 meso, 28 racemic) ; 107

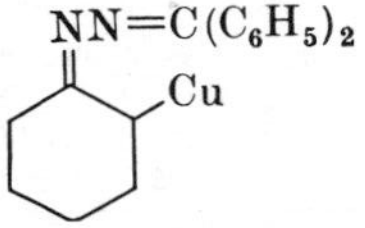

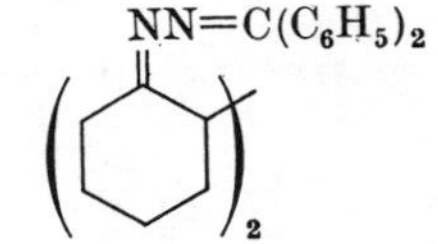

	CH₂=CH(CH₂)₄CuP(C₄H₉-n)₃	Heat	1-Hexene (44), 1,5-hexadiene (52)	11
	Z-n-C₄H₉CH=CHCu	20°, O₂	Z,Z-5,7-Dodecadiene (69)	86

Note: References 236–320 are on pp. 398–400.

^{a} When no oxidant is indicated, the reaction was performed under an inert atmosphere (nitrogen or argon).

309

Organocopper Species	Reaction Conditions[a] (Solvent)	Product(s) and Yield(s) (%)	Refs.
	A. Organocopper Compounds (RCu)		
C_7			
o-$CF_3C_6H_4Cu$	200–205°	o-$CF_3C_6H_4C_6H_4CF_3$-o (—)	102
m-$CF_3C_6H_4Cu$	Heat (benzene)	m-$CF_3C_6H_4C_6H_4CF_3$-m (I, high)	39
m-$CF_3C_6H_4Cu\cdot MgBr_2$	Air	I (37)	155
o-$CH_3C_6H_4Cu$	110–120° (neat)	o-$CH_3C_6H_4C_6H_4CH_3$-o (High)	73
m-$CH_3C_6H_4Cu$	100° (neat)	m-$CH_3C_6H_4C_6H_4CH_3$-m (High)	73
p-$CH_3C_6H_4Cu$	35°, 2 hr (ether)	p-$CH_3C_6H_4C_6H_4CH_3$-p (65–80)	242, 73, 29
4-CH_3-2,6-$(CH_3O)_2C_6H_2Cu$	80°, 2 hr (xylene) −78°, 2 hr, $CuCl_2$	[4-CH_3-2,6-$(CH_3O)_2C_6H_2$]$_2$ (I, 18) I (31)	109
$C_6H_5CH_2Cu$ *	−78°, 1 hr, O_2 25° (THF)	Bibenzyl (88)	38
C_7			
$C_6H_5SOCH_2Cu$ $C_6H_5SO_2CH_2Cu$	$CuCl_2$, O_2 (THF) 20° (1:3:4 ether:benzene:THF)	$C_6H_5SOCH_2CH_2SOC_6H_5$ * (45) $C_6H_5SO_2CH_2CH_2SO_2C_6H_5$ (20–25)	106b 246
(2-CH_2Cu-6-CH_3-pyridine structure)	60° (1:1 THF:ether)	(dimer structure, CH_2)$_2$ (71–77)	245
(2-$CH(CH_3)Cu$-pyridine structure)	60° (1:1 THF:ether)	[dimer structure, $CH(CH_3)$]$_2$ (63)	244
(4-$CH(CH_3)Cu$-pyridine structure)	60° (1:1 THF:ether)	[dimer structure, $CH(CH_3)$]$_2$ (40)	244
$C_6H_5(C_6H_5S)CHCHCu$	25° (THF)	$trans$-Stilbene (45)	237
$C_6H_5(C_6H_5S)_2CCu$	25° (THF)	$C_6H_5(C_6H_5S)C$=$C(SC_6H_5)C_6H_5$ (76)	237
$C_6H_5(2$-Pyridyl$S)CHCu$	25° (THF)	$trans$-Stilbene (86)	237
Z,n-C_4H_9CH=$C(CH_2OCH_3)Cu$	−10°, O_2	Z,Z-n-C_4H_9CH=$C(CH_2OCH_3)$]$_2$ (38)	246b

C_8			
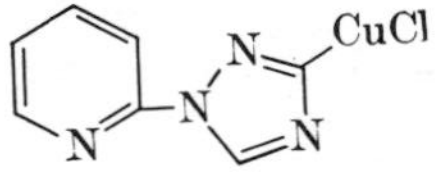 2-(imidazol-1-yl)pyridine–CuCl [structure]	$50°$, $CuCl_2$ (THF)	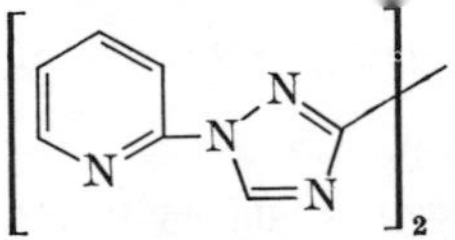 [dimer] (18)	247
$C_6H_5C{\equiv}CCu$	$125°$, 3 hr 4,6-dibromoresorcinol (DMF)	1,4-Diphenylbutadiyne (57)	133
"	$125°$, 2 hr 4,6-diiodoresorcinol (CH_3CO_2H)	" (22)	133
"	$15°$, O_2 (THF)	" (67)	40
$C_6H_5\overset{X}{C}CH_2Cu$ (I)		$C_6H_5\overset{X}{C}CH_2CH_2\overset{X}{C}C_6H_5$ (II)	246
I, X = O	$35°$	II, X = O (11)	
I, X = NC_6H_5	$35°$	II, X = NC_6H_5 (42)	
I, X = $NN{=}C(C_6H_5)CH_3$	$20–65°$, O_2 (THF)	II, X = $NN{=}C(C_6H_5)CH_3$ (52)	
I, X = $NN{=}C(C_6H_5)_2$	$20–65°$, O_2 (THF)	" (84)	
	$10–20°$, 12 hr (1:1 THF:ether)	" (96)	
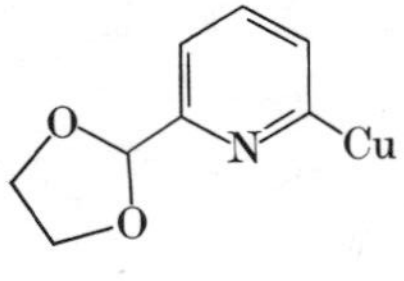 [2-pyridyl-dioxolane–Cu structure]	$-70°$, $CuCl_2$, O_2	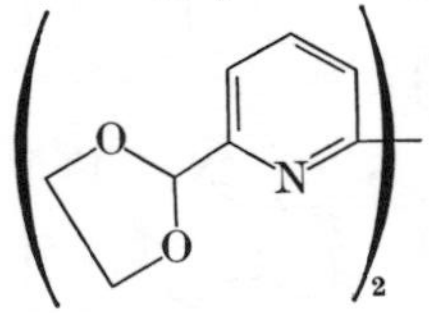 (46)	107
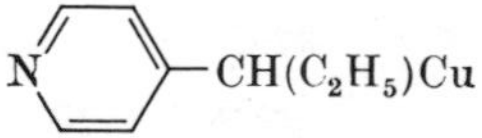 [4-pyridyl–CH(C_2H_5)Cu structure]	$60°$ (1:1 THF:ether)	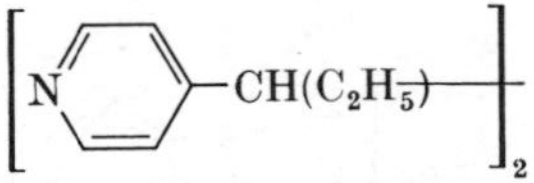 [$CH(C_2H_5)$]$_2$ (34)	245
$n\text{-}C_4H_9(C_2H_5)C{=}CHCu$ E-I, Z-I	$10°$, O_2, 1 hr	[$n\text{-}C_4H_9(C_2H_5)C{=}CH$]$_2$ (II) E-II (75), Z-II (68)	86
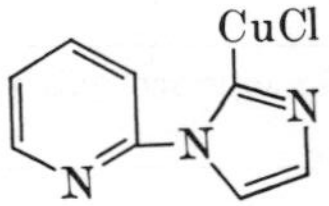 2-(imidazol-1-yl)pyridine–CuCl [structure]	$50°$, $CuCl_2$ (THF)	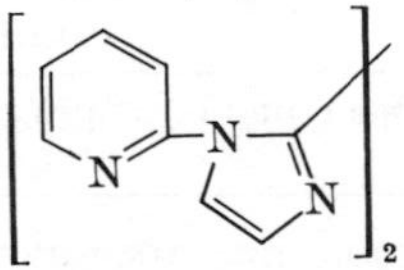 [dimer] (18)	247

Note: References 236–320 are on pp. 398–400.

[a] When no oxidant is indicated, the reaction was performed under an inert atmosphere (nitrogen or argon).

311

TABLE II. THERMAL AND OXIDATIVE DIMERIZATION OF ORGANOCOPPER AND ORGANOCUPRATE SPECIES (Continued)

Organocopper Species	Reaction Conditions^a (Solvent)	Product(s) and Yield(s) (%)	Refs.
A. Organocopper Compounds (RCu)			
C_8 (contd.)			
	50°, CuCl$_2$ (THF)	(21)	247
C_9			
	65°, CuCl$_2$ (THF)	(82)	247
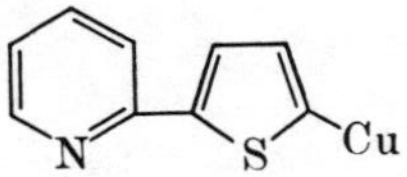	65°, 3 hr, O$_2$ (THF)	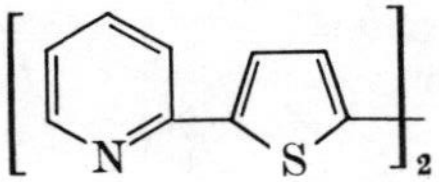 (63)	248
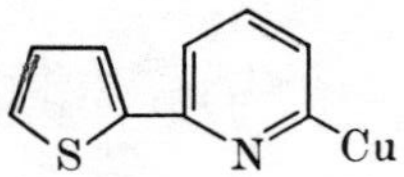	60°, 2.5 hr, O$_2$ (THF)	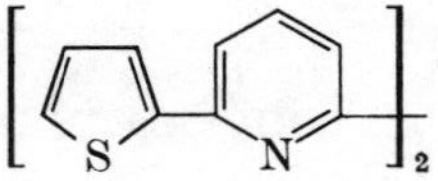 (5)	248
$C_6H_5PO(C_2H_5)CH_2Cu$ $NN=C(C_6H_5)_2$ *	CuCl$_2$, O$_2$ (THF)	$[C_6H_5PO(C_2H_5)CH_2]_2$ $NN=C(C_6H_5)_2$ * (34)	106b
$C_6H_5CCH(CH_3)Cu$	0–10°, 2.5 hr (1:1 THF:ether)	$[C_6H_5CH(CH_3)]_2$ (37 meso, 20 racemic)	107
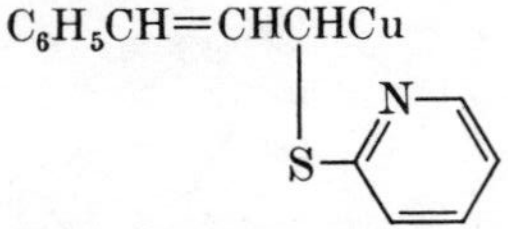 $C_6H_5CH=CHCHCu$	25° (THF)	1,6-Diphenylhexatriene (10)	237
C_{10} 1-C$_{10}$H$_7$Cu	Reflux (toluene)	Bi-1-naphthyl (Good)	249

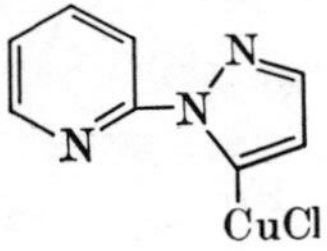

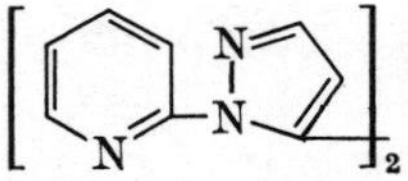

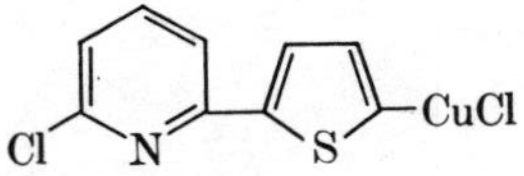

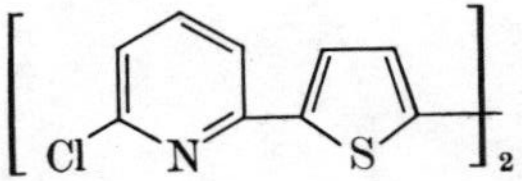

312

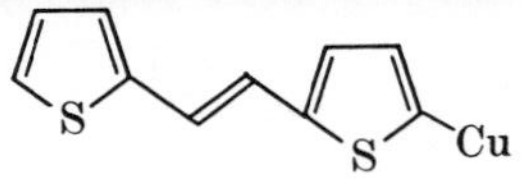

(28) 250

(32) 250

(7) 250

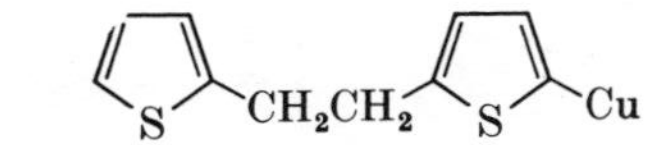

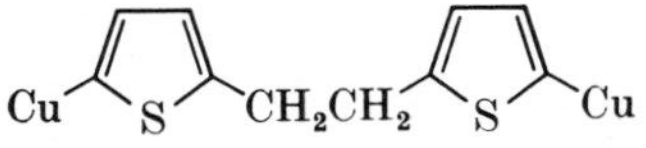

(67–69) 245

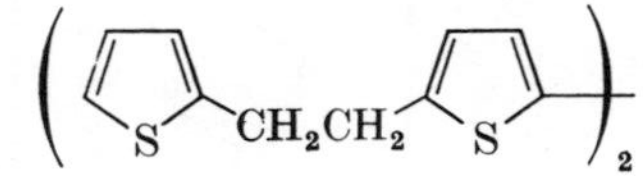

(58) 245

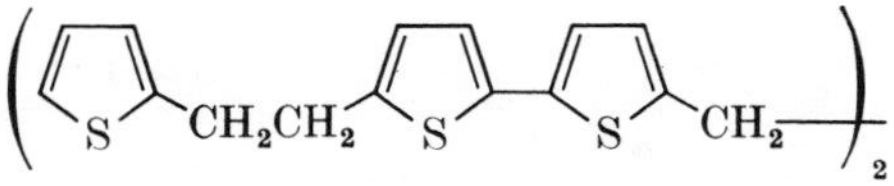

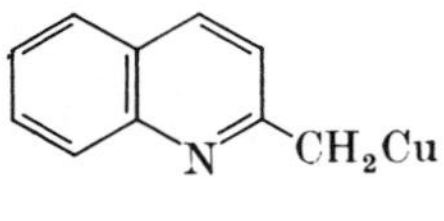

245

−65°, CuCl$_2$

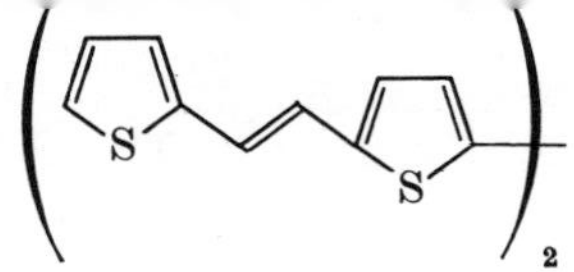

−65°, CuCl$_2$

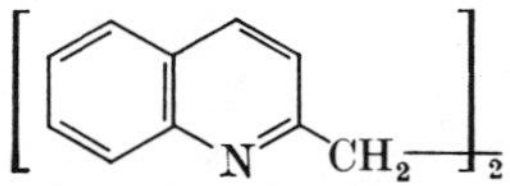

60–80°
(1:1 DME:ether)

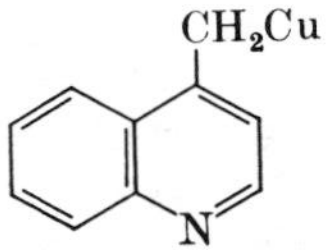

60–80°
(1:1 DME:ether)

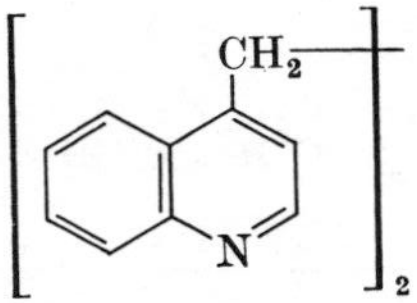

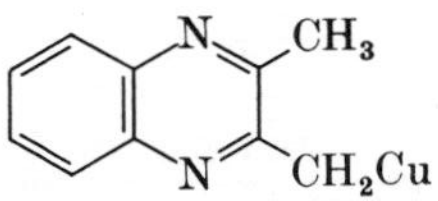

60–80°
(1:1 DME:ether)

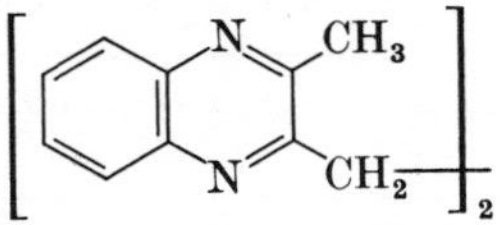

313

Note: References 236–320 are on pp. 398–400.

[a] When no oxidant is indicated, the reaction was performed under an inert atmosphere (nitrogen or argon).

TABLE II. THERMAL AND OXIDATIVE DIMERIZATION OF ORGANOCOPPER AND ORGANOCUPRATE SPECIES (*Continued*)

Organocopper Species	Reaction Conditions[a] (Solvent)	Product(s) and Yield(s) (%)	Refs.
		A. Organocopper Compounds (RCu)	
C_{10} (*contd.*) $NN=C(C_6H_5)_2$ $C_6H_5CCH(C_2H_5)Cu$ $E\text{-}C_6H_5(C_2H_5)C=CHCu$	$0\text{-}10°$, 2.5 hr (1:1 THF:ether) O_2	$NN=C(C_6H_5)_2$ $[C_6H_5CH(C_2H_5)]_2$ (49 meso, 27 racemic) $E,E\text{-}C_6H_5(C_2H_5)C=CHCH=C(C_2H_5)C_6H_5$ (55)	107 86
$C_6H_5C((CH_3)_2CH_2CuP(C_4H_9\text{-}n)_3$	$65°$	$C_6H_5C((CH_3)_3, C_6H_5CH_2CH(CH_3)_2,$ $C_6H_5CH=C(CH_3)_2, C_6H_5CHC(CH_3)=CH_2,$ 3 dimers	35, 251
C_{12}	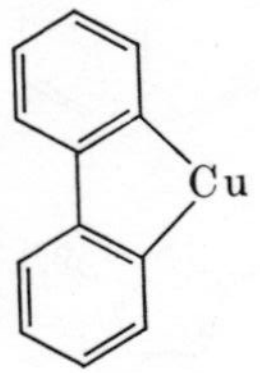 $-50°$, 20 min; then $25°$, 2 hr, $CuCl_2$ (1:1 THF:ether)	Biphenylene, (36) tetraphenylene (24)	252
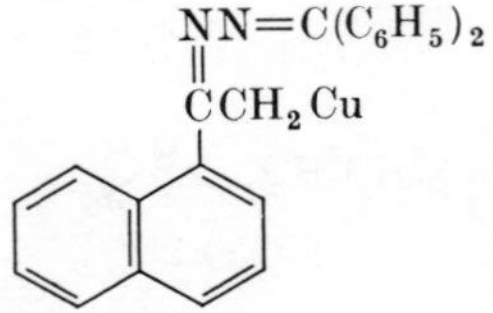 $NN=C(C_6H_5)_2$ $\|$ CCH_2Cu	$10\text{-}20°$, 14 hr (1:1 THF:ether)	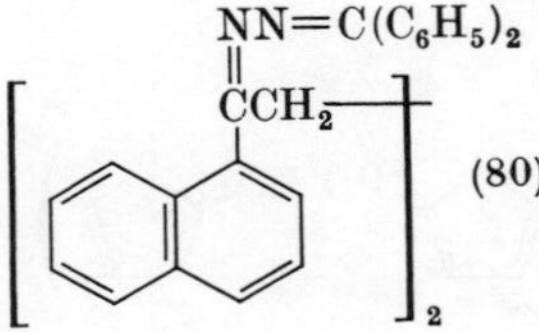 $\left[NN=C(C_6H_5)_2 \atop CCH_2 \right]_2$ (80)	107
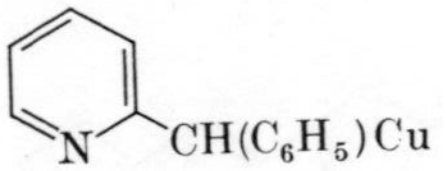 $CH(C_6H_5)Cu$	$60°$ (1:1 THF:ether)	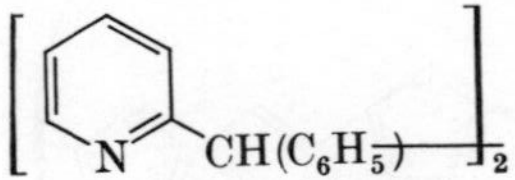 $\left[CH(C_6H_5) \right]_2$ (65)	245

314

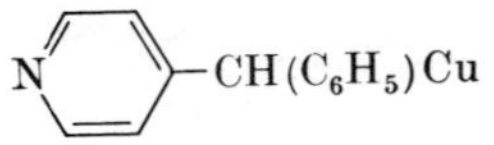

Reactant	Conditions	Product	Ref.

$$\text{4-NC}_5\text{H}_4\text{-CH(C}_6\text{H}_5)\text{Cu}$$

60° (1:1 THF:ether)

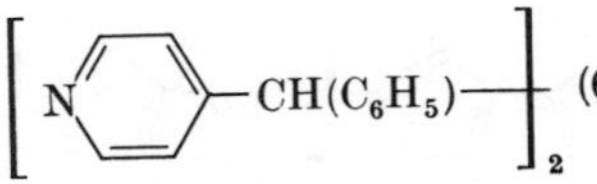

$$\left[\text{4-NC}_5\text{H}_4\text{-CH(C}_6\text{H}_5)\text{-}\right]_2 \quad (65)$$

245

$$Z\text{-}n\text{-C}_4\text{H}_9(\text{C}_6\text{H}_5)\text{C}=\text{CHCu}$$

$$\text{O}_2$$

$$Z,Z\text{-}n\text{-C}_4\text{H}_9(\text{C}_6\text{H}_5)\text{C}=\text{CHCH}=\text{C(C}_6\text{H}_5)\text{C}_4\text{H}_9\text{-}n \quad 86 \quad (-)$$

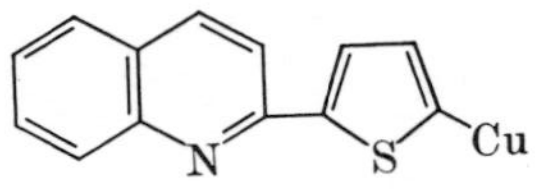

253

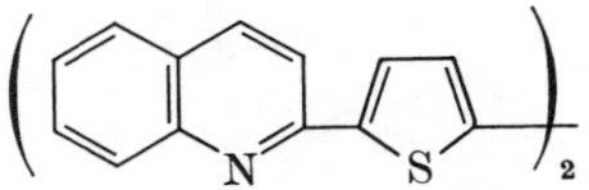

C$_{13}$

quinolyl–thienyl–Cu

35°, 3 hr, CuCl$_2$

(dithienyl-quinoline)$_2$ (14)

253

$$\text{(C}_6\text{H}_5)_2\text{PCH}_2\text{Cu}$$
$$\overset{\text{O}}{\|}$$

20°, O$_2$ (THF)

$$\underset{\text{O}}{\overset{\text{O}}{\|}}\text{(C}_6\text{H}_5)_2\text{PCH}_2\text{CH}_2\text{P(C}_6\text{H}_5)_2 \quad (43)$$

245

$$\text{(C}_6\text{H}_5)_2\text{CCu}$$
(2-thienyl-pyridyl-S)

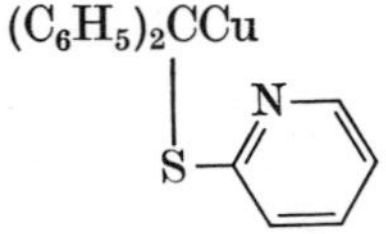

25°, (THF)

Tetraphenylethylene (33)

237

C$_{14}$

$$\text{C}_6\text{H}_5\text{CH}=\text{NCH(C}_6\text{H}_5)\text{Cu}$$

85° (DME)

$$[meso\text{-C}_6\text{H}_5\text{CH}=\text{NCH(C}_6\text{H}_5)]_2 \quad (58)$$

244

pyridine structure

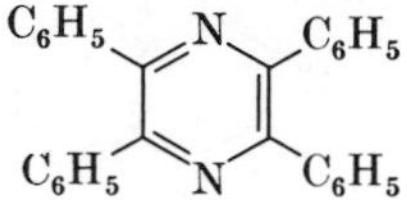

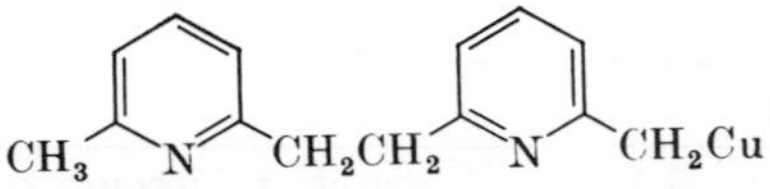

35°, CuBr$_2$

$$\underset{\text{C}_6\text{H}_5}{\overset{\text{C}_6\text{H}_5}{\text{N}}}\ \text{pyrazine}\ \underset{\text{C}_6\text{H}_5}{\overset{\text{C}_6\text{H}_5}{\text{N}}} \quad (18)$$

246

pyridine–CH$_2$Cu complex

60°

$$\left[\ \text{pyridine-CH}_2\text{-}\ \right]_2 \quad (48)$$

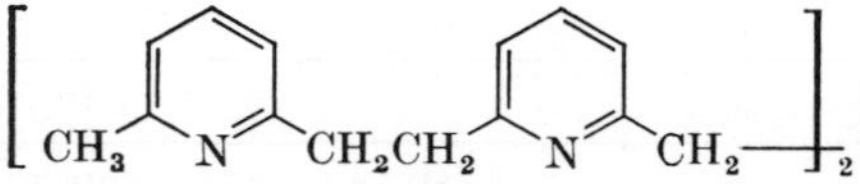

254

Note: References 236–320 are on pp. 398–400.

[a] When no oxidant is indicated, the reaction was performed under an inert atmosphere (nitrogen or argon).

TABLE II. THERMAL AND OXIDATIVE DIMERIZATION OF ORGANOCOPPER AND ORGANOCUPRATE SPECIES (Continued)

Organocopper Species	Reaction Conditions[a] (Solvent)	Product(s) and Yield(s) (%)	Refs.
A. Organocopper Compounds (RCu)			
C₁₄ (contd.)			
	60°	(1), (4)	254
	−65°, CuCl₂	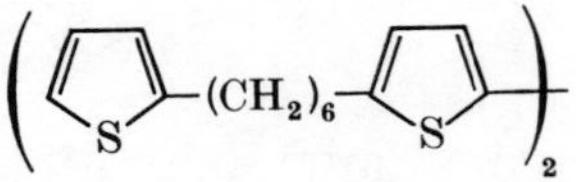 (40), 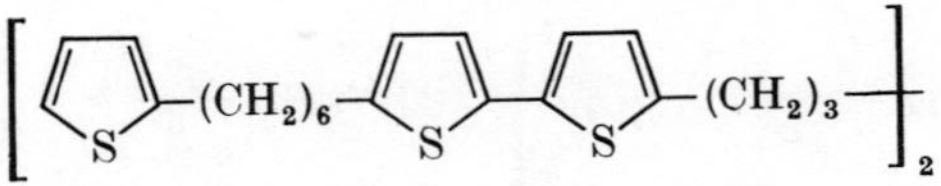 (7)	250

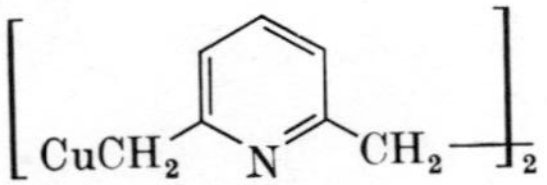

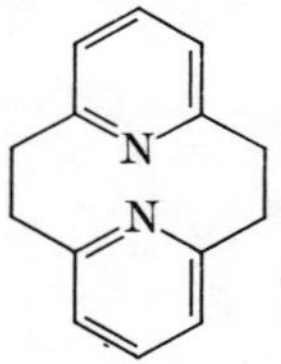

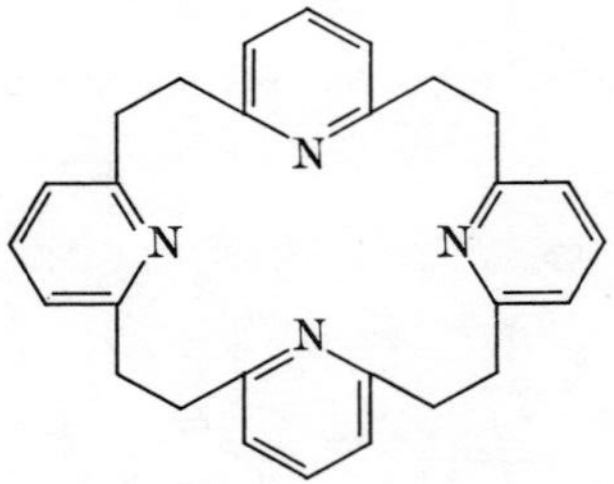

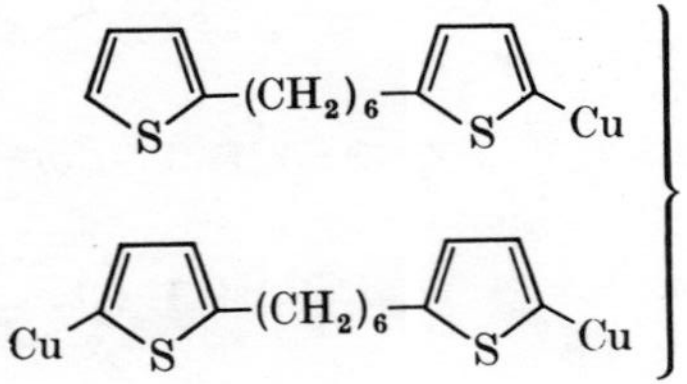

C_{15}	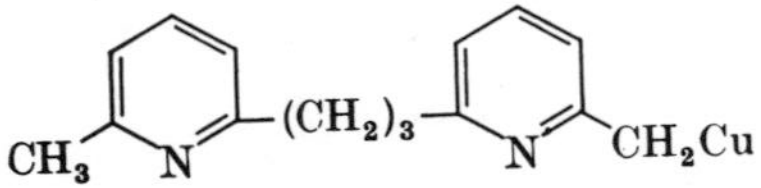	60°, 2 hr (THF)	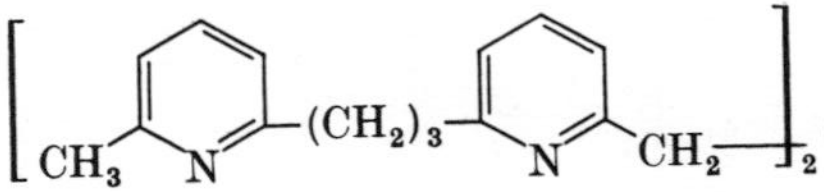 (40–42)	255
C_{16}	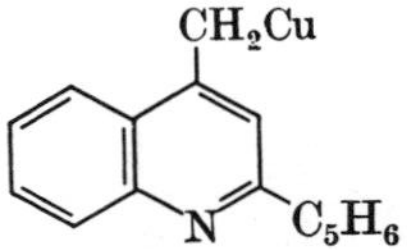	60–80° (1:1 DME:ether)	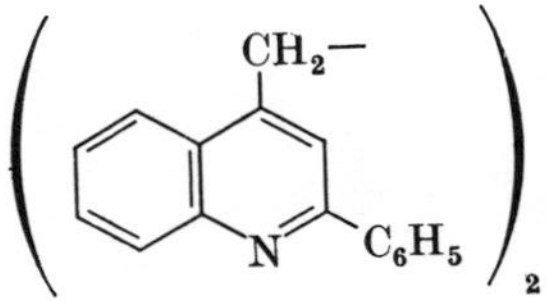 (24)	245
C_{17}	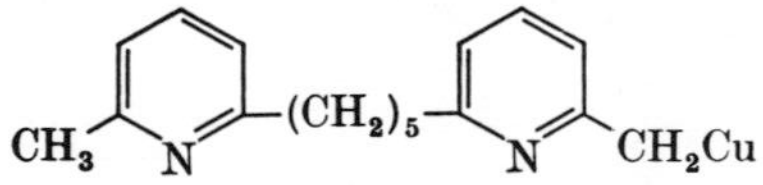	60°, 2 hr (THF)	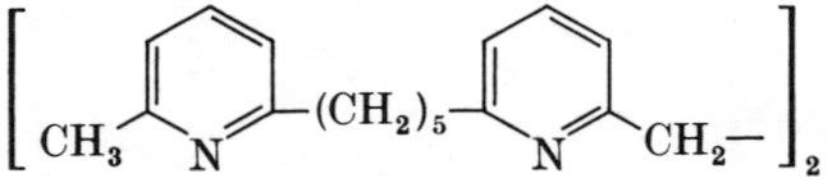 (36–39)	255
	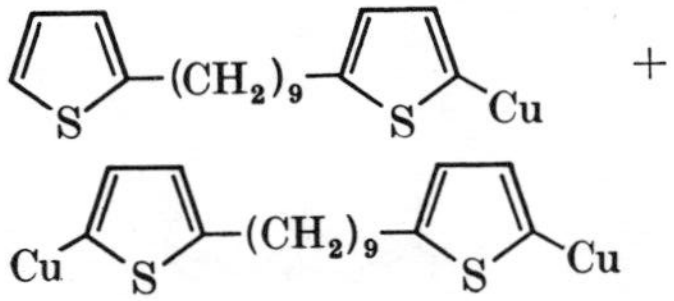	−65°, $CuCl_2$	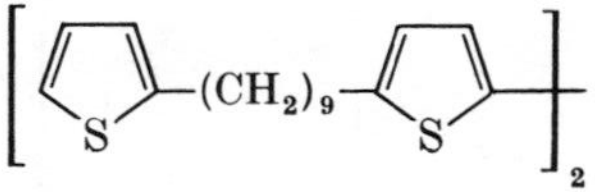 (33) 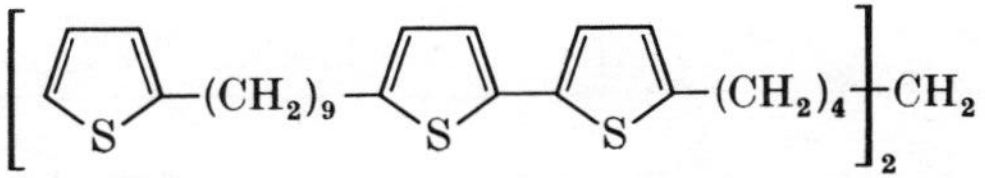 (6)	250

Note: References 236–320 are on pp. 398–400.

[a] When no oxidant is indicated, the reaction was performed under an inert atmosphere (nitrogen or argon).

TABLE II. THERMAL AND OXIDATIVE DIMERIZATION OF ORGANOCOPPER AND ORGANOCUPRATE SPECIES (*Continued*)

Organocopper Species	Reaction Conditions[a] (Solvent)	Product(s) and Yield(s) (%)	Refs.
	A. Organocopper Compounds (RCu)		
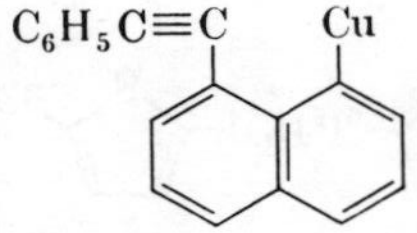 C₁₈	Reflux, 2 hr (THF)	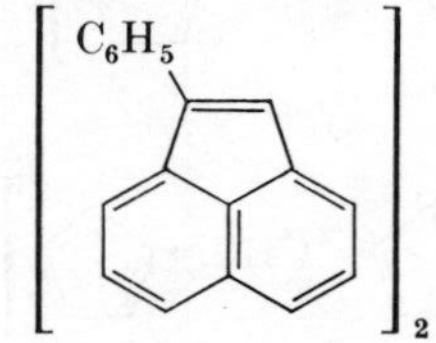 (—)	256
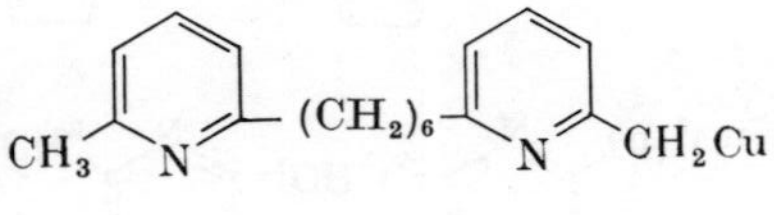 C₂₀	60°, 2 hr (THF)	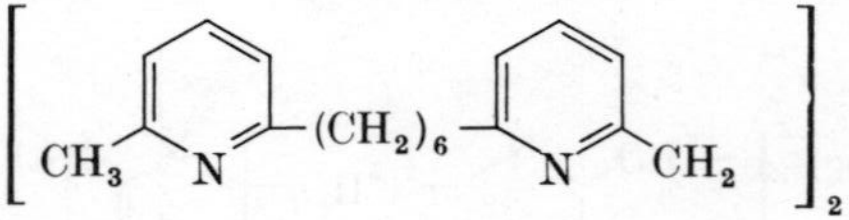	255 (32–35)
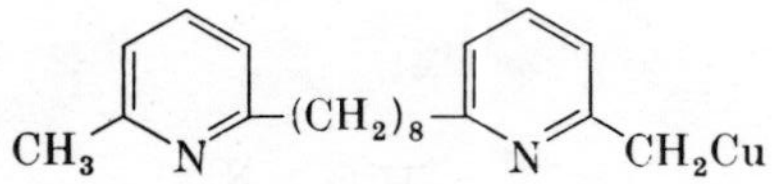	60° (THF)	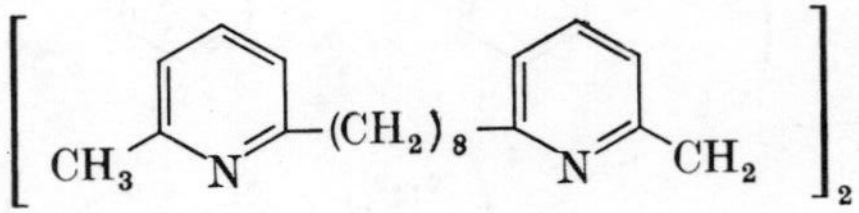	255 (38–40)
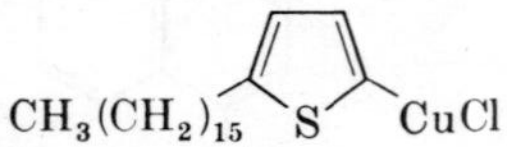	65° (THF)	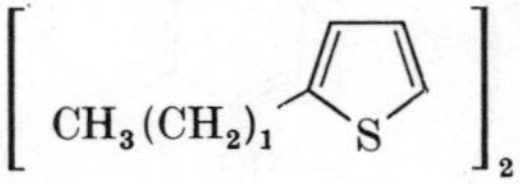 (66)	257a
C₂₂	140° [(*n*-C₄H₉)₂O]	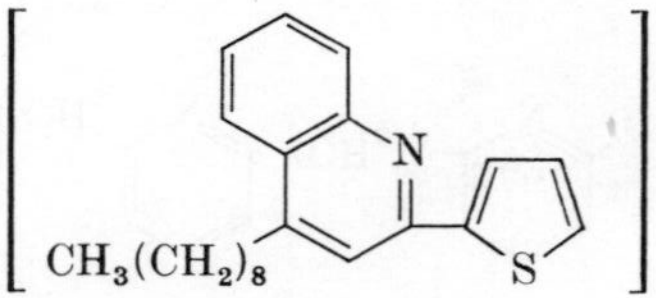 (24)	257a

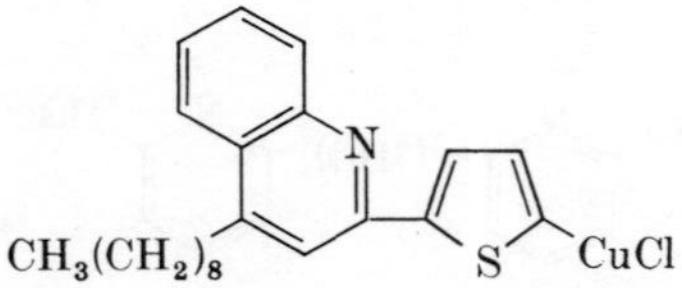

B. Homocuprates (RRCuMet)

C₃	(E—CH₃CH=CH)₂CuLi	−78°, O₂ (THF)	E,E-2,4-Hexadiene (78)	40
	(CH₃C=CH₂)₂CuLi	0°	2,3-Dimethyl-1,3-butadiene (63)	257b
C₄	(n-C₄H₉)₂CuLi	−78°, O₂ (THF)	n-Octane (84)	40
	(sec-C₄H₉)₂CuLi	−78°, O₂ (THF)	3,4-Dimethylhexane (82)	40
	(t-C₄H₉)₂CuLi	−78°, O₂ (THF)	2,2,3,3-Tetramethylbutane (14)	40
C₅	BrMgCu(CH₂)₄CuMgBr	−78°, O₂ (THF)	Cyclobutane (25)	40
	BrMgCu(CH₂)₅CuMgBr	−78°, O₂ (THF)	Cyclopentane (30)	40
C₆	(C₆H₅)₂CuLi	−78°, O₂ (THF)	Biphenyl (75)	40
C₈	(C₆H₅C≡C)₂CuLi	15°, O₂ (THF)	1,4-Diphenylbutadiyne (67)	40
C₁₀	[C₆H₅C(CH₃)₂CH₂]₂CuLi	−78°, O₂ (THF)	[C₆H₅C(CH₃)₂CH₂]₂ (88)	40

C. Mixed Homocuprates (RR'CuMet)

C₁, C₆	(CH₃)(C₆H₅)CuLi	−78°, O₂	Toluene (92)	33
C₁, C₇	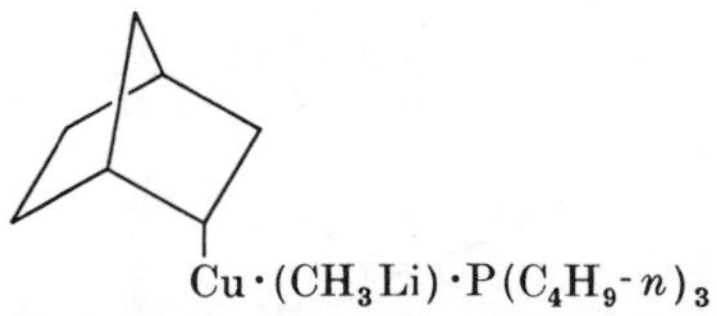 Cu·(CH₃Li)·P(C₄H₉-n)₃ (bicyclic structure)	−78°, C₆H₅NO₂ (THF-ether)	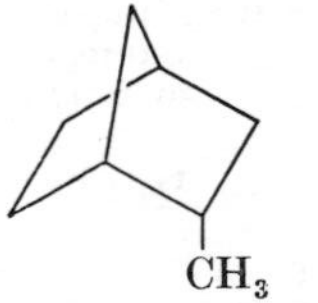 (bicyclic structure with CH₃) 50;98:2 endo:exo	147
C₁, C₁₀	(CH₃)(1-Naphthyl)CuLi	0°, C₆H₅NO₂	1-Methylnaphthalene (70)	33
C₁, C₁₁	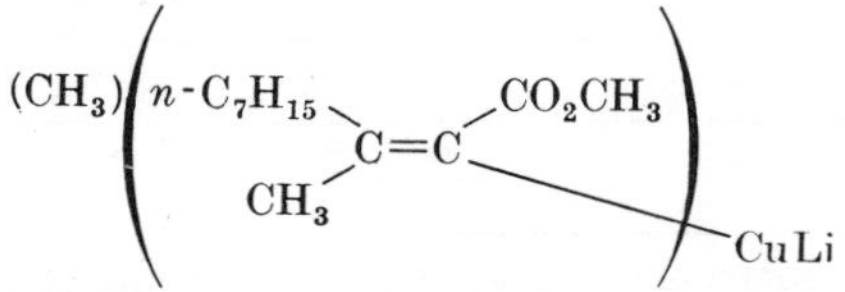 (CH₃) n-C₇H₁₅—C=C—CO₂CH₃ / CH₃ ...CuLi (structure)	−78°, O₂, 0.5 hr	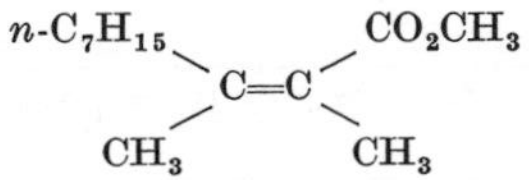 n-C₇H₁₅—C=C—CO₂CH₃ with CH₃, CH₃ (32–46)	144

Note: References 236–320 are on pp. 398–400.

ᵃ When no oxidant is indicated, the reaction was performed under an inert atmosphere (nitrogen or argon).

319

TABLE II. THERMAL AND OXIDATIVE DIMERIZATION OF ORGANOCOPPER AND ORGANOCUPRATE SPECIES (*Continued*)

	Organocopper Species	Reaction Conditions (Solvent)	Product(s) and Yield(s) (%)	Refs.
C_1,C_{11} (*contd.*)	$(CH_3)\left((n\text{-}C_7H_{15})(CH_3)C{=}C(CO_2CH_3)\right)CuLi$	−78°, O_2, 0.5 hr	$n\text{-}C_7H_{15}(CH_3)C{=}C(CH_3)CO_2CH_3$ (11)	144
C_4, C_5	$(n\text{-}C_4H_9)(n\text{-}C_5H_{11})CuLi$	−78°, O_2	*n*-Nonane (32–73)	33
C_4, C_6	$(n\text{-}C_4H_9)(C_6H_5)CuLi$	−78°, O_2 (THF)	*n*-Octane (33), biphenyl (28), *n*-butylbenzene (33)	40
		−78°, O_2 (THF)	*n*-Butylbenzene (55)	33
	$(sec\text{-}C_4H_9)(C_6H_5)CuLi\cdot P(C_4H_9\text{-}n)_3$	−78°, O_2	*sec*-Butylbenzene (20)	33
	$(sec\text{-}C_4H_9)(C_6H_5)CuLi$	−78°, O_2 (THF)	" (76)	33
	$(t\text{-}C_4H_9)(C_6H_5)CuLi$	−78°, O_2 (THF)	*t*-Butylbenzene (73)	33
C_4, C_7	$[CuLiHg]\cdot(C_4H_9\text{-}t)_3$ (I)		$C_4H_9\text{-}t$ (II)	66
	endo-I	−78°, $C_6H_5NO_2$ (THF)	II (45; 95:5 *endo*:*exo*)	
	exo-I	−78°, $C_6H_5NO_2$ (THF)	II (40; <2:>98 *endo*:*exo*)	

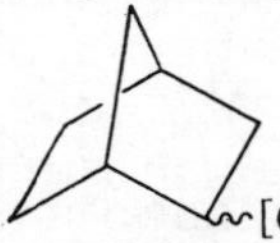

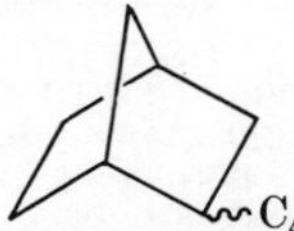

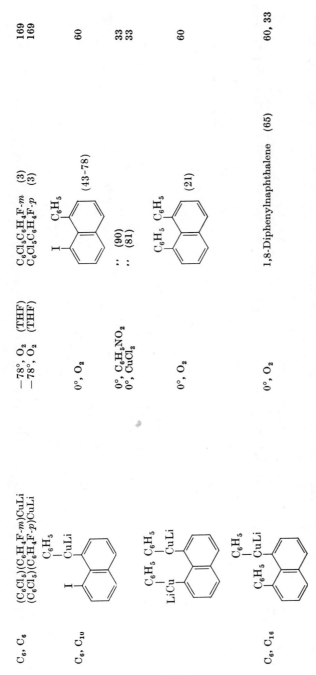

C_6, C_6	$(C_6Cl_5)(C_6H_4F\text{-}m)CuLi$	$-78°$, O_2 (THF)	$C_6Cl_5C_6H_4F\text{-}m$ (3)	169
	$(C_6Cl_5)(C_6H_4F\text{-}p)CuLi$	$-78°$, O_2 (THF)	$C_6Cl_5C_6H_4F\text{-}p$ (3)	169
C_6, C_{10}		$0°$, O_2	(43–78)	60
		$0°$, $C_6H_5NO_2$	:: (90)	33
		$0°$, $CuCl_2$	:: (81)	33
C_6, C_{10}		$0°$, O_2	(21)	60
C_6, C_{16}		$0°$, O_2	1,8-Diphenylnaphthalene (65)	60, 33

Note: References 236–320 are on pp. 398–400.

TABLE III. ORGANOCOPPER COUPLING WITH HALIDE SUBSTRATES

	Halide Substrate	Organocopper Reagent	Reaction Conditions (Solvent)	Product(s) and Yield(s) (%)	Refs.
			A. Alkyl Halides		
C₁	CH₂I₂	C₆F₅Cu (2 eq)	100°, 5 days (dioxane)	CH₂(C₆F₅)₂ (70)	156
	C₂H₅OCH₂Cl	Z-C₂H₅(CH₃)C=CHCu	[HMPA, (C₂H₅O)₃P]	Z-C₂H₅OCH₂CH=C(CH₃)C₂H₅ (68)	87
	ClC₂H₄OCH₂Br	Z-C₂H₅(CH₃)C=CHCu	—	Z-ClC₂H₄OCH₂CH=C(CH₃)C₂H₅ (82)	257c
		Z-i-C₄H₉(C₂H₅)C=CHCu	—	Z-ClC₂H₄OCH₂CH=C(C₂H₅)C₄H₉-i (78)	257c
	CH₃I	[CuLiHg]·(C₄H₉-t)₃ [norbornane structure]		[norbornane structure] (II) CH₃	66
		(I, 0.3 eq) endo-I [norbornane structure]	−78°, 30 min (THF)	endo-II (70)	
		exo-I	−78°, 30 min (THF)	exo-II (70)	
		Z-n-C₄H₉CH=CHCu	[HMPA, (C₂H₅O)₃P]	Z-2-Heptene (—)	87
		(n-C₄H₉)₂C=CHCu (0.5 eq)	−5°, 15 hr [HMPA, (C₂H₅O)₃P]	CH₃CH=C(C₄H₉-n)₂ (63)	87
C₂	ClCH₂CN	C₆F₅Cu	(—)	CH₃C₆F₅ (93)	103
	C₂H₅Br	C₆H₅Cu (10 eq)	0°, 23 hr (THF)	C₆H₅CH₂CN (45)	46
		C₂H₅Cu(C₂H₅MgBr)	2° (THF)	n-Butane (100)	258
		n-C₃H₇Cu	2° (THF)	n-Pentane (100)	38
		C₆H₅Cu	25°, 30 min (THF)	Ethylbenzene (—)	38
C₃	ClC₂H₄OCH(CH₃)Cl	Z-C₂H₅(CH₃)C=CHCu	—	ClC₂H₄OCH(CH₃)CH=C(CH₃)C₂H₅ (51)	257c
	Cl(CH₂)₃I	Z-C₂H₅(CH₃)C=CHCu	—	Z-Cl(CH₂)₃CH=C(CH₃)C₂H₅ (46)	257c
	1-Bromopropane	(C₆H₅C≡C)(n-C₄H₉)CuLi	(THF)	n-Heptane (38), C₆H₅C≡CC₄H₉-n (20)	259
C₄	2-Iodopropane	Z-C₂H₅(CH₃)C=CHCu	[HMPA, (C₂H₅O)₃P]	Z-(CH₃)₂CHCH=C(CH₃)C₂H₅ (15)	87
	CH₂=CHCH₂OCH₂CH₂CD₂I	(CH₃)₂CuLi	0°, 12 hr	CH₂=CHCH₂OCH₂CH₂CD₂CH₃ (—)	157
	n-C₄H₉Br	Z-C₂H₅(CH₃)C=CHCu	[HMPA, (C₂H₅O)₃P]	Z-n-C₄H₉CH=C(CH₃)C₂H₅ (39)	87

322

Substrate	Reagent	Conditions	Product (yield)	Refs.
(—)-(R)-2-Bromobutane	(C$_6$H$_5$)$_2$CuLi (0.39 M, 3 eq)	52°, 72 hr (3 eq LiBr, 3:2 THF:ether)	(+)-(S)-2-Phenylbutane (87)	33
C$_5$ n-C$_4$H$_9$I	Z-C$_2$H$_5$(CH$_3$)C=CHCu (1 eq)	[HMPA, (C$_2$H$_5$O)$_3$P] 100°, 36 hr (neat)	Z-n-C$_4$H$_9$CH=C(CH$_3$)C$_2$H$_5$ (58)	87
	C$_6$F$_5$Cu	25°, 26 hr	n-C$_4$H$_9$C$_6$F$_5$ (23)	156
1-Bromopentane	(n-C$_4$H$_9$)$_2$CuLi (0.4 M, 4 eq)	25°, 1 hr (THF)	n-Nonane (70)	33
	,,	25°, 1 hr	,, (98)	33
	(n-C$_4$H$_9$)$_2$CuLi· P(C$_4$H$_9$-n)$_2$ (0.5 M, 5 eq)	25°, 1 hr (5 eq LiI)	,, (93)	33
	(sec-C$_4$H$_9$)$_2$CuLi· P(C$_4$H$_9$-n)$_3$ (0.5 M, 5 eq)	25°, 1 hr (5 eq LiI)	3-Methyloctane (94)	33
	(t-C$_4$H$_9$)$_2$CuLi· P(C$_4$H$_9$-n)$_3$ (0.5 M, 5 eq)	25°, 1 hr (THF-pentane, 5 eq LiI)	2,2-Dimethylheptane (92)	33
	(CH$_3$C≡CCuC$_4$H$_9$-n)Li	−78°, then 0°, 2 hr (THF)	n-C$_5$H$_{11}$R (I) I, R = C$_4$H$_9$-n (79)	96b
	(C$_6$H$_5$CuC$_4$H$_9$-n)Li	−78°, then 0°, 2 hr (THF)	I, R = C$_4$H$_9$-n (84)	96b
	(CH$_3$C≡CCuC$_4$H$_9$-sec)Li	−78°, then 0°, 2 hr (THF)	I, R = C$_4$H$_9$-sec (80)	96b
	(C$_6$H$_5$CuC$_4$H$_9$-sec)Li	−78°, then 0°, 2 hr (THF)	I, R = C$_4$H$_9$-sec (75)	96b
	(CH$_3$C≡CCuC$_4$H$_9$-t)Li	−78° then 0°, 2 hr (THF)	I, R = C$_4$H$_9$-t (83)	96b
	(C$_6$H$_5$CuC$_4$H$_9$-t)Li	−78°, then 0°, 2 hr (THF)	I, R = C$_4$H$_9$-t (61)	96b
2-Bromopentane	(n-C$_4$H$_9$)$_2$CuLi (0.4 M, 4 eq)	25°, 26 hr (THF)	4-Methyloctane (12)	33
C$_2$H$_5$CBr(CH$_3$)$_2$	(n-C$_4$H$_9$)$_2$CuLi (0.4 M, 4 eq)	25°, 1 hr	3,3-Dimethylheptane (<10; <10^a)	33
1-Chloropentane	(n-C$_4$H$_9$)$_2$CuLi (0.4 M, 4 eq)	25°, 1 hr	n-Nonane (10; 32^a)	33
	,,	25°, 1 hr (THF)	n-Nonane (80; 86^a)	33

Note: References 236–320 are on pp. 398–400.

^{a} The reaction mixture was oxidized with oxygen at −78°.

TABLE III. Organocopper Coupling with Halide Substrates (*Continued*)

Halide Substrate	Organocopper Reagent	Reaction Conditions (Solvent)	Product(s) and Yield(s) (%)	Refs.
		A. Alkyl Halides (continued)		
C$_5$ (*contd.*)	(*n*-C$_4$H$_9$)$_2$CuLi (0.5 *M*, 5 eq)	25°, 1 hr (5 eq LiI)	*n*-Nonane (79; 80[a])	33
	(*sec*-C$_4$H$_9$)$_2$CuLi (0.5 *M*, 5 eq)	25°, 1 hr (5 eq LiI)	3-Methyloctane (64; 60[a])	33
1-Iodopentane	(CH$_3$)$_2$CuLi	25°, 3.5 hr (5 eq LiI)	*n*-Hexane (98)	33
	(*n*-C$_4$H$_9$)$_2$CuLi (0.5 *M*, 5 eq)	25°, 1 hr (ether-hexane, 5 eq LiI)	*n*-Nonane (53; 73[a])	33
	(*n*-C$_4$H$_9$)$_2$CuLi (0.4 *M*, 4 eq)	25°, 1 hr (THF)	*n*-Nonane (98)	33
	(*n*-C$_4$H$_9$)$_2$CuLi·P(C$_4$H$_9$-*n*)$_3$ (0.5 *M*, 5 eq)	25°, 1 hr (THF-hexane, 5 eq LiI)	'' (91)	33
	(*sec*-C$_4$H$_9$)$_2$CuLi (0.5 *M*, 5 eq)	25°, 1 hr (5 eq LiI)	3-Methyloctane (7; 7[a])	33
	(CH$_2$=CHCH$_2$)$_2$CuLi (0.3 *M*, 3 eq)	25°, 0.75 hr (6 eq LiI)	1-Octene (98)	33
C$_6$	4-Bromocyclohexanone	[CH$_2$=C(CH$_3$)]$_2$CuLi (5 eq)	0°, 6 hr (THF)	4-Isopropenylcyclohexanone (65)	114
1,1-Dichlorocyclohexane	(*n*-C$_4$H$_9$)$_2$CuLi (0.25 *M*, 10 eq)	0°, 2 hr	*n*-Butylcyclohexane (36), 1-*n*-butylcyclohexene (40)	159
Bromocyclohexane	(*n*-C$_4$H$_9$)$_2$CuLi·P(C$_4$H$_9$-*n*)$_3$ (0.5 *M*, 5 eq)	52°, 1 hr (THF-hexane, 5 eq LiI)	*n*-Butylcyclohexane (25; 25[a])	33
	(C$_6$H$_5$)$_2$CuLi (0.3 *M*, 3 eq)	52°, 96 hr (3 eq LiBr)	Phenylcyclohexane (10; 10[a])	33
Iodocyclohexane	(CH$_3$)$_2$CuLi (0.5 *M*, 5 eq)	0°, 10 hr	Methylcyclohexane (75)	54
trans-2-Iodocyclohexanol	(CH$_3$)$_2$CuLi (0.4 *M*, 10 eq)	0°, 18 hr, then 25°, 5 hr	*trans*-2-Methylcyclohexanol (75), cyclohexanone (10)	46
trans-4-Iodocyclohexanol	(CH$_3$)$_2$CuLi (0.4 *M*, 8 eq)	0°, 18 hr then 25°, 5 hr	*trans*-4-Methylcyclohexanol (35), cyclohexanol (60)	46

Substrate	Reagent	Conditions	Products (%)	Refs.
C₇				
7-Chloroquadricyclane	C₆F₅Cu	—	7-Pentafluorophenylquadricyclane (40)	155
7-Chloronorbornadiene	C₆F₅Cu	6 hr (n-Hexane)	7-Pentafluorophenylnorbornadiene (63)	155
4-Bromo-1-methylcyclohexene	$[CH_2=C(CH_3)]_2CuLi$ (5 eq)	0°, 6 hr (THF)	4-Isopropenyl-1-methylcyclohexene (80)	58, 114
1-Bromo-4-methylcyclohexane	$[CH_2=C(CH_3)]_2CuLi$ (5 eq)	0°, 6 hr (THF)	1-Isopropenyl-4-methylcyclohexane (80)	158, 114
1,1-Dichloroheptane	$(n\text{-}C_4H_9)_2CuLi$ (0.25 M, 10 eq)	0°, 4 hr	n-Undecane (20), 1-undecene (30), 7-butyltetradecane (35)	159
1-Chloroheptane	$(CH_3)_2CuLi$ (0.3 M, 10 eq)	25°, 5 days	n-Octane (70)	46
	$(n\text{-}C_4H_9)_2CuLi$ (0.25 M, 6 eq)	0°, 5 days	n-Undecane (75)	148
1-Iodoheptane	C_6H_5Cu (10 eq)	50–60°, 17 hr (THF)	1-Phenylheptane (61)	46
[bicyclic bromide structure]	C_6F_5Cu	20 hr (n-Hexane)	[bicyclic C_6F_5 structure] (83)	155
C₈				
5-Bromo-cis-cyclooctene	$(CH_3)_2CuLi$	25°	5-Methyl-cis-cyclooctene (23), cis-bicyclo[3,3,0] octane (50)	313
$(CH_3)_2C=CHCH_2CH_2CH(CH_3)Br$	$(p\text{-}CH_3C_6H_4)_2CuLi$ (5 eq)	−5°, 35 hr (THF)	$(CH_3)_2C=CHCH_2CH_2CH(CH_3)C_6H_4CH_3\text{-}p$ (65)	115
1,8-Dibromooctane	$(n\text{-}C_4H_9)_2CuLi$ (0.24 M, 5 eq)	0°, 21 hr	n-Dodecane (45)	46
1-Bromooctane	$(E\text{-}CH_3CH=CH)_2CuLi$ (1 eq)	−30 to +5°, 3 hr (4 eq HMPA)	E-2-Undecene (100)	232
	$(t\text{-}C_4H_9OCuC_4H_9\text{-}t)Li$ (0.25 M, 5 eq)	−50°, 4 hr (THF)	2,2-Dimethyldecane (83)	100
1-Chlorooctane	$(E\text{-}CH_3CH=CH)_2CuLi$ (1 eq)	25°, 48 hr (4 eq HMPA)	E-2-Undecene (80)	232
1-Iodooctane	$(CH_3)_2CuLi$ (1 eq)	0°, 17 hr	n-Nonane (88)	182
	$(CH_2=CH)_2CuLi\cdot P(C_4H_9\text{-}n)_3$ (0.5 M, 5 eq)	25°, 0.75 hr (5 eq LiI)	1-Decene (95; 91ᵃ)	33
	$(Z\text{-}CH_3CH=CH)_2CuLi\cdot P(OCH_3)_3$ (2 eq)	−30°, 15 min; then −30 to 25°, 1.5 hr	Z-2-Undecene (66; 98% retention)	231

Note: References 236–320 are on pp. 398–400.

ᵃ The reaction mixture was oxidized with oxygen at −78°.

TABLE III. Organocopper Coupling with Halide Substrates (Continued)

Halide Substrate	Organocopper Reagent	Reaction Conditions (Solvent)	Product(s) and Yield(s) (%)	Refs.
			A. Alkyl Halides (Continued)	
C_8 (contd.)	(E-$CH_3CH=CH)_2$CuLi-P(OCH$_3)_3$ (2 eq)	−30°, 15 min; then −30 to 25°, 1.5 hr	E-2-Undecene (73; 88% retention)	231
	(E-$CH_3CH=CH)_2$CuLi-O=P[N(CH$_3)_2]_2$ (1 eq)	−25°, 0.75 hr	E-2-Undecene (100)	232
	($C_6H_5)_2$CuLi (0.2 M, 2 eq)	25°, 2 hr (2 eq LiBr)	1-Phenyloctane (99)	33
	(t-C_4H_9OCuR)Li (0.25 M, 3 eq) R = C_4H_9-sec R = C_4H_9-t	−50°, 4 hr (THF)	n-C_8H_{17}R (I)	100
	(C_6H_5SCuR)Li (0.25 M, 2 eq) R = C_4H_9-sec R = C_4H_9-t	0° (THF)	I, R = C_4H_9-sec (52) I, R = C_4H_9-t (82)	101
C_{10}	1-Bromoadamantane	(CH$_3)_2$CuLi (4 eq)	35°, 96 hr	I, R = C_4H_9-sec (67) I, R = C_4H_9-t (98)	
	(CH$_3$CuCN)Li (4 eq)	35°, 96 hr	1-Methyladamantane (I, 80)	260
	C_2F_5Cu		I (35)	103
	m-$CF_3C_6H_4$Cu		1-Pentafluorophenyladamantane (93) 1-(m-Trifluoromethylphenyl)adamantane (48)	155
	1-Iododecane	CH$_3$Cu (10 eq)	25°, 12 hr (THF)	n-Undecane (68)	46
	(CH$_3)_2$CuLi (0.3 M, 5 eq)	0°, 6 hr	'' (90)	54, 46
	(n-$C_4H_9)_2$CuLi (0.2 M, 6 eq)	−45°, 0.5 hr; then n-C_4H_9I, 25°, 4.5 hr	n-Tetradecane (80)	148, 46
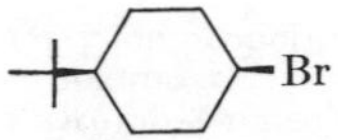	(CH$_3)_2$CuLi (0.25 M, 6 eq)	0°, 48 hr	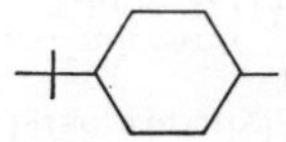 CH_3 (45 *trans*, 10 *cis*), 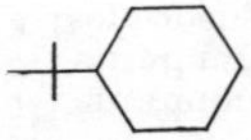 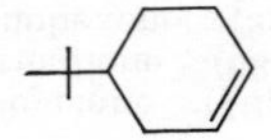 (5) (35)	54, 46

326

C$_{11}$	11-Bromoundecanoic acid	(CH$_3$)$_2$CuLi (0.25 M, 15 eq)	0°, 7 days	Dodecanoic acid (80)	46
	11-Iodoundecanoic acid	(n-C$_4$H$_9$)$_2$CuLi (0.25 M, 5 eq)	−40°, 8 hr; then n-C$_4$H$_9$I, 0°, 6 hr	Pentadecanoic acid (76)	148
	I(CH$_2$)$_{10}$CO$_2$CH$_3$	(CH$_3$)$_2$CuLi (0.2 M, 10 eq)	−20°, 4 hr	n-C$_{11}$H$_{23}$CO$_2$CH$_3$ (60), I(CH$_2$)$_{10}$CO$_2$CH$_3$ (35)	46
	I(CH$_2$)$_{10}$CON(CH$_3$)C$_6$H$_5$	(CH$_3$)$_2$CuLi (0.2 M, 8 eq)	−20°, 21 hr	n-C$_{11}$H$_{23}$CON(CH$_3$)C$_6$H$_5$ (70)	46
		(n-C$_4$H$_9$)$_2$CuLi (0.25 M, 5 eq)	−50°, 10 hr	n-C$_{14}$H$_{29}$CON(CH$_3$)C$_6$H$_5$ (82)	148

B. Alkenyl Halides

C$_2$	F$_2$C=CFI	C$_6$Br$_5$Cu (0.9 eq)	0°, 0.5 hr; 25°, 13 hr; reflux, 2 hr (THF)	F$_2$C=CFC$_6$Br$_5$ (50)	152
		C$_6$Cl$_5$Cu (1 eq)	50–60° (THF)	F$_2$C=CFC$_6$Cl$_5$ (32)	123
		C$_6$F$_5$Cu(MgBrCl) (0.25 M, 1 eq)	25–55°, 5 hr (THF)	F$_2$C=CFC$_6$F$_5$ (55)	138, 261
		C$_6$F$_5$Cu-dioxane (1 eq)	50–60° (THF)	'' (88)	123
		p-BrC$_6$F$_4$Cu (1 eq)	50–60° (THF)	F$_2$C=CFC$_6$F$_4$Br-p (60)	123
		C$_6$HF$_4$Cu (1 eq)	50–60° (THF)	F$_2$C=CFC$_6$HF$_4$ (45)	123
		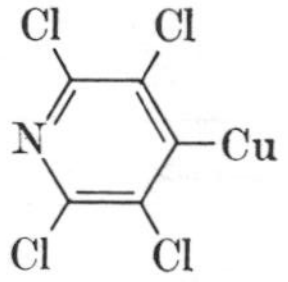 [tetrachloropyridyl]Cu (1 eq)	50–60° (THF)	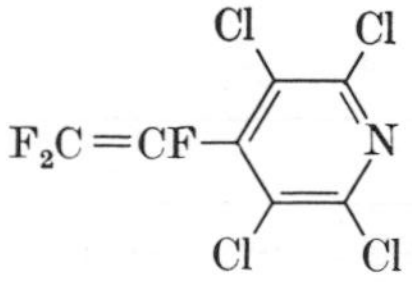 [tetrachloropyridyl]=CF—CF$_2$ (84)	123

Note: References 236–320 are on pp. 398–400.

TABLE III. ORGANOCOPPER COUPLING WITH HALIDE SUBSTRATES (*Continued*)

Halide Substrate	Organocopper Reagent	Reaction Conditions (Solvent)	Product(s) and Yield(s) (%)	Refs.
		B. Alkenyl Halides (Continued)		
C₂ (*contd.*)				
$CHBr=CBr_2$	[tetrafluoropyridyl]Cu (1 eq)	50–60° (THF)	$F_2C=CF$[tetrafluoropyridyl] (—)	123
	C_6F_5Cu (3 eq)	−10 to −30°, 3–6 hr (THF)	$C_6F_5C\!\equiv\!CC_6F_5$ (43)	151
	[trichlorothienyl]Cu	60°, 24 hr (THF)	[trichlorothienyl]$C\!\equiv\!C$[trichlorothienyl] (36)	140
$ClCH=CHI$	$Cu(CF_2)_3Cu^b$ (0.5 M, 0.9 eq)	100° (pyridine)	$ClCH=CH(CF_2)_3CH=CHCl$ (96)	82
$E\text{-}ClCH=CHI$	$n\text{-}C_7F_{15}Cu^b$ (0.4 M, 1 eq)	120°, 12 hr (DMF)	$E\text{-}ClCH=CHC_7F_{15}\text{-}n$ (65)	83, 82
$E\text{-}ICH=CHI$	$n\text{-}C_7F_{15}Cu^b$ (0.4 M, 1 eq)	120°, 12 hr (DMF)	$E\text{-}n\text{-}C_7F_{15}CH=CHC_7F_{15}\text{-}n$ (50)	83, 82
C₃ 2-Bromopropene	$(p\text{-}CH_3C_6H_4)_2CuLi$ (5 eq)	−5°, 24 hr (THF)	4-Isopropenyltoluene (55)	115
C₄ $Z\text{-}CH_3CH=C(Br)CO_2R$ (R = H, CH_3)	$(CH_3)_2CuLi$	−80°	$E\text{-}CH_3CH=CHCO_2R$ (R = H, CH_3) (—)	55
C₆ 1-Bromocyclohexene	$(CH_3)_2CuLi$ (0.25 M, 5 eq)	25°, 18 hr	1-Methylcyclohexene (25), 1-Bromocyclohexene (40)	46
	$(n\text{-}C_4H_9)_2CuLi$ (0.2 M, 6 eq)	−45°, 1 hr; then 0°, 3 hr	1-n-Butylcyclohexene (80)	148
1-Chlorocyclohexene	$(n\text{-}C_4H_9)_2CuLi$ (0.2 M, 5 eq)	0°, 62 hr	1-n-Butylcyclohexene (60)	148

Substrate	Reagent	Conditions	Product(s) (%)	Refs.
Z-CF$_3$CO$_2$(CH$_2$)$_3$-C(I)=CHCO$_2$CH$_3$	(CH$_3$)$_2$CuLi (1.1 eq)	−70° (THF)	E-HO(CH$_2$)$_3$C(CH$_3$)=CH=CHCO$_2$CH$_3$ (95)	261b
[cyclopentene with I and CH$_3$ substituents]	(CH$_3$)$_2$CuLi	—	(—)	262
C$_8$ n-C$_6$F$_{13}$CBr=CH$_2$	n-C$_4$F$_9$Cu^b	120° (DMF)	n-C$_6$F$_{13}$CH=CHR (I) I, R = C$_4$F$_9$-n (66)	262b
	n-C$_6$F$_{13}$Cu^b	120° (DMF)	I, R = C$_6$F$_{13}$-n (70)	
	n-C$_8$F$_{17}$Cu^b	120° (DMF)	I, R = C$_6$F$_{13}$-n (60)	
	(CH$_3$)$_2$CuLi	—	1,4-Dimethylcyclooctatetraene (95)	163
1,4-Dibromocyclo-octatetraene	(CH$_3$)$_2$CuLi (0.2 M, 1.7 eq)	−60°, 1 hr; then −10°	Methylcyclooctatetraene (93)	150, 263
Bromocyclooctatetraene	(C$_6$H$_5$)$_2$CuLi (0.5 M, 2 eq)	−35°, 1 hr; then reflux, 6 hr	Phenylcyclooctatetraene (58)	150
	CH$_3$Cu (20 eq)	0°, 27 hr; then 25°, 40 hr (THF)	E-C$_6$H$_5$CH=CHCR$_3$ (I, R = CH$_3$ 70)	46
E-C$_6$H$_5$CH=CHBr	(CH$_3$)$_2$CuLi (0.5 M, 5 eq)	0°, 2.5 hr	I, R=CH$_3$ (81)	54
	(C$_2$H$_5$)$_2$CuLi (0.1 M, 12 eq)	−78°, 3 hr, then −20°, 15 hr (pentane)	I, R = C$_2$H$_5$ (65)	148
	(n-C$_4$H$_9$)$_2$CuLi (0.25 M, 5 eq)	−95°, 1 hr, then −50°, 0.5 hr	I, R = C$_4$H$_9$-n (65)	46
	(n-C$_4$H$_9$)$_2$CuLi (0.3 M, 5 eq)	−78°, 2 hr then 0°, 14 hr (pentane)	'' (65)	46
	(C$_6$H$_5$)$_2$CuLi (0.24 M, 5 eq)	20°, 4 hr (15 eq LiBr)	E-Stilbene (90)	33
	(C$_6$H$_5$)$_2$CuLi (0.7 M, 5 eq)	−78°, 3 hr (ether-benzene)	Stilbene (84 E, 16 Z; 20 total)	110

Note: References 236–320 are on pp. 398–400.

^{b} The reagent was prepared in situ by heating the corresponding iodide with activated copper bronze in the indicated solvent.

329

TABLE III. ORGANOCOPPER COUPLING WITH HALIDE SUBSTRATES (Continued)

Halide Substrate	Organocopper Reagent	Reaction Conditions (Solvent)	Product(s) and Yield(s) (%)	Refs.
		B. Alkenyl Halides (continued)		
C$_8$ (contd.) Z-C$_6$H$_5$CH=CHBr	(CH$_3$)$_2$CuLi (0.5 M, 5 eq)	0°, 2.5 hr	Z-C$_6$H$_5$CH=CHCH$_3$ (—)	46
	(C$_6$H$_5$)$_2$CuLi (0.20 M, 5 eq)	25°, 2 hr (15 eq LiBr)	Z-Stilbene (73)	33
C$_6$H$_5$CH=CHBr	(C$_6$H$_5$)$_2$CuLi (0.7 M, 5 eq)	−78°, 3 hr (ether-benzene)	Stilbene (30 E, 70 Z)	110
	(i-C$_3$H$_7$)$_2$CuLi (0.7 M, 5 eq)	−78° 1 hr (1:2 ether:pentane)	C$_6$H$_5$CH=CHR (I) + C$_6$H$_5$CH$_2$CH$_2$R (II) R = C$_3$H$_7$-i, I (60), II (40)	110
	(t-C$_4$H$_9$)$_2$CuLi (0.7 M, 5 eq)	−78° 1 hr (2:5 ether:pentane)	R = C$_4$H$_9$-t, I (50), II (50)	110
	CF$_3$CH$_2$Cu (0.4 M, 1 eq)	120°, 12 hr (DMF)	I, R = CH$_2$CF$_3$ (18)	83, 82
	n-C$_3$H$_7$Cu^b (0.4 M, 1 eq)	120°, 12 hr (DMF)	I, R = C$_3$H$_7$-n (82)	83, 82
	n-C$_7$F$_{15}$Cu (0.4 M, 1 eq)	120°, 12 hr (DMF)	I, R = C$_7$F$_{15}$-n (95)	83
	Cu(CF$_2$)$_3$Cu (0.4 M, 1 eq)	120°, 12 hr (DMF)	C$_6$H$_5$CH=CH(CF$_2$)$_3$CH=CHC$_6$H$_5$ (95)	83, 82
[structure, I, I, OH]	(C$_2$H$_5$)$_2$CuLi (0.2 M, 8 eq)	−65 to −30°, 40 min, then −30°, 2.5 hr	[structure] OH (60^c)	161, 144
C$_9$ 1-Bromo-1-cycloocten-3-ol	(CH$_3$)$_2$CuLi (0.3 M, 3 eq)	0°, 4.5 hr	1-Methyl-1-cycloocten-3-ol (95)	192
Z-C$_6$H$_5$CH=C(Br)CO$_2$R (R = H, CH$_3$) E-C$_6$H$_5$CH=C(Br)CO$_2$R (R = H, CH$_3$)	(CH$_3$)$_2$CuLi	−80°	E-C$_6$H$_5$CH=C(CH$_3$)CO$_2$R (—), (R = H, CH$_3$) E-C$_6$H$_5$CH=CHCO$_2$R (—), (R = H, CH$_3$) Z-C$_6$H$_5$CH=C(CH$_3$)CO$_2$R (—), Z-C$_6$H$_5$CH=CHCO$_2$R (—) (R = H, CH$_3$)	55
[structure, OCH$_3$, I]	(CH$_3$)$_2$CuLi	—	[structure, OCH$_3$, H] (60)	263a
[structure, OTHP, HO]	(CH$_3$)CuLi		[structure, OTHP, HO] (30)	264, 265

Substrate	Reagent	Conditions	Product(s) (% Yield)	Refs.
	$(CH_3)_2CuLi$ (0.1 M, 12 eq)	5°, 30 hr	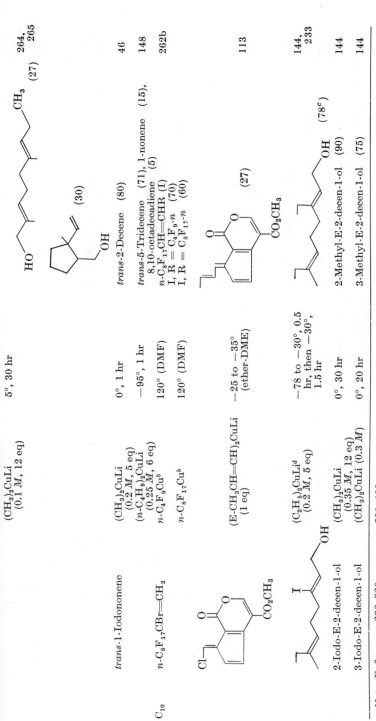 CH_3 (27) and vinylcyclopentane–CH_2OH (30)	264, 265
C_{10} *trans*-1-Iodononene	$(CH_3)_2CuLi$ (0.2 M, 5 eq)	0°, 1 hr	*trans*-2-Decene (80)	46
n-$C_8F_{17}CBr{=}CH_2$	$(n$-$C_4H_9)_2CuLi$ (0.25 M, 6 eq) n-$C_4F_9Cu^b$	−95°, 1 hr	*trans*-5-Tridecene (71), 1-nonene (15), 8,10-octadecadiene (5)	148
	n-$C_8F_{17}Cu^b$	120° (DMF)	n-$C_8F_{17}CH{=}CHR$ (I) I, R = C_4F_9-n (70) I, R = C_8F_{17}-n (60)	262b
[Cl-pyranone–CO_2CH_3]	n-$C_8F_{17}Cu^b$	120° (DMF)	[pyranone–CO_2CH_3] (27)	
	$(E$-$CH_3CH{=}CH)_2CuLi$ (1 eq)	−25 to −35° (ether-DME)		113
[I–E-2-decen-1-ol, OH]	$(C_2H_5)_2CuLi^a$ (0.2 M, 5 eq)	−78 to −30°, 0.5 hr, then −30°, 1.5 hr	[OH product] (78^c)	144, 233
2-Iodo-E-2-decen-1-ol	$(CH_3)_2CuLi$ (0.35 M, 12 eq)	0°, 30 hr	2-Methyl-E-2-decen-1-ol (90)	144
3-Iodo-E-2-decen-1-ol	$(CH_3)_2CuLi$ (0.3 M)	0°, 20 hr	3-Methyl-E-2-decen-1-ol (75)	144

Note: References 236–320 are on pp. 398–400.

[b] The reagent was prepared *in situ* by heating the corresponding iodide with activated copper bronze in the indicated solvent.

[c] Excess ethyl iodide was added to the reaction mixture before workup.

[a] The reagent was prepared at −50° for 1 hr.

331

TABLE III. ORGANOCOPPER COUPLING WITH HALIDE SUBSTRATES (Continued)

Halide Substrate	Organocopper Reagent	Reaction Conditions (Solvent)	Product(s) and Yield(s) (%)	Refs.
B. Alkenyl Halides (continued)				
C$_{10}$ (contd.) n-C$_3$H$_7$(CH$_3$)$\overset{6}{C}$(CH$_3$)=CH(CH$_2$)$_2$C(I)=CHCH$_2$OH (I), 1, 2E, 6Z	(n-C$_3$H$_7$)$_2$CuLi (4 eq)	−78°, 4 hr	n-C$_3$H$_7\overset{6}{C}$(CH$_3$)$_2$C(n-C$_3$H$_7$)=CHCH$_2$OH (II), II, 2E, 6Z (>50)	266b
I, 2E, 6Z	(n-C$_3$H$_7$)$_2$CuLi		II, 2E, 6E (—[c])	266c
C$_{11}$ 2-Iodo-3-methyl-Z-2-decen-1-ol	(CH$_3$)$_2$CuLi (0.2 M, 13 eq)	−25°, 2.5 hr	2,3-Dimethyl-Z-2-decen-1-ol (86[c])	144
	(C$_2$H$_5$)$_2$CuLi (0.2 M, 13 eq)	−35°, 6 hr	2-Ethyl-3-methyl-Z-2-decen-1-ol (—[c])	144
C$_{12}$	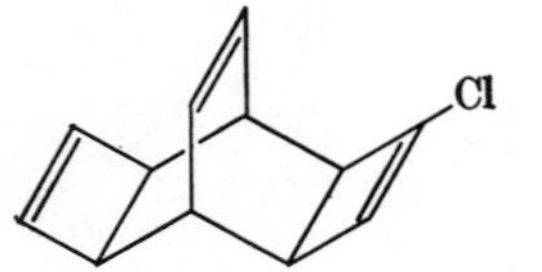 (CH$_3$)$_2$CuLi (0.3 M, 5 eq)	25°, 24 hr	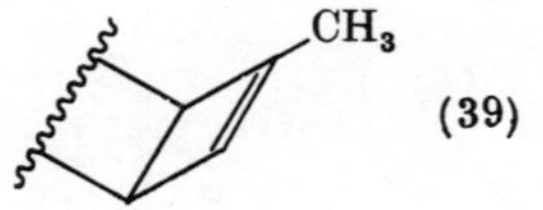 (39)	164
C$_{14}$ 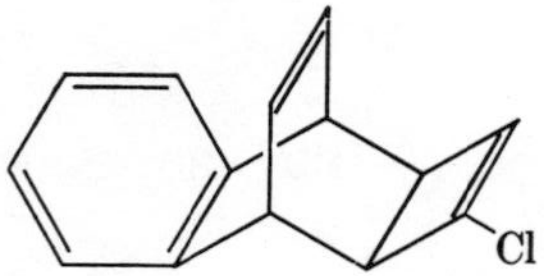	(CH$_3$)$_2$CuLi	—	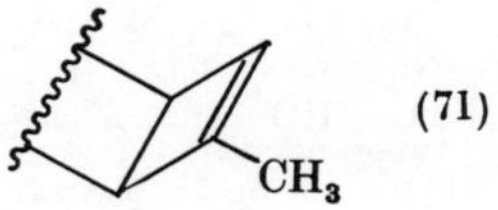 (71)	266
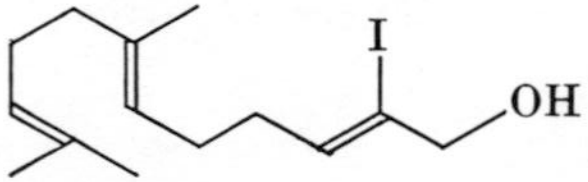	(CH$_3$)$_2$CuLi (0.1 M, 11 eq)	0°, 28 hr[e]	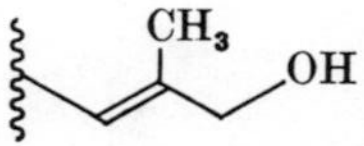 (58)	160, 46, 144
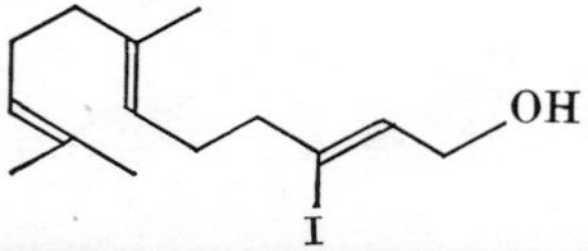	(CH$_3$)$_2$CuLi (0.2 M, 13 eq)	0°, 63 hr[e,f]	trans,trans-Farnesol (80)	160, 46, 144

332

C$_{15}$	Z-C$_6$H$_5$(I)C=CHCHOHC$_6$H$_5$	(CH$_3$)$_2$CuLi	0°	E-C$_6$H$_5$(CH$_3$)C=CHCHOHC$_6$H$_5$ (80[g])	267
C$_{16}$	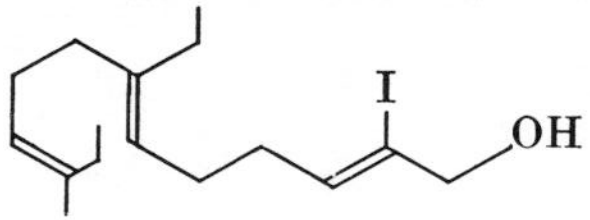	(CH$_3$)$_2$CuLi (0.2 M, 8 eq)	0°, 35 hr	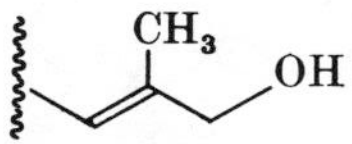 CH$_3$ OH (88)	144, 233
C$_{23}$	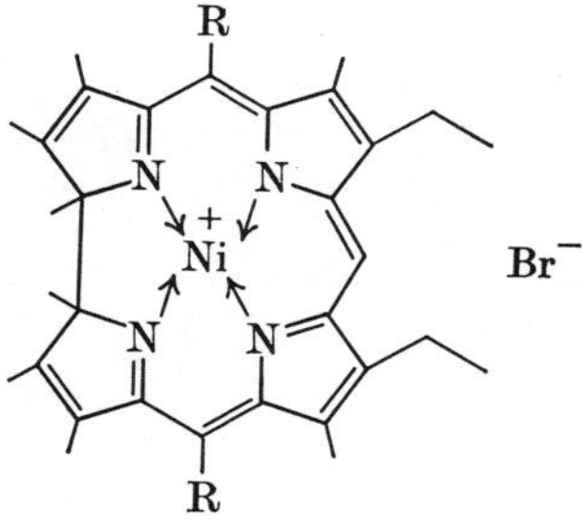 I, R = Br I, R = Cl	(CH$_3$)$_2$CuLi (CH$_3$)$_2$CuLi	— —	I, R = CH$_3$ (60) I, R = CH$_3$ (45)	222

C. Aryl Halides

C$_3$	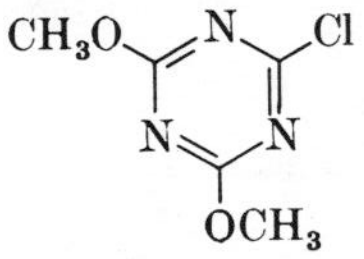	Cu(CF$_2$)$_3$Cu[b] (<1 eq)	115°, 2 hr (DMF)	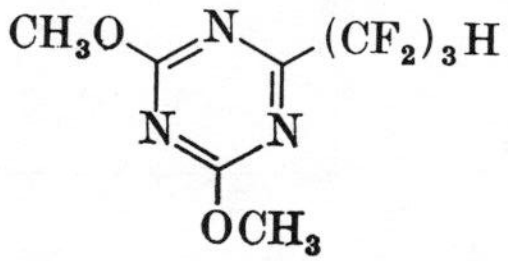 (18)	81

Note: References 236–320 are on pp. 398–400.

[b] The reagent was prepared *in situ* by heating the corresponding iodide with activated copper bronze in the indicated solvent.
[e] The halide was added slowly during 1 hr by a motor-driven syringe.
[f] After 42 hr at 0° an additional 13 eq of (CH$_3$)$_2$CuLi was added in one portion by syringe.
[g] Excess methyl iodide was added to the reaction mixture before workup.

333

TABLE III. ORGANOCOPPER COUPLING WITH HALIDE SUBSTRATES (Continued)

Halide Substrate	Organocopper Reagent	Reaction Conditions (Solvent)	Product(s) and Yield(s) (%)	Refs.
		C. Aryl Halides (Continued)		
C$_4$ 3-Bromothiophene	2-Thienyl-Cu	110° (pyridine)	2-Thienyl [thiophene] (14)	121
2-Iodothiophene	[N-CH$_3$ pyrrolyl]-Cu (3 eq)	100°, 3 hr (pyridine)	[N-CH$_3$ pyrrolyl-thienyl] (40)	122
	2-Thienyl-Cu	75° (pyridine)	Bi-2-thienyl (42)	121
	Cu(CF$_2$)$_3$Cu (1 eq)	100°, 35 min (pyridine)	[thienyl-(CF$_2$)$_3$-thienyl] (25)	81
3-Iodothiophene	[N-CH$_3$ pyrrolyl]-Cu (1 eq)	100–110°, 0.5 hr (quinoline)	3-Thienyl [N-CH$_3$ pyrrolyl] (8)	122
C$_5$ [2,3,5,6-tetrachloro-4-iodopyridine]	C$_6$Cl$_5$Cu	—	[tetrachloropyridine with R] (I)	268

Substrate	Reagent	Conditions	Product(s) (% Yield)	Refs.
	C_6F_5Cu	—	I, R = C_6Cl_5 (58); I, R = C_6F_5 (55)	81
(tetrachloro-4-pyridyl–Cu)		—	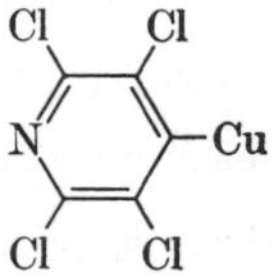 I, R = tetrachloro-4-pyridyl (72)	
3-Iodopyridine	$Cu(CF_2)_3Cu^b$ (<1 eq)	126°, 1.5 hr (DMF)	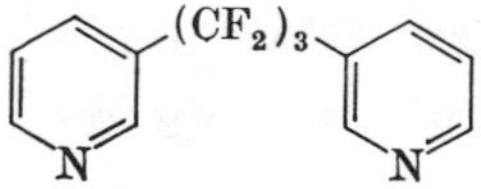 [3-pyridyl–$(CF_2)_3$–3-pyridyl] (52)	122
1-Methyl-2-iodopyrrole	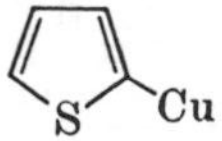 [2-thienyl–Cu] (0.7 M, 3 eq)	25° then 100°, 3 hr (pyridine)	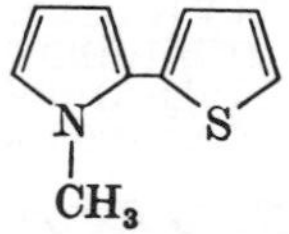 [1-methyl-2-(2-thienyl)pyrrole, CH_3] (42)	268
C_6 C_6Cl_5I	C_6F_5Cu	—	$C_6Cl_5C_6F_5$ (44)	268
	(tetrachloro-4-pyridyl–Cu)	—	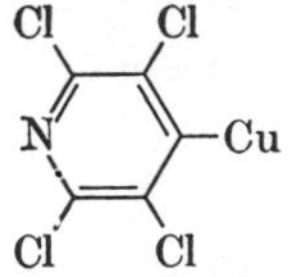 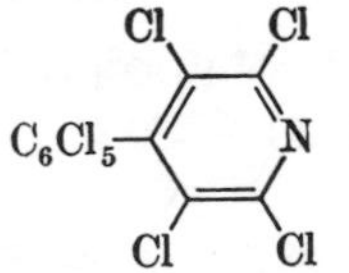 [C_6Cl_5–tetrachloropyridyl] (51)	268
C_6F_5Br	$Cu(CF_2)_3Cu^b$ (<1 eq)	125°, 95 min (DMF)	$C_6F_5C_6F_5$ (35), $C_6F_5(CF_2)_3C_6F_5$ (7), $C_6F_5C_7F_{15}\text{-}n$ (12)	81
	$n\text{-}C_7F_{15}Cu$ (1 eq)	120°, 1.5 hr (DMSO)		81
	C_6Cl_5Cu	—	$C_6F_5C_6Cl_5$ (80), (58)	268
	$C_6F_5Cu(MgBrCl)$ (0.25 M, 1 eq)	66°, 10 hr (THF)	$C_6F_5C_6F_5$ (I, 74)	138
C_6F_5I	$C_6F_5Cu(MgBrCl)$ (0.25 M, 1 eq)	60°, 18 hr (DMAC)	I (78)	138

Note: References 236–320 are on pp. 398–400.

b The reagent was prepared *in situ* by heating the corresponding iodide with activated copper bronze in the indicated solvent.

335

TABLE III. ORGANOCOPPER COUPLING WITH HALIDE SUBSTRATES (Continued)

Halide Substrate	Organocopper Reagent	Reaction Conditions (Solvent)	Product(s) and Yield(s) (%)	Refs.
		C. Aryl Halides (Continued)		
C_6 (contd.)				
	(2,3,5,6-tetrachloropyridin-4-yl)Cu	—	2,3,5,6-tetrachloro-4-(C_6F_5)pyridine (47)	268
p-$CH_3OC_6F_4I$	$C_6F_5Cu(MgBrCl)$ (0.25 M, 1 eq)	66°, 20 hr (THF)	p-$CH_3OC_6F_4C_6F_5$ (I, 70)	138
	$C_6F_5Cu(MgBrCl)$ (0.25 M, 1 eq)	70°, 48 hr (di-n-butyl ether)	I (45)	138
p-$C_6F_5OC_6F_4Br$	$C_6F_5Cu(MgBrCl)$ (0.25 M, 1 eq)	66°, 60 hr (THF)	p-$C_6F_5OC_6F_4C_6F_5$ (5)	138
$2,4,6$-$(O_2N)_3C_6H_2Cl$	$2,6$-$(CH_3O)_2C_6H_3Cu$	25°, 24 hr (DMF)	$2,4,6$-$(O_2N)_3C_6H_2C_6H_3(OCH_3)_2$-2,6 (36)	118
p-BrC_6H_4I	$H(CF_2)_6Cu^b$ (1 eq)	110°, 3 hr (DMSO)	p-$BrC_6H_4(CF_2)_6H$ (50)	81
o-ClC_6H_4I	C_6H_5Cu	50° (pyridine)	$C_6H_4C_6H_5$ (44)	81
p-ClC_6H_4I	$Cu(CF_2)_3Cu^b$ (<1 eq)	120°, 2 hr (DMSO)	$(p$-$ClC_6H_4CF_2)_2CF_2$ (80)	81
FC_6H_4I (I)			FC_6H_4R (II)	243, 153
o-I	C_6H_5Cu (1 eq)	40°, 40 hr (quinoline)	o-II, R = C_6H_5 (26)	
m-I	C_6F_5Cu (2 eq)	Reflux, 2 hr (benzene)	m-II R = C_6F_5 (73)	169
p-I	C_6F_5Cu (2 eq)	Reflux, 2 hr (benzene)	p-II R = C_6F_5 (78)	169
m- or p-$O_2NC_6H_4Br$	C_6F_5Cu (2 eq)	Reflux, 2 hr (benzene)	m- or p-$O_2NC_6H_4C_6F_5$ (85)	169
o-$O_2NC_6H_4Cl$	n-$C_3F_7Cu^b$ (<1 eq)	175°, 13 hr (DMF)	o-$O_2NC_6H_4C_3F_7$-n (17)	81
$O_2NC_6H_4I$ (I)			$O_2NC_6H_4R$ (II)	
o-I	C_6H_5Cu	25° (pyridine)	o-II, R = C_6H_5 (18)	153
	C_6F_5Cu	—	o-II, R = C_6F_5 (93)	155

336

Halide	Organocopper Reagent	Conditions	Products (% Yield)	Refs.
	m-CF₃C₆H₄Cu	—	o-II, R = C₆H₄CF₃-m (59)	155
	2-Thienyl-Cu	0° (pyridine)	o-II, R = 2-thienyl (56)	121
m-I	Cu(CF₂)₃Cu[b] (<1 eq)	130°, 1.5 hr (DMF)	m-II, R = (CF₂)₃C₆H₄NO₂-m (52)	81
	C₆F₅Cu	50° (pyridine)	m-II, R = C₆F₅ (85)	121
p-I	2-Thienyl-Cu	—	p-II, R = 2-thienyl (70)	155
	C₆F₅Cu	—	p-II, R = C₆F₅ (99)	169
(CH₃)₂NC₆H₄I (I)			(CH₃)₂NC₆H₄C₆F₅ (II)	
m-I	C₆F₅Cu (2 eq)	Reflux, 2 hr (benzene)	m-II (33)	147
p-I	C₆F₅Cu (2 eq)	Reflux, 2 hr (benzene)	p-II (26)	54
Bromobenzene	[CH₂=C(CH₃)]₂CuLi (5 eq)	0°, 6 hr (THF)	Isopropenylbenzene (85)	158
	Cu(CF₂)₃Cu[b] (<1 eq)	140°, 18 hr (DMF)	C₆H₅(CF₂)₃C₆H₅ (20)	81
p-DC₆H₄I	C₆H₅Cu (1 eq)	50° (pyridine)	p-DC₆H₄C₆H₅ (30)	243, 153
Iodobenzene	HCu·P(C₄H₉-n)₃ (CH₃)₂CuLi	25°, 14 hr (1 eq LiBr, 1 eq LiI)	Benzene (80) / Toluene (90)	147
		15 hr (no LiBr or LiI)	'' (95)	54
	(n-C₄H₉)₂CuLi (0.3 M)	0°, 3.5 hr, then C₄H₉I, 0°, 1.5 hr	n-Butylbenzene (75)	147
	(sec-C₄H₉)₂CuLi·P(C₄H₉-n)₃	-10°, 0.7 hr (5 eq LiI)	sec-Butylbenzene (<1; 20[a])	148
	CF₃Cu[b] (<1 eq)	150°, 12 hr (DMF)	C₆H₅CF₃ (45)	147
	n-C₃F₇Cu (<1 eq)	120°, 1 hr (DMSO)	C₆H₅C₃F₇-n (65)	81
	(CF₃)₂CFCu[b] (<1 eq)	125°, 1.5 hr (DMF)	C₆H₅CF(CF₃)₂ (40)	81
	Cu(CF₂)₃Cu[b] (<1 eq)	115°, 195 min (DMAC)	C₆H₅(CF₂)₃C₆H₅ (72)	81
	Cu(CF₂)₃Cu[b] (<1 eq)	80°, 22 hr (DMSO-C₆F₆)	'' (95)	81

Note: References 236–320 are on pp. 398–400.

[a] The reaction mixture was oxidized with oxygen at −78°.

[b] The reagent was prepared in situ by heating the corresponding iodide with activated copper bronze in the indicated solvent.

TABLE III. ORGANOCOPPER COUPLING WITH HALIDE SUBSTRATES (Continued)

Halide Substrate	Organocopper Reagent	Reaction Conditions (Solvent)	Product(s) and Yield(s) (%)	Refs.
		C. Aryl Halides (Continued)		
C_6 (contd.)	$Cu(CF_2)_4Cu$[b] (<1 eq)	120°, 85 min (DMF)	$C_6H_5(CF_2)_4C_6H_5$ (9)	81
	$n\text{-}C_7F_{15}Cu$[b] (1 eq)	110°, 2 hr (DMSO)	$C_6H_5C_7F_{15}\text{-}n$ (70)	81
	$H(CF_2)_{10}Cu$ (<1 eq)	120°, 1.5 hr (DMSO)	$C_6H_5(CF_2)_{10}H$ (65)	81
	$C_6H_5(CF_2)_3Cu$ (1 eq)	120°, 25 min (DMSO)	$C_6H_5(CF_2)_3C_6H_5$ (60)	81
	$HO_2C(CF_2)_3Cu$[b] (1 eq)	120°, 70 min (DMSO)	$C_6H_5(CF_2)_3CO_2H$ (60)	81
	$C_2H_5O_2C(CF_2)_3Cu$[b] (<1 eq)	135°, 95 min (DMF)	$C_6H_5(CF_2)_3CO_2C_2H_5$ (70)	81
	$(CH_3C{=}CH_2)_2CuLi$ (3 eq)	−78°, 10 hr	2-Phenylpropene (95)	257b
	C_6Cl_5Cu (1 eq)	100°, 30 min (neat)	$C_6H_5C_6Cl_5$ (55)	156, 268b
	C_6F_5Cu (1 eq)	100°, 44 hr (neat)	$C_6H_5C_6F_5$ (60)	156, 103
	$C_6F_5Cu(MgBrCl)$ (0.25 M, 1 eq)	66°, 168 hr (THF)	'' (71)	138
	$o\text{-}O_2NC_6H_4CO_2Cu$ (1 eq)	145–165° (quinoline)	$C_6H_5C_6H_4NO_2\text{-}o$ (39–50)	77
	2-Thienyl-Cu (1 eq)	Reflux, 30 min (pyridine)	2-Phenylthiophene (50)	269a, 121
	[tetrachlorothienyl-Cu structure] (2 eq)	110°, 38 hr (THF)	[trichloro-phenyl-thiophene structure, C_6H_5] (53)	140
	[1-methyl-2-pyrrolyl-Cu structure] (0.5 M, 0.9 eq)	100°, 8 hr (pyridine)	[1-methyl-2-phenylpyrrole structure, C_6H_5, CH_3] (41)	122

	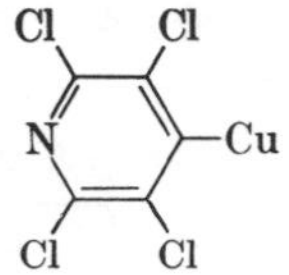	100°, 30 hr (neat)	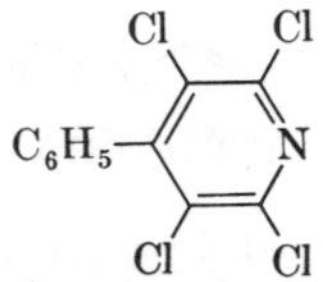 (56)	270, 156
m-Iodophenol	$(CH_3)_3SiCH_2Cu$	—	$C_6H_5CH_2Si(CH_3)_3$ (—)	271
Iodoanisole (I)	$Cu(CF_2)_3Cu^b$ (1 eq)	115°, 6 hr (DMSO)	$(m\text{-}HOC_6H_4CF_2)_2CF_2$ (69)	81
o-I	C_6F_5Cu (1 eq)	100°, 44 hr (neat)	$CH_3OC_6H_4R$ (II)	156
			o-II, R = C_6F_5 (60)	242,
		20°, 75 hr (quinoline)	o-II, R = C_6F_5 (30)	153
m-I	$Cu(CF_2)_3Cu^b$	145°, 2 hr (DMF)	m-II, R = $(CF_2)_3C_6H_4OCH_3$-m (55)	81
p-I	$[(CH_3)_2C=CH]_2CuLi$	35°, 12 hr (pyridine)	p-II, R = $CH=C(CH_3)_2$ (65)	269b
	C_6H_5Cu	50° (pyridine)	p-II, R = C_6H_5 (36)	153
	2-Furyl-Cu	150° (quinoline)	p-II, R = 2-furyl (—)	121
	2-Thienyl-Cu	115° (pyridine)	p-I, R = 2-thienyl (55)	121
	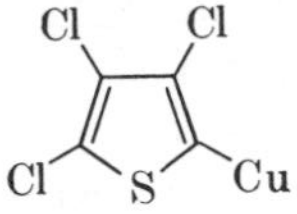	115°, 40 min (pyridine)	p-II, R = trichloro-2-thienyl (65)	238
	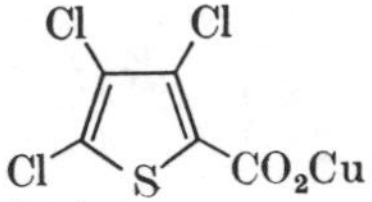	Reflux, 10 min (pyridine)	p-II, R = trichloro-2-thienyl (75)	238
2,6-$(CH_3O)_2C_6H_3I$	C_6H_5Cu	50°, 25 hr (quinoline)	2,6-$(CH_3O)_2C_6H_3C_6H_5$ (60)	243, 153
	2-Thienyl-Cu	110° (pyridine)	2,6-$(CH_3O)_2C_6H_3$-2-thienyl (56)	121

Note: References 236–320 are on pp. 398–400.

[b] The reagent was prepared *in situ* by heating the corresponding iodide with activated copper bronze in the indicated solvent.

[h] The reaction product, p-n-$C_7F_{15}C_6H_4SO_3Cu$, was converted via the free acid into the acid chloride with phosphorus pentachloride.

TABLE III. Organocopper Coupling with Halide Substrates (Continued)

Halide Substrate	Organocopper Reagent	Reaction Conditions (Solvent)	Product(s) and Yield(s) (%)	Refs.
		C. Aryl Halides (Continued)		
m-CH$_3$CO$_2$C$_6$H$_4$I	n-C$_3$F$_7$Cu^b (<1 eq)	130°, 50 min (DMF)	m-CH$_3$CO$_2$C$_6$H$_4$C$_3$F$_7$-n (65)	81
p-CH$_3$CO$_2$C$_6$H$_4$I	Cu(CF$_2$)$_3$Cu^b (<1 eq)	120°, 2 hr (DMF)	(m-CH$_3$CO$_2$C$_6$H$_4$CF$_2$)$_2$CF$_2$ (82)	81
	Cu(CF$_2$)$_3$Cu^b (<1 eq)	125°, 145 min (DMAc)	(p-CH$_3$CO$_2$C$_6$H$_4$CF$_2$)$_2$CF$_2$ (43)	81
p-CH$_3$O$_3$SC$_6$H$_4$I	n-C$_7$F$_{15}$Cu^b (1 eq)	120°, 290 min (DMF)	p-ClO$_2$SC$_6$H$_4$C$_7$F$_{15}$-n (14^h)	81
m-H$_2$NC$_6$H$_4$I	n-C$_9$F$_{19}$Cu^b (1 eq)	120°, 3.5 hr (DMSO)	m-H$_2$NC$_6$H$_4$C$_9$F$_{19}$-n (45)	81
m-CF$_3$C$_6$H$_4$I	C$_6$F$_5$Cu		m-CF$_3$C$_6$H$_4$C$_6$F$_5$ (43)	155
p-Iodobenzoic acid	n-C$_5$F$_{11}$Cu^b (1 eq)	11°, 3.5 hr (DMSO)	p-HO$_2$CC$_6$H$_4$C$_5$F$_{11}$-n (65)	81
Methyl o-bromobenzoate	2-Thienyl-Cu	110° (pyridine)	Methyl o-(2-thienyl)benzoate (13)	121
Methyl iodobenzoate (I)			CH$_3$O$_2$CC$_6$H$_4$R (II)	
o-I	cyclopentadienyl–Cu·P(C$_4$H$_9$-n)$_3$ 35°, 24 hr		o-II, R = cyclopentadienyl (34)	120
	C$_6$H$_5$Cu (1 eq)	20°, 20 hr (quinoline)	o-II, R = C$_6$H$_5$ (17–28)	243, 153
m-I	2-Thienyl-Cu	50° (pyridine)	o-II, R = 2-thienyl (50)	121
	C$_6$F$_5$Cu	Reflux, 2 hr (benzene)	m-II, R = C$_6$F$_5$ (96)	169
p-I	n-C$_7$F$_{15}$Cu^b (1 eq)	125°, 3 hr (DMF)	p-II, R = C$_7$F$_{15}$-n (70)	81
	Cu(CF$_2$)$_3$Cu^b	115°, 1 hr (DMF)	p-II, R = (CF$_2$)$_3$C$_6$H$_4$CO$_2$CH$_3$-p (63)	169, 120
	C$_6$F$_5$Cu (2 eq)	Reflux, 2 hr (benzene)	p-II, R = C$_6$F$_5$ (97)	81
Ethyl m-iodobenzoate	Cu(CF$_2$)$_3$Cu^b (<1 eq)	115°, 77 min (DMF)	(m-C$_2$H$_5$O$_2$CCF$_2$)$_2$CF$_2$ (76)	81
m-Iodotoluene	C$_6$F$_5$	—	m-CH$_3$C$_6$H$_4$C$_6$F$_5$ (85)	155
p-Iodotoluene	cyclopentadienyl–Cu·P(C$_4$H$_9$-n)$_3$ 35°, 50 hr		p-Tolylcyclopentadiene (50)	120
	C$_6$F$_5$Cu		p-CH$_3$C$_6$H$_4$C$_6$F$_5$ (85)	155
C$_8$ Dimethyl 4-iodophthalate	Cu(CF$_2$)$_7$Cu^b (1 eq)	120°, 12 hr (DMSO)	CH$_3$O$_2$C / CH$_3$O$_2$C · C$_6$H$_3$(CO$_2$CH$_3$)$_2$(CF$_2$)$_7$ (85)	81
o-HOCH(CH$_3$)C$_6$H$_4$Br	n-C$_3$F$_7$Cu^b (1 eq)	125°, 12 hr (DMSO)	o-HOCH(CH$_3$)C$_6$H$_4$C$_3$F$_7$-n (65)	81

340

	Substrate	Reagent	Conditions	Product(s) (% yield)	Refs.
C$_{10}$	1,8-Diiodonaphthalene	(C$_6$H$_5$)$_2$CuLi (0.6 M, 6 eq)	Reflux, 24 hr (LiBr, 18 eq)	1,8-Diphenylnaphthalene (47[i])	33
		(C$_6$H$_5$)$_3$CuLi$_2$ (0.2 M)	0°, 2 min	" (27[i])	60
	1-Iodonaphthalene	(CH$_3$)$_2$CuLi (0.38 M, 3 eq)	8 hr (no LiI present)	1-Methylnaphthalene (L, 33)	33
		CH$_3$Cu·P(C$_4$H$_9$-n)$_3$ (0.38 M, 3 eq)	23 hr (3 eq LiI); 2.5 hr (3 eq LiI); 25°, 2 hr then 60–70°, 2 hr (no LiI present)	I (20); I (70[j]); I (75)	
C$_{11}$	1-Bromo-2-methylnaphthalene	(C$_6$H$_5$)$_2$CuLi (0.16 M, 5 eq)	4 hr (LiBr, 15 eq)	1-Phenylnaphthalene (90[j])	81
		n-C$_7$F$_{15}$Cu[b] (1 eq)	160°, 40 min (HMPA)	1-n-C$_7$F$_{15}$-2-CH$_3$C$_{10}$H$_6$ (17)	60
C$_{12}$	1,8-Diiodo-9,10-anthraquinone	(C$_6$H$_5$)$_3$CuLi$_2$ (0.2 M, 3 eq)	0°, 10 sec, (5:1 THF:ether)	1,8-Diphenyl-9,10-anthraquinone (42[i])	60
	1-Iodo-9,10-anthraquinone	(C$_6$H$_5$)$_2$CuLi (0.2 M, 4 eq)	0°, 20 min (1:1 THF:ether)	1-Phenyl-9,10-anthraquinone (54[i])	60
C$_{14}$	(F-anthracene-Br structure)	(CH$_3$)$_2$CuLi (3 eq)	25°, 24 hr	(F-anthracene-R structure) (–) (mixture of R = H and CH$_3$)	272
C$_{15}$	I(CF$_2$)$_3$(phenyl)I structure	C$_6$H$_5$(CF$_2$)$_3$Cu[b] (<1 eq)	120°, 3 hr (DMF)	[C$_6$H$_5$(CF$_2$)$_3$—CF$_2$—CF$_2$—]$_2$ structure (70)	81
C$_{16}$	1-Iodo-8-phenylnaphthalene	(C$_6$H$_5$)$_2$CuLi (0.36 M, 3 eq)	0°, 3 min	1,8-Diphenylnaphthalene (65[i])	60

Note: References 236–320 are on pp. 398–400.

[b] The reagent was prepared *in situ* by heating the corresponding iodide with activated copper bronze in the indicated solvent.

[i] The reaction mixture was oxidized with oxygen at 0°.

[j] The reaction mixture was oxidized with nitrobenzene at 0°.

341

TABLE III. ORGANOCOPPER COUPLING WITH HALIDE SUBSTRATES (Continued)

Halide Substrate	Organocopper Reagent	Reaction Conditions (Solvent)	Product(s) and Yield(s) (%)	Refs.
		C. Aryl Halides (Continued)		
C_{16} (contd.) [structure: Cl-substituted tetrahydrobenzothieno-pyridine with C_6H_5]	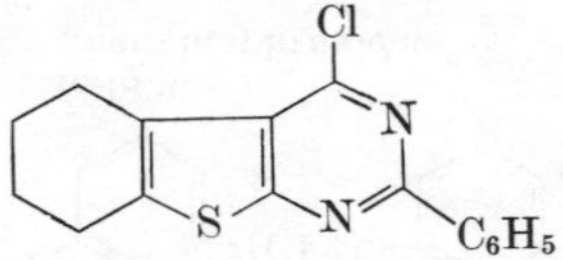 $(CH_3)_2CuLi$ (5 eq)	0–25°, 12 hr (THF-ether)	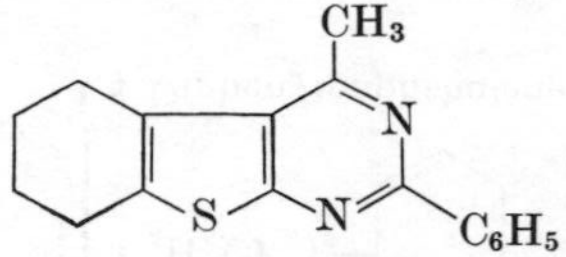 [structure: CH_3-substituted tetrahydrobenzothieno-pyridine with C_6H_5] (82)	170
C_{18} Bromo[18]annulene	$(CH_3)_2CuLi$	0°	Methyl[18]annulene (52)	168
		D. Benzylic, Allylic, and Propargylic Halides		
C_3 $CH_2=C(Br)CH_2Br$	$C_2H_5O_2CCH_2Cu$ (2 eq) $NCCH_2Cu$ (0.04 M, 3 eq)	−30°, 1 hr (THF) −25°, 1 hr (THF)	$CH_2=C(Br)CH_2CH_2CO_2C_2H_5$ (83) $CH_2=C(Br)CH_2CH_2CN$ (92)	143 142
Allyl bromide	$C_6H_5SCH_2Cu$ (1 eq)	−50 to +25°, 2 hr (THF)	$CH_2=CHCH_2CH_2SC_6H_5$ (—)	141
	Z-n-$C_4H_9CH=CHCu$	−10°, 15 hr (HMPA)	Z-1,4-Nonadiene (50)	86
	Z-$C_2H_5(CH_3)C=CHCu$ [HMPA, $(C_2H_5O)_3P$]		Z-$C_2H_5(CH_3)C=CHCH_2CH=CH_2$ (55)	87
	E-n-$C_4H_9(C_2H_5)C=CHCu$ −10° (HMPA)		E-n-$C_4H_9(C_2H_5)C=CHCH_2CH=CH_2$ (55)	86
	Z-n-$C_4H_9(C_2H_5)C=CHCu$ [HMPA, $(C_2H_5O)_3P$]		Z-n-$C_4H_9(C_2H_5)C=CHCH_2CH=CH_2$ (—)	87
	[structure: CO_2CH_3 / Cu vinyl]	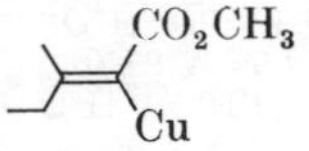 −78°, 0.5 hr (THF)	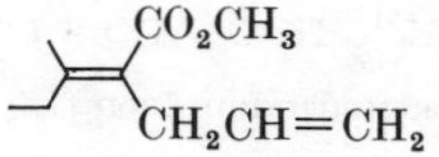 [structure: CO_2CH_3 / $CH_2CH=CH_2$ vinyl] (>80)	145
	[structure: dimethoxy-phenyl Cu, OCH_3/CH_3O/OCH_3] (0.5 eq) C_6H_5Cu (0.2 M, 1 eq)	—	Allylbenzene (31)	31
		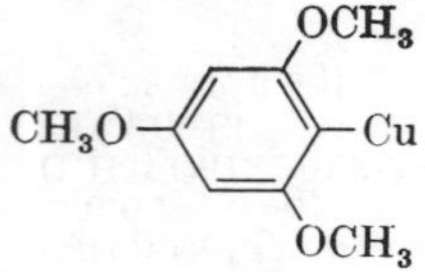 48 hr	$CH_2=CHCH_2C_6H_2(OCH_3)_3$-2,4,6 (75)	117, 119

Reagent	Conditions	Product (%)	Refs.
$o\text{-}CH_2N(CH_3)_2C_6H_4\text{-}Cu$	24 hr	$(CH_3)_2NCH_2$—CH_2=$CHCH_2$ (94)	117, 119
$m\text{-}FC_6H_4Cu\cdot MgBr_2$	—	$m\text{-}CH_2$=$CHCH_2C_6H_4F$ (67)	155
$p\text{-}FC_6H_4Cu\cdot MgBr_2$	—	$p\text{-}CH_2$=$CHCH_2C_6H_4F$ (69)	155
C_6Cl_5Cu (1 eq)	25°, 6 hr (THF)	CH_2=$CHCH_2C_6Cl_5$ (76)	156
C_6F_5Cu (1 eq)	25°, 12 hr (THF)	CH_2=$CHCH_2C_6F_5$ (68)	156
$o\text{-}CF_3C_6H_4Cu$	—	$o\text{-}CH_2$=$CHCH_2C_6H_4CF_3$ (71),	155
		I (51)	103
$m\text{-}CF_3C_6H_4Cu\cdot MgBr_2$	—	$m\text{-}CH_2$=$CHCH_2C_6H_4CF_3$ (I, 68)	155
$p\text{-}CF_3C_6H_4Cu\cdot MgBr_2$	—	$p\text{-}CH_2$=$CHCH_2C_6H_4CF_3$ (69)	155
thienyl-Cu (0.25 M, 0.9 eq)	25°, 6 hr (1:1 THF:ether)	CH_2=$CHCH_2$-thienyl (62)	71
thienyl-Cu	—	(66)	71
pyridyl-Cu	25°, several hr, (THF)	CH_2=$CHCH_2$-pyridyl (72–76)	270, 156
(0.5 M, 0.8 eq) $(CH_3)_3SiCH_2Cu$		CH_2=$CHCH_2CH_2Si(CH_3)_3$ (—)	271
$i\text{-}C_3H_7SCH$=$CHCH_2Cu$ (2 eq)	$-78°$, 4 hr	CH_2=$CHCH_2CH_2CH$=$CHSC_3H_{7}\text{-}i$ (86)	272b

Allyl chloride

Reagent	Conditions	Product (%)	Refs.
C_6Cl_5Cu (1 eq)	25°, 12 hr (THF)	CH_2=$CHCH_2C_6Cl_5$ (72)	156
C_6F_5Cu (1 eq)	25°, 12 hr (THF)	CH_2=$CHCH_2C_6F_5$ (60)	156

C_4 CH_3O_2CCH=$CHCH_2Br$

Reagent	Conditions	Product (%)	Refs.
$C_2H_5O_2CCH_2Cu$ (2 eq)	$-30°$, 1 hr (THF)	CH_3O_2CCH=$CHCH_2CH_2CO_2C_2H_5$ (89)	143

Note: References 236–320 are on pp. 398–400.

343

TABLE III. ORGANOCOPPER COUPLING WITH HALIDE SUBSTRATES (*Continued*)

Halide Substrate	Organocopper Reagent	Reaction Conditions (Solvent)	Product(s) and Yield(s) (%)	Refs.
			D. Benzylic, Allylic, and Propargylic Halides (Continued)	
C_4 (*contd.*) $C_2H_5O_2CC(=CH_2)CH_2Br$	$NCCH_2Cu$ (0.4 M, 3 eq)	$-78°$, 1 hr (THF)	$C_2H_5O_2CC(=CH_2)CH_2CH_2CN$ (89)	142
C_5 $CH_3CH=CHCH_2Br$	$m\text{-}CF_3C_6H_4Cu$	$-25°$, 1 hr (THF)	$CH_3CH=CHCH_2C_6H_4CF_3\text{-}m$ (57)	155
[structure: $BrCH_2$, CH_3, $C=C$, CO_2CH_3, H]	$NCCH_2Cu$ (0.4 M, 3 eq)		$NCCH_2CH_2C(CH_3)=CHCO_2CH_3$ (46 Z, 21 E)	142
C_6 $n\text{-}C_3H_7CH(Cl)C\equiv CH$	$(CH_3)_2CuLi$ $(C_2H_5)_2CuLi$ $(n\text{-}C_4H_9)_2CuLi$	$-5°$ $-30°$ $-60°$	$n\text{-}C_5H_7CH=C=CHR$ (I) I, R = CH_3 (62) I, R = C_2H_5 (66) I, R = $C_4H_9\text{-}n$ (71)	175b
$C_2H_5C(CH_3)ClC\equiv CH$	$(C_2H_5)_2CuLi$ $(n\text{-}C_4H_9)_2CuLi$	$-30°$ $-60°$	$C_2H_5C(CH_3)=C=CHR$ (I) I, R = CH_3 (66) I, R = C_2H_5 (63) I, R = $C_4H_9\text{-}n$ (64)	175b
3-Bromocyclohexene	$(CH_3)_2CuLi$ (0.5 M, 5 eq) $(n\text{-}C_4H_9)_2CuLi$ (0.25 M, 5 eq)	$0°$, 4 hr $-78°$, 1 hr	3-Methylcyclohexene (75), cyclohexene (10) 3-n-Butylcyclohexene (60)	54 148
	$C_2H_5O_2CCH_2Cu$ (2 eq)	$-30°$, 1 hr (THF)	[ring structure] $CH_2CO_2C_2H_5$ (69)	143
3-Chlorocyclohexene	$(CH_3)_2CuLi$ (0.2 M, 5 eq)	$0°$, 24 hr	3-Methylcyclohexene (65)	46
C_7 Benzotrichloride	$(CH_3)_2CuLi$ (0.25 M, 15 eq)	$-78°$, 1 hr	$C_6H_5(CH_3)C=C(Cl)C_6H_5$ (26), $C_6H_5C(CH_3)_2C(=CH_2)C_6H_5$ (36)	159a
Benzal chloride	$(CH_3)_2CuLi$ (0.25 M, 10 eq)	$0°$, 1 hr	Isopropylbenzene (40), 2,3-diphenylbutane (40; 1:1 d,1:meso)	172
Benzyl bromide	CH_3Cu (10 eq)	$0°$, 23 hr, then $25°$, 20 hr (THF)	Ethylbenzene (85)	46

Substrate	Reagent	Conditions	Product (yield)	Refs.
	$(CH_3)_2CuLi$ (0.4 M, 5 eq)	0°, 6.5 hr	" (89)	54
	$C_2H_5OCOCH_2Cu$	$-30°$, 1 hr (THF) [HMPA, $(C_2H_5O)_3P$]	$C_6H_5CH_2CH_2CO_2C_2H_5$ (62)	143
	$Z\text{-}C_2H_5(CH_3)C=CHCu$	—	$Z\text{-}C_6H_5CH_2CH_2CH=C(CH_3)C_2H_5$ (85)	87
	$m\text{-}CF_3C_6H_4Cu$	—	$C_6H_5CH_2C_6H_4CF_3\text{-}m$ (51)	155
	C_6F_5Cu		$C_6H_5CH_2C_6F_5$ (40)	103
	$(CH_3)_3SiCH_2Cu$	$-78°$, 2.5 hr (THF)	$C_6H_5CH_2CH_2Si(CH_3)_3$ (—)	271
	$(t\text{-}C_4H_9OCuC_4H_9\text{-}sec)Li$ (0.25 M, 3.0 eq)		$C_6H_5CH_2CH_2C_6H_5$ (>90)	100
	$(t\text{-}C_4H_9OCuC_4H_9\text{-}t)Li$ (0.25 M, 5.0 eq)	$-50°$, 4 hr (THF)	$C_6H_5CH_2C_4H_9\text{-}t$ (88)	100
Benzyl chloride	$(CH_3)_2CuLi$ (0.25 M, 5 eq)	0°, 1 hr	Ethylbenzene (80)	172
	$(p\text{-}CH_3OC_6H_4)_2CuLi$ (5 eq)	$-5°$, 25 hr (THF)	$C_6H_5CH_2C_6H_4OCH_3\text{-}p$ (50)	115
3,5-Dimethoxybenzyl chloride	$(n\text{-}C_4H_9)_2CuLi$ (1 eq)	0°, 4 hr	3,5-Dimethoxyphenylpentane (55)	221
3-Bromo-1-methylcyclohexene	$[CH_2=C(CH_3)]_2CuLi$ (5 eq)	0°, 6 hr (THF)	3-Isopropenyl-1-methylcyclohexene (75)	158
[structure: methylenecyclohexane-Br]	$i\text{-}C_3H_7SCH=CHCH_2Cu$ (2 eq)	$-78°$	[structure with $CH_2CH=CHSC_3H_{7}\text{-}i$] (89)	272b
[structure: THPO–chain–Br, $=CH_2$]	$(CH_3)_2CuLi$	—	[structure: THPO–chain, CH_2] (—)	216
[structure: dioxolane–THPO–chain–Br, CH_2]	$\left[(CH_3)_2NCH_2CH_2O\!-\!\overset{CH_2}{}\right]_2CuLi$ (0.2 M, 1.1 eq)	$-30°$, 6 hr	[structure: dioxolane–THPO–chain, $CH_2\,N(CH_3)_2$] (80–90)	146

Note: References 236–320 are on pp. 398–400.

TABLE III. Organocopper Coupling with Halide Substrates (*Continued*)

Halide Substrate	Organocopper Reagent	Reaction Conditions (Solvent)	Product(s) and Yield(s) (%)	Refs.
		D. Benzylic, Allylic, and Propargylic Halides (Continued)		
C₈ 1,2-Bis(bromomethyl)-benzene	$(CH_3)_2CuLi$ (0.25 M, 10 eq)	−78°, 3 hr	1,2-Diethylbenzene (87)	159
1,2-Bis(chloromethyl)-benzene	$(CH_3)_2CuLi$ (0.25 M, 10 eq)	0°, 1 hr	'' (77)	172
(1-Chloroethyl)benzene	$(CH_3)_2CuLi$ (0.25 M, 5 eq)	23°, 1 hr	Isopropylbenzene (40), 2,3-diphenylbutane (40; 1:1 d, 1:meso)	172
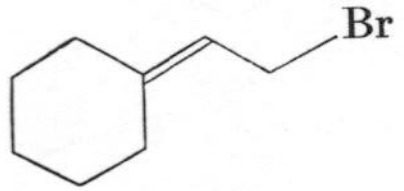	i-C_3H_7SCH=$CHCH_2Cu$ (2 eq)	−78°	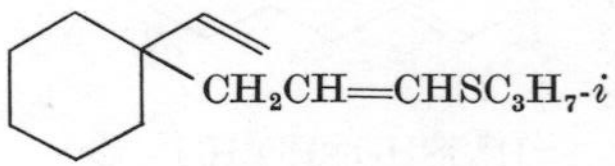 CH_2CH=$CHSC_3H_7$-i (87)	272b
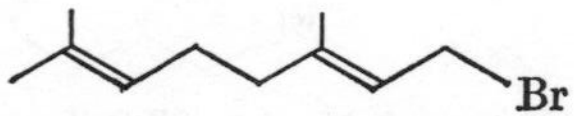 C₁₀	$C_2H_5O_2CCH_2Cu$ (2 eq)	−30°, 1 hr (THF)	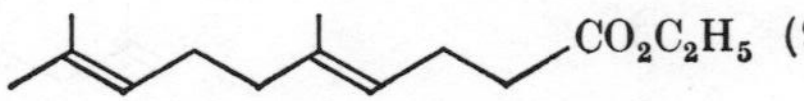 $CO_2C_2H_5$ (94)	143
	$C_6H_5SCH_2Cu$ (1 eq)	−20°, 12 hr (THF)	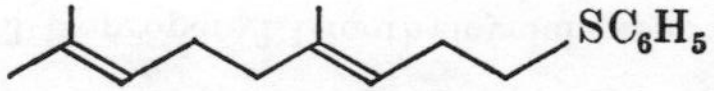 SC_6H_5 (76)	141
	$NCCH_2Cu$ (0.4 M, 3 eq)	−25°, 1 hr (THF)	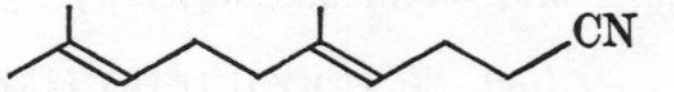 CN (92)	142
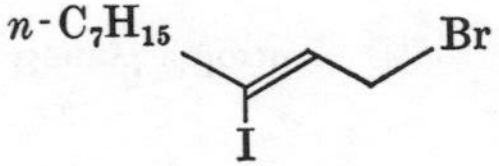 n-C_7H_{15} I	$NCCH_2Cu$ (0.4 M, 3 eq)	−25°, 1 hr (THF)	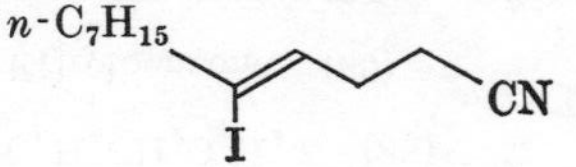 n-C_7H_{15} CN (87)	142
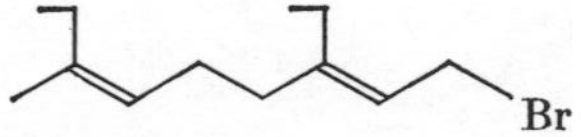 C₁₂ Br	$C_6H_5SCH_2Cu$ (0.25 M, 1.3 eq)	−55°, 1 hr, then −20°, 13 hr	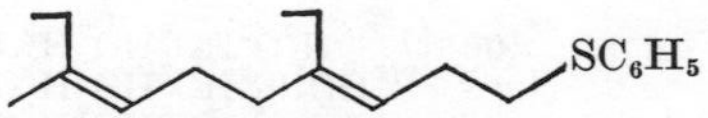 SC_6H_5 (92)	144
C₁₃ $(C_6H_5)_2CCl_2$	$(CH_3)_2CuLi$ (0.25 M, 10 eq)	0°, 1 hr	Tetraphenylethylene (81)	172
$(C_6H_5)_2CHCl$	m-$CF_3C_6H_4Cu$	—	$(C_6H_5)_2CHC_6H_4CF_3$-m (45)	155
	C_6F_5Cu	—	$(C_6H_5)_2CHC_6F_5$ (I, 85)	155
	$C_6F_5Cu·MgBr_2$	—	I (53)	

346

Substrate	Reagent	Conditions	Product (yield %)	Ref.
C_{15} [structure: tetraene chain bearing two Cl and an $OC(C_6H_5)_3$ group]	$(CH_3)_2CuLi$ (5 eq)	$-5°$, 1 hr	$C_2H_5(CH_3)C{=}CHCH_2CH_2C(C_2H_5){=}CHCH_2{-}\!(C_6H_5)_3COCH_2CH{=}C(CH_3)CH_2$ (—)	273

E. α-Halocarbonyl Substrates

Substrate	Reagent	Conditions	Product (yield %)	Ref.
C_2 Ethyl 2-bromoacetate	$[(CH_3)_2C{=}CH]_2CuLi$ (5 eq)	$-5°$, 4 hr (THF)	$(CH_3)_2C{=}CHCH_2CO_2C_2H_5$ (65)	158
C_5 $CH_3CHBrCOCHBrCH_3$	$(C_6H_5)_2CuLi$ (5 eq)	—	Ethyl phenylacetate (60)	115
	$(i\text{-}C_3H_7)_2CuLi$	—	$(CH_3)_2CHCOCH(CH_3)_2$ (22)	111
	$(CH_3)_2CuLi$	—	$CH_3CH(C_3H_7\text{-}i)COCH(C_3H_7\text{-}i)CH_3$ (8)	111
1,3-Dibromo-3-methyl-2-butanone	$(t\text{-}C_4H_9OCuR)Li$ (0.25 M, 5 eq); $R = C_4H_9\text{-}n$; $R = C_4H_9\text{-}sec$; $R = C_4H_9\text{-}t$	$-78°$, 0.5 hr (THF)	$(CH_3)_2CHCOCH_2R$ (I), $(CH_3)_2C(R)COCH_3$ (II); $R = C_4H_9\text{-}n$, I (55), II (25); $R = C_4H_9\text{-}sec$, I (42), II (8); $R = C_4H_9\text{-}t$, I (35)	274
$n\text{-}C_4H_9CCl_2CO_2C_3H_7\text{-}i$	$(CH_3)_2CuLi$	$20°$	$n\text{-}C_4H_9C(CH_3)(R)CO_2C_3H_7\text{-}i$ (I); I, R = Cl (32); I, R = H (30)	274b
C_6 2,6-Dibromocyclo-hexanone (I)			2-R-cyclohexanone (II)	
cis-I	$(CH_3)_2CuLi$ (0.25 M, 5 eq)	$-78°$, 0.5 hr	II, R = CH_3 (65)	99
trans-I	$(CH_3)_2CuLi$ (0.25 M, 5 eq)	$-78°$, 0.5 hr	II, R = CH_3 (98)	99
	$(n\text{-}C_4H_9)_2CuLi$ (0.25 M, 5 eq)	$-78°$, 0.5 hr	II, R = $C_4H_9\text{-}n$ (81)	274a
	$(sec\text{-}C_4H_9)_2CuLi$ (0.25 M, 5 eq)	$-78°$, 0.5 hr	II, R = $C_4H_9\text{-}sec$ (43)	274a
	$(t\text{-}C_4H_9CuCH_3)Li$ (0.25 M, 5 eq)	$-78°$, 0.5 hr	II, R = $C_4H_9\text{-}t$ (27)	274a
	$(t\text{-}C_4H_9OCuC_4H_9\text{-}n)Li$ (0.25 M, 5 eq)	$-78°$, 0.5 hr (THF)	II, R = $C_4H_9\text{-}n$ (93)	274a

Note: References 236–320 are on pp. 398–400.

TABLE III. ORGANOCOPPER COUPLING WITH HALIDE SUBSTRATES (*Continued*)

Halide Substrate	Organocopper Reagent	Reaction Conditions (Solvent)	Product(s) and Yield(s) (%)	Refs.
		E. α-Halocarbonyl Substrates (Continued)		
C$_6$ (*contd.*)	(t-C$_4$H$_9$OCuC$_4$H$_9$-*sec*)Li (0.25 M, 5 eq)	−78°, 0.5 hr (THF)	II, R = C$_4$H$_9$-*sec* (88)	274a
	(t-C$_4$H$_9$OCuC$_4$H$_9$-*t*)Li (0.25 M, 5 eq)	−78°, 0.5 hr (THF)	II, R = C$_4$H$_9$-*t* (78)	274a
	(C$_6$H$_5$SCuC$_4$H$_9$-*t*)Li (0.25 M, 5 eq)	−78°, 5 hr (THF)	II, R = C$_4$H$_9$-*t* (~75)	101

(I)′   (II)

Halide Substrate	Organocopper Reagent	Reaction Conditions (Solvent)	Product(s) and Yield(s) (%)	Refs.
C$_7$	(t-C$_4$H$_9$OCuR)Li (0.25 M, 5 eq) R = C$_4$H$_9$-*n* R = C$_4$H$_9$-*sec* R = C$_4$H$_9$-*t*	−78°, 0.5 hr	R = C$_4$H$_9$-*n*, I (48), II (16) R = C$_4$H$_9$-*sec*, I (60), II (8) R = C$_4$H$_9$-*t*, I (31), II (2)	99
2-Bromo-5-methylcyclo-hexanone	[CH$_2$=C(CH$_3$)]$_2$CuLi (5 eq)	−5°, 6 hr (THF)	2-Isopropenyl-5-methylcyclohexanone (58)	114
2-Chloro-5-methylcyclo-hexanone	[CH$_2$=C(CH$_3$)]$_2$CuLi (5 eq)	−5°, 4 hr (THF)	2-Isopropenyl-5-methylcyclohexanone (60)	158, 114
(CH$_3$)$_2$CBrCOCBr(CH$_3$)$_2$	(CH$_3$)$_2$CuLi (CH$_3$)$_2$CuLi (5 eq)	— −78°, 0.5 hr	(CH$_3$)$_3$CCOC(CH$_3$)$_3$ (33) (54)	111 99
n-C$_4$H$_9$CH(Br)COCH$_2$Br	(t-C$_4$H$_9$OCuR)Li (0.25 M, 5 eq) R = C$_4$H$_9$-*n* R = C$_4$H$_9$-*sec* R = C$_4$H$_9$-*t*	−78°, 0.5 hr (THF)	n-C$_4$H$_9$CH$_2$COCH$_2$R (I), n-C$_4$H$_9$CH(R)COCH$_3$ (II) R = C$_4$H$_9$-*n*, I (37), II (11) R = C$_4$H$_9$-*sec*, I (44), II (8) R = C$_4$H$_9$-*t*, I (53), II (6)	99
i-C$_3$H$_7$COCBr(CH$_3$)$_2$	(CH$_3$)$_2$CuLi (i-C$_3$H$_7$)$_2$CuLi (CH$_3$)$_2$CuLi (0.6 M, 3 eq)	−78°, 0.5 hr — 0°, 4 hr	R = CH$_3$, I (53), II (40) i-C$_3$H$_7$COC(CH$_3$)$_3$ (90) i-C$_3$H$_7$COC(CH$_3$)$_2$C$_3$H$_7$-*i* (33)	111 111 46
C$_8$ α-Bromoacetophenone			Acetophenone (95)	
	(t-C$_4$H$_9$OCuC$_4$H$_9$-*n*)Li (0.25 M, 2.5 eq)	−78°, 1 hr (THF)	C$_6$H$_5$COCH$_2$C$_4$H$_9$-*n* (56)	101

	Substrate	Reagent	Conditions	Product(s) (% yield)	Refs.
	2,8-Dibromocyclooctanone	$(CH_3)_2CuLi$ (0.25 M, 5 eq)	−78°, 0.5 hr	2-Methylcyclooctanone (80)	99
C_9	$n\text{-}C_3H_7CH(Br)COCH(Br)C_3H_7\text{-}n$	$(t\text{-}C_4H_9OCuR)Li$ (0.25 M, 5 eq); R = $C_4H_9\text{-}n$; R = $C_4H_9\text{-}sec$; R = $C_4H_9\text{-}t$; $(CH_3)_2CuLi$	−78°, 0.5 hr (THF); −78°, 0.5 hr, then CH_3I, 25°	$n\text{-}C_3H_7CH_2COCH(R)C_3H_7\text{-}n$ (I); I, R = $C_4H_9\text{-}n$ (75); I, R = $C_4H_9\text{-}sec$ (67); I, R = $C_4H_9\text{-}t$ (60); $n\text{-}C_3H_7CH(CH_3)COCH(CH_3)C_3H_7\text{-}n$ (62)	99
C_9	$i\text{-}C_3H_7CHBrCOCHBrC_3H_7\text{-}i$; $i\text{-}C_3H_7CH_2COCHBrC_3H_7\text{-}i$	$(i\text{-}C_3H_7)_2CuLi$; $(CH_3)_2CuLi$; $(i\text{-}C_3H_7)_2CuLi$; $(t\text{-}C_4H_9)_2CuLi$	—	$(i\text{-}C_3H_7)_2CHCOCH(C_3H_7\text{-}i)_2$ (12); $i\text{-}C_3H_7CH_2COCH(R)C_3H_7\text{-}i$ (I); I, R = CH_3 (63); I, R = $C_3H_7\text{-}i$ (45); I, R = $C_4H_9\text{-}t$ (16)	111, 111, 149
	$t\text{-}C_4H_9COCHBrC_3H_7\text{-}i$	$(CH_3)_2CuLi$; $(i\text{-}C_3H_7)_2CuLi$; $(t\text{-}C_4H_9)_2CuLi$	—	$t\text{-}C_4H_9COCH(R)C_3H_7\text{-}i$ (I); I, R = CH_3 (—); I, R = $C_3H_7\text{-}i$ (—); I, R = $C_4H_9\text{-}t$ (—)	149
C_{10}	$t\text{-}C_4H_9COCHBrC_4H_9\text{-}t$	$(CH_3)_2CuLi$; $(i\text{-}C_3H_7)_2CuLi$; $(t\text{-}C_4H_9)_2CuLi$	—	$t\text{-}C_4H_9COCH(R)C_4H_9\text{-}t$ (I); I, R = CH_3 (—); I, R = $C_3H_7\text{-}i$ (—); I, R = $C_4H_9\text{-}t$ (—)	149
C_{11}	$t\text{-}C_5H_{11}COCHBrC_3H_7\text{-}i$	$(CH_3)_2CuLi$; $(i\text{-}C_3H_7)_2CuLi$; $(t\text{-}C_4H_9)_2CuLi$	—	$t\text{-}C_5H_{11}COCH(R)C_3H_7\text{-}i$ (I); I, R = CH_3 (—); I, R = $C_3H_7\text{-}i$ (—); I, R = $C_4H_9\text{-}t$ (—)	149
C_{11}	$t\text{-}C_4H_9CH_2COCHBrC_4H_9\text{-}t$	$(CH_3)_2CuLi$; $(i\text{-}C_3H_7)_2CuLi$; $(t\text{-}C_4H_9)_2CuLi$	—	$t\text{-}C_4H_9CH_2COCH(R)C_4H_9\text{-}t$ (I); I, R = CH_3 (—); I, R = $C_3H_7\text{-}i$ (—); I, R = $C_4H_9\text{-}t$ (—)	149
C_{12}	cis-2,12-Dibromocyclododecanone	$(CH_3)_2CuLi$ (0.25 M, 5 eq)	−78°, 0.5 hr	2-Methylcyclododecanone (97)	99
C_{13}	$(C_2H_5)_3CCOCHBrC_3H_7\text{-}i$; 2,12-Dibromo-2-methylcyclododecanone	$(i\text{-}C_3H_7)_2CuLi$; $(CH_3)_2CuLi$	−78°, 0.5 hr, then CH_3I, 25°; −78°	2,12-Dimethylcyclododecanone (94); $(C_2H_5)_3CCOCH(C_3H_7\text{-}i)_2$ (—); 2,2-Dimethylcyclododecanone (54)	149, 274a
C_{15}	$C_6H_5CHBrCOCHBrC_6H_5$	$(CH_3)_2CuLi$ (0.25 M, 5 eq)	−78°, 0.5 hr	$C_6H_5CH_2COCH(CH_3)C_6H_5$ (72)	99

Note: References 236–320 are on pp. 398–400.

TABLE III. ORGANOCOPPER COUPLING WITH HALIDE SUBSTRATES (*Continued*)

Halide Substrate	Organocopper Reagent	Reaction Conditions (Solvent)	Product(s) and Yield(s) (%)	Refs.
		F. Acyl Halides		
C₂ Oxalyl chloride	C_6Cl_5Cu (2 eq)	0°, several hr	$C_6Cl_5COCOC_6Cl_5$ (71)	76
	C_6F_5Cu (2 eq)	0°, several hr	$C_6F_5COCOC_6F_5$ (71)	76
	(tetrachloropyridyl)Cu (2 eq)	0°, several hr	(tetrachloropyridyl–CO)₂ (57)	76
Chloroacetyl chloride	C_6Cl_5Cu	−10°, 6 hr	$ClCH_2COC_6Cl_5$ (78)	275
	C_6F_5Cu	−10°, 6 hr	$ClCH_2COC_6F_5$ (48)	275
Acetyl bromide	C_6Cl_5Cu (<1 eq)	0°, several hr (ether-hexane)	$CH_3COC_6Cl_5$ (65)	70
	C_6F_5Cu (<1 eq)	0°, several hr (ether-hexane)	$CH_3COC_6F_5$ (83)	70
	(tetrachloropyridyl)Cu	0°, several hr (THF-hexane)	(tetrachloropyridyl–COCH₃) (64)	70
Acetyl chloride	C_6H_5Cu (0.3 eq)	−18°	Acetophenone (66)	31
	C_6H_5Cu (10 eq)	25°, 1 hr	" (74)	46
	$(C_6H_5)_2CuLi$ (1 eq)	−5°	" (55)	176
	$p\text{-}CH_3OC_6H_4Cu$		*p*-Methoxyacetophenone (52)	31
	$2,4,6\text{-}(CH_3O)_3C_6H_2Cu$	24 hr	$CH_3COC_6H_2(OCH_3)_3\text{-}2,4,6$ (55)	119
	$4\text{-}(CH_3)_3SiC_6H_4Cu$		$4\text{-}(CH_3)_3SiC_6H_4COCH_3$ (—)	276
	C_6F_5Cu (1 eq)	−10 to −30°, 3-6 hr (THF)	$CH_3COC_6F_5$ (82)	151, 277a, 70
	$o\text{-}CF_3C_6H_4Cu\cdot MgBr_2$	—	$CH_3COC_6H_4CF_3\text{-}o$ (60)	155
	$m\text{-}CF_3C_6H_4Cu$	—	$CH_3COC_6H_4CF_3\text{-}m$ (Good)	155

Substrate	Reagent	Conditions	Product (Yield %)	Refs.
	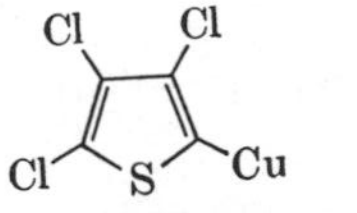 thiophene–Cu structure (Cl, Cl) (0.5 M, 0.9 eq)	−78°, 5 min, then 25°, 5 hr (THF)	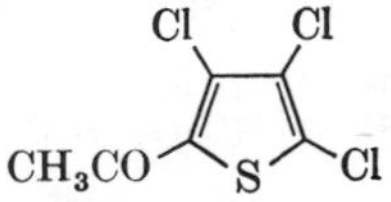 CH$_3$CO–thiophene(Cl,Cl)–Cl (76)	140
	Cu–thiophene(Cl,Cl)–Cu	—	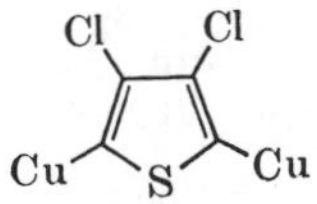 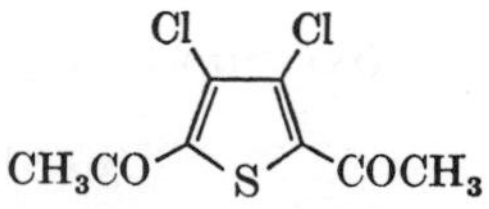 CH$_3$CO–thiophene(Cl,Cl)–COCH$_3$ (60–64)	71
C$_3$	n-C$_4$H$_9$C≡CCu	(HMPA)	CH$_3$COC≡CC$_4$H$_9$-n (—)	277b
	(C$_6$H$_5$)$_3$SiCu	—	CH$_3$COSi(C$_6$H$_5$)$_3$ (74)	278
Malonyl chloride	C$_6$H$_5$	—	C$_6$F$_5$COCH$_2$COC$_6$F$_5$ (—)	279
α-Chloropropionyl chloride	(C$_6$H$_5$)$_2$CuLi (1 eq)	−5°	α-Chloropropiophenone (15)	176
C$_4$	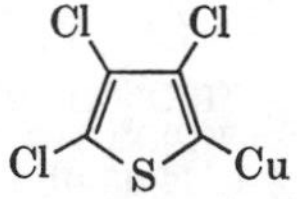 Cu–thiophene(Cl,Cl)–Cu (0.25 M, 0.9 eq)	25°, 18 hr (1:1 THF:ether)	E-RCOCH=CHCOR (I) I, R = trichloro-2-thienyl (55)	140
E-ClCOCH=CHCOCl	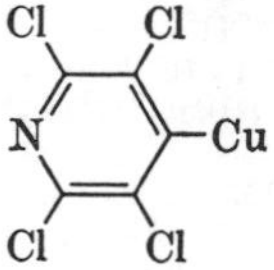 tetrachloropyridyl–Cu (2 eq)	0°, several hr	I, R = tetrachloro-4-pyridyl (30)	76
	C$_6$Cl$_5$Cu (2 eq)	0°, several hr	I, R = C$_6$Cl$_5$ (79)	76
	C$_6$F$_5$Cu (2 eq)	0°, several hr	I, R = C$_6$F$_5$ (75)	76
ClCO(CH$_2$)$_2$COCl	C$_6$Cl$_5$Cu (2 eq)	0°, several hr	RCO(CH$_2$)$_2$COR (I) I, R = C$_6$Cl$_5$ (45)	76
	C$_6$F$_5$Cu (2 eq)	0°, several hr	I, R = C$_6$F$_5$ (43)	76
	C$_6$F$_5$Cu (2 eq)	−10 to −30°,	I, R = C$_6$F$_5$ (71)	151

Note: References 236–320 are on pp. 398–400.

TABLE III. Organocopper Coupling with Halide Substrates (Continued)

Halide Substrate	Organocopper Reagent	Reaction Conditions (Solvent)	Product(s) and Yield(s) (%)	Refs.
F. Acyl Halides (Continued)				
C_4 (contd.)				
	[tetrachloropyridyl]–Cu (2 eq)	3–6 hr (THF)	I, R = tetrachloro-4-pyridyl (59)	76
$CH_3O_2C(CH_2)_2COCl$	$(n\text{-}C_4H_9)_2CuLi$ (0.25 M, 3 eq)	0°, several hr	$CH_3O_2C(CH_2)_2COR$ (I)	234
	$(t\text{-}C_4H_9OCuR)Li$ (0.25 M, 1.2 eq) R = C_4H_9-sec	−78°, 15 min (THF)	I, R = C_4H_9-n (84)	100
	R = C_4H_9-t $(C_6H_5SCuC_4H_9\text{-}t)Li$ (0.25 M, 1.2 eq)	−78°, 15 min (THF)	I, R = C_4H_9-sec (66) I, R = C_4H_9-t (61) I, R = C_4H_9-t (66)	101
Isobutyryl chloride	$(CH_3)_2CuLi$ (1 eq) $(C_6H_5)_2CuLi$ (1 eq)	−5° −5°	$i\text{-}C_3H_7COCH_3$ (45) Isobutyrophenone (67)	176 176
C_5				
$ClCO(CF_2)_3COCl$	[tetrachlorothiophene]–Cu	25°, 12 hr (1:1 THF:ether)	[tetrachlorothiophene]–$CO(CF_2)_3CO$–[tetrachlorothiophene] (22)	140
[cyclopropane diacyl chloride]	$(CH_3)_2CuLi$ (2.2 eq.)	−60°, 15 min	[cyclopropane di-$COCH_3$] (60)	279a
$t\text{-}C_4H_9COCl$	$(CH_3)_2CuLi$ (0.25 M, 0.9 eq) $(CH_3)_2CuLi$ (0.25 M, 3 eq)	−78°, 15 min	$t\text{-}C_4H_9COCH_3$ (84)	28
	$i\text{-}C_3H_7Cu$ $(n\text{-}C_4H_9)_2CuLi$ (0.25 M, 3 eq)	−78° −78°, 15 min	$t\text{-}C_4H_9COC_3H_7\text{-}i$ (94) $t\text{-}C_4H_9COC_4H_9\text{-}n$ (90)	281 280
$i\text{-}C_4H_9COCl$	$t\text{-}C_4H_9Cu$ $[(CH_3)_2C{=}CH]_2CuLi$ (5 eq)	−5° −5°, 4 hr (THF)	$t\text{-}C_4H_9COC_4H_9\text{-}t$ (88) $i\text{-}C_4H_9COCH{=}C(CH_3)_2$ (70)	281 158
$(CH_3)_3SiCH(n\text{-}C_3H_7)COCl$	$(n\text{-}C_4H_9)_2CuLi$ (3 eq.)	−78°	$(CH_3)_3SiCH(n\text{-}C_3H_7)COC_3H_7\text{-}n$ (60)	281a

	Reactant	Reagent	Conditions	Product(s) (% Yield)	Refs.
C₆	ClCO(CH₂)₄COCl	(CH₃)₂CuLi (0.25 M, 3 eq)	−78°, 15 min	ROC(CH₂)₄COR (I) I, R = CH₃ (92)	96
		(n-C₄H₉)₂CuLi (0.25 M, 3 eq)	−78°, 15 min	I, R = C₄H₉-n (90)	96
		C₆F₅Cu (2 eq)	0°	I, R = C₆F₅ (76)	76
		C₆Cl₅Cu (2 eq)	0°	I, R = C₆Cl₅ (78)	76
			0°	I, R = tetrachloro-4-pyridyl (66)	76
	n-C₄H₉O₂C(CH₂)₄COCl	(CH₃)₂CuLi (0.25 M, 3 eq)	−78°, 15 min	n-C₄H₉O₂C(CH₂)₄COR (I) I, R = CH₃ (83)	96
		(n-C₄H₉)₂CuLi (0.25 M, 3 eq)	−78°, 15 min	I, R = C₄H₉-n (93)	100
		(t-C₄H₉OCuR)Li (0.25 M, 1.3 eq) R = C₄H₉-sec R = C₄H₉-t	−78°, 15 min (THF)	I, R = C₄H₉-sec (89) I, R = C₄H₉-t (73)	
	n-C₅H₁₁COCl	(CH₃)₂CuLi (0.25 M, 3 eq)	−78°, 15 min	2-Heptanone (81)	280
		(n-C₄H₉)₂CuLi (0.25 M, 3 eq)	−78°, 15 min	5-Decanone (79)	280
C₇	p-Nitrobenzoyl chloride	(CH₃)₂CuLi (0.25 M, 3 eq)	−78°, 15 min	p-Nitroacetophenone (50)	280
		C₆F₅Cu	25°	p-O₂NC₆H₄COC₆F₅ (83)	155
			48 hr	p-O₂NC₆H₄COC₆H₂(OCH₃)₃-2,4,6 (75)	117, 119

Note: References 236–320 are on pp. 398–400.

353

TABLE III. ORGANOCOPPER COUPLING WITH HALIDE SUBSTRATES (Continued)

Halide Substrate	Organocopper Reagent	Reaction Conditions (Solvent)	Product(s) and Yield(s) (%)	Refs.
		F. Acyl Halides (Continued)		
C₇ (contd.)	(CH₃)₂N–[C₆H₄]–Cu·CuBr	48 hr	p-O₂NC₆H₄COC₆H₄N(CH₃)₂-p (60)	119
	[o-C₆H₄(CH₂N(CH₃)₂)]–Cu	48 hr	p-O₂NC₆H₄COC₆H₄CH₂N(CH₃)₂-o (75)	117, 119
p-Iodobenzoyl chloride	m-CF₃C₆H₄Cu	—	p-O₂NC₆H₄COC₆H₄CF₃-m (72)	103
	m-CF₃C₆H₄Cu·MgBr₂		p-O₂NC₆H₄COC₆H₄CF₃-m (83)	155
	(CH₃)₂CuLi (0.25 M, 3 eq)	−78°, 15 min	p-IC₆H₄COR (I)	96
			I, R = CH₃ (98)	
	(n-C₄H₉)₂CuLi (0.25 M, 3 eq)	−78°, 15 min	I, R = C₄H₉-n (85)	
Benzoyl chloride	HCu·P(C₄H₉-n)₃	25°, 2 hr	Benzaldehyde (50)	147
	CH₃Cu (1 eq)	25°, 20 min (THF)	Acetophenone (53)	46
	CH₃Cu·P(C₄H₉-n)₃		" (60)	35, 18
	(CH₃)₂CuLi (0.25 M, 3 eq)	−78°, 15 min	" (94)	101, 35, 176
	C₂H₅Cu (1 eq)	−78°	Propiophenone (24)	31
	i-C₃H₇Cu	−25°, 36 hr	Isobutyrophenone (41)	282
	[cyclopentadienyl]–CuP(C₄H₉-n)₃	0°, 23 hr	[cyclopentadiene structure with C₆H₅ and O₂CC₆H₅] (67)	283
	C₆H₅Cu (0.1 M, 0.7 eq)	12 hr	Benzophenone (55)	31
	(C₆H₅)₂CuLi (1 eq)	−5°	" (59)	176, 115
	[C₆H₅C₄H₈]CuMgBr	0°		224

354

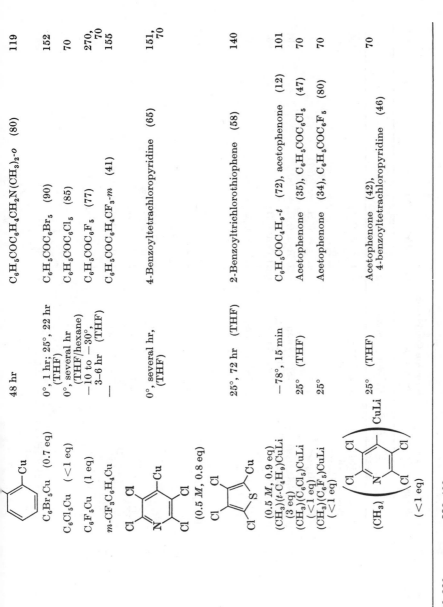

(o-CH$_3$C$_6$H$_4$Cu)	48 hr	C$_6$H$_5$COC$_6$H$_4$CH$_2$N(CH$_3$)$_2$-o (80)	119
C$_6$Br$_5$Cu (0.7 eq)	0°, 1 hr; 25°, 22 hr (THF)	C$_6$H$_5$COC$_6$Br$_5$ (90)	152
C$_6$Cl$_5$Cu (<1 eq)	0°, several hr (THF/hexane)	C$_6$H$_5$COC$_6$Cl$_5$ (85)	70
C$_6$F$_5$Cu (1 eq)	−10 to −30°, 3–6 hr (THF)	C$_6$H$_5$COC$_6$F$_5$ (77)	270, 70
m-CF$_3$C$_6$H$_4$Cu	—	C$_6$H$_5$COC$_6$H$_4$CF$_3$-m (41)	155
(tetrachloropyridyl-Cu) (0.5 M, 0.8 eq)	0°, several hr, (THF)	4-Benzoyltetrachloropyridine (65)	151, 70
(trichlorothienyl-Cu)	25°, 72 hr (THF)	2-Benzoyltrichlorothiophene (58)	140
(CH$_3$)(t-C$_4$H$_9$)CuLi (0.5 M, 0.9 eq) (3 eq)	−78°, 15 min	C$_6$H$_5$COC$_4$H$_9$-t (72), acetophenone (12)	101
(CH$_3$)(C$_6$Cl$_5$)CuLi (<1 eq)	25° (THF)	Acetophenone (35), C$_6$H$_5$COC$_6$Cl$_5$ (47)	70
(CH$_3$)(C$_6$F$_5$)CuLi (<1 eq)	25°	Acetophenone (34), C$_6$H$_5$COC$_6$F$_5$ (80)	70
(tetrachloropyridyl)(CH$_3$)CuLi (<1 eq)	25° (THF)	Acetophenone (42), 4-benzoyltetrachloropyridine (46)	70

Note: References 236–320 are on pp. 398–400.

355

TABLE III. Organocopper Coupling with Halide Substrates (Continued)

F. Acyl Halides (Continued)

Halide Substrate	Organocopper Reagent	Reaction Conditions (Solvent)	Product(s) and Yield(s) (%)	Refs.
C_7 (contd.)	$(t\text{-}C_4H_9OCuR)Li$ (0.25 M, 1.2 eq) R = C_3H_7-iso R = C_4H_9-n R = C_4H_9-sec R = C_4H_9-t	$-78°$, 15 min (THF)	C_6H_5COR (I) I, R = C_3H_7-iso (80) I, R = C_4H_9-n (66) I, R = C_4H_9-sec (87) I, R = C_4H_9-t (82)	100
	$(C_6H_5SCuC_4H_9\text{-}t)$ (0.25 M, 1.10 eq)	$-78°$, 15 min (THF)	I, R = C_4H_9-t (85)	101
Benzoyl fluoride	$(CH_3)_2CuLi$ (0.25 M, 3 eq)	$-78°$, 15 min	C_6H_5COR (I) I, R = CH_3 (72)	96
	$(n\text{-}C_4H_9)_2CuLi$ (0.25 M, 3 eq)	$-78°$, 15 min	I, R = C_4H_9-n (87)	96
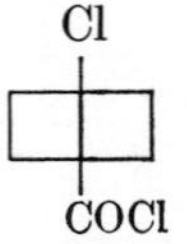 Cl—☐—COCl	$(CH_3)_2CuLi$	—	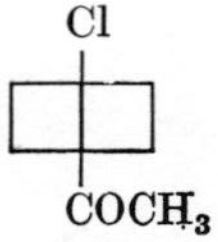 Cl—☐—$COCH_3$ (—)	285
cyclo-$C_6H_{11}COCl$	$(CH_3)_2CuLi$ (0.25 M, 3 eq)	$-78°$, 15 min	cyclo-$C_6H_{11}COR$ (I) I, R = CH_3 (86)	280
	$(C_2H_5)_2CuLi$ (0.25 M, 3 eq)	$-78°$, 15 min	I, R = C_2H_5 (71)	280
	$(n\text{-}C_4H_9)_2CuLi$ (0.25 M, 3 eq)	$-78°$, 15 min	I, R = C_4H_9-n (80)	280
	$(C_6H_5CuI)MgBr$ (1 eq)	$0°$	I, R = C_6H_5 (—)	284
C_8 Phthaloyl chloride	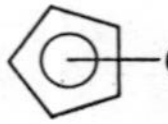 ⬡—$CuP(C_4H_9\text{-}n)_3$	$0°$, 2.5 hr	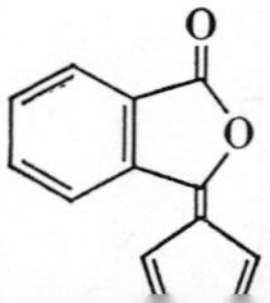 (II)	283

Acid Chloride	Reagent	Conditions	Product(s) (% Yield)	Refs.
	C_6F_5Cu (2 eq)	0°, several hr	I, R = C_6Cl_5 (61) I, R = C_6F_5 (65)	285b
(1-methylcyclohexane-1-carbonyl chloride, COCl)	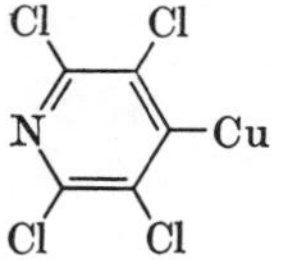 tetrachloro-4-pyridyl–Cu (2 eq)	0°, several hr	I, R = tetrachloro-4-pyridyl (62)	285b
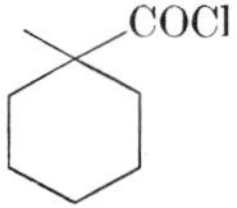	CH_3Cu (4 eq)	—	 (cyclohexyl)COCH$_3$ (65)	
	$(CH_3)_2CuLi$ (4 eq)	—	 (cyclohexyl)COCH$_3$ (65)	
$(i\text{-}C_3H_7)_2CHCOCl$	CH_3Cu	−78°	$(i\text{-}C_3H_7)_2CHCOR$, R = CH_3 (95)	281
	$C_2H_5Cu(MgI_2)$	−5°	R = C_2H_5 (84–97)	281, 286a
	$t\text{-}C_4H_9Cu$	−5°	R = $C_4H_9\text{-}t$ (80–86)	281
	$(C_2H_5)_3CCu$	−5°	R = $C(C_2H_5)_3$ (77)	281
$3,5\text{-}(CF_3)_2C_6H_3COCl$	$(CH_3)_2CuLi$ (0.25 M, 3 eq)	−78°, 15 min	$3,5\text{-}(CF_3)_2C_6H_3COCH_3$ (92)	280
	$(n\text{-}C_4H_9)_2CuLi$ (0.25 M, 3 eq)	−78°, 15 min	$3,5\text{-}(CF_3)_2C_6H_3COC_4H_9\text{-}n$ (75–80)	
C₉				
$C_6H_5C{\equiv}CCOCl$	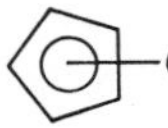 (cyclopentadienyl)$CuP(C_4H_9\text{-}n)_3$	−20°, 3 hr	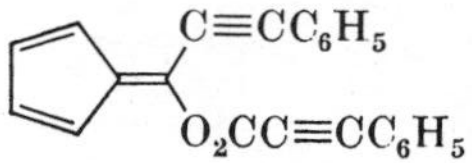 (fulvenyl) $C{\equiv}CC_6H_5$ / $O_2CC{\equiv}CC_6H_5$ (20)	283
$trans\text{-}C_6H_5CH{=}CHCOCl$	 (cyclopentadienyl)$CuP(C_4H_9\text{-}n)_3$	0°, 6.5 hr	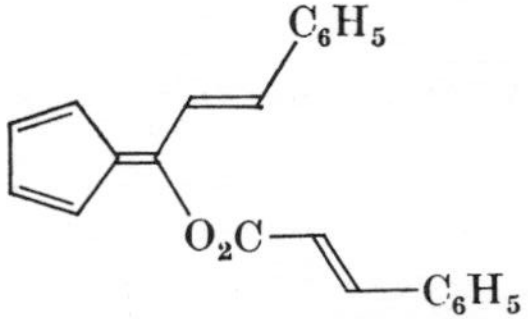 (fulvenyl) C_6H_5 / O_2C ... C_6H_5 (26)	283

Note: References 236–320 are on pp. 398–400.

TABLE III. Organocopper Coupling with Halide Substrates (*Continued*)

	Halide Substrate	Organocopper Reagent	Reaction Conditions (Solvent)	Product(s) and Yield(s) (%)	Refs.
			F. Acyl Halides (Continued)		
C_{10}	n-$C_8H_{17}COCl$	(t-$C_4H_9OCuC_3H_7$-i)Li	−78°, 15 min (3:1 THF/pentane)	n-$C_8H_{17}COC_3H_7$-i (95)	286b
	n-$C_4H_9CO(CH_2)_4COCl$	(CH₃)₂CuLi (0.25 M, 3 eq)	−78°, 15 min	n-$C_4H_9CO(CH_2)_4COR$ (I) I, R = CH₃ (95)	96
		(n-C_4H_9)₂CuLi (0.25 M, 3 eq)	−78°, 15 min	I, R = C_4H_9-n (83)	
C_{11}	t-C_4H_9——COCl			t-C_4H_9——COR (II)	67
	I-*cis*	(t-C_4H_9)₂CuMgX (0.6 eq)	0°	II-*cis*, R = C_4H_9-t (60)	
		(C₆H₅)₂CuLi (0.6 eq)	0°	II-*cis*, R = C₆H₅ (85)	
		(p-CH₃OC₆H₄)₂CuMgX (0.6 eq)	0°	II-*cis*, R = $C_6H_4OCH_3$-p (>50)	
		[p-(CH₃)₂NC₆H₄]₂CuLi (0.6 eq)	0°	II-*cis*, R = $C_6H_4N(CH_3)_2$-p (85)	
	I-*trans*	(C₆H₅)₂CuLi (0.6 eq)	0°	II-*trans*, R = C₆H₅ (62)	
		(p-CH₃OC₆H₄)₂CuMgX (0.6 eq)	0°	II-*trans*, R = $C_6H_4OCH_3$-p (55)	
		[p-(CH₃)₂NC₆H₄]₂CuLi (0.6 eq)	0°	II-*trans*, R = $C_6H_4N(CH_3)_2$-p (50)	
	Br(CH₂)₁₀COCl	(t-C_4H_9OCuR)Li (0.25 M, 1.2 eq) R = C_4H_9-sec R = C_4H_9-t	−78°, 15 min (THF)	Br(CH₂)₁₀COR (I) I, R = C_4H_9-sec (82) I, R = C_4H_9-t (78)	100
	I(CH₂)₁₀COCl	(CH₃)₂CuLi (0.25 M, 3 eq)	−78°, 15 min	I(CH₂)₁₀COR (I) I, R = CH₃ (91)	96
		(n-C_4H_9)₂CuLi (0.25 M, 3 eq)	−78°, 15 min	I, R = C_4H_9-n (93)	
C_{12}	NC(CH₂)₁₀COCl	(CH₃)₂CuLi (0.25 M, 3 eq)	−78°, 15 min	NC(CH₂)₁₀COR (I) I, R = CH₃ (80)	96
		(n-C_4H_9)₂CuLi (0.25 M...)	−78°, 15 min	I, R = C_4H_9-n (>95)	96

Substrate	Reagent	Conditions	Product(s) (%)	Refs.
(cut) cyclopentane: COCl, (CH₂)₆CO₂C₂H₅, CO₂CH₃, O₂CCH₃	(...M, 3 eq) / (C₂H₅)₂CuLi (0.25 M, 3 eq)	(cut) 15 min / −78°, 15 min	I, R = CH₃ (93) / I, R = C₂H₅ (80)	286, 176 / 280
	(n-C₄H₉)₂CuLi (0.25 M, 3 eq)	−78°, 15 min	I, R = C₄H₉-n (90)	280
	(C₆H₅)₂CuLi (1 eq)	−5°	I, R = C₆H₅ (88)	176
	(CH₃)₂CuLi	−78°	cyclopentane: COCH₃, (CH₂)₆CO₂C₂H₅, CO₂CH₃, O₂CCH₃ (—)	177
C₁₉ (C₆H₅)₂C=CHC(CH₃)-(C₂H₅)COCl	(CH₃)₂CuLi	−78°, 25 min	(C₆H₅)₂C=CHC(CH₃)(C₂H₅)COCH₃ (60)	287
C₂₂ (steroid with FCO–CH=, CH₃CO₂)	(CH₃)₂CuLi (1.5 eq)	0°, 4 hr	CH₃COCH₂=C–CH₃ cyclopentane fused structure (60)	288
C₂₄ (steroid with COCl side chain, AcO)	(t-C₄H₉OCuC₃H₇-i)Li (2.2 eq)	−78°, 15 min (3:1 THF:pentane)	cyclopentane with COC₃H₇-i side chain (80)	286b

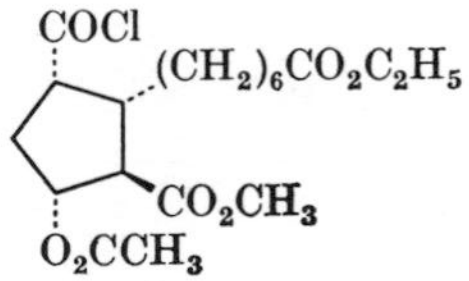

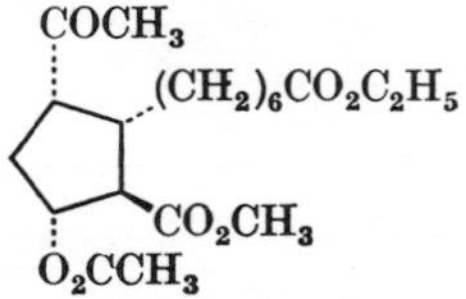

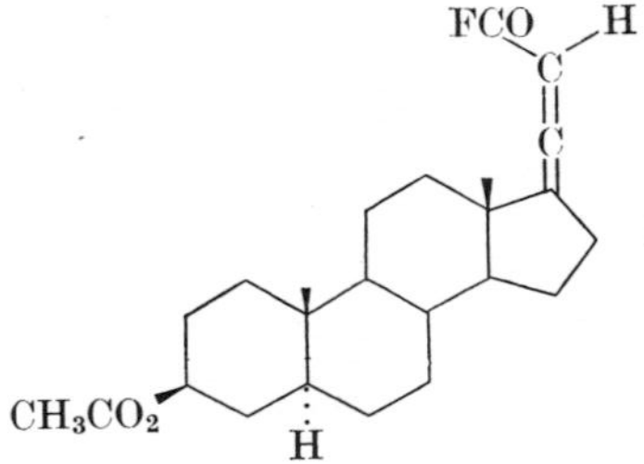

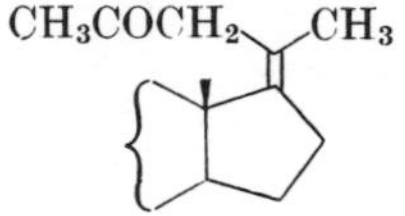

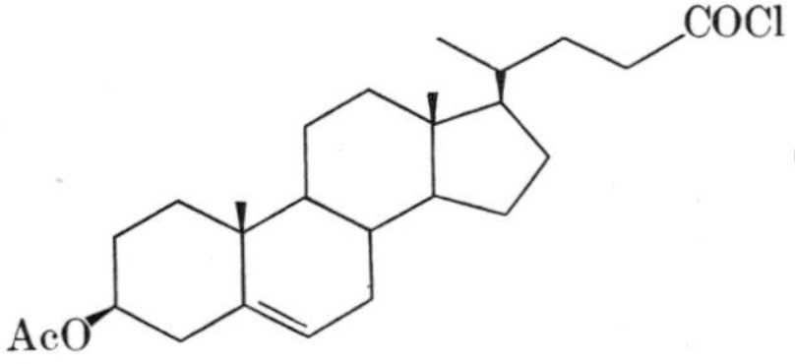

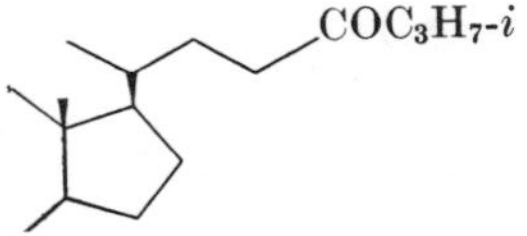

359

Note: References 236–320 are on pp. 398–400.

TABLE III. ORGANOCOPPER COUPLING WITH HALIDE SUBSTRATES (*Continued*)

Halide Substrate	Organocopper Reagent	Reaction Conditions (Solvent)	Product(s) and Yield(s) (%)	Refs.
		F. Acyl Halides (Continued)		
C_{27}	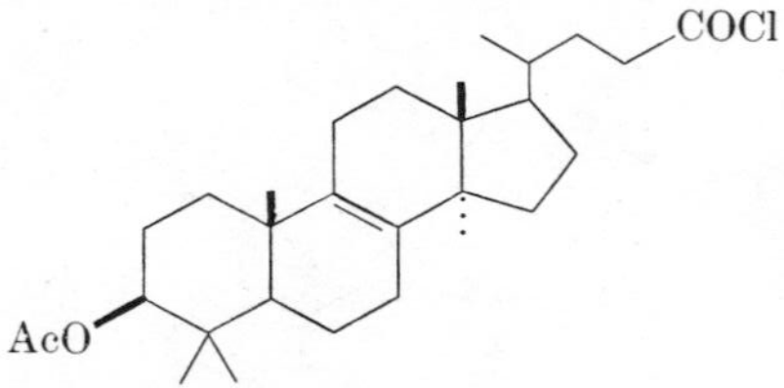 $(C_6H_5)_2CuLi$ (3 eq)	$-20°$, 20 min (THF: ether)	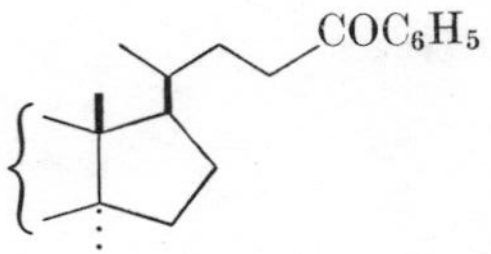 COC_6H_5 (50)	288d
C_{30}	CH_3Cu (4 eq) $n\text{-}C_4H_9Cu$ (4 eq) C_6H_5Cu (4 eq) $2\text{-Pyridyl-}CH_2Cu$ (4 eq)	$-20°$, 2 hr (THF) $-20°$, 2 hr (THF) $-20°$, 2 hr (THF) $-60°$, 2 hr then $-20°$, 2 hr (THF)	(I) $I, R = CH_3$ (87) $I, R = C_4H_9\text{-}n$ (82) $I, R = C_6H_5$ (75) $I, R = 2\text{-Pyridyl-}CH_2$ (38)	285b
		G. Miscellaneous Halides		
C_2 $(CH_3)_3SiC\!\equiv\!CBr$	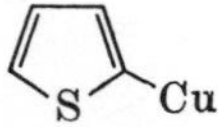	$0°$, 3 hr, then $20°$, 6 hr	$(CH_3)_3SiC\!\equiv\!C$ 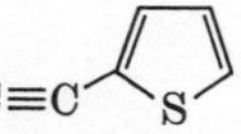 (32)	166
$(C_2H_5)_3SiC\!\equiv\!CBr$ $(CH_3)_3SiC\!\equiv\!CCl$	C_6F_5Cu $m\text{-}FC_6H_4Cu$	Reflux, 10 hr (THF) $0°$, 3 hr, then $20°$, 6 hr	$(C_2H_5)_3SiC\!\equiv\!CC_6F_5$ (85) $(CH_3)_3SiC\!\equiv\!CAr$ (I)	134 166
		$-25°$, then $20°$, 6 hr	$I, Ar = C_6H_4F\text{-}m$ (29)	
	C_6H_5Cu $o\text{-}CH_3OC_6H_4Cu$	$0°$, 3 hr, then $20°$, 6 hr	$I, Ar = C_6H_5$ (64) $I, Ar = C_6H_4OCH_3\text{-}o$ (45)	
	$m\text{-}CH_3OC_6H_4Cu$	$0°$, 3 hr, then $20°$,	$I, Ar = C_6H_4OCH_3\text{-}m$ (33)	

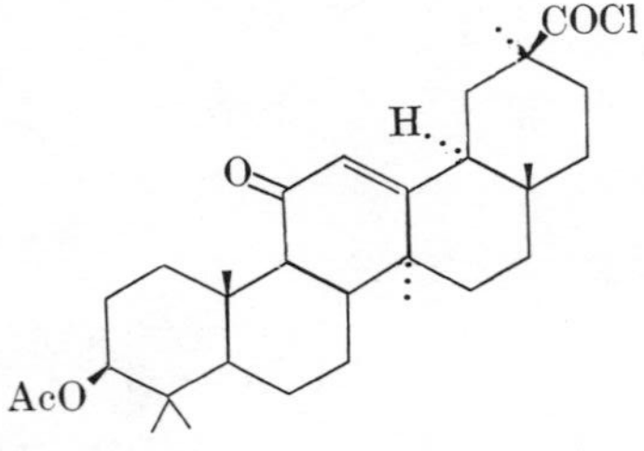

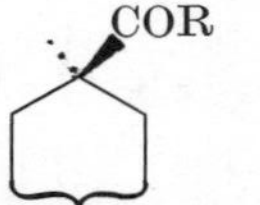

Reagent	Conditions	Product (%)	Refs.
o-CF₃C₆H₄Cu	6 hr	I, Ar = C₆H₄CF₃-o (51)	
m-CF₃C₆H₄Cu	0°, 3 hr, then 20°, 6 hr	I, Ar = C₆H₄CF₃-m (30)	
p-CF₃C₆H₄Cu	0°, 3 hr, then 20°, 6 hr	I, Ar = C₆H₄CF₃-p (38)	
o-CH₃C₆H₄Cu	0°, 3 hr, then 20°, 6 hr	I, Ar = C₆H₄CH₃-o (48)	
m-CH₃C₆H₄Cu	0°, 3 hr, then 20°, 6 hr	I, Ar = C₆H₄CH₃-m (80)	
o-(CH₃)₂NCH₂C₆H₄Cu	0°, 3 hr, then 20°, 6 hr	I, Ar = C₆H₄CH₂N(CH₃)₂-o (38)	
2-(CH₃)₂N-4-CH₃C₆H₃Cu	0°, 3 hr, then 20°, 6 hr	I, Ar = C₆H₃N(CH₃)-2-CH₃-4 (20)	
2-Naphthyl-Cu	0°, 3 hr, then 20°, 6 hr	I, Ar = 2-Naphthyl (47)	
2-Furyl-Cu	−25° then 20°, 6 hr	I, Ar = 2-Furyl (47)	

$$\text{I, Ar} = \overset{CH_3}{\underset{S}{\bigcirc}} \quad (61)$$

Reagent	Conditions	Product (%)	Refs.
(thienyl-Cu structure, 2,5-dimethyl)	−25° then 20°, 6 hr		
3-Furyl-Cu	—	E-THPOCH₂C(CH₃)=CHC≡C-Furyl-3 (—)	288e

C₆′   E-THPOCH₂C(CH₃)=CHC≡CCl

$$\underset{n\text{-}C_3H_7}{\overset{H}{C}}=C=\underset{X}{\overset{H}{C}} \quad (I)$$

I, X = Br

I, X = I

Reagent	Conditions	Product (%)	Refs.
(CH₃)₂CuLi (2.5 eq)	−5°, 1–3 hr	$\underset{n\text{-}C_3H_7}{\overset{H}{C}}=C=\underset{R}{\overset{H}{C}}$ (II) — II, R = CH₃ (85)	178, 175b
(C₂H₅)₂CuLi (2.5 eq)	−30°, 1–3 hr	II, R = C₂H₅ (68)	
(CH₃)₂CuLi (2.5 eq)	−5°, 1–3 hr	II, R = CH₃ (68)	
(n-C₄H₉)₂CuLi (2.5 eq)	−60°, 1–3 hr	II, R = C₄H₉-n (64)	

C₆

$$\underset{C_2H_5}{\overset{CH_3}{C}}=C=\underset{Br}{\overset{H}{C}}$$

Reagent	Conditions	Product (%)	Refs.
(CH₃)₂CuLi (2.5 eq)	−5°, 1–3 hr	$\underset{C_2H_5}{\overset{CH_3}{C}}=C=\underset{R}{\overset{H}{C}}$ (I) — I, R = CH₃ (85)	175b, 178
(n-C₄H₉)₂CuLi (2.5 eq)	−60°, 1–3 hr	I, R = C₄H₉-n (87)	

Note: References 236–320 are on pp. 398–400.

ᵉ Excess ethyl iodide was added to the reaction mixture before workup.

TABLE III. Organocopper Coupling with Halide Substrates (Continued)

Halide Substrate	Organocopper Reagent	Reaction Conditions (Solvent)[a]	Product(s) and Yield(s) (%)	Refs.
		G. Miscellaneous Halides (continued)		
C$_7$ [dibromonorcarane structure]	(CH$_3$)$_2$CuLi (0.3 M, 10 eq)	$-15°$, 96 hr	7,7-Dimethylnorcarane (65), exo-7-methylnorcarane (25)	54
	(CH$_3$)$_2$CuLi (0.3 M, 10 eq)	0°, 20 hr (pentane)	7,7-Dimethylnorcarane (80), exo-7-methylnorcarane (10)	46
	(C$_2$H$_5$)$_2$CuLi (0.3 M, 9 eq)	$-45°$, 1 hr, then 0°, 16 hr	7,7-Diethylnorcarane (60°), 7-ethylnorcarane (20°)	148
[bicyclic OH structure]	(CH$_3$)$_2$CuLi (2 eq)	$-15°$, 96 hr	[product structure] (22)	180
C$_8$ [dibromo OH structure]	(CH$_3$)$_2$CuLi (0.7 M, 10 eq)	0°, 5 hr	[OH product structure] (low)	179
[dibromo bicyclic diene structure]	(CH$_3$)$_2$CuLi	—	[product structure] (—)	288b
C$_{10}$ n-C$_9$H$_{19}$COSC$_2$H$_5$	(CH$_3$)$_2$CuLi (1.2 eq)	$-78°$, 2 hr	n-C$_9$H$_{19}$COR (I), I, R = CH$_3$ (75)	288f
	(i-C$_3$H$_7$)$_2$CuLi (1.1 eq)	$-40°$, 4 hr (THF)	I, R = i-C$_3$H$_7$ (66)	
	(n-C$_4$H$_9$)$_2$CuLi (1.0 eq)	$-40°$, 2 hr	I, R = n-C$_4$H$_9$ (87)	
	(C$_6$H$_5$)$_2$CuLi (1.2 eq)	$-40°$, 2 hr	I, R = C$_6$H$_5$ (74)	
C$_{19}$ [Si—naphthyl-α, X tetralin structure] X = Cl or F	(CH$_3$)$_2$CuLi	20	[Si—naphthyl-α, CH$_3$ product structure] (high)	288c

TABLE IV. CUPROUS ACETYLIDE COUPLING WITH HALIDE SUBSTRATES

	Halide Substrate	Organocopper Reagent	Reaction Conditions (Solvent)[a]	Product(s) and Yield(s) (%)	Refs.
			A. Alkenyl Halides		
C_2	$I_2C=CI_2$	$C_6H_5C≡CCu$	—	$(C_6H_5C≡C)_2C=C(C≡CC_6H_5)_2$ (40)	82
	E-ClCH=CHI	$C_6H_5C≡CCu$ (0.6 M, 1 eq)	100°, 2 hr	$ClCH=CHC≡CC_6H_5$ (94)	83, 82
		$C_6H_5OCH_2C≡CCu$ (0.6 M, 1 eq)	100°, 1 hr	E-$ClCH=CHC≡CCH_2OC_6H_5$ (—)	83, 82
	ICH=CHI	$C_6H_5OCH_2C≡CCu$	—	$C_6H_5OCH_2C≡CCH=CHC≡CCH_2OC_6H_5$ (40)	82, 83
	E-ICH=CHI	n-$C_4H_9C≡CCu$ (1 eq)	10 min	E-n-$C_4H_9C=CCH=CH$—CHR (II) II, R = I (40)	289
		n-$C_4H_9C≡CCu$ (2 eq)	2 hr	II, R = $C≡CC_4H_9$-n (60)	289,
		$C_6H_5C≡CCu$ (1 eq)	10 min	E-$C_6H_5C≡CCH=CHR$ (II) II, R = I (30)	289,
		$C_6H_5C≡CCu$ (2 eq)	90°, 4 hr (DMF)	II, R = $C≡CC_6H_5$ (55)	289
C_8	$C_6H_5CH=CHBr$	$C_6H_5C≡CCu$	15 hr	$C_6H_5CH=CHC≡CC_6H_5$ (75)	82
C_{10}	$C_6H_5SO_2CH=C(C_2H_5)I$	$C_6H_5C≡CCu$ (1 eq)	15 hr	$C_6H_5SO_2CH=C(C_2H_5)C≡CC_6H_5$ (56)	290
C_{15}	p-$CH_3C_6H_4SO_2CH=C(R)I$ I, R = C_6H_5 I, R = cyclo-C_6H_{11}	$C_6H_5C≡CCu$		p-$CH_3C_6H_4SO_2CH=C(R)C≡CC_6H_5$ (II) II, R = C_6H_5 (15) II, R = cyclo-C_6H_{11} (80)	290
			B. Alkynyl Halides		
C_3	$HOCH_2C≡CBr$	$CH_3C≡CCu$	25°	$HOCH_2C≡CC≡CCH_3$ (60)	129
		$C_6H_5C≡CCu$	25°	$HOCH_2C≡CC≡CC_6H_5$ (65)	129
		(bithiophene)—C≡CCu	25°	$HOCH_2C≡CC≡C$—(bithiophene) (71)	129

Note: References 236–320 are on pp. 398–400.

[a] Pyridine was used as solvent at reflux unless noted otherwise.

TABLE IV. Cuprous Acetylide Coupling with Halide Substrates (Continued)

	Halide Substrate	Organocopper Reagent	Reaction Conditions (Solvent)[a]	Product(s) and Yield(s) (%)	Refs.
			B. Alkynyl Halides (Continued)		
C$_8$	C$_6$H$_5$C≡CBr	p-BrC$_6$H$_4$C≡CCu (1 eq)	(DMF)	1,4-Diphenylbutadiyne (—), 1,4-di-p-bromophenylbutadiyne (—), 1-p-bromophenyl-4-phenylbutadiyne (—)	291
	C$_6$H$_5$C≡CI	C$_6$H$_5$C≡CCu	25°	1,4-Diphenylbutadiyne (96)	129
			C. Aryl Halides		
C$_3$	2-Iodothiazole	C$_6$H$_5$C≡CCu	125°, 12 hr	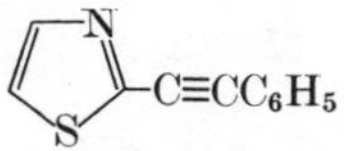 (thiazole)—C≡CC$_6$H$_5$ (58)	292
C$_4$	2,5-Diiodothiophene	CH$_3$C≡CCu	120°, 6 hr	I—(thiophene)—C≡CR (I)	125
		CH$_2$=CHC≡CCu; (C$_2$H$_5$O)$_2$CHC≡CCu; THP—OCH$_2$C≡CCu	120°, 4 hr; 8 hr; —	I, R = CH$_3$ (18); I, R = CH=CH$_2$ (29); I, R = CHO (—)[b]; I, R = CH$_2$O—THP (—)	125, 293, 294
	O$_2$N—(furan)—I	C$_6$H$_5$C≡CCu	3–8 hr	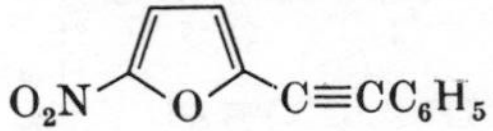 O$_2$N—(furan)—C≡CC$_6$H$_5$ (48)	128
	2-Iodofuran	HOCH$_2$C≡CCu	4 hr	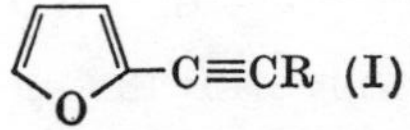 I, R = CH$_2$OH (67)	127, 293
		C$_6$H$_5$C≡CCu; C$_6$H$_5$CH$_2$C≡CCu	125°, 12 hr; 6 hr	(furan)—C≡CR (I); I, R = C$_6$H$_5$ (82); I, R = CH$_2$C$_6$H$_5$ (58)	292; 127, 293
	2-Iodothiophene	C$_6$H$_5$C≡CCu	—	2-Phenylethynylthiophene (74)	295
		o-O$_2$NC$_6$H$_4$C≡CCu (0.9 eq)	8 hr	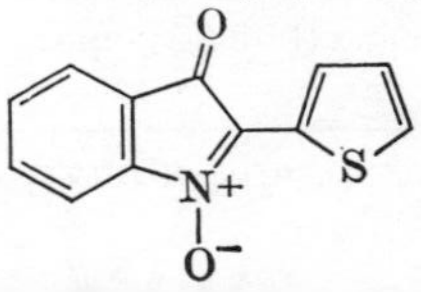 2-(thiophen-2-yl) indolone N-oxide (69)	131

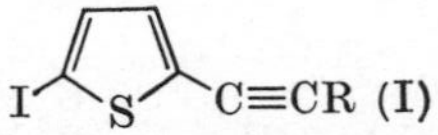

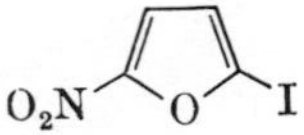

HOCH₂C≡CCu	11 hr	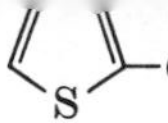 $\left[\text{thienyl}\right]$—C≡CR (I)	127, 329
		I, R = CH₂OH (67)	
THP—OCH₂C≡CCu	7 hr	I, R = CH₂O—THP (—)	293
(C₂H₅O)₂CHC≡CCu	8 hr	I, R = CH(OC₂H₅)₂ (—)	293
2-Thienyl-C≡CCu	125°, 12 hr	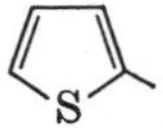 [thienyl–thienyl] (90)	292
2-FurylCH=CHC≡CCu	—	I, R = CH=CHfuryl-2 (60)	127
Ferrocenyl-C≡CCu	—	I, R = ferrocenyl (80)	295

C₅

[3,5-diiodo-4-hydroxypyridine structure]

		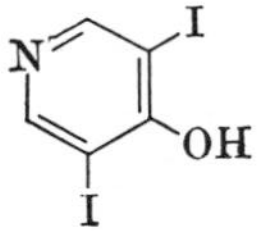 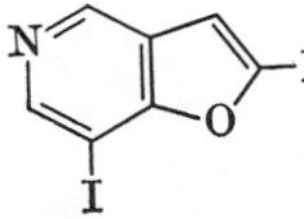 [furopyridine]—R (I)	126
n-C₃H₇C≡CCu (1 eq)	70°	I, R = C₃H₇-n (24)	
n-C₆H₁₃C≡CCu	100°	I, R = C₆H₁₃-n (37)	
C₆H₅C≡CCu	100°	I, R = C₆H₅ (86)	
2-PyridylC≡CCu	100°	I, R = 2-Pyridyl (37)	
HOCH₂CH₂C≡CCu	100°	I, R = CH₂CH₂OH (61)	

2-Iodopyridine	n-C₄H₉C≡CCu	120°, 1 hr	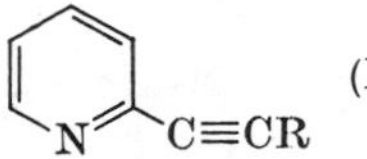 [pyridine]—C≡CR (I)	296

C₆H₅C≡CCu	120°, 18 hr	I, R = C₄H₉-n (95)	133
p-O₂NC₆H₄C≡CCu	120°, 7 hr	I, R = C₆H₅ (25)	296
p-CH₃OC₆H₄C≡CCu	120°, 2 hr	I, R = C₆H₄NO₂-p (52)	296
p-C₆H₅C₆H₄C≡CCu	120°, 1 hr	I, R = C₆H₄OCH₃-p (68)	296
p-C₆H₅C₆H₄C≡CC≡CCu	120°, 2 hr	I, R = C₆H₄C₆H₅-p (54)	296
		I, R = C≡CC₆H₄C₆H₅-p (7)	296

Note: References 236–320 are on pp. 398–400.

[a] Pyridine was used as solvent at reflux unless noted otherwise.
[b] The initial acetal product was hydrolyzed with acid.

365

TABLE IV. CUPROUS ACETYLIDE COUPLING WITH HALIDE SUBSTRATES (Continued)

Halide Substrate	Organocopper Reagent	Reaction Conditions (Solvent)[a]	Product(s) and Yield(s) (%)	Refs.
		C. Aryl Halides (Continued)		
C$_4$ (contd.) 3-Iodopyridine	C$_6$H$_5$C≡CCu (1 eq)	120°, 9 hr	3-Phenylethynylpyridine (47)	133
	o-O$_2$NC$_6$H$_4$C≡CCu (0.9 eq)	Reflux, 8 hr	(22)	131
	n-C$_3$H$_7$C≡CCu	125° (DMF)	I, R = C$_3$H$_7$-n (85)	126
	n-C$_4$H$_9$C≡CCu	125° (DMF)	I, R = C$_4$H$_9$-n (78)	126
	n-C$_6$H$_{13}$C≡CCu	125° (DMF)	I, R = C$_6$H$_{13}$-n (30)	126
	C$_6$H$_5$C≡CCu	125° (DMF)	I, R = C$_6$H$_5$ (57) (I)	126
	n-C$_3$H$_7$C≡CCu	100°	I, R = C$_3$H$_7$-n (82)	126
	n-C$_4$H$_9$C≡CCu	100°	I, R = C$_4$H$_9$-n (88)	126
	n-C$_6$H$_{13}$C≡CCu	100°	I, R = C$_6$H$_{13}$-n (92)	126
	HOCH$_2$CH$_2$C≡Cu	100°	I, R = CH$_2$CH$_2$OH (61)	126
	C$_6$H$_5$C≡CCu (0.2 M, 1 eq)	110–120°, 9 hr	I, R = C$_6$H$_5$ (75–82) (I)	235, 126
	o-O$_2$NC$_6$H$_4$C≡CCu (0.9 eq)	8 hr	(6.5)	131

Reactant	Reagent	Time	Product	Refs.
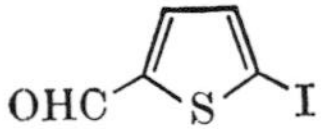 5-I-thiophene-2-CHO (OHC⟨S⟩I)	CH₃C≡CCu	10 hr	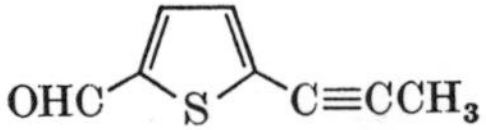 5-(propynyl)-thiophene-2-CHO (OHC⟨S⟩C≡CCH₃) (61)	127, 293
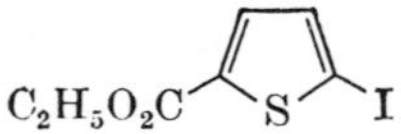 5-I-thiophene-2-CO₂C₂H₅ (C₂H₅O₂C⟨S⟩I)	2-Furyl-CH=CHC≡CCu	—	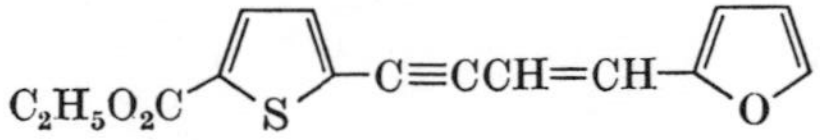 C₂H₅O₂C⟨S⟩C≡CCH=CH-furan (55)	127
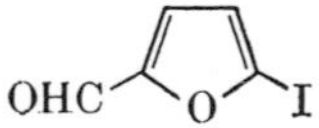 5-I-furan-2-CHO (OHC⟨O⟩I)			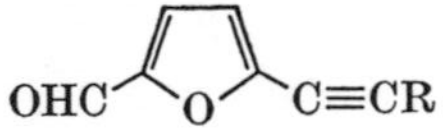 OHC⟨O⟩C≡CR (I)	127
	n-C₄H₉C≡CCu	3–8 hr	I, R = C₄H₉-n (74)	128
	C₆H₅C≡CCu	3–8 hr	I, R = C₆H₅ (65)	128
	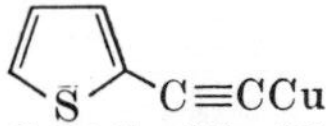 thienyl-C≡CCu (⟨S⟩C≡CCu)	3–8 hr	I, R = 2-thienyl (80)	128
	C₆H₅C≡C—C≡CCu	3–8 hr	I, R = C₆H₅C≡C (69)	128
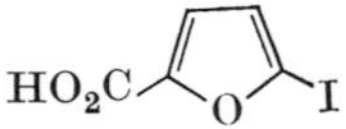 5-I-furan-2-CO₂H (HO₂C⟨O⟩I)	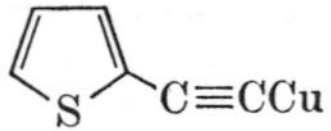 thienyl-C≡CCu (⟨S⟩C≡CCu)	3–8 hr	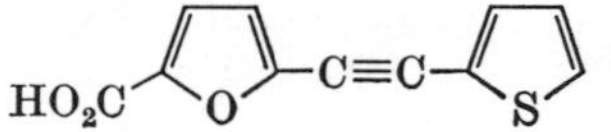 HO₂C⟨O⟩C≡C⟨S⟩ (65)	128
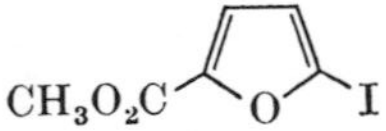 5-I-furan-2-CO₂CH₃ (CH₃O₂C⟨O⟩I)	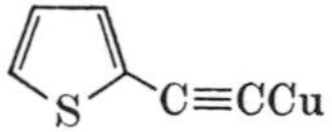 thienyl-C≡CCu (⟨S⟩C≡CCu)	3–8 hr	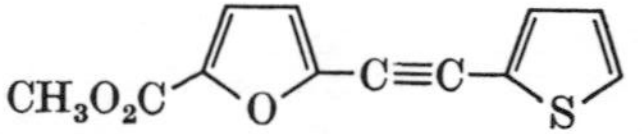 CH₃O₂C⟨O⟩C≡C⟨S⟩ (90)	128
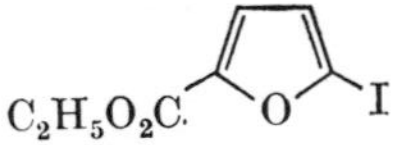 5-I-furan-2-CO₂C₂H₅ (C₂H₅O₂C⟨O⟩I)	CH₃C≡CCu	5 hr	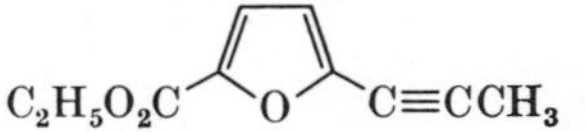 C₂H₅O₂C⟨O⟩C≡CCH₃ (60)	127, 293
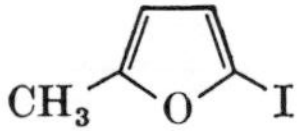 5-I-2-CH₃-furan (CH₃⟨O⟩I)	HOCH₂C≡CCu	4 hr	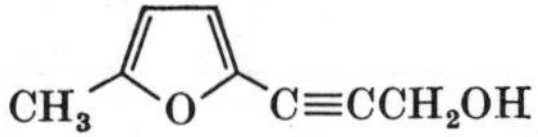 CH₃⟨O⟩C≡CCH₂OH (65)	127, 293
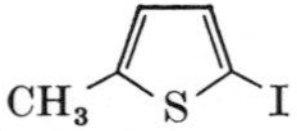 5-I-2-CH₃-thiophene (CH₃⟨S⟩I)	(C₂H₅O)₂CHC≡CCu	6 hr	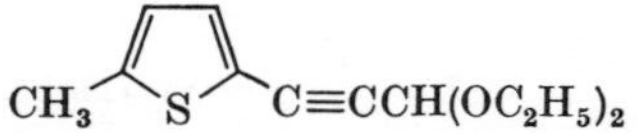 CH₃⟨S⟩C≡CCH(OC₂H₅)₂ (85)	127, 293

Note: References 236–320 are on pp. 398–400.

[a] Pyridine was used as solvent at reflux unless noted otherwise.

367

TABLE IV. CUPROUS ACETYLIDE COUPLING WITH HALIDE SUBSTRATES (Continued)

Halide Substrate	Organocopper Reagent	Reaction Conditions (Solvent)[a]	Product(s) and Yield(s) (%)	Refs.
		C. Aryl Halides (Continued)		
C$_6$ C$_6$Cl$_5$I	C$_6$H$_5$C≡CCu	—	C$_6$Cl$_5$C≡CC$_6$H$_5$ (49)	135
C$_6$F$_5$Br	C$_6$H$_5$C≡CCu (0.8 eq)	10 hr	C$_6$F$_5$C≡CC$_6$H$_5$ (33)	135
C$_6$F$_5$I	C$_6$H$_5$C≡CCu (0.8 eq)	10 hr	C$_6$F$_5$C≡CC$_6$H$_5$ (55)	135, 297
2,4,6-(O$_2$N)$_3$C$_6$H$_2$Cl	C$_6$H$_5$C≡CCu (1 eq)	100°, 3 hr (DMF)	2,4,6-(O$_2$N)$_3$C$_6$H$_2$C≡CC$_6$H$_5$ (34)	133
2,4-Dibromophenol	C$_6$H$_5$C≡CCu (1 eq)	120°, 22 hr	I, R = C$_6$H$_5$ (55)	133
	2-PyridylC≡CCu (1 eq)	—	I, R = 2-pyridyl (38)	
o-, m-, or p-BrC$_6$H$_4$I	n-C$_3$H$_7$C≡CCu (1 eq)	120°, 22 hr	I, R = C$_3$H$_7$-n (40)	298
	C$_6$H$_5$C≡CCu (0.4 M, 1 eq)	12 hr	o-, m-, or p-BrC$_6$H$_4$C≡CC$_6$H$_5$ (70–80)	
o-, m-, or p-ClC$_6$H$_4$I	C$_6$H$_5$C≡CCu (0.4 M, 1 eq)	12 hr	o-, m-, or p-ClC$_6$H$_4$C≡CC$_6$H$_5$ (70–80)	298, 256
o-, m-, or p-FC$_6$H$_4$I	C$_6$F$_5$C≡CCu (0.4 M, 1 eq)	12 hr	o-, m-, or p-FC$_6$H$_4$C≡CC$_6$H$_5$ (70–80)	298
IC$_6$H$_4$I (I) o-I m-I p-I	C$_6$H$_5$C≡CCu (2 eq)	16 hr	C$_6$H$_5$C≡CC$_6$H$_4$C≡CC$_6$H$_5$ (II) o-II (61) m-II (42) p-II (45)	133, 299
o-O$_2$NC$_6$H$_4$I	C$_6$H$_5$C≡CCu	100°, 8 hr	o-O$_2$NC$_6$H$_4$C≡CC$_6$H$_5$ (84)	300, 84
o-O$_2$NC$_6$H$_4$I	o-O$_2$NC$_6$H$_4$C≡CCu (0.9 eq)	8 hr	o-O$_2$NC$_6$H$_4$C≡CC$_6$H$_4$NO$_2$-o (93)	131

Aryl halide	Cuprous acetylide (eq)	Conditions	Product (% yield)	Refs.
	2,4-(O₂N)₂C₆H₃C≡CCu (0.9 eq)	8 hr	(3-oxo-indolenine 1-oxide, 5-O₂N); C₆H₄NO₂-o (75)	131
p-O₂NC₆H₄I	o-O₂NC₆H₄C≡CCu (0.9 eq)	8 hr	(3-oxo-indolenine 1-oxide); C₆H₄NO₂-p (71)	131
	C₆H₅C≡CCu	80°, 10 hr	p-O₂NC₆H₄C≡CC₆H₅ (74)	131, 84, 300
2-O₂N-3-CH₃OC₆H₃I	o-O₂NC₆H₄C≡CCu (0.9 eq)	8 hr	(3-oxo-indolenine 1-oxide); C₆H₃OCH₃-3-NO₂-2 (71)	131
2-O₂N-4-CH₃OC₆H₃I	4-CH₃OC₆H₄C≡CCu (0.9 eq)	3 hr	2-NO₂-4-CH₃OC₆H₃C≡CC₆H₄OCH₃-4 (95)	131
o-HOC₆H₄Br	C₆H₅C≡CCu (1 eq) 2-PyridylC≡CCu (1 eq)	120°, 22 hr 120°, 22 hr	(benzofuran) R (I) I, R = C₆H₅ (53) I, R = 2-pyridyl (50)	133

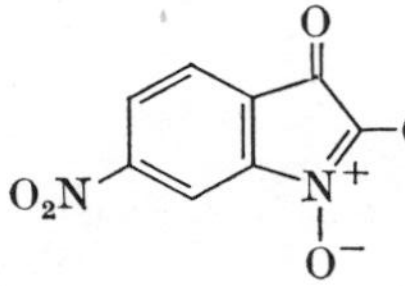

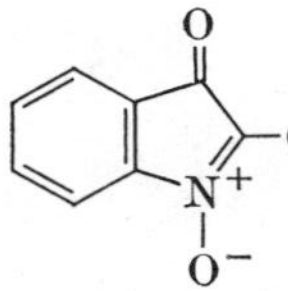

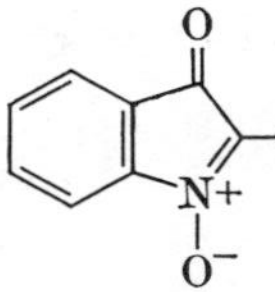

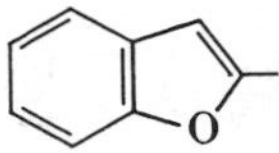

$$
\begin{array}{l}
\text{In the products the indolone structures are drawn as:} \\
\text{O=C—C(=N$^+$—O$^-$)—aryl fused benzo ring}
\end{array}
$$

Note: References 236–320 are on pp. 398–400.

ᵃ Pyridine was used as solvent at reflux unless noted otherwise.

TABLE IV. CUPROUS ACETYLIDE COUPLING WITH HALIDE SUBSTRATES (*Continued*)

Halide Substrate	Organocopper Reagent	Reaction Conditions (Solvent)[a]	Product(s) and Yield(s) (%)	Refs.
		C. Aryl Halides (Continued)		
C$_6$ (*contd.*) o-HOC$_6$H$_4$I	C$_6$H$_5$C≡CCu (1 eq)	120°, 22 hr (DMF)	I, R = C$_6$H$_5$ (88)	133
	n-C$_3$H$_7$C≡CCu (1 eq)	120°, 22 hr	I, R = C$_3$H$_7$-n (60)	133
	Ferrocenyl-C≡CCu (0.15 M, 1 eq)	8 hr	I, R = ferrocenyl (80)	130
p-HOC$_6$H$_4$I	C$_6$H$_5$C≡CCu	125°, 8 hr	p-HOC$_6$H$_4$C≡CC$_6$H$_5$ (82)	84, 299
o-CH$_3$OC$_6$H$_4$I	C$_6$H$_5$C≡CCu (0.4 M, 1 eq)	12 hr	o-CH$_3$OC$_6$H$_4$C≡CC$_6$H$_5$ (70–80)	298, 300, 84
	o-O$_2$NC$_6$H$_4$C≡CCu (0.9 eq)	8 hr	o-CH$_3$OC$_6$H$_4$C≡CC$_6$H$_4$NO$_2$-o (83)	131
p-CH$_3$OC$_6$H$_4$I	C$_6$H$_5$C≡CCu (1 eq)	10 hr	p-CH$_3$OC$_6$H$_4$C≡CC$_6$H$_5$ (83–99)	300, 84
	o-O$_2$NC$_6$H$_4$C≡CCu	8 hr	(37)	131
Iodobenzene	t-C$_4$H$_9$C≡CCu (1 eq)	16 hr	C$_6$H$_5$C≡CC$_4$H$_9$-t (84)	301
	n-C$_5$H$_{11}$C≡CCu	100°, several hr (HMPA)	C$_6$H$_5$C≡CC$_5$H$_{11}$-n (36)	302
	THP-OCH$_2$C≡CCu	—	C$_6$H$_5$C≡CCH$_2$O-THP (—)	294
	C$_6$H$_5$C≡CCu	100°, several hr (HMPA)	Diphenylacetylene (65)	302
	C$_6$H$_5$C≡CCu (1 eq)	1 hr	'' (87–90)	300, 84
	C$_6$F$_5$C≡CCu	8 hr	C$_6$H$_5$C≡CC$_6$F$_5$ (74)	135

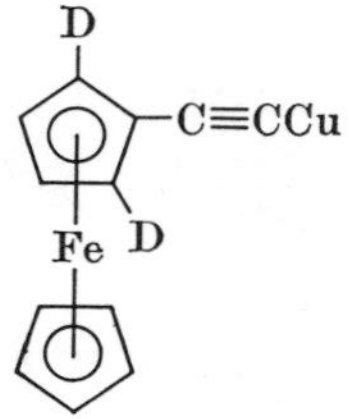

Reactant	Conditions	Product	Ref.
C≡CCu ferrocene-D; o-HSC$_6$H$_4$Br^c	8 hr	C≡CC$_6$H$_5$ ferrocene-D (77)	303
n-C$_3$H$_7$C≡CCu (1 eq) n-C$_4$H$_9$C≡CCu (1 eq) C$_6$H$_5$C≡CCu (1 eq) C$_2$H$_5$OCOC≡CCu^d HOCH$_2$C≡CCu^d	— — (DMF) (DMF) 110° (DMF)	benzothiophene R (I) I, R = C$_3$H$_7$-n (80) I, R = C$_4$H$_9$-n (80) I, R = C$_6$H$_5$ (90) I, R = CO$_2$C$_2$H$_5$ (35) I, R = CH$_2$OH (10)	171
C$_6$H$_5$C≡CCu	3–8 hr	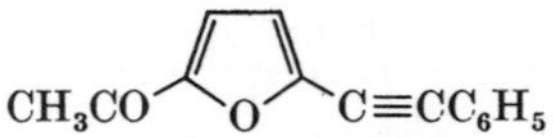 CH$_3$CO—furan—C≡CC$_6$H$_5$ (72)	128
C$_6$H$_5$C≡C—C≡CCu	3–8 hr	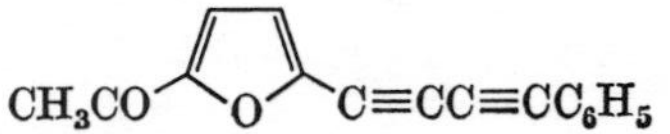 CH$_3$CO—furan—C≡C—C≡CC$_6$H$_5$ (71)	128
o-H$_2$NC$_6$H$_4$I		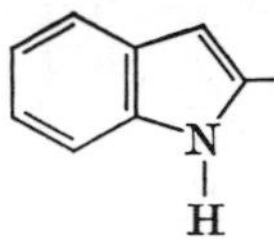 indole-R (I)	128

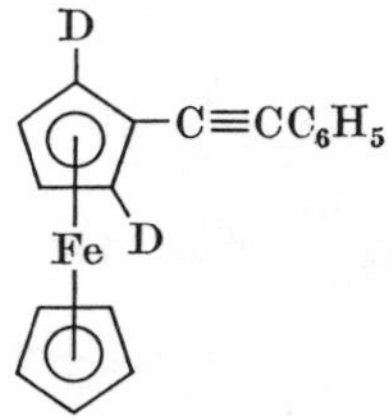

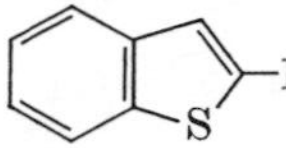

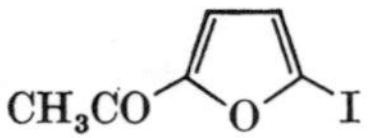

Note: References 236–320 are on pp. 398–400.

[a] Pyridine was used as solvent at reflux unless noted otherwise.

[c] This reactant was added slowly (24 hr) to a pyridine solution of the organocopper reagent.

[d] The organocopper reagent was generated *in situ* from the corresponding acetylene, N-ethylpiperidine, and (presumably) a Cu source.

TABLE IV. CUPROUS ACETYLIDE COUPLING WITH HALIDE SUBSTRATES (Continued)

Halide Substrate		Organocopper Reagent	Reaction Conditions (Solvent)[a]	Product(s) and Yield(s) (%)	Refs.
			C. Aryl Halides (Continued)		
C_6 (contd.)	o-$H_2NC_6H_4I$ (contd.)	$C_2H_5C{\equiv}CCu$ (1 eq)	120°, 8 hr	I, R = C_2H_5-n (12)	133
		n-$C_3H_7C{\equiv}CCu$ (1 eq)	120°, 8 hr	I, R = C_3H_7-n (70)	133, 84
		n-$C_4H_9C{\equiv}CCu$ (1 eq)	120°, 8 hr	I, R = C_4H_9-n (35)	133
		$C_6H_5C{\equiv}CCu$ (1 eq)	115° (DMF)	I, R = C_6H_5 (89)	133
		2-Pyridyl$C{\equiv}CCu$		o-$H_2NC_6H_4C{\equiv}C$- 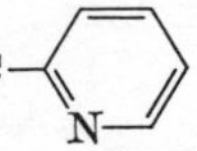 (pyridyl) (50)	133
	p-$H_2NC_6H_4I$	$C_6H_5C{\equiv}CCu$ (1 eq)	120°, 8 hr	o-$H_2NC_6H_4C{\equiv}CC_6H_5$ (59)	133
		$C_6H_5C{\equiv}CCu$ (1 eq)	125°, 24 hr	p-$H_2NC_6H_4C{\equiv}CC_6H_5$ (76)	84
	o-$C_2H_5NHC_6H_4I$	n-$C_3H_7C{\equiv}CCu$ (1 eq)	120°, 22 hr (DMF)	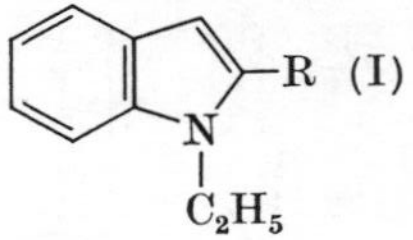 (indole, N-C_2H_5) R (I) I, R = C_3H_7-n (50)	133
		$C_6H_5C{\equiv}CCu$ (1 eq)	120°, 22 hr (DMF)	I, R = C_6H_5 (50)	133
	p-HO-o-$H_2NC_6H_3I$	$C_6H_5C{\equiv}CCu$ (1 eq)	120°, 22 hr (DMF)	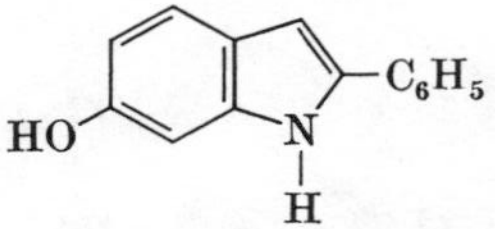 (hydroxyindole, C_6H_5) (57)	133
C_7	2,4-$Cl_2C_6H_3CO_2H$	$C_6H_5C{\equiv}CCu$ (1 eq)	125°, 3 hr	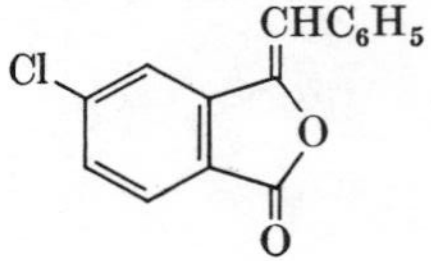 (chloro-benzofuranone, CHC_6H_5) (69)	133

372

Substrate	Organocopper Reagent	Conditions	Product (Yield %)	Refs.
CH₃C≡C–[2-iodothiophene] ($CH_3C{\equiv}C$–thienyl–I)	$CH_3OCOC{\equiv}CCu$	120°, 5 hr	$CH_3C{\equiv}C$–thienyl–S–$C{\equiv}CCO_2CH_3$ (35)	125
$HO_2CCH{=}CH$–[5-iodofuran]	$C_6H_5C{\equiv}CCu$	3–8 hr	$HO_2CCH{=}CH$–furyl–O–$C{\equiv}CC_6H_5$ (64)	128
$o\text{-}HO_2CC_6H_4Br$	$C_6H_5C{\equiv}CCu^a$	(DMF)	[CHR isobenzofuranone] (I)	133
$o\text{-}HO_2CC_6H_4Cl$	$C_6H_5C{\equiv}CCu^a$	(DMF)	I, R = C_6H_5 (53)	
$o\text{-}HO_2CC_6H_4I$	$C_6H_5C{\equiv}CCu$	—	I, R = C_6H_5 (39)	
	$C_6H_5C{\equiv}CCu^a$	125°, 6 hr (DMF)	I, R = C_6H_5 (65)	
	$n\text{-}C_3H_7C{\equiv}CCu$		I, R = C_6H_5 (90)	
	$C_2H_5OCOC{\equiv}CCu^a$		I, R = $C_3H_7\text{-}n$ (22)	
			I, R = $CO_2C_2H_5$ (39)	
$p\text{-}HO_2CC_6H_4I$	$C_2H_5C{\equiv}CCu$ (1 eq)	25°, 21 hr	$p\text{-}HO_2CC_6H_4C{\equiv}CC_6H_5$ (85)	84
$o\text{-}RO_2CC_6H_4I$	$o\text{-}O_2NC_6H_4C{\equiv}CCu$ (1 eq)	110°, 8 hr	$o\text{-}RO_2CC_6H_4C{\equiv}CC_6H_4NO_2\text{-}o$ (I)	131
R = CH₃		8 hr	I, R = CH_3 (93)	
R = C₂H₅			I, R = C_2H_5 (92)	
$o\text{-}H_2NCOC_6H_4I$	$n\text{-}C_3H_7C{\equiv}CCu$ (1 eq)	—	$o\text{-}H_2NCOC_6H_4C{\equiv}CC_3H_7\text{-}n$ (50)	133
	$C_6H_5C{\equiv}CCu$ (1 eq)	120°, 8 hr	$o\text{-}H_2NCOC_6H_4C{\equiv}CC_6H_5$ (47)	133
	$o\text{-}O_2NC_6H_4C{\equiv}CCu$ (0.9 eq)	8 hr	[aminoindenone], $C_6H_4NO_2\text{-}o$ (66)	131
	$1\text{-}C_{10}H_7C{\equiv}CCu$ (0.9 eq)	8 hr	$o\text{-}H_2NCOC_6H_4C{\equiv}CC_{10}H_7\text{-}1$ (87)	131

Note: References 236–320 are on pp. 398–400.

ᵃ Pyridine was used as solvent at reflux unless noted otherwise.

ᵃ The organocopper reagent was generated *in situ* from the corresponding acetylene, N-ethylpiperidine, and (presumably) a Cu source.

TABLE IV. CUPROUS ACETYLIDE COUPLING WITH HALIDE SUBSTRATES (Continued)

Halide Substrate	Organocopper Reagent	Reaction Conditions (Solvent)[a]	Product(s) and Yield(s) (%)	Refs.
		C. Aryl Halides (Continued)		
C_7 (contd.) $CH_3C_6H_4I$ (I)			$CH_3C_6H_4C{\equiv}CR$ (II)	
o-I	$o\text{-}O_2NC_6H_4C{\equiv}Cu$ (0.9 eq)	8 hr	o-II, R = $C_6H_4NO_2\text{-}o$ (67)	131
m-I	$C_6H_5C{\equiv}CCu$ (0.4 M, 1 eq)	12 hr	m-II, R = C_6H_5 (70–80)	298
p-I	$C_6H_5C{\equiv}CCu$ (0.4 M, 1 eq)	12 hr	p-II, R = C_6H_5 (70–80)	298
o-Iodobenzyl alcohol	$n\text{-}C_3H_7C{\equiv}CCu$	—	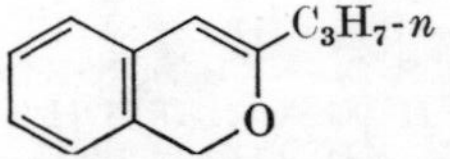 $C_3H_7\text{-}n$ (50)	61
	$C_6H_5C{\equiv}CCu$ (1 eq)	7 hr	$o\text{-}HOCH_2C_6H_4C{\equiv}CC_6H_5$ (50)	133
	$C_6H_5C{\equiv}CCu$ (1 eq)	48 hr	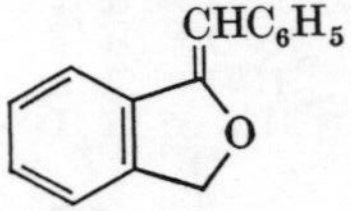 CHC_6H_5 (80)	61
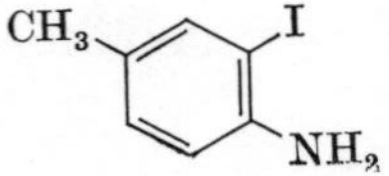	$C_6H_5C{\equiv}CCu$ (1 eq)	120°, 22 hr (DMF)	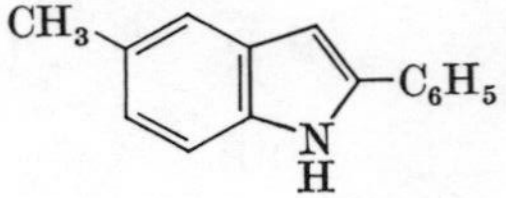 (90)	133
	$C_6H_5C{\equiv}CCu$ (1 eq)	120°, 6 hr	$2\text{-}H_2N\text{-}5\text{-}CH_3C_6H_3C{\equiv}CC_6H_5$ (92)	133
C_8	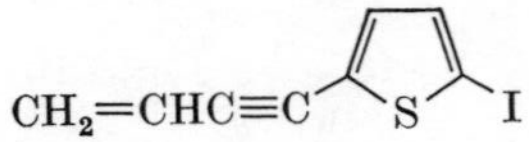 $THP\text{-}OCH_2\text{-}CH{=}CHC{\equiv}CCu$	120°, 3 hr	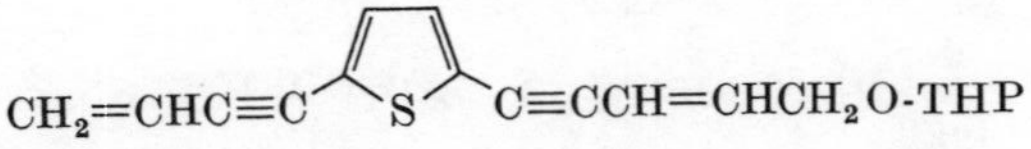 (57)	125

374

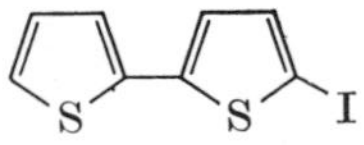

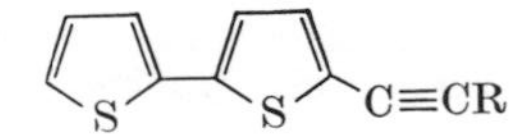

Substrate	Reagent	Conditions	Product	Refs.
5-Iodo-2,2'-bithiophene	CH₂=CHC≡CCu	7 hr	I, R = CH=CH₂ (18)	127, 304
	HOCH₂CH₂C≡CCu	9 hr	I, R = CH₂CH₂OH (51)	127, 305
	CH₃C≡CCu	6 hr	I, R = CH₃ (—)	305
	(C₂H₅O)₂CHC≡CCu	12 hr	I, R = CH(OC₂H₅)₂ (—)	305
	HOCH₂C≡CCu	4 hr	I, R = CH₂OH (—)	305
	THP-OCH₂C≡CCu	—	I, R = CH₂O-THP (—)	294
	CH₃COC≡CCu	6 hr	I, R = COCH₃ (—)	305
	ClCH₂CHOHC≡CCu	—	I, R = CHOHCH₂Cl (Low)	306
	CH₃CO₂CH₂-CH(O-THP)C≡CCu	110°, 1 hr	I, R = CH(O-THP)CH₂O₂CCH₃ (54)	307

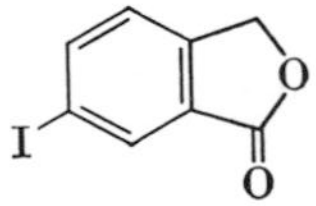

Product: $\text{(thiophene)}_2\text{—C}{\equiv}\text{CR}$ (I)

Substrate	Reagent	Conditions	Product	Refs.
(5-iodophthalide)	C₆H₅C≡CCu (1.5 eq)	Reflux, 10 hr (DMSO)	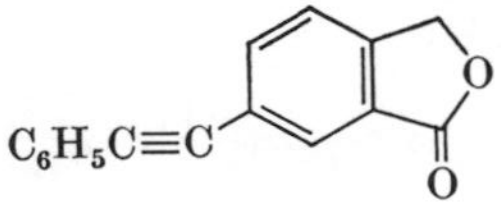 (78)	308
o-Iodophenylacetic acid	n-C₃H₇C≡CCu	—	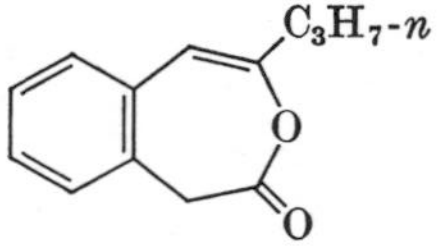 (60)	61
p-Iodoacetophenone	C₆H₅C≡CCu	Reflux (DMF)	p-CH₃COC₆H₄C≡CC₆H₅ (—)	299
o-, m-, or p-C₂H₅C₆H₄I	C₆H₅C≡CCu	12 hr	o-, m-, or p-C₂H₅C₆H₄C≡CC₆H₅ (70–80)	298
2,5-Dimethyliodobenzene	C₆H₅C≡CCu (0.4 M, 1 eq)	12 hr	2,5-(CH₃)₂C₆H₃C≡CC₆H₅ (70–80)	298

Note: References 236–320 are on pp. 398–400.

a Pyridine was used as solvent at reflux unless noted otherwise.

TABLE IV. Cuprous Acetylide Coupling with Halide Substrates (*Continued*)

Halide Substrate	Organocopper Reagent	Reaction Conditions (Solvent)[a]	Product(s) and Yield(s) (%)	Refs.
C. Aryl Halides (Continued)				
C$_9$ thiophene (I), R = CH$_3$; R = CO$_2$C$_2$H$_5$; *o*-, *m*-, or *p*-*i*-C$_3$H$_7$C$_6$H$_4$I	CH$_2$=CHC≡CCu	7–9 hr	thiophene (II) C≡CCH=CH$_2$ II, R = CH$_3$ (53) II, R = CO$_2$C$_2$H$_5$ (43)	127, 305
o-, *m*-, or *p*-*i*-C$_3$H$_7$C$_6$H$_4$I	C$_6$H$_5$C≡CCu (0.4 *M*, 1 eq)	12 hr	*o*-, *m*-, or *p*-*i*-C$_3$H$_7$C$_6$H$_4$C≡CC$_6$H$_5$ (70–80)	298
2,4,6-(CH$_3$)$_3$C$_6$H$_2$I	C$_6$H$_5$C≡CCu	12 hr	2,4,6-(CH$_3$)$_3$C$_6$H$_2$C≡CC$_6$H$_5$ (70–80)	298
C$_{10}$ 1-Bromo-8-iodonaphthalene	C$_6$H$_5$C≡CCu		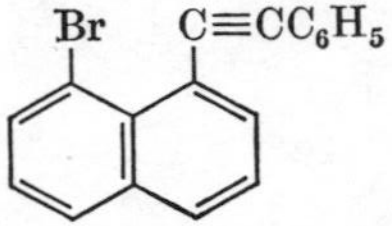 Br / C≡CC$_6$H$_5$ naphthalene (75)	309, 256
1,8-Diiodonaphthalene	C$_6$H$_5$C≡CCu *o*-CH$_3$C$_6$H$_4$C≡CCu *p*-CH$_3$C$_6$H$_4$C≡CCu 2,4-(CH$_3$)$_2$C$_6$H$_3$C≡CCu 2,4,6-(CH$_3$)$_3$C$_6$H$_2$C≡CCu	5 hr	1,8-C$_{10}$H$_6$(C≡CR)$_2$ (I) I, R = C$_6$H$_5$ (62) I, R = C$_6$H$_4$CH$_3$-*o* (~60) I, R = C$_6$H$_4$CH$_3$-*p* (~60) I, R = C$_6$H$_2$(CH$_3$)$_2$-2,4 (~60) I, R = C$_6$H$_2$(CH$_3$)$_3$-2,4,6 (80)	132
1-Iodonaphthalene	THP-OCH$_2$C≡CCu	—	naphthalene C≡CCH$_2$O–THP (—)	294
	C$_6$H$_5$C≡CCu	10 hr	1-Naphthylphenylacetylene (75)	309
	Ferrocenyl-C≡CCu	8 hr	Ferrocenyl-1-naphthylacetylene (83)	130

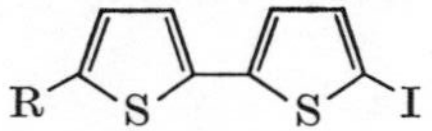

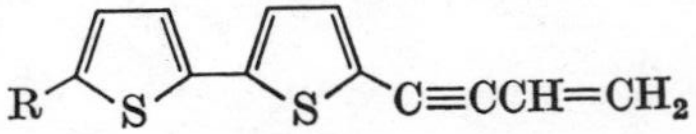

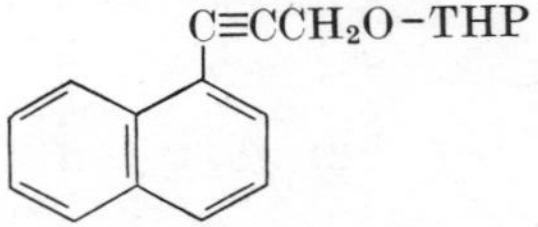

376

C_n	Substrate	Reagent	Conditions	Product (Yield)	Ref.
	[thiophene, C_6H_5— and —I substituted]	$o\text{-}O_2NC_6H_4C{\equiv}CCu$ (0.9 eq)	8 hr	1-Naphthyl-2-nitrophenylacetylene (88)	131
	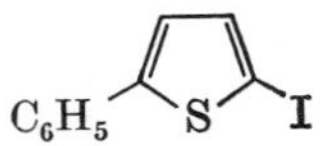	$CH_3C{\equiv}CCu$	7 hr	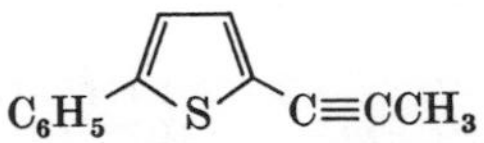 [thiophene, C_6H_5— and S—$C{\equiv}CCH_3$] (51)	127, 293
	1,1′-Diiodoferrocene	$C_6H_5C{\equiv}CCu$	—	$1,1'\text{-}(C_6H_5C{\equiv}C)_2$-ferrocene (60)	295
	Bromoferrocene	$C_6H_5C{\equiv}CCu$	Reflux, 3 hr (DMF)	$C_6H_5C{\equiv}C$-ferrocene (48)	171
	Iodoferrocene	$C_6H_5C{\equiv}CCu$ (1 eq)	—	$C_6H_5C{\equiv}C$-ferrocene (84)	295
		Ferrocenyl-$C{\equiv}C$—Cu	—	Diferrocenylacetylene (85)	295
		Ruthenocenyl-$C{\equiv}C$—Cu (0.6 M, 1 eq)	8 hr	Ruthenocenyl-$C{\equiv}C$-ferrocene (60)	130
	$o\text{-},\ m\text{-},$ or $p\text{-}t\text{-}C_4H_9C_6H_4I$	$C_6H_5C{\equiv}CCu$ (0.4 M, 1 eq)	12 hr	$o\text{-},\ m\text{-},$ or $p\text{-}t\text{-}C_4H_9C_6H_4C{\equiv}CC_6H_5$ (70–80)	298
C_{12}	o-Iodobiphenyl	$o\text{-}O_2NC_6H_4C{\equiv}CCu$ (0.9 eq)	8 hr	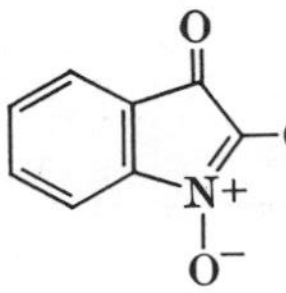 [isatogen N-oxide] $C_6H_4C_6H_5\text{-}o$ (73)	131
C_{18}	[steroidal iodophenol structure]	$n\text{-}C_3H_7C{\equiv}CCu$ $n\text{-}C_4H_9C{\equiv}CCu$ $n\text{-}C_6H_{13}C{\equiv}CCu$ $HOCH_2CH_2C{\equiv}CCu$ $C_6H_5C{\equiv}CCu$	100–110°	(I) [furan-fused ring structure] I, R = $C_3H_7\text{-}n$ (~70) I, R = $C_4H_9\text{-}n$ (~70) I, R = $C_6H_{13}\text{-}n$ (~70) I, R = CH_2CH_2OH (~70) I, R = C_6H_5 (~70)	136

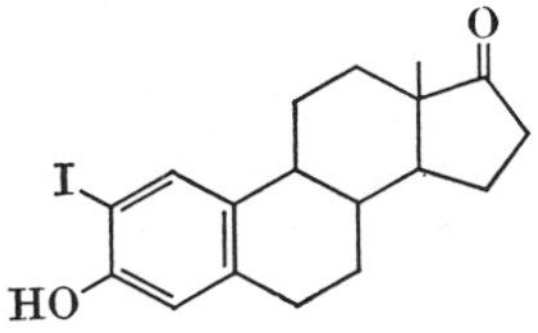

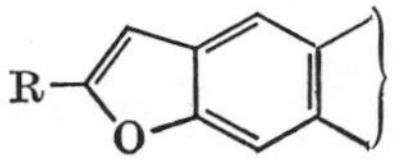

Note: References 236–320 are on pp. 398–400.

^{a} Pyridine was used as solvent at reflux unless noted otherwise.

TABLE IV. CUPROUS ACETYLIDE COUPLING WITH HALIDE SUBSTRATES (Continued)

Halide Substrate	Organocopper Reagent	Reaction Conditions (Solvent)[a]	Product(s) and Yield(s) (%)	Refs.
		C. Aryl Halides (Continued)		
C_{18} (contd.)	$n\text{-}C_3H_7C\equiv CCu$	$100\text{--}110°$	$(—)$	136
		D. Benzylic, Allylic, and Propargylic Halides		
C_3				
Propargyl chloride	$(CH_3)_2C(OH)C\equiv CCu$	(H_2O)	$HC\equiv CCH_2C\equiv CC(OH)(CH_3)_2$ $(—)$	175a
2,3-Dibromopropene	$(CH_3)_2C(OH)C\equiv CCu$	$25°$, 36 hr (H_2O)	$CH_2=CBrCH_2C\equiv CC(OH)(CH_3)_2$ (50)	175a
Allyl bromide	$HOCH_2C\equiv CCu$	$20°$, 1 hr (H_2O)	$CH_2=CHCH_2C\equiv CCH_2OH$ (13)	175a
	$n\text{-}C_5H_{11}C\equiv CCu$	$100°$, 16 hr (HMPA)	$CH_2=CHCH_2C\equiv CC_5H_{11}\text{-}n$ $(37\text{-}44)$	85
	'' (1 eq)	$80°$, 1 hr (DMF, excess NaCN)	(96)	302
	$n\text{-}C_7F_{15}C\equiv CCu^e$	$25°$, 10 min (THF)	$CH_2CH=CHC_7F_{15}\text{-}n$ $(—)$	310
	$(n\text{-}C_5H_{11}C\equiv C)\text{-}(n\text{-}C_4H_9)CuLi$ (0.5 eq)	$-10°$ (ether-HMPA)	$CH_2=CHCH_2C\equiv CC_5H_{11}\text{-}n$ (89)	18
	$C_6H_5C\equiv CCu$	Reflux (DMF)	$CH_2=CHCH_2C\equiv CC_6H_5$ (40)	299, 173, 299
	$C_6H_5C\equiv CCu$	$240°$ $(C_6H_5NO_2)$	(83)	
Allyl chloride	$(CH_3)_2C(OH)C\equiv CCu$	$40°$ (H_2O)	$CH_2=CHCH_2C\equiv CC(OH)(CH_3)_2$ (50)	175a
	$n\text{-}C_5H_{11}C\equiv CCu$ (1 eq)	$25°$ (HMPA, 1 eq) NaCN	$CH_2=CHCH_2C\equiv CC_5H_{11}\text{-}n$ (60)	302
Allyl iodide	$(C_2H_5)_2C(OH)C\equiv CCu$	$25°$, 45 min (H_2O)	$CH_2=CHCH_2C\equiv CC(OH)(C_2H_5)_2$ (30)	175a
	$n\text{-}C_5H_{11}C\equiv CCu$ (1 eq)	$25°$ (HMPA, 1 eq) NaCN	$CH_2=CHCH_2C\equiv CC_5H_{11}\text{-}n$ (74)	302

	Substrate	Reagent	Conditions	Product (% yield)	Ref.
C₄	1,4-Dichloro-2-butene 1-Bromo-2-butene	$(CH_3)_2C(OH)C\equiv CCu$ (1 eq) $n\text{-}C_5H_{11}C\equiv CCu$	(H_2O) 25° (HMPA, 1 eq) NaCN	$ClCH_2CH=CHCH_2C\equiv CC(OH)(CH_3)_2$ (28) $CH_3CH=CHCH_2C\equiv CC_5H_{11}\text{-}n,$ $CH_2=CHCH(CH_3)C\equiv CC_5H_{11}\text{-}n$ (76, total)	175a 302
	2-Bromomethylpropene	$n\text{-}C_5H_{11}C\equiv CCu$ (1 eq)	25° (HMPA, 1 eq) NaCN)	$CH_2=C(CH_3)CH_2C\equiv CC_5H_{11}\text{-}n$ (80)	302
	2-Chloromethylpropene	$HOCH_2C\equiv CCu$ $n\text{-}C_5H_{11}C\equiv CCu$ (1 eq)	30°, 1 hr (H₂O) 25° (HMPA, 1 eq) NaCN)	$CH_2=C(CH_3)CH_2C\equiv CCH_2OH$ (13) $CH_2=C(CH_3)CH_2C\equiv CC_5H_{11}\text{-}n$ (63)	175a 302
C₅	$(CH_3)_2C=CHCH_2Br$	$(CH_3)_2C(OH)C\equiv CCu$ (1 eq) $n\text{-}C_5H_{11}C\equiv CCu$ (1 eq)	(H_2O) 25° (HMPA, 1 eq) NaCN)	$CH_2=C(CH_3)CH_2C\equiv CC(OH)(CH_3)_2$ (50) $(CH_3)_2C=CHCH_2C\equiv CC_5H_{11}\text{-}n$ (50)	175a 302
	$(CH_3)_2C=CHCH_2Cl$	$n\text{-}C_5H_{11}C\equiv CCu$ (1 eq)	25° (HMPA, 1 eq) NaCN)	$(CH_3)_2C=CHCH_2C\equiv CC_5H_{11}\text{-}n$ (60)	302
C₇	Benzyl bromide	$C_6H_5C\equiv CCu$	245°, 5 min (N-methyl-pyrrolidone)	$C_6H_5CH_2C\equiv CC_6H_5$ (90)	173
C₉	$C_6H_5CH=CHCH_2Br$	$HOCH_2C\equiv CCu$ $(CH_3)_2C(OH)C\equiv CCu$	40°, 3 hr (H₂O)	$C_6H_5CH=CHCH_2C\equiv CR$ (I) I, R = CH_2OH (8) I, R = $C(OH)(CH_3)_2$ (45)	175a

E. α-Halocarbonyl Substrates

	Substrate	Reagent	Conditions	Product (% yield)	Ref.
C₈	$C_6H_5COCH_2Br$	$n\text{-}C_3H_7C\equiv CCu$ $C_6H_5C\equiv CCu$	140°, 5 min (neat) 140°, 16 hr (HOCH₂CH₂OH)	2,5-Di-n-propylfuran (29) Acetophenone (47)	173 133
		$C_6H_5C\equiv CCu$	240°, 5 min (C₆H₅NO₂)	2,5-Diphenylfuran (54)	173
	(bromo diketone structure)	$C_6H_5C\equiv CCu$	—	(furanone structure) C_6H_5 (43)	173

Note: References 236–320 are on pp. 398–400.

ᵃ Pyridine was used as solvent at reflux unless noted otherwise.
ᵉ The reagent was generated *in situ* from the corresponding iodide and activated copper bronze in DMSO at 110°.

TABLE IV. CUPROUS ACETYLIDE COUPLING WITH HALIDE SUBSTRATES (Continued)

Halide Substrate	Organocopper Reagent	Reaction Conditions (Solvent)[a]	Product(s) and Yield(s) (%)	Refs.
		E. α-Halocarbonyl Substrates (Continued)		
C_{15} $C_6H_5COCHBrCOC_6H_5$	$C_6H_5C\equiv CCu$	—	[furan ring with C_6H_5CO and C_6H_5 substituents] (24)	173
		F. Acyl Halides		
C_2 Acetyl bromide	$n\text{-}C_5H_{11}C\equiv CCu$	80°, 1 hr	$CH_3COC\equiv CR$ (I); I, R = C_5H_{11}-n (40)	302
Acetyl chloride	$n\text{-}C_3H_7C\equiv CCu$ (0.5 eq)	25°, 24 hr (neat)	I, R = C_3H_7-n (75)	61, 311
	$n\text{-}C_4H_9C\equiv CCu$ (1 eq)	25°, 20 hr (LiI)	I, R = C_4H_9-n (70)	98
	$n\text{-}C_5H_{11}C\equiv CCu$ (1 eq)	25°, 20 hr (LiI)	I, R = C_5H_{11}-n (82)	98
	$(CH_3)(n\text{-}C_5H_{11}C\equiv C)\text{-}CuLi$	25°, 20 hr (HMPA)	I, R = C_5H_{11}-n (58)	98
C_3 $CH_2=CHCOCl$	$C_6H_5C\equiv CCu$ (1 eq)	25°, 20 hr (LiI)	I, R = C_6H_5 (82)	98
C_4 $E\text{-}CH_3CH=CHCOCl$	$n\text{-}C_4H_9C\equiv CCu$ (1 eq)	25°, 20 hr (LiI)	$CH_2=CHCOC\equiv CC_4H_9$-n (80)	98
	$(CH_3)(n\text{-}C_5H_{11}C\equiv C)\text{-}CuLi$	25°, 20 hr (HMPA)	$E\text{-}CH_3CH=CHCOC\equiv CC_5H_{11}$-$n$ (78)	98
C_5 $n\text{-}C_4H_9COCl$	$n\text{-}C_4H_9C\equiv CCu$ (1 eq)	25°, 20 hr (LiI)	$n\text{-}C_4H_9COC\equiv CC_4H_9$-$n$ (96)	98
C_7 Benzoyl chloride	$(CH_3)(n\text{-}C_5H_{11}C\equiv C)\text{-}CuLi$	25°, 20 hr (HMPA)	$C_6H_5COC\equiv CR$ (I); I, R = C_5H_{11}-n (71)	98
	$C_6H_5C\equiv CCu$ (1 eq)	70°, 2 hr (benzene) then 70°, 6 hr (pyridine)	I, R = C_6H_5 (42)	312
	$C_6H_5C\equiv CCu$ (1 eq)	25°, 20 hr (LiI)	I, R = C_6H_5 (90)	98
$C_6H_5\overset{\text{N}C_6H_5}{\underset{\shortparallel}{C}}Cl$	$C_6H_5C\equiv CCu$	25° (LiI)	$\overset{\text{N}C_6H_5}{}C_6H_5COC\equiv CC_6H_5$ (50)	18
C_9 $C_6H_5CH_2CH_2CH_2COCl$	$n\text{-}C_3H_7C\equiv CCu$ (0.5 eq)	25°, 24 hr (neat)	$C_6H_5CH_2CH_2COC\equiv CC_3H_7$-$n$ (—)	61

Note: References 236–320 are on pp. 398–400.

[a] Pyridine was used as solvent at reflux unless noted otherwise.

	Alcohol Derivative	Organocopper Reagent	Reaction Conditions (Solvent)	Product(s) and Yield(s) (%)	Refs.
			A. p-Toluenesulfonates (ROTs)		
C_1	CH_3OTs (with structure CuC_4H_9-n)	$LiP(C_4H_9$-$n)_3$	$-20°$	[structure with CH_3] (50)	147
C_2	$BrCH_2CH_2OTs$	$(n$-$C_4H_9)_2CuLi$ (3 eq)	$-78, 1$ hr	1-Bromohexane (90)	181
C_3	$CH_2{=}CHCH_2OTs$	$(C_6H_5)_2CuLi$ (0.4 M, 2 eq)	$-20°$, 6 hr	Allylbenzene (89)	181
C_4	$(+)$-(S)-sec-C_4H_9OTs	$(C_6H_5)_2CuLi$ (3 eq)	—	$(-)$-(R)-sec-$C_4H_9C_6H_5$ (45)	45
	$(+)$-(S)-sec-$C_4H_9OSO_2CH_3$	$(C_6H_5)_2CuLi$	$-78°$, 2 hr, then $-20°$, 5 hr	$(-)$-(R)-sec-$C_4H_9C_6H_5$ (33)	182
C_5	$cyclo$-C_5H_9OTs	$(CH_3)_2CuLi$ (2 eq)	$0°$	Methylcyclopentane (65)	45
		$(C_6H_5)_2CuLi$ (0.15 M, 2 eq)	$0°$, 12 hr	Phenylcyclopentane (61)	181
	n-$C_5H_{11}OTs$	$(n$-$C_4H_9)_2CuLi$ (0.4 M, 4 eq)	$25°$, 1 hr (THF)	n-Nonane (98)	33
		$(n$-$C_4H_9)_2CuLi$ (0.04 M, 5 eq)	$-78°$, 0.5 hr	n-Nonane (98)	181
C_6	$(CH_3)_3CCH_2OTs$	$(C_6H_5)_2CuLi$ (0.14 M, 3 eq)	$25°$, 72 hr	Neopentylbenzene (80)	181
	$cyclo$-$C_6H_{11}OTs$	$(CH_3)_2CuLi$ (2 eq)	$0°$, 5 hr	Methylcyclohexane (20), cyclohexene (60)	45
C_7	[bicyclic structure] OTs	$(CH_3)_2CuLi$ (5 eq)	$25°$, 6 hr	[bicyclic structure R] (I)	313
		n-$C_4H_9)_2CuLi$ (5 eq)	$0°$, 3 hr	I, R = CH_3 (62)	
		t-$C_4H_9CuOC_4H_9$-$t)Li$	$-50°$, 3 hr (THF)	I, R = C_4H_9-n (84); I, R = C_4H_9-t (87)	

Note: References 236–320 are on pp. 398–400.

TABLE V. ORGANOCOPPER COUPLING WITH ALCOHOL DERIVATIVES (*Continued*)

Alcohol Derivative	Organocopper Reagent	Reaction Conditions (Solvent)	Product(s) and Yield(s) (%)	Refs.
		A. p-Toluenesulfonates (ROTs) (*Continued*)		
C_7 (*contd.*)				
OTs (bicyclic structure)	$(CH_3)_2CuLi$ (2 eq)	0°, 6 hr	2-Methylnorbornane (58), 77% *exo*, 23% *endo*	182
OTs (bicyclic structure)	$(CH_3)_2CuLi$ (2 eq)	0°, 6 hr	*exo*-2-Methylnorbornane (65)	182
$C_6H_5CH_2CH_2OTs$	$(C_6H_5)_2CuLi$ (3 eq)	—20°, 18 hr	*exo*-2-Phenylnorbornane (35) *n*-Hexylbenzene (73)	45
	$(t\text{-}C_4H_9)_2CuLi$ (5 eq)	—20°, 18 hr	$C_6H_5CH_2CH_2C_4H_9\text{-}t$ (70)	
	$(C_6H_5)_2CuLi$ (2 eq)	—78 to 0°, 18 hr	Bibenzyl (100)	
	$(C_2H_5)_2CuLi$ (0.06 M, 5 eq)	—20°, 5 hr	2-Phenylbutane (33)	
$C_6H_5(CH_3)CHOTs$				181
OTs (cyclooctene structure)	$(CH_3)_2CuLi$ (0.2 M, 5 eq)	25°, 18 hr	5-Methylcyclooctene (64)	313
$n\text{-}C_8H_{17}OTs$	$n\text{-}C_4H_9)_2CuLi$ (0.2 M, 5 eq)	0°, 12 hr	5-*n*-Butylcyclooctene (81)	
	$(CH_3)_2CuLi$ (2 eq)	0°, 1 hr	*n*-Nonane (95)	45
	$(C_2H_5)_2CuLi$ (2 eq)	0°, 5 hr	*n*-Decane (96)	232
	$E\text{-}CH_3CH=CH_2CuLi$	—10°, 2.5 hr (1:1 THF:ether)	E-2-Undecene (95)	
	$(sec\text{-}C_4H_9)_2CuLi$ (5 eq)	—15°, 16 hr	3-Methylundecane (85)	45
C_8				

382

Substrate	Organocopper reagent	Conditions	Product(s) (%)	Refs.
	(t-C₄H₉)₂CuLi (5 eq)	−15°, 16 hr	2,2-Dimethyldecane (90)	45
	(C₆H₅)₂CuLi (0.06 M, 2 eq)	0°, 24 hr	1-Phenyloctane (81)	181
n-C₆H₁₃CH(CH₃)OTs	(CH₃)₂CuLi (2 eq)	−78°, 4 hr	2-Methyloctane (87)	45
	(n-C₄H₉)₂CuLi (3 eq)	0°, 3 hr	5-Methylundecane (65)	
	(n-C₄H₉)₂CuLi (0.04 M, 2 eq)	−20°, 0.5 hr	5-Methylundecane (63)	118
	(C₆H₅)₂CuLi (0.04 M, 2 eq)	−20°, 3 hr	2-Phenyloctane (60)	
C₉ C₆H₅CH=CHCH₂OTs	(CH₃)₂CuLi	0°, 7 hr	1-Phenyl-1-butene (45)	181
C₉ C₆H₅COCH₂CH₂OTs	(CH₃)₂CuLi (0.1 M, 3 eq)	−78°, 2 hr	Butyrophenone (83)	181
 C₁₀ t-C₄H₉—⟨cyclohexane⟩—OTs (I) (I-cis, I-trans)	(CH₃)₂CuLi (2 eq)		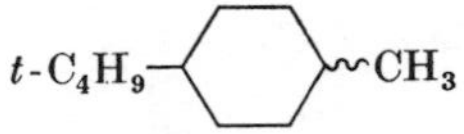 t-C₄H₉—⟨cyclohexane⟩—CH₃ (II)	45, 182
	(CH₃)₂CuLi (8 eq)	−78°, 8 hr, then 0°	II-trans (6)	
	(CH₃)₂CuLi (8 eq)	−78°, 6 hr, then −10°	II-cis (36)	
C₁₁ H₂C—CH(CH₂)₉OTs (epoxide)	(CH₃)₂CuLi (2 eq)	−20°	CH₃CH₂CHOH(CH₂)₉CH₃ (—)	181
C₁₆ 9-CH₃, 10-CH₂OTs anthracene derivative	(CH₃)₂CuLi (0.5 M, 5 eq)	25°, 24 hr	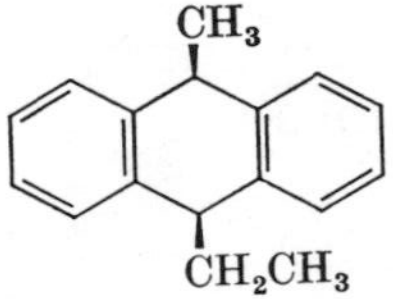 9-CH₃, 10-CH₂CH₃ anthracene derivative (79ᵃ)	314

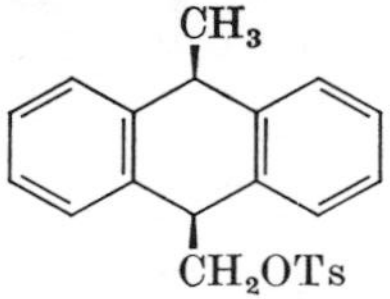

B. Allylic Acetates (ROAc)

Substrate	Organocopper reagent	Conditions	Product(s)	Refs.
C₇ CH₃—⟨cyclohexene⟩—OAc	(CH₃)₂CuLi	—	  CH₃—⟨cyclohexene⟩—CH₃ (Major product)	185

Note: References 236–320 are on pp. 398–400.

ᵃ Excess methyl iodide was added to the reaction mixture before workup.

TABLE V. ORGANOCOPPER COUPLING WITH ALCOHOL DERIVATIVES (Continued)

Alcohol Derivative	Organocopper Reagent	Reaction Conditions (Solvent)	Product(s) and Yield(s) (%)	Refs.
B. Allylic Acetates (ROAc) (Continued)				
C_7 (contd.) OSi(CH$_3$)$_2$C$_4$H$_9$-t (cyclopentene lactone, =O, O)	(n-C$_5$H$_{11}$···CuLi)$_2$ OSi(CH$_3$)$_2$C$_4$H$_9$-t	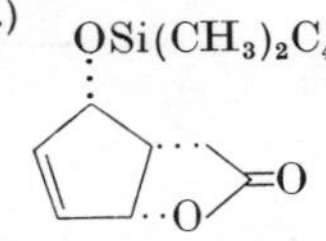 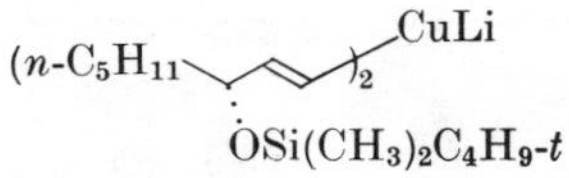 (1 eq) −78°	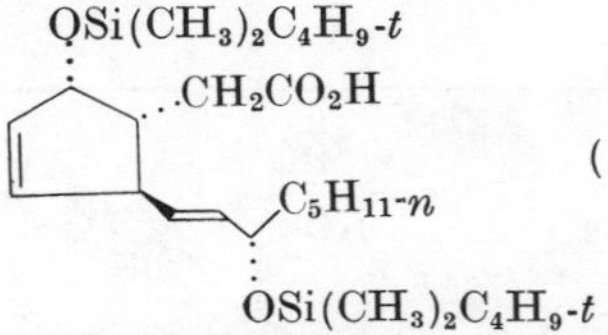 OSi(CH$_3$)$_2$C$_4$H$_9$-t ···CH$_2$CO$_2$H ···C$_5$H$_{11}$-n OSi(CH$_3$)$_2$C$_4$H$_9$-t (−)	314b
C_8 CH$_2$=C(CH$_3$)CH(OAc)C$_4$H$_9$-n	(CH$_3$)$_2$CuLi	−10°, 15 min	E-CH$_3$CH$_2$(CH$_3$)C=CHCH$_4$C$_4$H$_9$-n (78)	44
CH$_2$=C(CH$_3$)CH(OAc)CH$_2$CH$_2$C(H$_3$C dioxolane)	(CH$_3$)$_2$CuLi	−10°, 0.5 hr	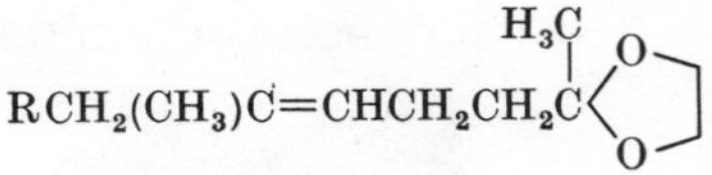 RCH$_2$(CH$_3$)C=CHCH$_2$CH$_2$C (H$_3$C dioxolane) (I) E-I, R = CH$_3$ (77)	44
	(n-C$_4$H$_9$)$_2$CuLi (C$_6$H$_5$)$_2$CuLi	−10°, 1 hr −10°, 0.5 hr	E-I, R = C$_4$H$_9$-n (67) E-I, R = C$_6$H$_5$ (54), Z-I, R = C$_6$H$_5$ (15)	44 44
C_{10} CO$_2$CH$_3$ (with OAc)	(CH$_3$)$_2$CuLi	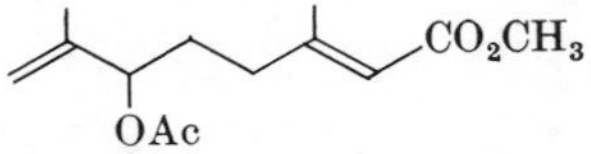 −10°, 0.5 hr	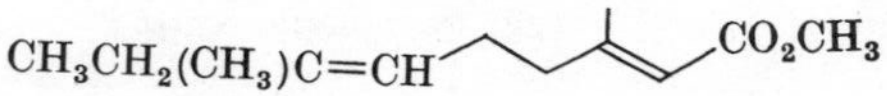 CH$_3$CH$_2$(CH$_3$)C=CH···CO$_2$CH$_3$ (I) E-I (66) Z-I (6)	44
E-t-C$_4$H$_9$CH$_2$CH=CHC(CH$_3$)$_2$OAc	(n-C$_4$H$_9$)$_2$CuLi	−10°, 0.5 hr	E-t-C$_4$H$_9$CH$_2$CH$_2$CH=CHC(CH$_3$)$_2$R (I), t-C$_4$H$_9$CH$_2$CH$_2$CH(R)CH=C(CH$_3$)$_2$ (II) R = C$_4$H$_9$-n, I (75), II (15)	44
t-C$_4$H$_9$CH$_2$CH(OAc)CH=C(CH$_3$)$_2$	(CH$_3$)$_2$CuLi	−10°, 10 hr	R = CH$_3$, I (48), II (9)	44
	(n-C$_4$H$_9$)$_2$CuLi	−10°, 4 hr	R = C$_4$H$_9$-n, I (70), II (15)	44

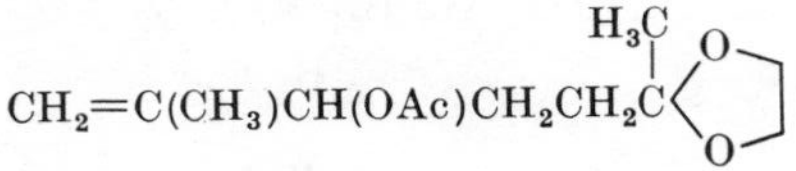

Reactant	Reagent	Conditions	Product(s) and Yield(s) (%)	Refs.
C$_{15}$ [structure: polyene with OAc groups and CO$_2$C$_2$H$_5$]	(CH$_3$)$_2$CuLi	$-10°$, 0.5 hr	C$_2$H$_5$(CH$_3$)C=CHCH$_2$CH$_2$C(C$_2$H$_5$)=CHCH$_2$...CHCH$_2$ C$_2$H$_5$O$_2$CCH=C(CH$_3$)CH$_2$ (I) E,E,E-I (14), E,Z,Z-I (8), E,Z,E-I (76)	44 184
C$_{20}$ [steroid structure with CH=CH$_2$ / OAc; CH$_3$O]	(CH$_3$)$_2$CuLi (1.5 eq)	$-10°$, then 25°, 12 hr	[structure: $\overset{H}{\underset{}{}}$ CH$_2$R, C= (I)] I, R = CH$_3$ (33)	183
	(n-C$_4$H$_9$)$_2$CuLi (C$_6$H$_5$)$_2$CuLi	— —	I, R = C$_4$H$_9$-n (40) I, R = C$_6$H$_5$ (29)	

C. Propargylic Acetates (ROAc)

Reactant	Reagent	Conditions	Product(s) and Yield(s) (%)	Refs.
C$_5$ HC≡CCH(OAc)CH=CH$_2$	(CH$_3$)$_2$CuLi (n-C$_4$H$_9$)$_2$CuLi (n-C$_8$H$_{17}$)$_2$CuLi (n-C$_8$H$_{17}$)$_2$CuLi	$-20°$ -20 to $-30°$ $-30°$ $-30°$	RCH=C=CHCH=CH$_2$ (I), HC≡C-CH=CHCH$_2$R (II) R = CH$_3$, I (40), II (3) R = C$_4$H$_9$-n, I (48), II (12) R = C$_8$H$_{17}$-n, I (47), II (5) n-C$_8$H$_{17}$CH=C=CCH=CHCH(OCH$_3$)$_2$ (25)	315
C$_6$ HC≡CCH(OAc)- CH=CH(OCH$_3$)$_2$	(CH$_3$)$_2$CuLi (1.2 eq)	$-10°$, then 25°, 5 hr	[structure] C=CHCH$_3$ (81)	315
C$_7$ [cyclopentane with OAc and C≡CH]	(CH$_3$)$_2$CuLi (1.2 eq)	$-10°$ then 25°, 5 hr	[cyclohexane] =C=CHCH$_3$ (81)	186, 43
C$_8$ [cyclohexane with OAc and C≡CH]	(n-C$_4$H$_9$)$_2$CuLi (2 eq)	$-10°$ then 25°, 5 hr	[cyclohexane] =C=CHR (I) I, R = CH$_3$ (85) I, R = C$_4$H$_9$-n (65)	186, 43

Note: References 236–320 are on pp. 398–400.

TABLE V. ORGANOCOPPER COUPLING WITH ALCOHOL DERIVATIVES (Continued)

	Alcohol Derivative	Organocopper Reagent	Reaction Conditions (Solvent)	Product(s) and Yield(s) (%)	Refs.
			C. Propargylic Acetates (ROAc) (Continued)		
C_9	$AcOCH(C_6H_5)C{\equiv}CH$	$(CH_3)_2CuLi$ (1.2 eq)	$-10°$ then $25°$, 5 hr	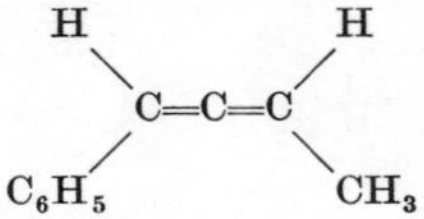 (38)	186, 43
	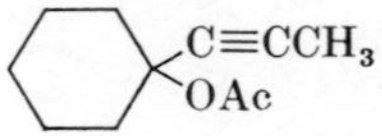 cyclohexane–$C{\equiv}CCH_3$ / OAc	$(CH_3)_2CuLi$ (5 eq)	$-10°$ then $25°$, 5 hr	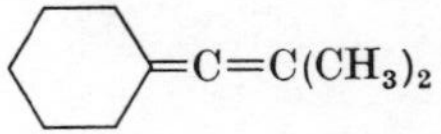 cyclohexylidene$=C=C(CH_3)_2$ (71)	186, 43
		$(n\text{-}C_4H_9)_2CuLi$	$0°$, 5 hr	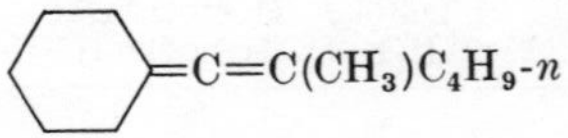 cyclohexylidene$=C=C(CH_3)C_4H_9\text{-}n$	43
	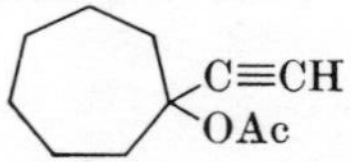 cycloheptane–$C{\equiv}CH$ / OAc	$(CH_3)_2CuLi$ (1.2 eq)	$-10°$ then $25°$, 5 hr	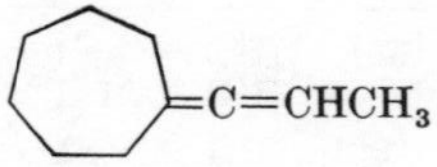 cycloheptylidene$=C=CHCH_3$ (82)	186, 43
C_{12}	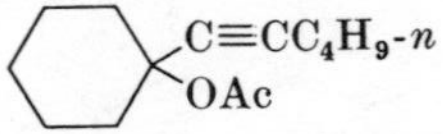 cyclohexane–$C{\equiv}CC_4H_9\text{-}n$ / OAc	$(CH_3)_2CuLi$	$0°$, 5 hr	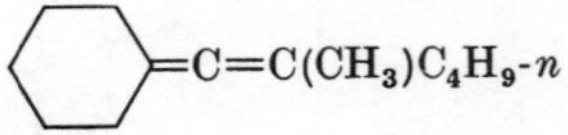 cyclohexylidene$=C=C(CH_3)C_4H_9\text{-}n$ (47)	43
C_{20}	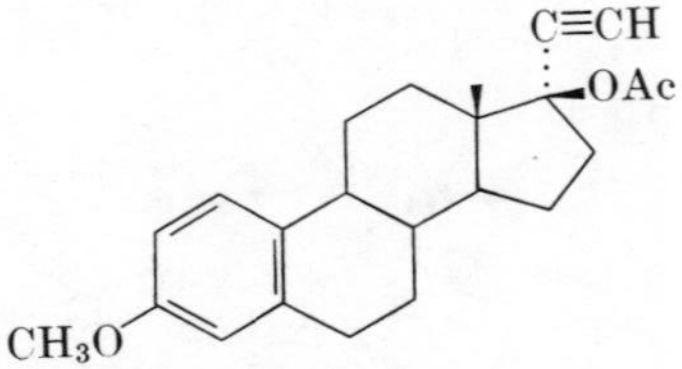	$(CH_3)_2CuLi$ (0.1 M, 1 eq)	$0°$, 1 hr	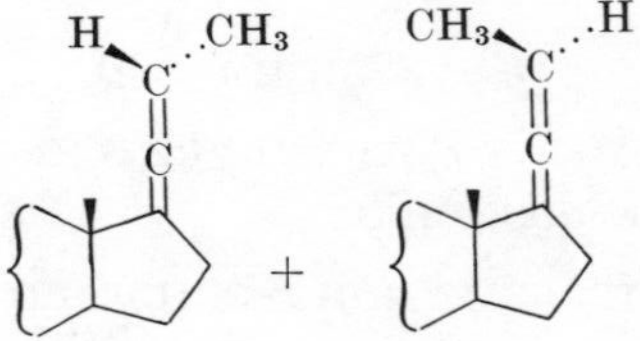 (1:4, ~90 total)	187, 43

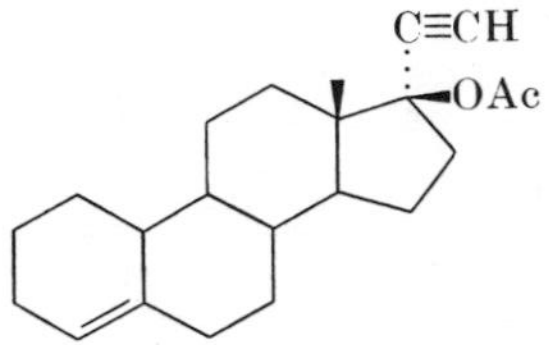

$C\equiv CH$

OAc

$(CH_3)_2CuLi$
(0.1 M, 1 eq)

0°, 1 hr

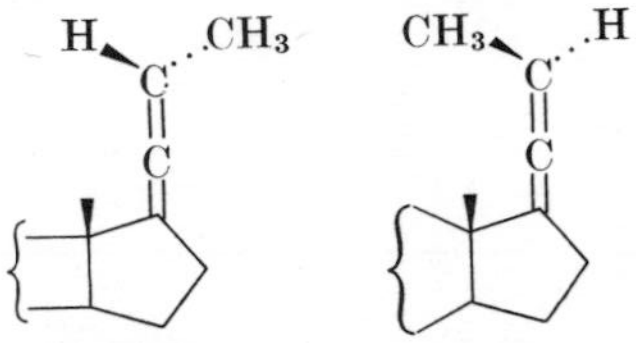

CH_3 H
C=C

(minor)

CH_3 H
C=C

(major)

(~50)

187

OAc

$CH_2CH(OAc)C\equiv C(CH_2)_3OAc$

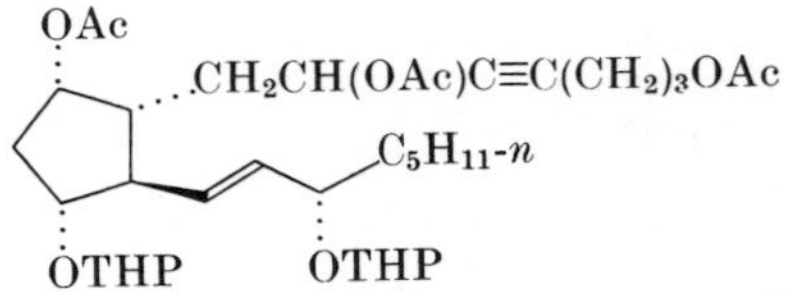

$C_5H_{11}\text{-}n$

OTHP

OTHP

$(CH_3)_2CuLi$ (4 eq)

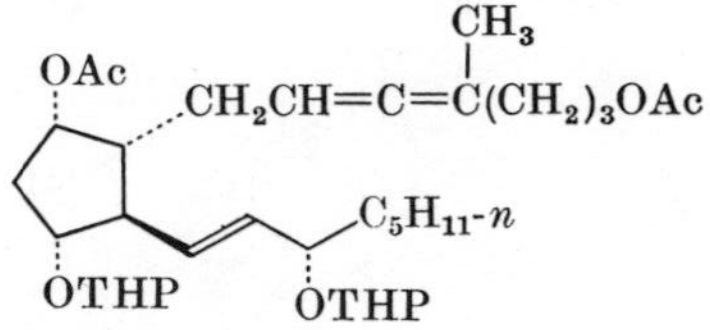

−78°

CH_3

$CH_2CH=C=C(CH_2)_3OAc$

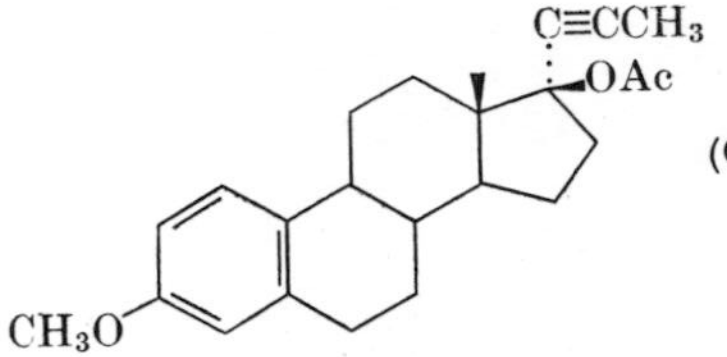

$CH_2CH=C$

$C_5H_{11}\text{-}n$

OTHP

OAc

OTHP

(75)

316

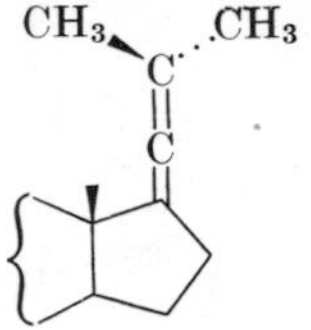

C_{21}

$C\equiv CCH_3$

OAc

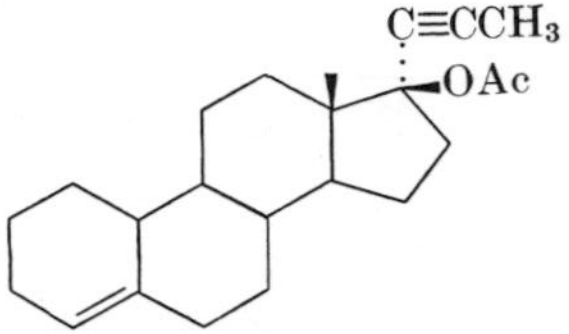

CH_3O

$(CH_3)_2CuLi$
(0.1 M, 1 eq)

0°, 1 hr

CH_3 CH_3
C=C

(~60)

187

$C\equiv CCH_3$

OAc

$(CH_3)_2CuLi$
(0.1 M, 1 eq)

0°, 1 hr

CH_3 CH_3
C=C

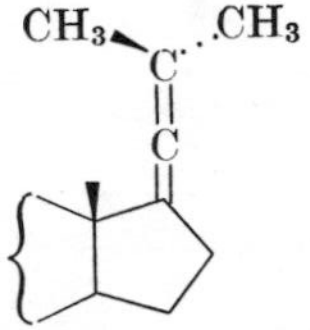

(~85)

187

Note: References 236–320 are on pp. 398–400.

387

TABLE V. ORGANOCOPPER COUPLING WITH ALCOHOL DERIVATIVES (*Continued*)

Alcohol Derivative	Organocopper Reagent	Reaction Conditions (Solvent)	Product(s) and Yield(s) (%)	Refs.
C. Propargylic Acetates (ROAc) (*Continued*)				

Note: References 236–320 are on pp. 398–400.

b The crude product was treated with acetic anhydride in pyridine.

388

TABLE II. ORGANOCOPPER COUPLING WITH EPOXIDES

	Substrate	Organocopper Reagent	Reaction Conditions (Solvent)	Product(s) and Yield(s) (%)	Refs.
C_3	Propylene oxide	$(CH_3)_2CuLi$ (2 eq)	0°, 13.5 hr	$CH_3CHOHCH_2CH_3$ (89), $(CH_3)_2CHCH_2OH$ (4), $(CH_3)_3COH$ (3)	50
C_4	CH_3—(epoxide)—$CO_2C_2H_5$	$(CH_3)_2CuLi$ (2 eq)	0°, 3 hr	$CH_3CHOHCH(CH_3)CO_2C_2H_5$ (67)	50
	3,4-Epoxy-1-butene	$(CH_3)_2CuLi$ (0.17 M, 2 eq)	Reflux, 0.5 hr	2-Penten-1-ol (94; 3.8:1 E:Z)	190
		$(n\text{-}C_4H_9)_2CuLi$	—	2-Octen-1-ol (93; 86:14 E:Z)	189
		$(C_6H_5)_2CuLi$	—	4-Phenyl-2-buten-1-ol (85; 90:10 E:Z)	189
			(ether:benzene 5:1)		
C_5	1,2-Epoxybutane	$(CH_3)_2CuLi$ (0.13 M, 2 eq)	0°, 13.5 hr	3-Pentanol (88)	50
	3,4-Epoxy-2-methyl-1-butene	$(CH_3)_2CuLi$ (0.3 M, 5 eq)	−10°	3-Methyl-2-penten-1-ol (93; 92:8 E:Z)	190, 189
	(2,3-epoxy ester, $CO_2C_2H_5$)	$(CH_3)_2CuLi$ (3 eq)	−5°	HO⟨$CO_2C_2H_5$⟩ (75), + OH⟨$CO_2C_2H_5$⟩ (25)	188
	Cyclopentene 1,2-oxide	$(CH_3)_2CuLi$ (0.1 M, 5 eq)	25°, 6 hr	*trans*-2-Methylcyclopentanol (75), cyclopentanone (10)	154
		$(n\text{-}C_4H_9)_2CuLi$ (5 eq)	25°	*trans*-2-*n*-Butylcyclopentanol (55), cyclopentanone (10), cyclopentanol (19)	154
	(epoxide with CO_2CH_3, $CH_2CO_2CH_3$; $C_6H_5CH_2O$)	$(i\text{-}C_4H_9)_2CuLi$	−40°	*i*-C_4H_9⟨CO_2H⟩ / HO⟨CO_2CH_3⟩ (60)	316b
	(bicyclic epoxide, $C_6H_5CH_2O$ groups)	$(CH_2{=}CHCH_2)_2CuLi$ (0.3 M, 2 eq)	−78 to 25°, 2 hr	($CH_2CH{=}CH_2$, OH product; $C_6H_5CH_2O$) (95)	137

Note: References 236–320 are on pp. 398–400.

TABLE VI. ORGANOCOPPER COUPLING WITH EPOXIDES (*Continued*)

Substrate	Organocopper Reagent	Reaction Conditions (Solvent)	Product(s) and Yield(s) (%)	Refs.
C_6	$(CH_3)_2CuLi$	—, 0°, 0.5 hr	$R = CH_3$, I (35), II (42)	191
	$(CH_3)_2CuLi$ (0.1 M, 2 eq)	0°, 0.5 hr	$R = CH_3$, I (42), II (49)	192
	$(C_6H_5)_2CuLi$ (0.25 M, 5 eq)	0°, 0.5 hr	$R = C_6H_5$, I (60), II (27)	192
	$(t\text{-}C_4H_9)_2CuLi$	−40°, 4 hr	$R = t\text{-}C_4H_9$, I (8; 17 *cis*), II (41; 6 *cis*)	192
	$(CH_3)_2CuLi$	—	(−)	191, 317
	$(CH_3)_2CuLi$	—	No reaction	191
	$(CH_3)_2CuLi$ (4.7 eq)	—	(76) + ketonic products (24)	188

390

1,2-Epoxycyclohexane	$(CH_3)_2CuLi$ (0.4 *M*, 5 eq)	0°, 18 hr then 25°, 5 hr	*trans*-2-Methylcyclohexanol (75) cyclohexanone (15)	46, 50
	$(n\text{-}C_4H_9)_2CuLi$ (0.1 *M*, 5 eq)	25°	*trans*-2-*n*-Butylcyclohexanol (60)	154
	$(t\text{-}C_4H_9)_2CuLi$	—	*trans*-2-*t*-Butylcyclohexanol (—)	192
	$(C_6H_5)_2CuLi$ (2 eq)	0°, 5 hr	*trans*-2-Phenylcyclohexanol (81)	50
1,4-Epoxycyclohexane	$(CH_3)_2CuLi$ (0.2 *M*, 5 eq)	0°, 4 hr then 25°, 18 hr	No reaction	46
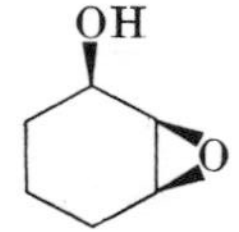	$(CH_3)_2CuLi$ (3 eq)	−5°	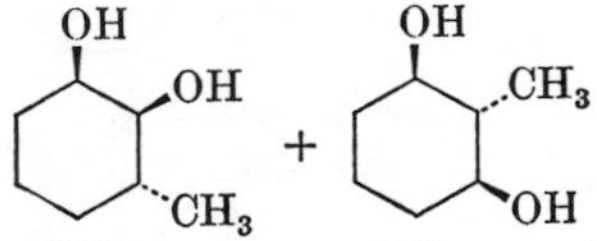 (83) (14)	188
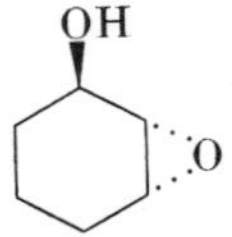	$(CH_3)_2CuLi$ (3 eq)	−5°	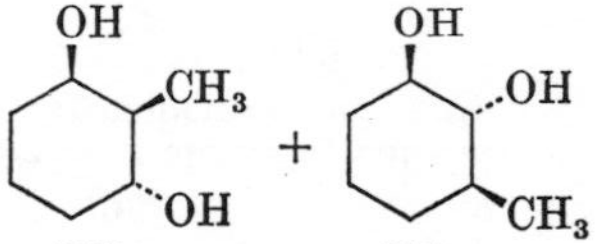 (85) (15)	188
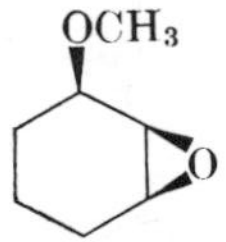	$(CH_3)_2CuLi$ (3 eq)	−5°	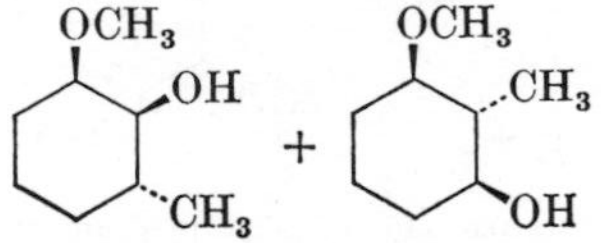 (49) (45)	188
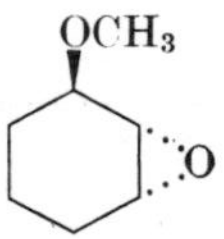	$(CH_3)_2CuLi$ (3 eq)	−5°	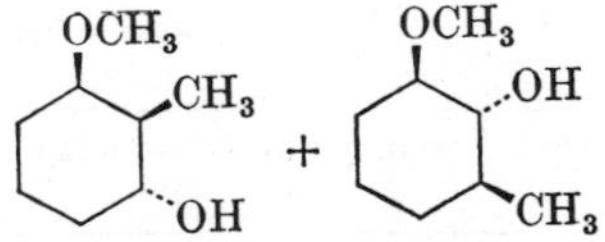 (29) (69)	188

Note: References 236–320 are on pp. 398–400.

TABLE VI. ORGANOCOPPER COUPLING WITH EPOXIDES (*Continued*)

Substrate	Organocopper Reagent	Reaction Conditions (Solvent)	Product(s) and Yield(s) (%)	Refs.
C_6 (*contd.*)	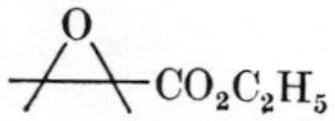 $(CH_3)_2CuLi$	—	No reaction	188
C_7 *exo*-Norbornene 2,3-oxide	$(CH_3)_2CuLi$ (0.1 M, 5 eq)	55°, 5 hr (1,2-dimethoxy-ethane)	*trans*-3-Methyl-2-norborneol (58)	154
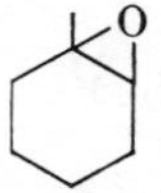	$(CH_3)_2CuLi$	—	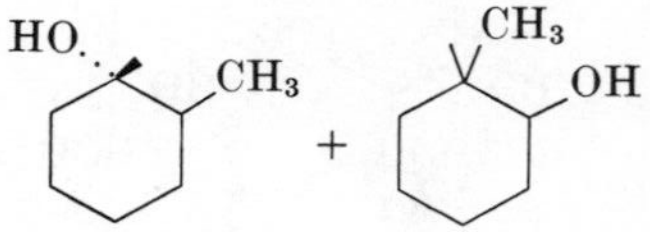 (96) + (4)	188
Cycloheptene 1,2-oxide	$(CH_3)_2CuLi$ (0.1 M, 5 eq)	25°, 48 hr	*trans*-2-Methylcycloheptanol (60), cycloheptanone (20)	154
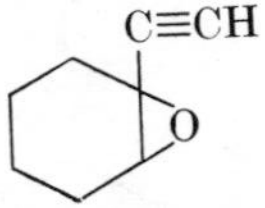	$(CH_3)_2CuLi$ (6 eq) $(n\text{-}C_4H_9)_2CuLi$ (6 eq)	−25° −25°	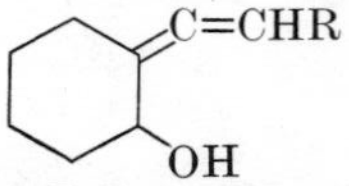 (I) I, R = CH$_3$ (40) I, R = $C_4H_9\text{-}n$ (34)	317b
C_8	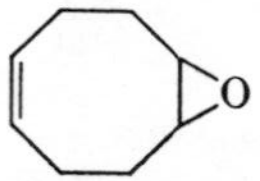 $(CH_3)_2CuLi$	25°, 7 hr	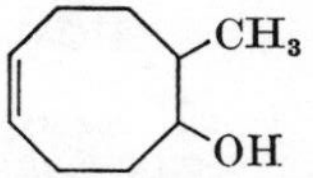 (45)	313
cis-1,2-Epoxycyclooctane	$(CH_3)_2CuLi$ (0.1 M, 5 eq)	70°, 30 hr (1,2-dimethoxy-ethane)	*trans*-2-Methylcyclooctanol (21), cyclooctanone (15)	154

392

C_9	$n\text{-}C_4H_9C\!\equiv\!C$— [epoxide]	$(CH_3)_2CuLi$ (6 eq) $(n\text{-}C_4H_9)_2CuLi$ (6 eq)	$-25°$ $-25°$	$n\text{-}C_4H_9C(R)\!=\!C\!=\!C(CH_3)CH_2OH$ (I) I, $R = CH_3$ (75) I, $R = C_4H_9\text{-}n$ (34)	317b
C_{10}	$n\text{-}C_4H_9C\!\equiv\!C$— [epoxide] —$C_2H_5$	$(CH_3)_2CuLi$ (6 eq)	$-25°$	$n\text{-}C_4H_9C(CH_3)\!=\!C\!=\!CHCHOHCH_3$ (60)	317b
C_{11}	[epoxide]$(CH_2)_8CO_2CH_3$	$(CH_3)_2CuLi$ (2 eq)	$0°$, 13.5 hr	$CH_3CH_2CHOH(CH_2)_8CO_2CH_3$ (37)	50
C_{12}	[epoxide]$(CH_2)_8COCH_3$	$(CH_3)_2CuLi$ (2 eq)	$-50°$, 0.5 hr	$CH_3CH_2CHOH(CH_2)_8COCH_3$ (68)	50
C_{13}	(I) I, α-oxide I, β-oxide			(II) II, $R_1 = OH$ $R_2 = CH_3$ (75) II, $R_1 = CH_3$ $R_2 = OH$ (70)	317c
C_{15}	$C_6H_5C\!\equiv\!C$— [epoxide]	$(CH_3)_2CuLi$ (6 eq)	$-25°$ (Hexane)	$C_6H_5C(CH_3)\!=\!C\!=\!C(C_3H_7\text{-}i)C(CH_3)_2OH$ (68)	317b
C_{27}	$2\alpha,3\alpha$-Epoxy-5α-cholestane	$(CH_3)_2CuLi$ $(0.1\,M,\ 5\ eq)$	$25°$, 24 hr	$2\text{-}\beta$-Methyl-5α-cholestan-$3\text{-}\alpha$-ol (62)	154

Note: References 236–320 are on pp. 398–400.

393

TABLE VII. ORGANOCOPPER COUPLING WITH MISCELLANEOUS SUBSTRATES

	Substrate	Organocopper Reagent	Reaction Conditions (Solvent)	Product(s) and Yield(s) (%)	Refs.
C_2	$N_2CHCO_2C_2H_5$	C_6F_5Cu (1.1 eq)	0° (THF)	$C_6F_5CH_2CO_2C_2H_5$ (43)	103
		C_6H_5Cu (1.4 eq)	−15°	Ethyl phenylacetate (52)[a]	116
		$p\text{-}CH_3C_6H_4Cu$ (1.4 eq)	−15°	Ethyl p-tolylacetate (46)[a]	116
C_4		$o\text{-}C_6F_5C_6F_4Cu$	(—)	$o\text{-}C_6F_5C_6F_4CH_2CO_2C_2H_5$ (—)	139
	$C_2H_5C(OC_2H_5)_2CO_2C_2H_5$	$(CH_3)_2CuLi$ (6 eq)	35°, 3.5 hr	$C_2H_5C(OC_2H_5)_2COCH_3$ (95)	196
	$n\text{-}C_4H_9SC(CH_3){=}CHCO_2C_2H_5$	$(CH_3)_2CuLi$ (10 eq)	0°, 2 hr	$(CH_3)_2C{=}CHCO_2C_2H_5$ (I, 70)	317d
	$C_6H_5SC(CH_3){=}CHCO_2C_2H_5$	$(CH_3)_2CuLi$ (10 eq)	0°, 1 hr	I (80)	317c
	$CH_3CO_2C(CH_3){=}CHCO_2C_2H_5$	$(CH_3)_2CuLi$ (1.1 eq)	−78°	I (91)	317e
	$Z\text{-}CH_3CO_2C(CH_3){=}CHCO_2CH_3$	$(C_2H_5)_2CuLi$	—	$C_2H_5C(CH_3){=}CHCO_2CH_3$ (II, 52) (1:1 Z:E)	317e
C_5	$Z\text{-}CH_3CO_2C(C_2H_5){=}CHCO_2CH_3$	$(CH_3)_2CuLi$	−78°	II (92; 1:10. 8 Z:E)	317e
	$CH_3CO_2C(CH_3){=}CHCOCH_3$	$(CH_3)_2CuLi$ (1.0 eq)	−78°	$(CH_3)_2C{=}CHCOCH_3$ (76)	317e
	[cyclohexenone with O_2CCH_3 and CH_3 substituents]	$(CH_3)_2CuLi$ (1.0 eq)	−78°	[cyclohexanone with $(CH_3)_2C$] (99)	317e
C_6	[γ-butyrolactone $CH_3CO_2C'(CH_3){=}$]	$(CH_3)_2CuLi$ (1.0 eq)	−78°	[γ-butyrolactone with $(CH_3)_2C{=}$] (88)	317e
	1,3,5-Trinitrobenzene	[2-thienyl]–Cu	−10 to 0°, 5 hr (pyridine)	[benzene ring: R, NO_2, NO_2, O_2N] (I) (—)	318, 195

Substrate	Reagent	Conditions	Product(s)	Refs.
	2-Furyl·Cu	−10 to 0°, 5 hr (pyridine)	I, R = 2-furyl (45)[b]	195
	2,6-(CH₃O)₂C₆H₃Cu	—	I, R = 2,6-(CH₃O)₂C₆H₃ (46–70)	118
	C₆H₅C≡CCu	25°, 15 min (pyridine)	I, R = C₆H₅C≡C (5) +	194

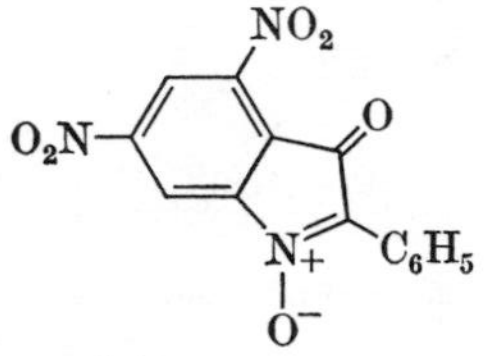

$$I, R = C_6H_5C\!\equiv\!C \quad (6)$$

Substrate	Reagent	Conditions	Product(s)	Refs.
m-Dinitrobenzene	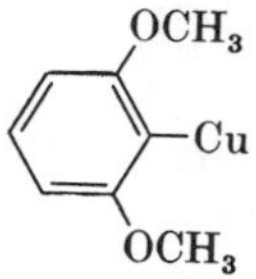	50°, 20 hr (pyridine)	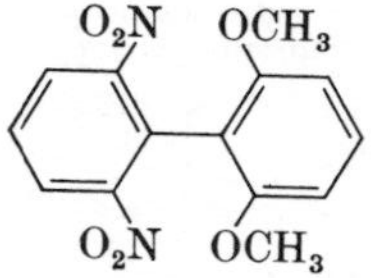	118

reagent: 2,6-dimethoxyphenyl copper structure (OCH₃ / Cu / OCH₃)

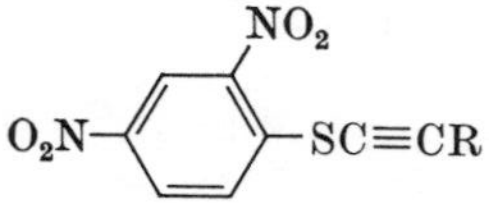

product (42): O₂N / OCH₃ — OCH₃ / O₂N biphenyl structure

Substrate	Reagent	Conditions	Product(s)	Refs.
(2-nitro-4-nitrophenyl SCl)	n-C₃H₇C≡CCu (1 eq)	Reflux, 48 hr (CH₃CN)	SC≡CR (I)	61
	C₆H₅C≡CCu (1 eq)	Reflux, 48 hr (CH₃CN)	I, R = C₃H₇-n (70) I, R = C₆H₅ (80)	

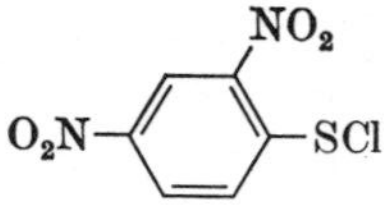

product I structure: O₂N / NO₂ / SC≡CR

C₇	C₆H₅CO₂C₂H₅	(CH₃)₂CuLi (3 eq)	—	C₆H₅COCH₃ (85), C₆H₅COH(CH₃)₂ (5–15)	196
	C₆H₅CH₂N⁺(CH₃)₃Br⁻	(n-C₄H₉)₂CuLi	0°, 24 hr	C₆H₅CH₂CH₂C₄H₉-n (67)	313
	(γ-lactone structure)	(CH₃)₂CuLi	−10°, 0.5 hr	C₂H₅(CH₃)C=CCH₂CH₂CO₂H (E 60, Z 24)	44

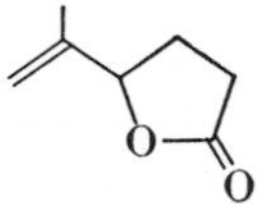

Note: References 236–320 are on pp. 398–400.

[a] The reaction mixture was hydrolyzed with dilute hydrochloric acid.

[b] This product was formed by acidification (H₂SO₄) and oxidation (p-benzoquinone) of the reaction mixture.

TABLE VII. ORGANOCOPPER COUPLING WITH MISCELLANEOUS SUBSTRATES (*Continued*)

	Substrate	Organocopper Reagent	Reaction Conditions (Solvent)	Product(s) and Yield(s) (%)	Refs.
C_8	$N_2CHCOC_6H_4NO_2$-p	C_6H_5Cu (1.4 eq)	$-15°$	$C_6H_5CH_2COC_6H_4NO_2$-p (87a)	116
	$N_2CHCOC_6H_5$	C_6H_5Cu (1.4 eq)	$-15°$	$C_6H_5CH_2COC_6H_5$ (35a)	116
		p-$CH_3C_6H_4Cu$ (1.4 eq)	$-15°$	p-$CH_3C_6H_4CH_2COC_6H_5$ (31a)	116
	E-$C_6H_5CH{=}CHSCH_3$	$(n$-$C_4H_9)_2CuLi$ (0.25 M, 10 eq)	$-40°$, 10 min, then 25°, 15 hr	E-1-Phenylhexene (50)	165
	E-$C_6H_5CH{=}CHSOC_6H_4Cl$-p	$(n$-$C_4H_9)_2CuLi$ (0.25 M, 10 eq)	0°, 2 hr	E-1-Phenylhexene (50)	159
	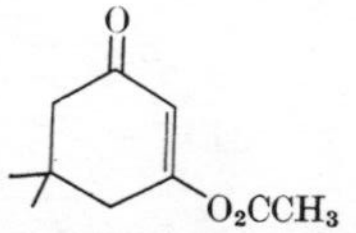	$(CH_3)_2CuLi$ (1.0 eq)	$-78°$	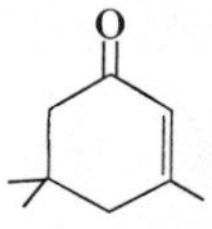 (91)	317e
C_9	Z-$(CH_3)_2C{=}CH(CH(CH_3)_2)$-$C(OAc){=}CHCO_2CH_3$	$(CH_3)_2CuLi$ (1.1 eq)	$-78°$	$(CH_3)_2C{=}CH(CH_2)_2C(CH_3){=}CHCO_2CH_3$ (83.5; 1:10.3 Z:E)	318b
	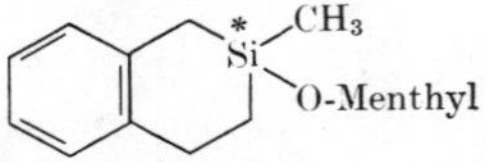	$(CH_3)_2CuLi$	20°	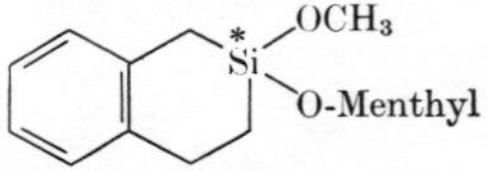 (high)	288c
C_{10}	Bromobullvalene	$(CH_3)_2CuLi$	$-10°$, 24 hr	Methylbullvalene (92)	319
	E-t-$C_4H_9CH_2CH_2CH{=}CHC$-$(CH_3)_2OCH_3$	$(CH_3)_2CuLi$		No reaction	44
	E-$C_6H_5C(CH_3){=}$ $CHCH_2SO_2C_6H_4Cl$-p	$(CH_3)_2CuLi$ (0.25 M, 5 eq)	0°, 1 hr	E-2-Phenyl-2-pentene (86)	116
C_{13}	$N_2C(C_6H_5)_2$	C_6H_5Cu (1.4 eq)	$-15°$ then 25°	$(C_6H_5)_3CH$ (41a)	116
		C_6H_5Cu (1.4 eq)	$-15°$ then 25°	$(C_6H_5)_3COH$ (39c)	116
	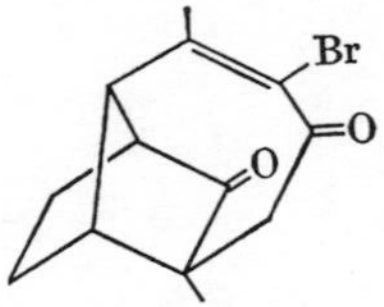	$(CH_3)_2CuLi$ (3 eq)	$-40°$, then 0°, 15 min	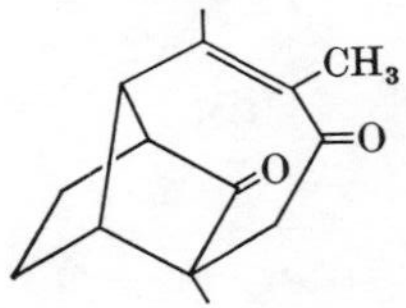 (24)	320

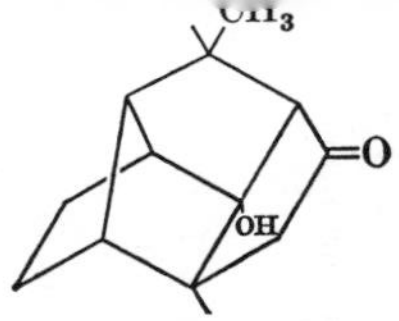

(61)

C$_{16}$	C$_6$H$_5$-α-Naphthyl SiH$_2$	(CH$_3$)$_2$CuLi	20°	C$_6$H$_5$-α-Naphthyl Si(CH$_3$)$_2$ (high)	288c
	C$_6$H$_5$-α-Naphthyl SiOCH$_3$	(CH$_3$)$_2$CuLi	20°	C$_6$H$_5$-α-Naphthyl SiCH$_3$ (high)	288c
	C$_6$H$_5$-α-Naphthyl Si(H)O-menthyl	(CH$_3$)$_2$CuLi	20°	C$_6$H$_5$-α-Naphthyl Si(H)CH$_3$ (high, retention)	288c
	(C$_6$H$_5$)$_3$SiH	(CH$_3$)$_2$CuLi	20°	(C$_6$H$_5$)$_3$SiCH$_3$ (high)	288c

C$_{18}$ (structure, Si–X, Naphthyl-α)   (CH$_3$)$_2$CuLi   20°   (structure Si–CH$_3$, Naphthyl-α) (high, retention)   288c

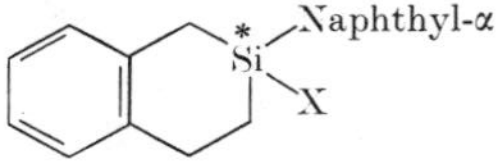

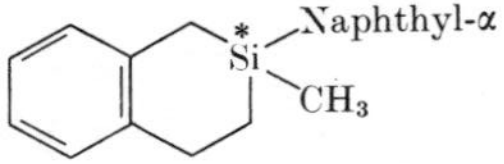

C$_{19}$ X = H or OCH$_3$

C$_{21}$ (steroid, CH$_3$O, OAc, CF$_2$ cyclopropene)   (CH$_3$)$_2$CuLi (10 eq)   —   (74)   183

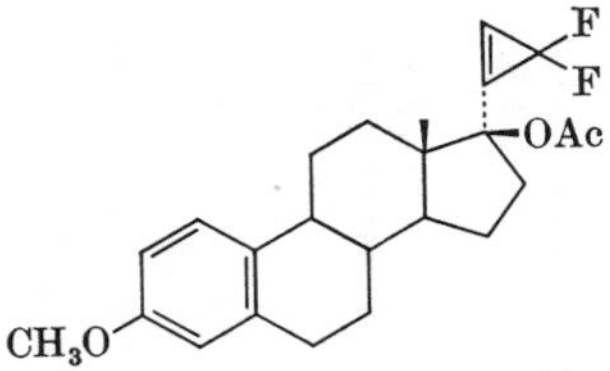

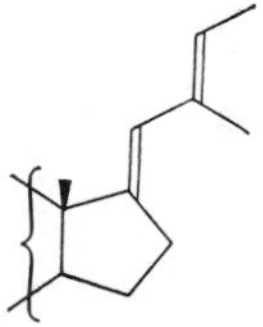

C$_{22}$ (steroid, AcO, OAc, CF$_2$ cyclopropene, H)   (CH$_3$)$_2$CuLi (10 eq)   —   (low)   183

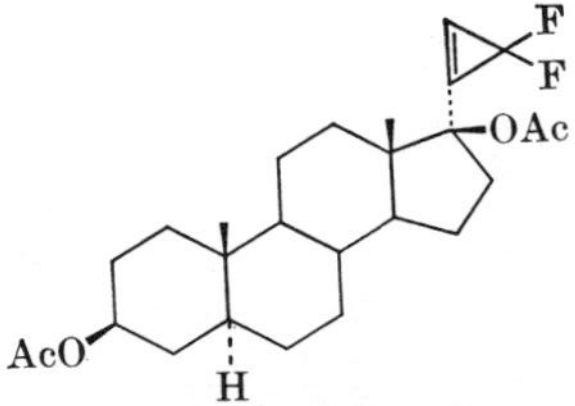

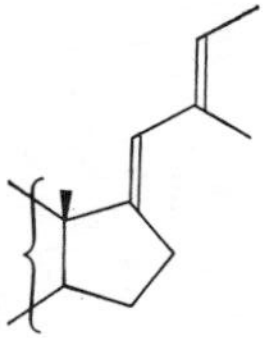

Note: References 236–320 are on pp. 398–400.

[a] The reaction mixture was hydrolyzed with dilute hydrochloric acid.
[b] This product was formed by acidification (H$_2$SO$_4$) and oxidation (p-benzoquinone) of the reaction mixture.
[c] Oxygen was bubbled into the reaction mixture before hydrolysis with dilute hydrochloric acid.

[236] M. Tamura and J. Kochi, *Bull. Chem. Soc. Japan*, **44**, 3063 (1971).

[237] T. Makaiyama, K. Narasaka, and M. Furusato, *Bull. Chem. Soc. Jap.*, **45**, 652 (1972).

[238] (a) M. Nilsson and C. Ullenius, *Acta Chem. Scand.*, **25**, 2428 (1971).

[238] (b) J. E. Parks, B. E. Wagner, and R. H. Holm, *J. Organometal. Chem.*, **56**, 53 (1973).

[239] R. J. DePasquale and C. Tamborski, *J. Organometal. Chem.*, **13**, 273 (1968).

[240] R. Reich, *C.R. Acad. Sci.*, **177**, 322 (1923).

[241] M. Ryang, K. Yoshida, H. Yokoo, and S. Tsutsumi, *Bull. Chem. Soc. Jap.*, **38**, 636 (1965).

[242] H. Hashimoto and T. Nakano, *J. Org. Chem.*, **31**, 891 (1966).

[243] M. Nilsson and O. Wennerström, *Tetrahedron Lett.*, **1968**, 3307.

[244] T. Kauffmann, G. Beissner, E. Köppelmann, D. Kuhlmann, A. Schott, and H. Schrecken, *Angew. Chem.*, **80**, 117 (1968); *Angew. Chem., Int. Ed. Engl.*, **7**, 131 (1968).

[245] T. Kauffmann, D. Kuhlman, W. Sahm, and H. Schrecken, *Angew. Chem.*, **80**, 566 (1968); *Angew. Chem., Int. Ed. Engl.*, **7**, 541 (1968).

[246] (a) T. Kauffmann, G. Beissner, H. Berg, E. Köppelmann, J. Legler, and M. Schönfelder, *Angew. Chem.*, **80**, 565 (1968).

[246] (b) J. F. Normant, A. Alexakis, and J. Villieras, in press.

[247] T. Kauffmann, J. Legler, E. Ludorff, and H. Fischer, *Angew. Chem., Int. Ed. Engl.*, **11**, 846 (1972).

[248] T. Kauffmann, E. Wienhöfer, and A. Woltermann, *Angew. Chem.*, **83**, 796 (1971); *Angew. Chem., Int. Ed. Engl.*, **10**, 741 (1971).

[249] R. G. R. Bacon and H. A. O. Hill, *Quart. Rev.* (London), **19**, 103 (1965).

[250] T. Kauffmann, J. Jackisch, A. Woltermann, and P. Röwemeier, *Angew. Chem., Int. Ed. Engl.*, **11**, 844 (1972).

[251] G. M. Whitesides, E. J. Panek, and E. R. Stedronsky, *J. Amer. Chem. Soc.*, **94**, (1972).

[252] G. Wittig and G. Klar, *Ann.*, **704**, 91 (1967).

[253] T. Kauffmann, J. Jackisch, H. J. Streitberger, and E. Wienhöfer, *Angew. Chem.*, **83**, 799 (1971); *Angew. Chem., Int. Ed. Engl.*, **10**, 744 (1971).

[254] T. Kauffmann, G. Beissner, W. Sahm, and A. Woltermann, *Angew. Chem.*, **82**, 815 (1970); *Angew. Chem., Int. Ed. Engl.*, **9**, 808 (1970).

[255] T. Kauffmann, G. Beissner, and R. Maibaum, *Angew. Chem.*, **83**, 795 (1971); *Angew. Chem., Int. Ed. Engl.*, **10**, 740 (1971).

[256] S. A. Kandel and R. E. Dessy, *J. Amer. Chem. Soc.*, **88**, 3207 (1966).

[257] (a) T. Kauffmann and A. Woltermann, *Angew. Chem., Int. Ed. Engl.*, **11**, 842 (1972).

[257] (b) J. F. Normant, G. Cahiez, C. Chuit, and J. Villieras, *Tetrahedron Lett.*, **1973**, 2407.

[258] M. Tamura and J. Kochi, *J. Amer. Chem. Soc.*, **93**, 1485 (1971).

[259] L. I. Zakharkin and L. P. Sorokina, *Zh. Org. Khim.*, **6**, 2490 (1970); *J. Org. Chem. USSR*, **6**, 2482 (1970).

[260] J-P. Gorlier, L. Hamon, J. Levisalles, and J. Wagnon, *Chem. Commun.*, **1973**, 88.

[261] (a) C. Tamborski, E. J. Soloski, and R. J. DePasquale, *J. Organometal. Chem.*, **15**, 494 (1968).

[261] (b) T. A. Bryson, *Tetrahedron Lett.*, **1973**, 4923.

[262] (a) G. Zweifel, G. M. Clark, and R. Lynd, *Chem. Commun.*, **1971**, 1593.

[262] (b) G. Santini, M. LeBlanc, and J. Riess, *Tetrahedron*, **29**, 2411 (1973).

[263] M. Brookhart, E. R. Davis, and D. L. Harris, *J. Amer. Chem. Soc.*, **94**, 7853 (1972).

[263] (a) M. S. Baird, *J. Chem. Soc. Chem. Commun.*, **1974**, 196.

[264] J. A. Katzenellenbogen and E. J. Corey, *J. Org. Chem.*, **37**, 1441 (1972).

[265] E. J. Corey, K. Achiwa, and J. A. Katzenellenbogen, *J. Amer. Chem. Soc.*, **91**, 4318. (1969).

[266] (a) L. A. Paquette and J. C. Stowell, *Tetrahedron Lett.*, **1970**, 2259.

[266] (b) S. B. Bowlus and J. A. Katzenellenbogen, *J. Org. Chem.*, **38**, 2733 (1973).

[266] (c) M. P. Cooke, Jr., *Tetrahedron Lett.*, **1973**, 1983.

[267] W. T. Borden, Harvard University, personal communication.

[268] (a) A. E. Jukes, S. S. Dua, and H. Gilman, Iowa State University, personal communication.

[268] (b) L. A. Levy, *Synthesis*, **1973**, 107.

[269] (a) A. M. Nilsson, *Tetrahedron Lett.*, **1966**, 679; (b) M. Nilsson and R. Wahren, *J. Organometal. Chem.*, **16**, 515 (1969).

[270] S. S. Dua, A. E. Jukes, and H. Gilman, *Org. Prep. Proc.*, **1**, 187 (1969).

[271] M. F. Lappert and R. Pearce, *Chem. Commun.*, **1973**, 24.

[272] (a) G. L. Anderson and L. M. Stock, *J. Org. Chem.*, **36**, 1140 (1971).

[272] (b) K. Oshima, H. Yamamoto, and H. Nozaki, *J. Amer. Chem. Soc.*, **95**, 7927 (1973).

[273] E. E. van Tamelen and J. P. McCormick, *J. Amer. Chem. Soc.*, **92**, 737 (1970).

[274] (a) G. H. Posner and J. J. Sterling, The Johns Hopkins University, unpublished results.

[274] (b) J. Villieras, J-R. Disnar, D. Masure, and J. F. Normant, *J. Organometal. Chem.*, **57**, C95 (1973).

[275] A. C. Ranade and H. Gilman, *J. Heterocycl. Chem.*, **6**, 253 (1969).

[276] H. Gilman and A. F. Webb, Iowa State University, unpublished results.

[277] (a) A. E. Jukes and H. Gilman, *J. Organometal. Chem.*, **17**, 145 (1969).

[277] (b) M. Bourgain and J. F. Normant, *Bull. Soc. Chim. France*, **1973**, 2137.

[278] A. E. Jukes, S. S. Dua, and H. Gilman, *J. Organometal. Chem.*, **12**, p44 (1968).

[279] R. Filler, personal communication as reported by S. S. Dua, A. E. Jukes, and H. Gilman, *Organometal. Chem. Rev.*, **1**, 87 (1970).

[279] (a) C. Frejaville and R. Julien, *Tetrahedron Lett.*, **1974**, 1021.

[280] G. H. Posner and C. E. Whitten, *Tetrahedron Lett.*, **1970**, 4647.

[281] J. E. Dubois, M. Boussu, and C. Lion, *Tetrahedron Lett.*, **1971**, 829.

[281] (a) P. F. Hudrlik and D. Petersen, *Tetrahedron Lett.*, **1974**, 1133.

[282] E. J. Corey and I. Kuwajima, Harvard University, unpublished results.

[283] R. Lundin, C. Moberg, R. Wahren, and O. Wennerström, *Acta Chem. Scand.*, **26**, 2045 (1972).

[284] N. T. Luong-Thi and H. Rivière, *Tetrahedron Lett.*, **1971**, 587.

[285] (a) E. N. Cain, *Tetrahedron Lett.*, **1971**, 1865.

[285] (b) S. Rozen, I. Shahak, and E. D. Bergmann, *Synthesis*, **1972**, 701.

[286] (a) J. E. Dubois and M. Boussu, *C.R. Acad. Sci., Ser. C*, **273**, 1101 (1971).

[286] (b) Unpublished results of J. I. Shulman, The Proctor and Gamble Co., Miami Valley Laboratories.

[287] W. G. Dauben and J. Seeman, University of California (Berkeley), unpublished results.

[288] (a) P. Crabbé and E. Velarde, *Chem. Commun.*, **1972**, 241.

[288] (b) S. W. Staley and N. J. Pearl, *J. Amer. Chem. Soc.*, **95**, 2731 (1973).

[288] (c) G. Chauviere and R. Corriu, *J. Organometal. Chem.*, **50**, C5 (1973).

[288] (d) B. Ganem and M. S. Kellogg, *J. Org. Chem.*, **39**, 575 (1974).

[288] (e) D. W. Knight and G. Pattenden, *J. Chem. Soc. Chem. Commun.*, **1974**, 188.

[288] (f) R. J. Anderson, C. A. Henrick, and L. D. Rosenblum, *J. Amer. Chem. Soc.*, manuscript submitted.

[289] L. Yu. Ukhin, A. M. Sladkov, and U. N. Gorshkov, *Zh. Org. Khim.*, **4**, 25 (1968); *J. Org. Chem. USSR*, **4**, 21 (1968).

[290] W. E. Truce and G. C. Wolf, *J. Org. Chem.*, **36**, 1727 (1971).

[291] A. M. Sladkov and L. Yu. Ukhin, *Russ. Chem. Rev.*, **37**, 748 (1968).

[292] T. Teitei, P. J. Collin, and W. H. F. Sasse, *Austr. J. Chem.*, **25**, 171 (1972).

[293] R. E. Atkinson, R. F. Curtis and J. A. Taylor, *J. Chem. Soc., C*, **1967**, 578.

[294] R. E. Atkinson, R. F. Curtis, D. M. Jones, and J. A. Taylor, *Chem. Commun.*, **1967**, 718.

[295] M. D. Rausch, A. Seigel, and L. P. Klemann, *J. Org. Chem.*, **31**, 2703 (1966).

[296] I. L. Kotlyarevskii, V. N. Andrievskii, and M. S. Shvartsberg, *Chem. Heterocycl. Compounds*, **3**, 236 (1967).

[297] R. Filler and E. W. Heffern, *J. Org. Chem.*, **32**, 3249 (1967).

[298] G. Martelli, P. Spagnolo, and M. Tiecco, *J. Chem. Soc., B*, **1970**, 1413.

[299] A. M. Sladkov, L. Ukhin, and V. V. Korshak, *Isv. Akad. Nauk SSSR*, **1963**, 2213; *Bull. Acad. Sci. USSR, Div. Chem. Sci.*, **1963**, 2043.

[300] C. E. Castro and R. D. Stephens, *J. Org. Chem.*, **28**, 2163 (1963).

[301] J. J. Eisch and M. W. Foxton, *J. Org. Chem.*, **36**, 3520 (1971).

[302] J. F. Normant, M. Bourgain, and A. N. Rone, *C.R. Acad. Sci., Ser. C*, **270**, 354 (1970).

[303] M. D. Rausch and A. Siegel, *J. Organometal. Chem.*, **17**, 117 (1969).

400 ORGANIC REACTIONS

304 R. E. Atkinson, R. E. Curtis, and G. T. Phillips, *Tetrahedron Lett.*, **1964**, 3159.
305 R. E. Atkinson, R. F. Curtis, and G. T. Phillips, *J. Chem. Soc.*, *C*, **1967**, 2011.
306 R. E. Atkinson, R. F. Curtis, and G. T. Phillips, *J. Chem. Soc.*, *C*, **1966**, 1101.
307 F. Bohlmann, C. Zdero, and W. Gordon, *Chem. Ber.*, **100**, 1193 (1967).
308 M. P. Briede and O. Ya. Neiland, *Zh. Org. Khim.*, **6**, 1701 (1970); *J. Org. Chem. USSR* **6**, 1706 (1970).
309 R. E. Dessy and S. A. Kandil, *J. Org. Chem.*, **30**, 3857 (1965).
310 P. L. Coe and N. E. Milner, *J. Organometal. Chem.*, **39**, 395 (1972).
311 H. O. House, L. E. Huber, and M. J. Umen, *J. Amer. Chem. Soc.*, **94**, 8471 (1972).
312 A. M. Sladkov and I. R. Gol'ding, *Zh. Org. Khim.*, **3**, 1338 (1967) [*C.A.*, **67**, 99804h (1967).
313 G. H. Posner and J-S. Ting, The Johns Hopkins University, unpublished results; G. H. Posner and J-S. Ting, *Tetrahedron Lett.*, **1974**, 683.
314 (a) D. J. Schaeffer and H. E. Zieger, *J. Org. Chem.*, **34**, 3958 (1969).
314 (b) E. J. Corey and J. Mann, *J. Amer. Chem. Soc.*, **95**, 6832 (1973).
315 C. Descoins, C. A. Henrick, and J. B. Siddall, *Tetrahedron Lett.*, **1972**, 3777.
316 (a) P. Crabbé and H. Carpio, *Chem. Commun.*, **1972**, 904.
316 (b) J. Auerbach, T. Ipaktchi, and S. M. Weinreb, *Tetrahedron Lett.*, **1973**, 4561.
317 (a) J. A. Staroscik and B. Rickborn, *J. Org. Chem.*, **37**, 738 (1972).
317 (b) P. R. Ortiz de Montellano, *J. Chem. Soc.*, *Chem. Commun.*, **1973**, 709.
317 (c) D. R. Hicks, R. Ambrose, and B. Fraser-Reid, *Tetrahedron Lett.*, **1973**, 2507.
317 (d) G. H. Posner and D. J. Brunelle, *Chem. Commun.*, 907 (1973).
317 (e) C. P. Casey, D. F. Marten, and R. A. Boggs, *Tetrahedron Lett.*, **1973**, 2071.
318 (a) O. Wennerström, Ph.D. Dissertation, Royal Institute of Technology, Stockholm, Sweden, 1971.
318 (b) C. P. Casey and D. F. Marten, *Synthetic Commun.*, **3**, 321 (1973).
319 G. Schröder, U. Prange, and J. F. M. Oth, *Chem. Ber.*, **105**, 1854 (1972).
320 J. E. McMurry and S. J. Isser, *J. Amer. Chem. Soc.*, **94**, 7132 (1972).

CHAPTER 3

CLEMMENSEN REDUCTION OF KETONES IN ANHYDROUS
ORGANIC SOLVENTS

E. Vedejs

University of Wisconsin, Madison, Wisconsin

CONTENTS

INTRODUCTION

The Clemmensen reduction of ketones and aldehydes using zinc and
hydrochloric acid is the simplest direct method for converting the carbonyl
group into a methylene group. Procedures and results with acid-stable
compounds were reviewed by Martin in *Organic Reactions* in 1942 and
more recently by Staschewski.[1, 2] Typically, the carbonyl compound is

[1] E. L. Martin, *Org. Reactions*, **1**, 155 (1942).
[2] D. Staschewski, *Angew. Chem.*, **71**, 726 (1959).

401

refluxed for several hours with 40% aqueous hydrochloric acid, amalgamated zinc, and a water-immiscible organic cosolvent such as toluene. Because of these harsh conditions, reports of successful Clemmensen reduction of polyfunctional ketones have been rare. However, the milder procedure described by Yamamura and his collaborators using dry hydrogen chloride in organic solvents extends the synthetic potential of Clemmensen reduction to acid- and heat-sensitive compounds; this procedure is summarized later (pp. 412–414) in this review. Other developments that define the scope of both aqueous and anhydrous reduction conditions are discussed, and an effort is made to compare the properties of possible reduction intermediates with other organozinc species.

MECHANISM

Because the mechanism of the Clemmensen reduction is poorly understood, much additional information is necessary before the effect of experimental variables on results can be rationalized. Studies by Nakabayashi,[3] Brewster,[4] and numerous earlier workers have established several general characteristics of the reaction that suggest a stepwise reduction involving organozinc intermediates. It has been shown that reduction occurs with zinc but not with other metals of comparable reduction potential. The rate-determining step does not involve an electrochemical process (i.e., two one-electron transfer steps) because the rate of reduction is not sufficiently sensitive to changes in the zinc reduction potential.[3] An electrochemical reduction (pinacol coupling) often competes with the Clemmensen reduction, but the two reactions do not have a common intermediate in the only case studied.[3] The reduction rate is relatively insensitive to acid concentration but responds sufficiently to changes in halide concentration to suggest involvement of halide in the initial step.[3] Intermediates have not been identified conclusively, but older mechanisms involving alcohols as intermediates have been ruled out because alcohols are generally not reduced under Clemmensen conditions.[2, 4]

The experimental and kinetic data are rationalized by the rate-limiting attack of zinc and chloride ion on the carbonyl group with subsequent rapid protonation to afford the α-hydroxyalkylzinc chloride 1.[3] The nature of any further intermediates is highly speculative, but a second reductive step is ultimately necessary in order to form an organozinc species capable of undergoing protolysis to the hydrocarbon product. One possibility involves acid-catalyzed reduction of the carbon-oxygen bond

---

[3] T. Nakabayashi, *J. Amer. Chem. Soc.*, **82**, 3900, 3906, 3909 (1960).

[4] J. H. Brewster, *J. Amer. Chem. Soc.*, **76**, 6364 (1954); J. H. Brewster, J. Patterson, and D. A. Fidler, *ibid.*, **76**, 6368 (1954).

Scheme 1

$$\left[R-\underset{\underset{ZnCl}{|}}{\overset{\overset{Cl}{|}}{C}}-R' \right] \xrightarrow{H^+} RCHClR'$$

$$R-\overset{O}{\underset{}{\overset{\|}{C}}}-R' \xrightarrow{\underset{HCl}{Zn}} \left[R-\underset{\underset{ZnCl}{|}}{\overset{\overset{OH}{|}}{C}}-R' \right] \xrightarrow{\underset{Zn}{H^+}} \left[R-\underset{\underset{ZnCl}{|}}{\overset{\overset{ZnCl}{|}}{C}}-R' \right] \xrightarrow{H^+} RCH_2R'$$

1

$HCl\uparrow$ Zn $\overset{Zn(?)}{HCl}$

$\underset{R \quad R'}{\overset{\overset{ZnCl}{|}}{\overset{+}{C}}}$ $\underset{R \quad R'}{\overset{\overset{O}{\|}}{C}}$

$$\left[\underset{RR'C-CRR'}{\overset{HO \quad ZnCl}{|\qquad|}} \right] \longrightarrow RR'C\!\!=\!\!CRR' \longleftarrow \left[\underset{RR'C-CRR'}{\overset{ClZnO \quad RR'C}{|\qquad\quad|}} \right]$$

$$R-\overset{Zn}{\underset{}{\overset{\|}{C}}}-R' \underset{2}{\longleftrightarrow} R-\overset{+Zn}{\underset{}{\overset{|}{C}}}-R'$$

of **1** to a bis(chlorozinc)alkyl structure. An alternative reduced species **2** has been suggested,[3] but this unprecedented molecule involves either charge separation or unlikely carbon $2p$-zinc $4p\pi$-bonding.

The proposed bis(chlorozinc)alkyl also lacks precedent, although a related substance appears to be formed in the zinc-copper couple reduction of iodomethyl tosylate, as evidenced by the formation of methane upon hydrolysis and of the corresponding amount of methylene iodide upon treatment with iodine.[5] Methane is also formed in low yield upon hydrolysis of the reduction product of methylene iodide with zinc-copper couple, but Blanchard and Simmons have shown that methane results from stepwise reduction to methyl iodide and then to methylzinc iodide, and not from bis(iodozinc)methane as suggested previously.[6, 7]

The proposed reduction of **1** by zinc in the presence of an acid catalyst is an example of the high electrophilic reactivity of α-heteroatom-substituted /zinc alkyls. Thus hydrolysis of bis(chloromethyl)zinc in

[5] M. Jautelat, Ph.D. Thesis, Heidelberg, 1965.
[6] E. P. Blanchard and H. E. Simmons, *J. Amer. Chem. Soc.*, **86**, 1337 (1964).
[7] E. Emschwiller, *Compt. Rend.*, **188**, 1555 (1929).

aqueous zinc iodide affords 80 % of methyl iodide and only 1 % of methyl chloride, indicating that nucleophilic displacement of chloride by iodide is faster than protolysis of the carbon-zinc bond.[8] This observation raises the possibility that acid-catalyzed nucleophilic displacement of hydroxide by chloride may be faster than reduction under Clemmensen conditions; if it is, the result would be conversion of the α-hydroxyalkylzinc chloride 1 into an α-chloroalkylzinc chloride. Protolysis of the zinc-carbon bond of the latter would explain the occasional appearance of alkyl chlorides as side products of Clemmensen reduction. It is also conceivable that some alkyl chlorides are reduced to hydrocarbons and may serve as Clemmensen intermediates. Few experimental data about this point are available since it has been assumed that chlorides would be formed from alcohols which are definitely not reduced under Clemmensen conditions. It is reported that cyclohexyl chloride is not reduced under conditions which convert cyclohexanone into cyclohexane,[4] but reduction of the exocyclic methylene group of the gibberellin 3 to a methyl group using zinc and dry hydrogen chloride suggests that an intermediate tertiary chloride would be reduced.[9]

Other side reactions accompanying Clemmensen reduction can be explained on the basis of a polar carbon-heteroatom bond in intermediates at the same reduction stage as 1, represented for simplicity by chlorozinc-carbonium ions in the following discussion. Migration of an adjacent substituent (hydride, alkyl, aryl) to the positive center would afford monomeric olefin, as illustrated for the formation of cyclohexene from cyclohexanone. The relative yield of cyclohexene increases from 6 to 47 % as the concentration of hydrogen chloride in the aqueous reduction

[8] H. Hoberg, *Ann.*, **656**, 15 (1962); G. Wittig and F. Wingler, *ibid.*, **656**, 18 (1962).

[9] B. E. Cross and J. C. Stewart, *J. Chem. Soc.*, *C*, **1971**, 245.

medium is decreased from 20 to 3 %.[10a] Actual yields were not reported, however, so it is unclear whether the dependence of product ratio on acid concentration has any bearing on the reduction mechanism or merely reflects selective destruction of cyclohexene under strongly acidic conditions. Synthetically useful yields of alkenes can be obtained by reduction of ketones with zinc in the presence of chlorotrimethylsilane (aprotic conditions, ether solution).[10b]

Reduction of medium-sized ring ketones affords the transannular insertion products.[11] Thus bicyclo[3.3.0]octane is formed in addition to cyclooctene and cyclooctane upon reduction of cyclooctanone. Related carbonium ion-like rearrangements are observed upon zinc reduction of α,α-diiodoalkanes, probably via α-iodoalkylzinc iodides which are closely related to the proposed Clemmensen intermediates.[12]

Dimeric olefins are often formed as side products of Clemmensen reduction, especially from aryl ketones.[2] These products may result from self-condensation of the hydroxyalkylzinc chloride 1, followed by elimination of zinc chlorohydroxide (Scheme 1, p. 403). This mechanism has a precedent in the for mation of ethylene from iodomethylzinc iodide or of stilbene from α-iodobenzylzinc iodide.[6, 12a, 13]

Alternatively, condensation of unreacted ketone with a bis(chlorozinc) species derived from the second reductive step and subsequent elimination could be invoked. Analogous reactions between aldehydes or ketones and methylene iodide in the presence of excess zinc have been reported to give methylene derivatives.[14] However, formation of the methylene derivatives

[10] (a) G. E. Risinger, E. E. Mach, and K. W. Barnett, *Chem. Ind.* (London), **1965**, 679; (b) W. B. Motherwell, *Chem. Commun.*, **1973**, 935.

[11] E. Muller, G. Fiedler, H. Huber, B. Narr, H. Suhr, and K. Witte, *Z. Naturforsch.*, **18B**, 5 (1963).

[12] (a) H. E. Simmons, E. P. Blanchard, and R. D. Smith, *J. Amer. Chem. Soc.*, **86**, 1347 (1964); (b) R. Newman, *Tetrahedron Lett.*, **1964**, 2541.

[13] L. Y. Goh and S. H. Goh, *J. Organometal. Chem.*, **23**, 5 (1970).

[14] H. Hashimoto, M. Hida, and S. Miyano, *J. Organometal. Chem.*, **10**, 518 (1967); *ibid.*, **12**, 263 (1968); I. T. Harrison, R. J. Rawson, P. Turnbull, and J. H. Fried, *J. Org. Chem.*, **36**, 3515 (1971).

as well as the dimeric olefins from Clemmensen reduction can also be explained by condensation of an α-halozinc halide with a carbonyl group to form the zinc salt of a β-haloalcohol which would afford olefin upon further reduction.

The preceding discussion relates Clemmensen intermediates to α-haloalkylzinc halides, the carbenoid reagents of the Simmons-Smith cyclopropane synthesis.[15] Another similarity between the two types of reactions is apparent in the formation of 7-phenylnorcarane by aprotic reduction of benzaldehyde with zinc and boron trifluoride etherate in the presence of cyclohexene.[16] The scope of this reaction is not known, although it has been reported that zinc does not reduce ketones under similar conditions.

In summary, it is possible to rationalize the principal side products of Clemmensen reduction on the basis of hypothetical intermediates derived from the first reductive step. In general, the second reductive step (resulting in formation of hydrocarbon) competes effectively with protolysis, rearrangement, or intermolecular coupling of the intermediate **1**. To explain the observation that protolysis after the first reductive step is a minor reaction pathway, it has been argued that both reductive steps involve species which are bound to the solid zinc surface. However, electron transfer from the metal surface to dissolved intermediates could be more efficient than various possible side reactions. Experimental support for either argument is lacking, but there is no compelling reason to assume unusual bonding properties for the great variety of zinc surfaces (liquid 2% zinc amalgam, solid zinc amalgam, ordinary zinc dust, etc.) which can be used and generally afford similar product mixtures.

SCOPE AND LIMITATIONS

In most instances the success of the Clemmensen reduction depends on the stability of a given ketone to acid. This is not a severe limitation, especially under the conditions reported by Yamamura and co-workers (pp. 412–414).

1,3-Diketones

1,3- and 1,4-Diketones seldom give useful yields of Clemmensen reduction products. Intramolecular pinacol coupling to cyclopropanediols is the favored initial reaction of acyclic 1,3-diones, cyclohexane-1,3-diones, and 2-acylcycloalkanones. It is possible to trap the cyclopropanediols as

---

[15] H. E. Simmons, T. L. Cairns, S. A. Vladuchick, and C. M. Hoiness, *Org. Reactions*, **20**, 1 (1973).

[16] I. Elphimoff-Felkin and P. Sarda, *Chem. Commun.*, **1969**, 1065.

the diacetates by using acetic anhydride as the reduction medium; but, under ordinary Clemmensen conditions, cleavage of the cyclopropane ring is rapid and a mixture of α- and β-hydroxyketones is formed. Further reduction of the hydroxyketones is then possible, and the ultimate products may include rearranged and unrearranged monoketones and hydrocarbons derived from reduction of the initial products.[17]

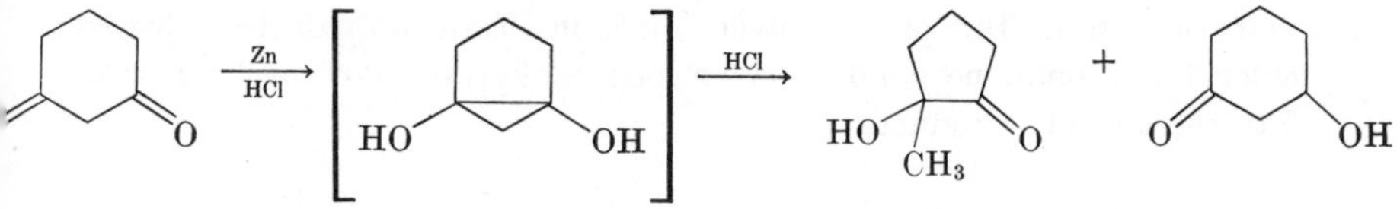

1,4-Diketones

Reduction of 1,4-diketones is complex and unpredictable. Occasionally, normal reduction is observed as in the conversion of 1-phenylpentane-1,4-dione into 5-phenylpentan-2-one or of cholestane-3,6-dione into cholestan-6-one.[18] More commonly the initial process is reductive fragmentation of the C_2–C_3 bond. Depending on conditions, cyclohexane-1,4-dione affords as many as twenty-six products, beginning with cleavage to hexane-2,5-dione followed by numerous reduction processes, rearrangements, acid-catalyzed cyclization, etc.[18, 19] Cleavage of a strained cyclobutane bond is more easily controlled, and the diketone 4 is reduced smoothly to a dihydro derivative. However, internal pinacol coupling to a cyclobutane diol occurs upon prolonged treatment with zinc.[19]

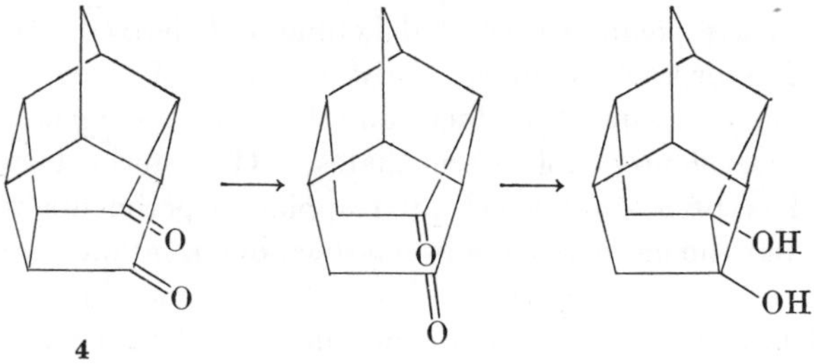

1,5-Diketones

Internal pinacol coupling is a general reaction of diketones such as cyclooctane-1,5-dione and bicyclo[3.3.1]nonane-3,7-dione which can

[17] J. G. St. C. Buchanan and P. D. Woodgate, *Quart. Rev.*, **23**, 522 (1969).
[18] J. G. St. C. Buchanan and B. R. Davis, *J. Chem. Soc., C*, **1967**, 1340.
[19] E. Wenkert and J. E. Yoder, *J. Org. Chem.*, **35**, 2986 (1970).

adopt conformations with the two carbonyl groups in close proximity.[20] Other intramolecular interactions are responsible for the unusual behavior of the related 1,5-diketone 1,5-dimethylbicyclo[3.3.0]octane-3,7-dione. Zinc in acetic anhydride-hydrogen chloride affords the acetate 6 in 50% yield, while reduction in aqueous hydrochloric acid results in 1,5-dimethylbicyclo[3.3.0]octane as the major product.[20] The solvent effect indicates that a reduction intermediate such as 5 is converted primarily into hydrocarbon in the protic solvent but, in acetic anhydride-hydrogen chloride, carbon-zinc bonds survive long enough to allow cyclization of 5 to the tricyclic product.

α,β-Unsaturated Ketones

Reduction of simple α,β-unsaturated ketones affords mixtures containing the corresponding saturated ketone and derived hydrocarbons, ketonic and hydrocarbon dimers derived from radical coupling at the β position, pinacol coupling products, and skeletal rearrangement products derived from cyclopropanol intermediates.[17] The intermediacy of cyclopropanols has been established by trapping experiments with acetic anhydride, but the mechanism of cyclopropanol formation is not known.[21]

Certain steroidal enones can be reduced in acceptable yield with the result that first the enone double bond and then the carbonyl group are reduced (see Table I). Reduction of the double bond is especially facile in systems such as 8 owing to activation by a second carbonyl group, and the resulting γ-ketoacid is reduced normally.[22, 23] Analogous reduction

[20] W. T. Borden and T. Ravindranathan, *J. Org. Chem.*, **36**, 4125 (1971).

[21] I. Elphimoff-Felkin and P. Sarda, *Tetrahedron Lett.*, **1969**, 3045.

[22] J. A. Marshall and S. F. Brady, *J. Org. Chem.*, **35**, 4068 (1970); D. L. Dreyer, *ibid.*, **36**, 3719 (1971).

[23] K. Ohkata and T. Hanafusa, *Bull. Chem. Soc. Jap.*, **43**, 2204 (1970).

of the enedione **9** to the dihydro derivative also occurs without rearrangement or fragmentation.[24]

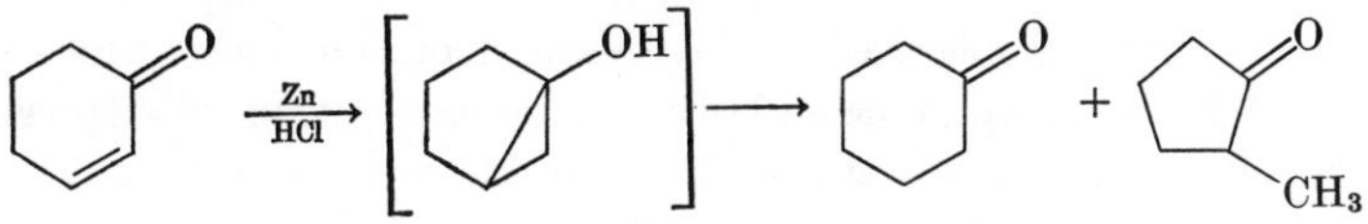

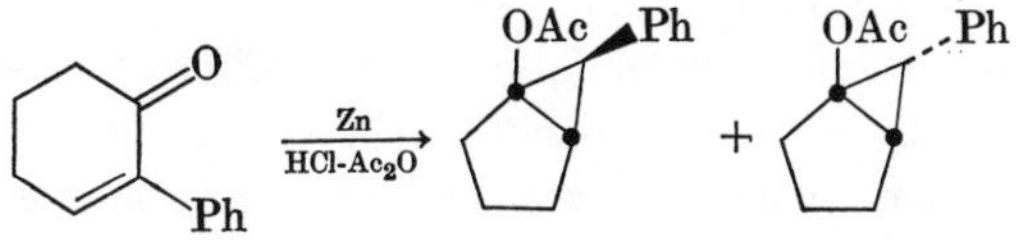

Ketones with α-Heteroatom Substituents

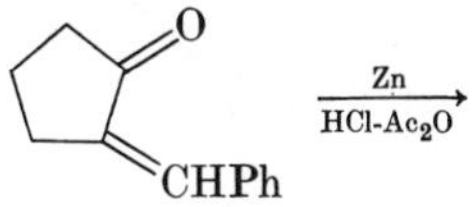

Heteroatoms attached to carbon atoms alpha to the carbonyl function are subject to reductive elimination under Clemmensen conditions.[17] Similar reactions occur with other reducing metals; an electrochemical mechanism is probably involved. Transfer of two electrons from the metal

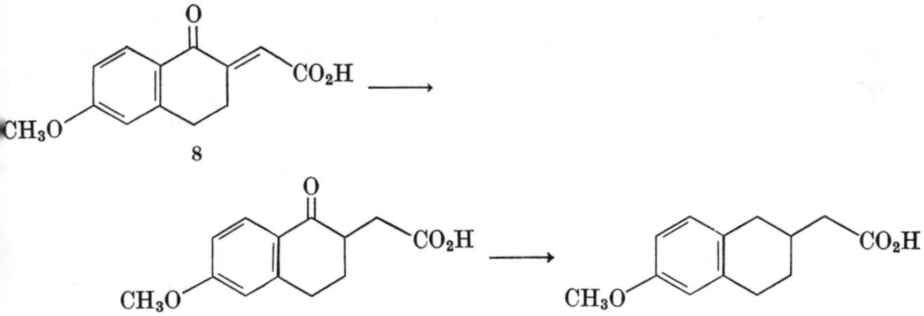

[24] E. Vedejs, unpublished results.

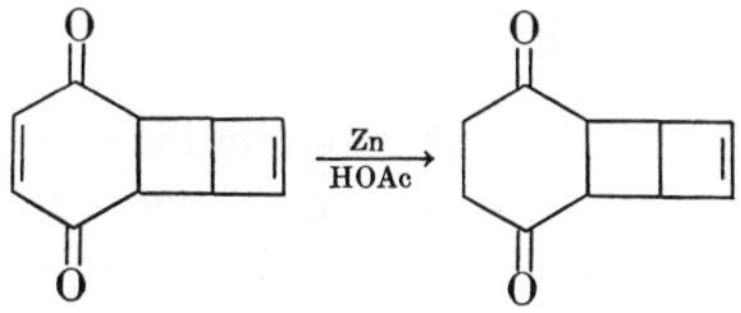

to the carbonyl group, followed by departure of the heteroatom as the anion affords an enolate which is converted into the corresponding ketone by acid. Clemmensen reduction of α-dicarbonyl compounds occurs by way of a related electrochemical mechanism. Thus the Diels-Alder adduct of 1,2-naphthoquinone and cyclopentadiene is reduced stepwise by zinc in acetic acid, first to an α-hydroxyketone, more slowly to a mono-ketone, and ultimately to the alkene.[24] Reduction of oxalylcyclopentanone (an α-ketoester as well as a 1,3-diketone) may also involve an electro-chemical reduction to the α-hydroxyester followed by reductive elimina-tion of hydroxide and hydrolysis to the ketoacid **10**. Clemmensen reduction of the isolated cyclopentanone carbonyl group of **10** occurs only under forcing conditions.[25]

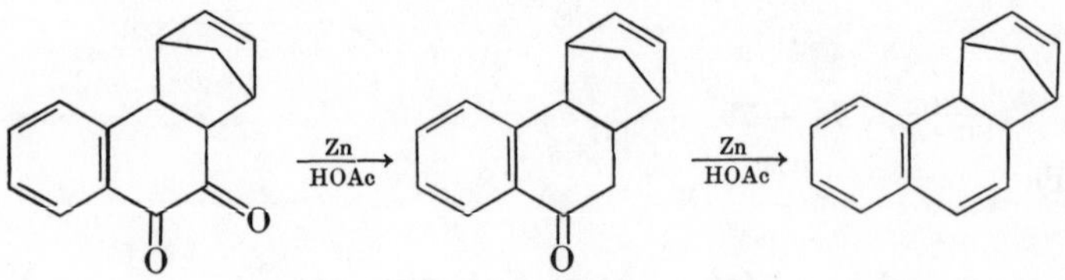

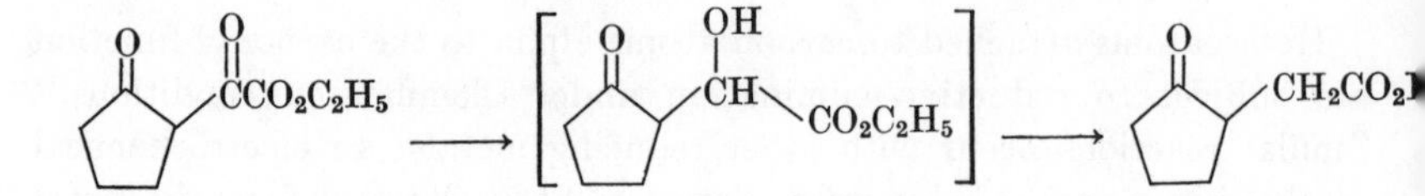

Reductive elimination of the α-chloro substituents in 1,4-dichloro-bicyclo[2.2.1]heptan-7-one does not occur because the intermediate enolate would have to violate Bredt's rule.[26]

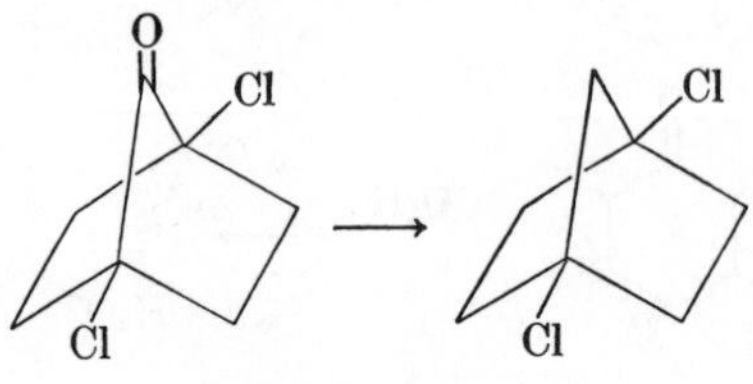

Hindered Ketones

The rate of Clemmensen reduction is sensitive to steric hindrance, as expected for a heterogeneous reaction. Ketones having adjacent t-butyl- or neopentyl-like substituents are reduced slowly, and in extreme cases

[25] R. Mayer, H. Burger, and B. Matauschek, *J. Prakt. Chem.* [IV] **14**, 261 (1961).
[26] A. P. Marchand and W. R. Weimar, Jr., *J. Org. Chem.*, **34**, 1109 (1969).

such as **11** and **12** reduction fails completely.[27, 28] Substantial differences in reduction rate due to steric factors permit selective reduction of the 3-keto group of androsta-3,17-dione in 67 % yield (Table I).

COMPARISON WITH OTHER METHODS OF REDUCTION

In view of the limitations of the Clemmensen reduction, other reasonably general methods for conversion of carbonyl into $-CH_2-$, such as Raney nickel desulfurization of the derived thioketal, or Wolff-Kishner reduction, may be preferred.[29, 30]

Desulfurization is particularly useful for selective reduction of enones and ketones having α-heteroatom substituents. Bisthioketals derived from 1,3- or 1,4-diketones are reduced without rearrangement, and selective reduction of one carbonyl group is possible if the corresponding monothioketal can be prepared.[27] The most common limitation of the desulfurization method is the hydrogenation of alkenes by active forms of Raney nickel.

The Wolff-Kishner reduction is a useful alternative to either the Clemmensen reduction or the desulfurization procedure, both of which employ a reducing metal capable of cleaving N–O bonds, reducing imines, hydrazines, azo compounds, and other electron-deficient functional groups. The Wolff-Kishner method is especially suited for reduction of medium-ring or strained-ring ketones to the corresponding hydrocarbons without rearrangement. There are few specific reports of the Clemmensen reduction applied to strained ketones, and the lack of positive results is discouraging.

In other respects the Wolff-Kishner reduction is more limited in scope than the Clemmensen reduction since enones, 1,3- and 1,4-dicarbonyl

[27] H. A. P. DeJongh and H. Wynberg, *Tetrahedron*, **20**, 2553 (1964).
[28] A. T. Blomquist and B. H. Smith, *J. Org. Chem.*, **32**, 1684 (1967).
[29] G. R. Pettit and E. E. van Tamelen, *Org. Reactions*, **12**, 356 (1962).
[30] H. H. Szmant, *Angew. Chem. Int.*, *Ed. Engl.*, **7**, 120 (1968).

compounds of various kinds, and α-heteroatom-substituted ketones behave anomalously. Furthermore, the presence of base-sensitive substituents precludes use of the Wolff-Kishner reductions in which potassium hydroxide is employed in typical experiments at temperatures between 100 and 200°. The tosylhydrazone modification avoids strong base and requires temperatures no higher than 80°.[31] This technique employs sodium borohydride or sodium cyanoborohydride to convert the tosylhydrazone into a tosylhydrazine which decomposes to the hydrocarbon at 80°.

REDUCTIONS WITH HYDROGEN CHLORIDE IN APROTIC ORGANIC SOLVENTS

Clemmensen reduction in organic solvents (alcohols, acetic acid) has been known for some time,[2] but it was generally found that a homogeneous liquid phase favored the formation of dimeric products (pinacols).[32] However, Yamamura and his associates have shown that anhydrous hydrogen chloride and zinc dust in organic solvents (ether, tetrahydrofuran, acetic anhydride, benzene) affords hydrocarbons in high yield.[33-39] Optimum results are obtained when a large excess of activated zinc dust in diethyl ether saturated with hydrogen chloride at ice-bath temperatures is used. In contrast to the original Clemmensen method, typical reductions are complete within an hour at 0° (Procedure A, p. 414). Activation of the zinc dust is recommended for hindered ketones, but commercial zinc dust may be used in most instances. A large excess of hydrogen chloride is generally used, but as little as 2 moles of acid per mole of substrate is sufficient for reduction of unhindered ketones.[40, 41] Slow addition of a small excess of deuterium chloride to the ketone and zinc dust in tetrahydrofuran (see Procedure B, p. 415) is convenient for reduction to gem-dideutero hydrocarbons, typically 75–80% d₂.[40]

[31] L. Cagliotti, *Tetrahedron*, **22**, 487 (1966); R. O. Hutchins, B. E. Maryanoff, and C. A Milewski, *J. Amer. Chem. Soc.*, **93**, 1793 (1971).

[32] G. E. Risinger and J. A. Thompson, *J. Appl. Chem.*, **13**, 346 (1963).

[33] S. Yamamura, S. Ueda, and Y. Hirata, *Chem. Commun.*, **1967**, 1049.

[34] S. Yamamura and Y. Hirata, *J. Chem. Soc.*, *C*, **1968**, 2887.

[35] S. Yamamura, *Chem. Commun.*, **1968**, 1494.

[36] S. Yamamura, H. Irikawa, and Y. Hirata, *Tetrahedron Lett.*, **1967**, 3361.

[37] M. Toda, Y. Hirata, and S. Yamamura, *Chem. Commun.*, **1969**, 919.

[38] M. Toda, Y. Hirata, H. Irikawa, and S. Yamamura, *Nippon Kagaku Zashi*, **91**, 103 (1970) [*C.A.*, **73**, 22137j (1970)].

[39] M. Toda, M. Hayashi, Y. Hirata, and S. Yamamura, *Bull. Chem. Soc. Jap.*, **45**, 264 (1972).

[40] R. P. Steiner, Ph.D. Thesis, University of Wisconsin, 1972 [*Diss. Abstr.*, **33**, 3563-B (1973)].

[41] I. Felkin, personal communication.

Successful reduction of α,β-unsaturated ketones may require a large excess of acid. At low acid concentration it appears that partially reduced organozinc intermediates survive long enough in solution to undergo intermolecular condensation. Thus, treatment of 4,4-diphenylcyclohex-2-en-1-one according to Procedure A (excess zinc dust in diethyl ether

Excess HCl
Et₂O, Zn

Ph   Ph   Ph   Ph
                     (60%)

3 Equiv
Zn   HCl, THF

Ph   Ph

Ph   Ph
              (46%)

saturated with hydrogen chloride) affords 1,1-diphenylcyclohexane,[24] while Procedure B (3 equivalents of hydrogen chloride in tetrahydrofuran) results in dimeric triene, but no diphenylcyclohexane.[40] Reduction of α-tetralone according to Procedure B also leads to dimeric hydrocarbons.[40]

Benzylidenecyclohexanone, however, affords monomeric products using 2 equivalents of hydrogen chloride in ether.[41] The benzylidenecyclohexane

CHPh   CH₂Ph   CHPh   CH₂Ph

          Et₂O/HCl
          Zn, −15°
O   O   +   +

and 1-benzylcyclohexene may be formed via the allylic alcohol 13, because in control experiments the alcohol 13 furnished these two products in a

CHPh

OH
13

CH₃   OH   CH₃   CH₃

          75%
                                +

                                        (minor)

Ph   OH   60%
              Ph   +   Ph
              (4:1)

combined yield of 95 %. Other typical allylic alcohols are reduced smoothly under the same conditions, as shown for 3-methylcyclohex-2-enol and cinnamyl alcohol.[41]

From available data there is no reason to believe that reduction with zinc and dry hydrogen chloride differs mechanistically from the original method. The same side reactions are observed, including formation of monomeric and dimeric alkenes, transannular insertion products from cyclooctanone, and alkyl chlorides from certain steroidal ketones. Reduction in ether generally proceeds directly to the hydrocarbon, but alkyl chlorides become significant products when benzene is the solvent.[39] This observation again raises the possibility that alkyl chlorides are precursors of hydrocarbons under Clemmensen reduction conditions.

Yamamura and co-workers initially observed reduction with hydrogen chloride in acetic anhydride (Procedure C).[33, 34, 36] This procedure is less convenient than Procedure A or B, and it yields enol acetates and saturated acetates as side products. Other solvents such as methanol or ethyl acetate are unsatisfactory. As in aqueous Clemmensen reduction, hydrogen chloride (or hydrogen bromide) is necessary for good results. Sulfuric acid, toluenesulfonic acid, and fluoroboric acid have been tried without success.[24, 38, 39]

EXPERIMENTAL PROCEDURES

Activated zinc dust. Commercial zinc dust (16 g, 325 mesh) was activated by stirring for 3–4 minutes with 100 ml of 2 % hydrochloric acid. The zinc was immediately filtered under suction, washed to neutrality with water, and then washed with 50 ml ethanol, 100 ml of acetone, and diethyl ether. The resulting powder was dried at 90° under vacuum (10 minutes) and was used within 10 hours of preparation.

Cholestane (*Example of Procedure A. Zinc, Diethyl Ether, Excess Hydrogen Chloride*).[39, 42] Cholestan-3-one (0.5 g, 1.30 mmol) was dissolved in 75 ml of dry ether saturated with hydrogen chloride at 0°. Activated zinc dust (5.0 g, 0.076 g-at) was slowly added to the cooled mixture with vigorous stirring at a rate such that the temperature was maintained below 5°. The reaction was exothermic and considerable hydrogen evolution occurred. The reaction mixture was stirred for 1 hour at 0° and then filtered. The filtrate was shaken with 500 ml of ice water and then washed to neutrality with aqueous sodium carbonate. The aqueous washings were extracted with additional ether, the combined ether extracts were dried over sodium sulfate and evaporated under vacuum. Chromatography of

[42] S. Yamamura, M. Toda, and Y. Hirata, *Org. Syntheses*, **53**, 86 (1973).

the residual oil over silica gel (Mallinckrodt, 100 mesh, 25 g) using benzene as eluant afforded 0.43 g (89%) of cholestane, mp 77.5–79°. A slightly modified procedure on a preparative scale gave a yield of 80%.[44]

1,1-Diphenylcyclohexane (*Example of Procedure B. Zinc, Tetrahydrofuran, 3 Equivalents of Hydrogen Chloride*).[40] 4,4-Diphenylcyclohexanone (0.5 g, 2 mmol) was dissolved in dry tetrahydrofuran (10 ml, distilled from lithium aluminum hydride) at 0° and was stirred vigorously with 2 g of commercial zinc dust. A previously titrated solution of dry hydrogen chloride (3 equiv, 6 mmol) in tetrahydrofuran (*ca*. 5 ml) was added dropwise over 20 minutes while the reaction temperature was maintained below 10°. The mixture was then stirred overnight at room temperature, diluted with 75 ml of ether, and worked up according to Procedure A. Chromatography of the crude product over 20 g of silica gel using 2:1 hexane-benzene as eluant afforded 0.345 g (74%) of 1,1-diphenylcyclohexane, mp 40–41°.

4,4-Dideuterio-1,1-diphenylcyclohexane (*Preparation of gem-Dideuterated Hydrocarbons Using Procedure B*).[40] Freshly distilled trimethylchlorosilane (1.63 g, 15 mmol) was added by syringe with gentle agitation to deuterium oxide (0.11 g, 6 mmol) in 5 ml of dry tetrahydrofuran in a dropping funnel. After 5 minutes the solution of deuterium chloride was added to 4,4-diphenylcyclohexanone (0.5 g, 2 mmol) as described under Procedure B above. 4,4-Dideuterio-1,1-diphenylcyclohexane (0.33 g, 70%) was isolated as before, mp 39–40.5°. Analysis by nmr and low-voltage mass spectroscopy indicated 81% d_2, 10% d_1, 6% d_3, 3% d_4.

Cholestane (*Example of Procedure C. Zinc, Acetic Anhydride, Excess Hydrogen Chloride*.)[39] Cholestan-3-one (0.25 g, 0.65 mmol) was dissolved with stirring in 10 ml of acetic anhydride saturated with hydrogen chloride at 0°. Activated zinc powder (2.5 g) was added slowly at a rate such that the temperature did not exceed 5°. After the mixture was stirred at 0° for 6 hours, it was poured with vigorous stirring into a large volume of water and made basic with sodium carbonate. After carbon dioxide evolution ceased, the product was extracted with diethyl ether and purified by the same method as described under Procedure A to yield cholestane (0.21 g, 87%), mp 77.5–79°.

TABULAR SURVEY

Ketones that have been reduced by zinc and hydrogen chloride in aprotic organic solvents are listed in the tables. All examples which make reference to the general reduction conditions of Yamamura and collaborators through December 1973 (*Science Citation Index*) are included. For comparison, reduction of androstane-3,17-dione with zinc amalgam in aqueous hydrochloric acid is included, but no effort has been made to survey the numerous other examples of reduction by the original method.

The reaction conditions are specified unless one of the three general procedures described under Experimental Procedures was employed. These are A (zinc dust, diethyl ether, excess hydrogen chloride, 1 hour at 0°), B (zinc dust, tetrahydrofuran, slow addition of 3 equiv of hydrogen chloride at 0°), and C (zinc dust, acetic anhydride, excess hydrogen chloride, x hours at 0°). Commercial zinc dust is used unless specified otherwise.

Many of the Clemmensen reductions using anhydrous hydrogen chloride employ ketones in the steroid series. These examples are surveyed in Tables I and II. Table III deals with α,β-unsaturated ketones and includes a number of examples in which undesirable side reactions play a major role. Table IV includes simple cyclic ketones as well as several complex natural products.

TABLE I. Steroidal Ketones and Steroidal Enones

Formula	Ketone	Conditions	Product(s) and Yield(s) (%)	Refs.
$C_{19}H_{28}O_2$	Androstane-3,17-dione	A	Androstane (75)	39
		Zinc amalgam, aq. HCl, heat	" (35)	38
		C, 6 hr; *freshly* activated Zn	" (66), 17β-acetoxyandrostane (26)	34
		C, 6 hr; 1 d activated Zn	Androstan-17-one (67), 17β-acetoxyandrostane (6)	34
$C_{21}H_{32}O_3$	17β-Acetoxyandrostan-3-one	C, 10 hr, 25°	17β-Acetoxyandrostane (79)	34
$C_{23}H_{36}O_3$	3β-Acetoxy-5α-pregnan-20-one	C, 10 hr, 25°	3β-Acetoxy-5α-pregnane (70)	34
$C_{27}H_{44}O$	Cholest-1-en-3-one	A	Cholestane (88)	39
		C, 2 hr	" (30–32), cholestan-3-one (30–40), 3-acetoxycholest-2-ene (10–24)	
	Cholest-4-en-3-one	A	Cholestane (48), coprostane (40)	39
$C_{27}H_{46}O$	Cholestan-3-one	A	Cholestane (89)	39
		Excess HCl in tetrahydrofuran, 1 hr, 0°	" (44)	
		C, 2 hr	" (87)	34
		Excess HBr in Ac₂O, 2 hr	" (66)	39
		Excess HCl in hexane, 1 hr, 0°	" (57), 3-chlorocholestane (8)	
		Excess HCl in benzene, 1 hr, 0°	Cholestane (64), 3-chlorocholestane (21)	39
$C_{29}H_{48}O_3$	3-β-Acetoxycholestan-6-one	C, 10 hr, 25°	3-β-Acetoxycholestane (54)	34

TABLE II. STEROIDAL KETONES WITH α-HETEROATOM SUBSTITUENTS[a]

Formula	Ketone	Conditions	Product(s) and Yield(s) (%)
$C_{25}H_{36}O_5$	3β,17α-Diacetoxypregn-5-en-20-one	C, 6 hr	3β-Acetoxypregn-5-ene (62)
$C_{27}H_{45}BrO$	2-α-Bromocholestan-3-one	A	Cholestane (85)
		C, 6 hr	,, (86), 3-acetoxycholest-2-ene (8)
$C_{29}H_{48}O_3$	α-Acetoxycholestan-3-one (1:1 mixture of 2- and 4-acetoxy isomers)	A	Cholestane (79)
		C, 6 hr	,, (90), unidentified acetates (2)
$C_{31}H_{49}BrO_5$	3β,5α-Diacetoxy-7α-bromo-cholestan-6-one	C, 6 hr	3β-Acetoxycholestane (73)

[a] All data are from reference 39.

TABLE III. ENONES AND ARYL KETONES

Formula	Ketone	Conditions	Product(s) and Yield(s) (%)	Refs
$C_7H_{10}O$	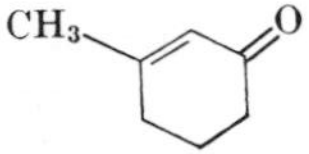	Et_2O, −15°, 1.5 hr, 2 mol HCl	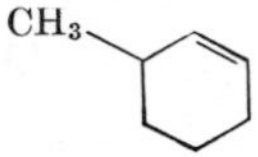 (Low)	41
C_9H_8O	1-Indanone	A	Indane (42)	39
		C, 2 hr, 0°	(22);	39
$C_{10}H_{10}O$	α-Tetralone	B	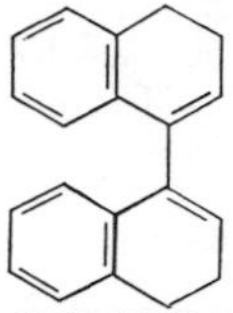 (19) + 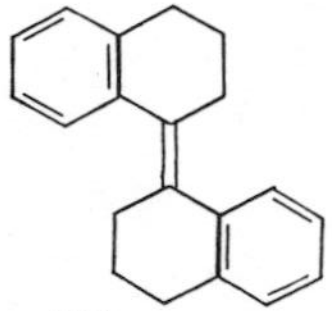 (5)	40
$C_{11}H_{12}O_3$	$C_6H_5CO(CH_2)_2CO_2CH_3$	A	$C_6H_5(CH_2)_3CO_2CH_3$ (41)	39
		C, 2 hr, 0°	,, (45)	39
$C_{11}H_{13}BrO$	$C_6H_5COC(CH_3)_2CH_2Br$	1:5 Ac_2O:Et_2O, −15°, 1.5 hr, 2 mol HCl	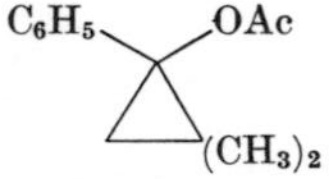 + $C_6H_5CH_2C(CH_3)_2CH_2OAc$ (1:2.5) (—)	41
$C_{12}H_{12}O$	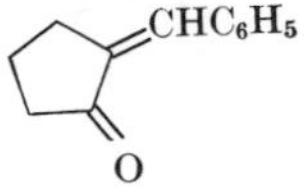	1:5 Ac_2O:Et_2O, 1.5 hr, −15° 2 mol HCl	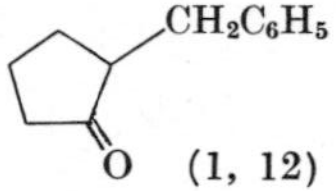 (48) + (1, 12) 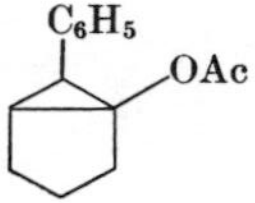	21, 41
		Et_2O, 1.5 hr, −15°, 2 mol HCl	I, (40) + 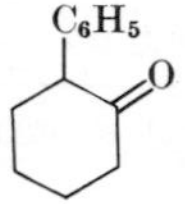 (10)	21, 41

419

TABLE III. Enones and Aryl Ketones (*Continued*)

Formula	Ketone	Conditions	Product(s) and Yields(s) (%)	Refs.
$C_{12}H_{12}O$	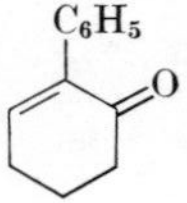	Et_2O, 1.5 hr, $-15°$, 2 mol HCl	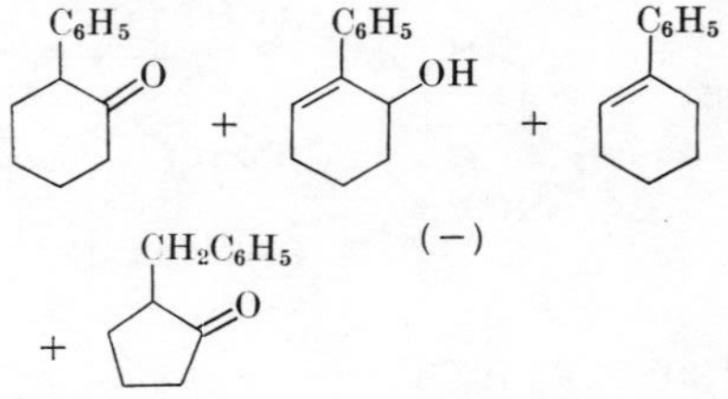 (−)	21, 41
		1:5 $Ac_2O:Et_2O$, 1.5 hr, $-15°$, 2 mol HCl	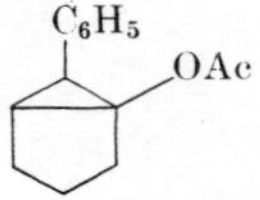 (35) 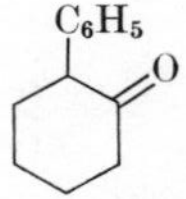 + (18) 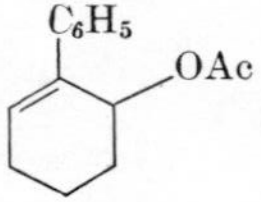 + (18)	21, 41
$C_{16}H_{22}O_3$	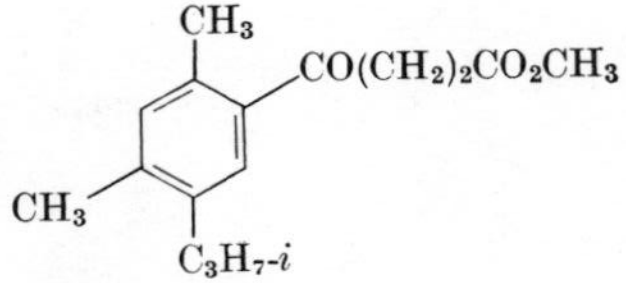	C, 2 hr	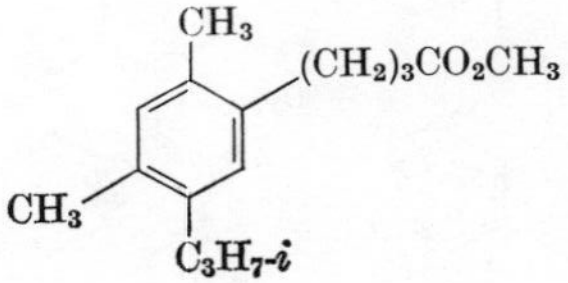 (87)	38
$C_{18}H_{16}O$	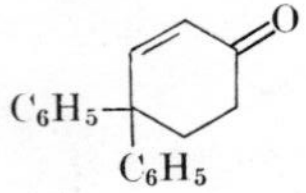	A	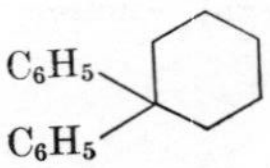 (60)	24
		B	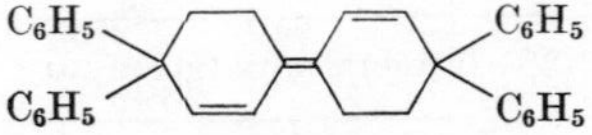 (46)	40

420

TABLE IV. SATURATED KETONES

Formula	Ketone	Conditions	Product(s) and Yield(s) (%)	Refs.
$C_7H_{10}O$	Norbornan-2-one	B	Norbornane (80)	40
		B, DCl	,, (80) (75% d_2, 17% d_1, 7% d_3)	40
$C_7H_{12}O$	2-Methylcyclohexanone	B	Methylcyclohexane (77)	40
$C_8H_{14}O$	2,6-Dimethylcyclohexanone	B	1,3-Dimethylcyclohexane (18)	40
$C_8H_{14}O$	Cyclooctanone	B	Cyclooctane (19), cyclooctene (7), bicyclo[3.3.0]octane (31)	40
$C_{10}H_{14}O_2$	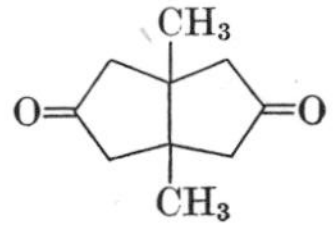	Ac_2O, −5°, 2 hr, excess HCl; basic hydrolysis	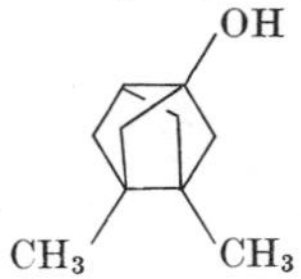 (50) 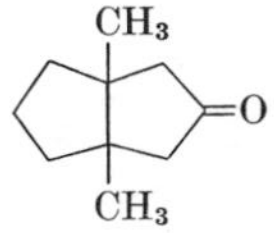 (10)	20
		Zinc amalgam, 25°, 2 hr, H_2O—HCl	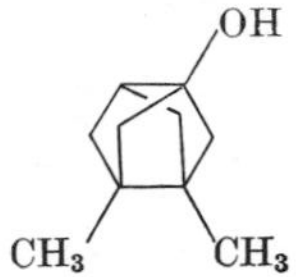 (10) 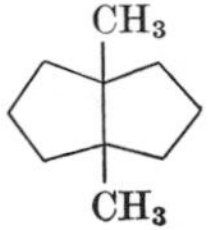 (major)	20
$C_{17}H_{22}O$	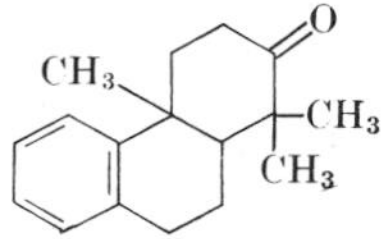	B	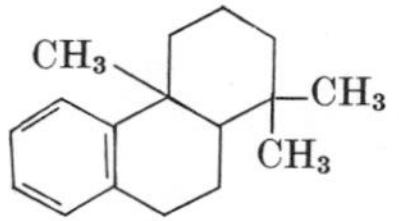 (50)	40
$C_{18}H_{18}O$	4,4-Diphenylcyclohexanone	B	1,1-Diphenylcyclohexane (74)	40
		B, DCl	,, (75) (81% d_2, 10% d_1, 6% d_3, 3% d_4)	40

421

TABLE IV. Saturated Ketones (*Continued*)

Formula	Ketone	Conditions	Product(s) and Yield(s) (%)	Refs.
$C_{19}H_{23}O_4$	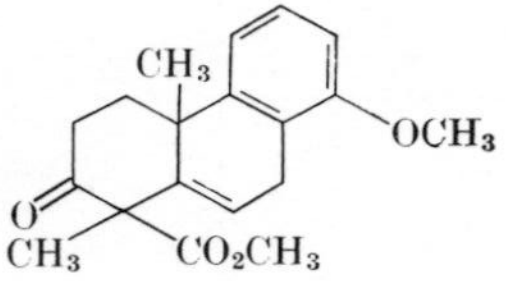	C	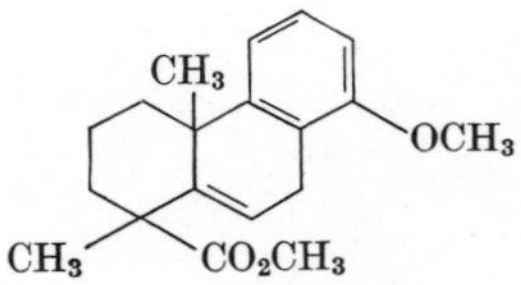 (high)	43
$C_{19}H_{26}O_4$	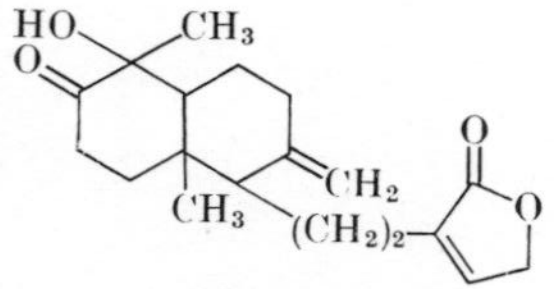	A. C	No useful result[a]	44
$C_{21}H_{26}O_5$	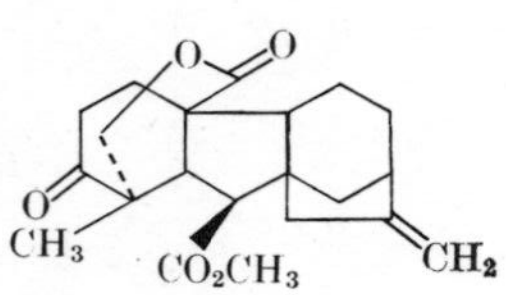	B	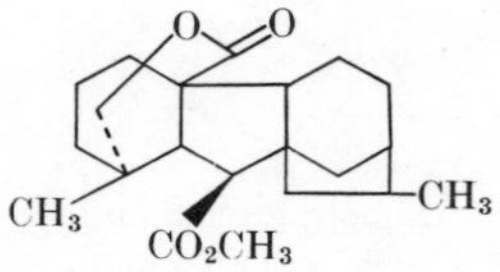 (40)	9
$C_{31}H_{47}NO_5$	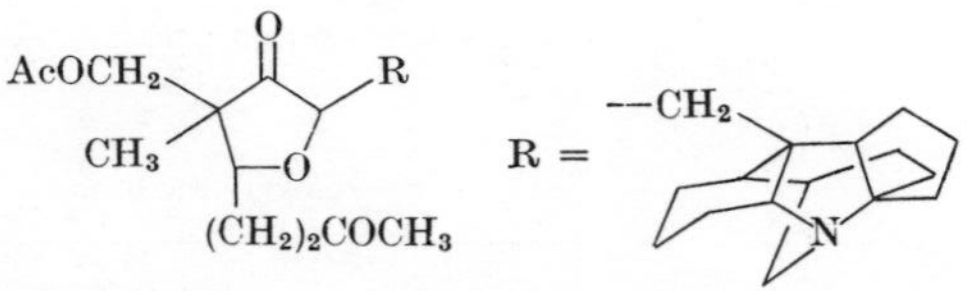	C, 18 hr	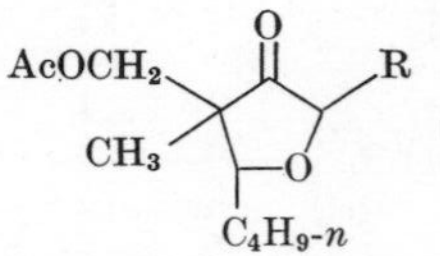 (46)	36

$R =$

[43] H. Kakisawa *et al.*, personal communication cited in reference 34.

[44] W. R. Chan, D. R. Taylor, C. R. Willis, R. L. Bodden, and H. W. Fehlhaber, *Tetrahedron*, **27**, 5081 (1971).

CHAPTER 4

THE REFORMATSKY REACTION

MICHAEL W. RATHKE

Michigan State University, East Lansing, Michigan

CONTENTS

INTRODUCTION

The Reformatsky reaction is the reaction of a carbonyl compound, usually an aldehyde or ketone, with an α-haloester in the presence of zinc metal to furnish, after hydrolysis, a β-hydroxyester.[1] Subsequent dehydration of the hydroxyester is commonly carried out to form an α,β-unsaturated ester. Summaries of certain aspects of the reaction have

$$\begin{array}{c} R_1 \\ \diagdown \\ \diagup \\ R_2 \end{array} C{=}O \ + \ X{-}\overset{|}{\underset{|}{C}}CO_2R \ \xrightarrow[\text{2. H}_3O^+]{\text{1. Zn}} \ \begin{array}{c} R_1 \ OH \\ \diagdown \diagup \\ \diagup \ \overset{|}{\underset{|}{C}}{-}\overset{|}{\underset{|}{C}}CO_2R \\ R_2 \end{array} + \ \begin{array}{c} R_1 \\ \diagdown \\ \diagup \\ R_2 \end{array} C{=}\overset{|}{C}CO_2R$$

been included in review articles,[2] and an interesting paper on the history and background of the reaction has appeared.[3] A useful review of the synthetic aspects of the Reformatsky reaction also has been included in a recent book.[4] A review in Russian is available.[4a]

This chapter summarizes some of the more important advances in the understanding and use of the Reformatsky reaction since the original chapter was written for Volume 1 of this series.[5] Studies on the nature of intermediates, side reactions, stereochemistry, and variations of the original reaction are discussed. Procedures used for the reaction and a discussion of other methods presently available are included. Finally, there is a table of representative examples of the reaction that have appeared between 1941 and 1971.

INTERMEDIATES IN THE REACTION

The course of the Reformatsky reaction with carbonyl compounds is usually formulated as shown in the accompanying equations. Ample

[1] S. N. Reformatsky, *Chem. Ber.*, **20**, 1210 (1887).

[2] (a) D. G. M. Diaper and A. Kuksis, *Chem. Rev.*, **59**, 89 (1959). (b) M. Gaudemar, *Organometal. Chem. Rev.*, *A*, **8**, 183 (1972).

[3] A. Sementsov, *J. Chem. Educ.*, **34**, 530 (1957).

[4] Herbert O. House, *Modern Synthetic Reactions*, 2nd ed., W. A. Benjamin, New York, 1972, pp. 671–682.

[4a] N. S. Vul'fson and L. Kh. Vinograd, *Reactions and Research Methods for Organic Compounds, Book 17: Reformatskii Reaction*, Khimiya, Moscow, 1967.

[5] R. L. Shriner, *Org. Reactions*, **1**, 1 (1942).

$$X\overset{|}{\underset{|}{C}}CO_2R + Zn \longrightarrow (\overset{|}{\underset{|}{C}}CO_2R)ZnX \xrightarrow{R_1R_2CO} R_1R_2\overset{ZnX}{\underset{\underset{|}{O}}{\overset{\|}{C}}}CCO_2R \xrightarrow{H_3O^+}$$

$$\underset{\textbf{1}}{} \qquad\qquad \underset{\textbf{2}}{}$$

$$R_1R_2\overset{HO}{\underset{|}{\overset{|}{C}}} \overset{}{\underset{|}{C}}CO_2R + HOZnX$$

evidence exists for the formation of zinc alkoxides corresponding to **2**. A number of these salts, obtained from reactions of ethyl bromoacetate or ethyl α-bromopropionate with aldehydes and ketones, have been isolated as pale-yellow solids with satisfactory analyses.[6]

A more difficult problem has been to determine the nature of the first intermediate in the reaction, the Reformatsky reagent **(1)**. The reagent is usually formulated with a zinc-carbon bond **(3)**, because it reacts as a carbon rather than an oxygen nucleophile. However, another formulation as a zinc-oxygen bonded enolate is possible **(4)**.

$$X Zn\overset{|}{\underset{|}{C}}CO_2R \qquad\qquad \underset{/}{\overset{\backslash}{}}C=C\overset{OZnX}{\underset{\backslash}{\overset{/}{}}OR}$$

$$\underset{\textbf{3}}{} \qquad\qquad\qquad \underset{\textbf{4}}{}$$

The Reformatsky reaction is normally conducted in a single step,[5] combining haloester and carbonyl substrate in a solvent with zinc metal. Recently it has been shown that a two-step process involving initial formation of the Reformatsky reagent is possible.[7,8] This development has permitted a number of detailed studies of the nature of the reagent.

The infrared spectrum of the reagent obtained from ethyl α-bromoisobutyrate and zinc metal in an ether-benzene solvent possesses a strong band at 1525 cm^{-1}, suggesting a zinc-oxygen bond.[9] Only a relatively weak band is observed in the ester carbonyl region (1730 cm^{-1}) and it is attributed to unreacted bromoester. In addition, the chemical behavior of the Reformatsky reagent **(5)** is completely analogous to that of the corresponding magnesium reagent **(6)**, assumed to exist in enolate form.[10, 11] These results are most consistent with formulation **4** for the Reformatsky reagent.

[6] J. F. J. Dippy and J. C. Parkins, *J. Chem. Soc.*, **1951**, 1570.
[7] A. Siegel and H. Keckeis, *Monatsh. Chem.*, **84**, 910 (1953).
[8] C. A. Grob and P. Brenneisen, *Helv. Chim. Acta.*, **41**, 1184 (1958).
[9] W. R. Vaughan, S. C. Bernstein, and M. E. Lorber *J. Org. Chem.*, **30**, 1790 (1965).
[10] W. R. Vaughan and H. P. Knoess, *J. Org. Chem.*, **35**, 2394 (1970).
[11] H. E. Zimmerman and M. D. Traxler, *J. Amer. Chem. Soc.*, **79**, 1920 (1957).

$$[(C_6H_5)_2CHCHCO_2CH_3]ZnBr \qquad (C_6H_5)_2CHCH=C\begin{matrix}OMgBr\\ \\OCH_3\end{matrix}$$

6

On the other hand, from nuclear magnetic resonance and infrared data obtained in a variety of solvents, carbon-bonded structures analogous to 3 are postulated for the reagent obtained from zinc and ethyl bromoacetate.[12] Especially interesting are the chemical shifts of the α-methylene protons that range from 2.0 δ in dimethoxymethane to 1.21 δ in hexamethylphosphoramide. The low values appear inconsistent with an enolate structure.

The reaction of optically active methyl α-bromopropionate with zinc and aromatic aldehydes produces optically active esters.[13] The structures of the products correspond to inversion at the asymmetric carbon. The preservation of optical activity, estimated to be about 5%, is most simply explained by carbon-bonded structures analogous to 3 for the intermediate.

$$\underset{\overset{|}{CO_2CH_3}}{\overset{H}{\underset{|}{CH_3\blacktriangleright C}\blacktriangleleft Br}} + Zn + ArCHO \xrightarrow[\text{2. } H_3O^+]{\text{1. } C_6H_6, \text{ reflux}} \underset{\overset{|}{OH}\ \overset{|}{CO_2CH_3}}{\overset{H\ \ H}{\underset{|}{Ar}\blacktriangleright C-C}\blacktriangleleft CH_3} + \underset{\overset{|}{H}\ \ \overset{|}{CO_2CH_3}}{\overset{OH\ H}{\underset{|}{Ar}\blacktriangleright C-C}\blacktriangleleft CH_3}$$

Clearly, further investigations concerning the nature of the Reformatsky reagent would be useful.

SCOPE AND LIMITATIONS

Side Reactions

The Reformatsky reaction is subject to a number of side reactions as described in Volume 1.[5] The most common side reactions are probably those of the reagent with the carbonyl component to generate an aldehyde or ketone enolate[14] or reaction to form β-ketoesters[15] derived from the starting haloester. The stoichiometry of the latter reaction was determined to be as shown in the accompanying equation.[15] The importance

$$2\,R_2CBrCO_2C_2H_5 + 2\,Zn \xrightarrow{C_6H_6} [R_2CCOCR_2CO_2C_2H_5]ZnBr + BrZnOC_2H_5$$

[12] M. Gaudemar and M. Martin, *C. R. Acad. Sci.*, Ser. C, **267**, 1053 (1968).
[13] J. Canceill, J. Gabard, and J. Jacques, *Bull. Soc. Chim. Fr.*, **1968**, 231.
[14] M. S. Newman, *J. Amer. Chem. Soc.*, **64**, 2131 (1942).
[15] A. S. Hussey and M. S. Newman, *J. Amer. Chem. Soc.*, **70**, 3024 (1948).

of this condensation reaction is reported to increase in the order:

$$BrCH_2CO_2C_2H_5 < CH_3BrCHCO_2C_2H_5 < (CH_3)_2BrCCO_2C_2H_5$$

Discrete Reformatsky reagents prepared from ethyl α-bromoiso-butyrate[9] and from methyl 2-bromo-3,3-diphenylpropanoate[10] undergo a slow "dimerization" on heating or prolonged standing. The hydrolysis products are the corresponding condensed esters, ethyl isobutyryliso-butyrate (7) and methyl 5,5-diphenyl-2-(diphenylmethyl)-3-oxopentanoate (8). The reaction may proceed via a ketene intermediate which reacts with

$$(CH_3)_2CHCOC(CH_3)_2CO_2C_2H_5 \qquad (C_6H_5)_2CHCH_2COCH[CH(C_6H_5)_2]CO_2CH_3$$
$$7 \qquad\qquad\qquad\qquad\qquad\qquad\qquad 8$$

a second molecule of the reagent to furnish the condensation product.[9]

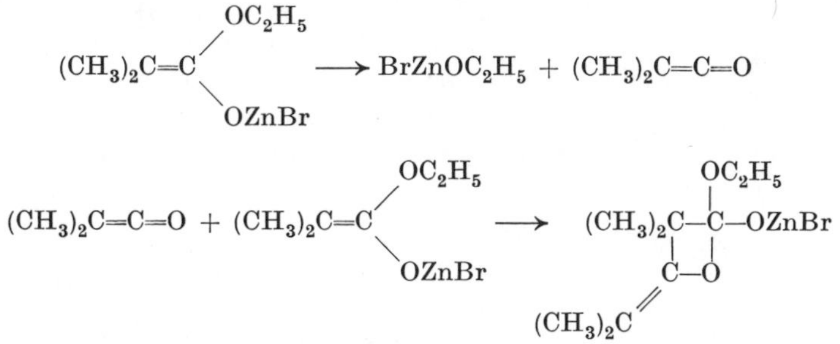

The Reformatsky reaction often gives poor yields with aliphatic alde-hydes. This result may be attributed to the ready self-condensation of aldehydes under the basic reaction conditions.[16] However, in at least one case, traces of acid on the zinc metal caused acid-catalyzed condensation of aldehydes to trioxanes.[16]

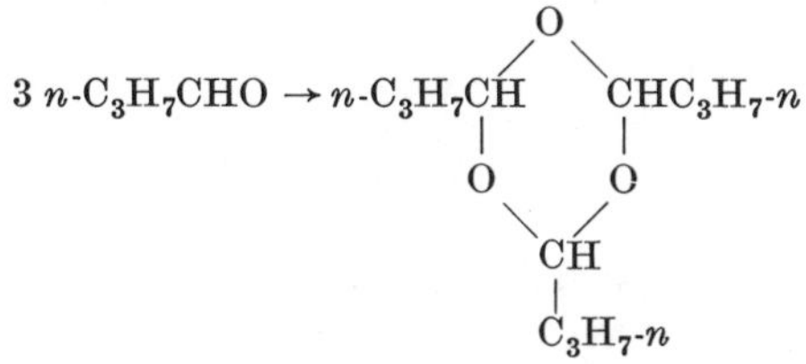

Reversal of the Reformatsky reaction[9, 17] after extended refluxing may also lead to decreased yields by eventual self-condensation of the starting carbonyl substrate.[17]

[16] J. W. Frankenfeld and J. J. Werner, J. Org. Chem., 34, 3689 (1969).
[17] C. R. Hauser and W. H. Puterbaugh, J. Amer. Chem. Soc., 75, 4756 (1953).

$$BrCH_2CO_2C_2H_5 + Zn + C_6H_5COCH_3 \begin{cases} \xrightarrow[\text{reflux, 1.25 hr}]{C_6H_6} C_6H_5C(OH)(CH_3)CH_2CO_2C_2H_5 \\ \qquad\qquad\qquad (60\%) \\ \xrightarrow[\text{reflux, 29 hr}]{C_6H_6-C_6H_5CH_3} C_6H_5C(OH)(CH_3)CH_2CO_2C_2H_5 \\ \qquad\qquad\qquad (1\%) \end{cases}$$

0.5   0.62   0.61
mol   g-at.   mol

$$+ \underset{\underset{CH_3}{|}}{C_6H_5C}=CHCOC_6H_5$$
$$(49\%)$$

Stereochemistry

The Reformatsky reaction of α-haloesters such as ethyl α-bromopropionate with aldehydes or ketones normally produces a mixture of two diastereomeric hydroxyesters.[18] The effect of reaction conditions on the ratio of erythro and threo isomers has been studied by a number of workers.[19-25] Most of the results can be rationalized with metalchelated structures of minimum steric interactions. The solvent appears to have a major influence on the ratio of isomers obtained.[25] In some reactions a thermodynamic mixture of diastereomers is obtained at higher temperatures.[24]

$$(C_2H_5CHCO_2C_2H_5)ZnBr + C_6H_5COCH_3 \rightarrow$$
$$C_6H_5C(OH)(CH_3)CH(C_2H_5)CO_2C_2H_5$$
Dimethoxymethane (69%) (25% erythro)
Dimethyl sulfoxide (55%) (52% erythro)

With rigid carbonyl systems, attack usually occurs from the less hindered side, as indicated from the predominance of *exo* isomer obtained from norbornanone.[26]

$$BrCH_2CO_2C_2H_5 + Zn +$$

$$\xrightarrow{C_6H_6}$$

$$\sim CH_2CO_2C_2H_5$$
$$HO$$

0.1   0.11   0.1
mol   g-at.   mol
(48%) (90% *endo* hydroxyl)

[18] L. Canonica and F. Pelizzoni. *Gazz. Chem. Ital.*, **84**, 553 (1954).

[19] J. Canceill, J. J. Basselier, and J. Jacques, *Bull. Soc. Chim. Fr.*, **1963**, 1906.

[20] M. Mousseron, M. Mousseron-Canet, J. Neyrolles, and Y. Beziat, *Bull. Soc. Chim. Fr.* **1963**, 1483.

[21] M. Mousseron-Canet and Y. Beziat, *Bull. Soc. Chim. Fr.*, **1968**, 2572.

[22] F. Gaudemar-Bardone and M. Gaudemar, *C. R. Acad. Sci., Ser. C*, **266**, 403 (1968).

[23] Y. Beziat and M. Mousseron-Canet, *Bull. Soc. Chim. Fr.*, **1968**, 1187.

[24] J. Canceill, J. Gabard, and J. Jacques, *Bull. Soc. Chim. Fr.*, **1968**, 231.

[25] F. Gaudemar-Bardone and M. Gaudemar, *Bull. Soc. Chim. Fr.*, **1969**, 2088.

[26] F. Lauria, V. Vecchietti, W. Logemann, G. Tosolini, and E. Dradi, *Tetrahedron*, **25**, 3989 (1969).

Partial asymmetric syntheses of β-hydroxyesters have been obtained by the use of haloesters of optically active alcohols.[27, 28] For example, reaction of $(-)$-menthyl bromoacetate (9) in benzene solution with acetophenone and zinc followed by saponification produces $(+)$-β-hydroxy-β-phenylbutyric acid of 30% optical purity.[27] Variations in the reaction conditions have little effect on the specific rotation of the product.

$$\underset{\substack{0.11 \text{ mol} \\ \mathbf{9}}}{BrCH_2CO_2C_{10}H_{19}} + \underset{0.11 \text{ mol}}{C_6H_5COCH_3} + \underset{0.165 \text{ g-at.}}{Zn} \rightarrow \rightarrow \underset{(53\%) \ (30\% \text{ optical purity})}{C_6H_5C(OH)(CH_3)CH_2CO_2H}$$

Variations

Variations in reactants and substrates that have increased the usefulness of the Reformatsky reaction are considered in the following subsections.

Metals Other Than Zinc

Attempts to use the more reactive magnesium in place of zinc in the Reformatsky reaction normally result in self-condensation of the haloester. However, the use of t-butyl haloesters overcomes this difficulty and sometimes gives better yields.[29] In the accompanying example, zinc

$$\underset{0.55 \text{ mol}}{CH_3BrCHCO_2C_4H_9\text{-}t} + \underset{0.08 \text{ g-at.}}{Mg} + \underset{0.05 \text{ mol}}{(C_6H_5)_2CO} \xrightarrow{(C_2H_5)_2O}$$

$$\underset{(81\%)}{(C_6H_5)_2C(OH)CH(CH_3)CO_2C_4H_9\text{-}t}$$

gave none of the desired β-hydroxyester. A similar use of magnesium was reported earlier for reactions of substituted haloesters.[30] With highly hindered acids, even the ethyl esters can be used.

$$\underset{0.1 \text{ mol}}{} + Mg + \underset{0.1 \text{ mol}}{} \xrightarrow{C_6H_6, \ (C_2H_5)_2O}$$

Reformatsky reactions with lithium,[31] aluminum,[31, 32] or cadmium[32, 33] have also been reported. In some cases, the solvent choice appears to be

[27] J. A. Reid and E. E. Turner, J. Chem. Soc., 1949, 3365.
[28] M. H. Palmer and J. A. Reid, J. Chem. Soc., 1960, 931.
[29] T. Moriwake, J. Org. Chem., 31, 983 (1966).
[30] J. Jacques and C. Weidmann, Bull. Soc. Chim. Fr., 1958, 1478.
[31] T. Moriwake, Mem. Sch. Eng., Okayama Univ., 1967, 93 [C.A., 69, 86567 (1970)].
[32] M. Gaudemar, C. R. Acad. Sci., Ser. C, 268, 1439 (1969).
[33] J. Cason and R. J. Fessenden, J. Org. Chem., 22, 1326 (1957).

critical. For example, the following reaction did not proceed satisfactorily with ether, dimethoxymethane, or tetrahydrofuran as the solvent.[32]

$$\text{BrCH}_2\text{CO}_2\text{C}_4\text{H}_9\text{-}t + \text{Cd} + \text{C}_6\text{H}_5\text{CHO} \xrightarrow{(\text{CH}_3)_2\text{SO}} \text{C}_6\text{H}_5\text{CHOHCH}_2\text{CO}_2\text{C}_4\text{H}_9\text{-}t$$
$$(65\%)$$

Unsaturated Haloesters

4-Bromocrotonate esters may undergo the Reformatsky reaction with attack at either the 4 position to produce the normal product or at the 2 position to produce the abnormal product. Steric effects appear to be

$$\text{BrCH}_2\text{CH}=\text{CHCO}_2\text{R} + \text{Zn} + \text{R}_2\text{CO} \rightarrow$$
$$\text{R}_2\text{C(OH)CH}_2\text{CH}=\text{CHCO}_2\text{R} + \text{R}_2\text{C(OH)CHCO}_2\text{R}$$
$$\underset{\text{Normal product}}{} \qquad \underset{\text{Abnormal product}}{\overset{|}{\text{CH}=\text{CH}_2}}$$

important, and hindered methyl ketones give predominantly the normal product.[34] In a similar study aliphatic aldehydes gave a mixture of both isomers with branching in the alkyl portion of the aldehyde favoring the normal product.[35]

$$\underset{\text{0.1 mol}}{\text{BrCH}_2\text{CH}=\text{CHCO}_2\text{C}_2\text{H}_5} + \underset{\text{0.4 g-at.}}{\text{Zn}} + \underset{\text{0.17 mol}}{\text{CH}_3\text{COR}}$$

$$\nearrow \quad 100\% \text{ abnormal, R} = \text{CH}_3$$
$$(\text{C}_2\text{H}_5)_2\text{O}$$
$$\searrow \quad 100\% \text{ normal, R} = \text{C(CH}_3)_3$$

In some cases the product obtained depends on the reaction conditions. Cyclohexanone reacts to give predominantly the normal product in refluxing benzene and the abnormal product in refluxing ether.[36] With

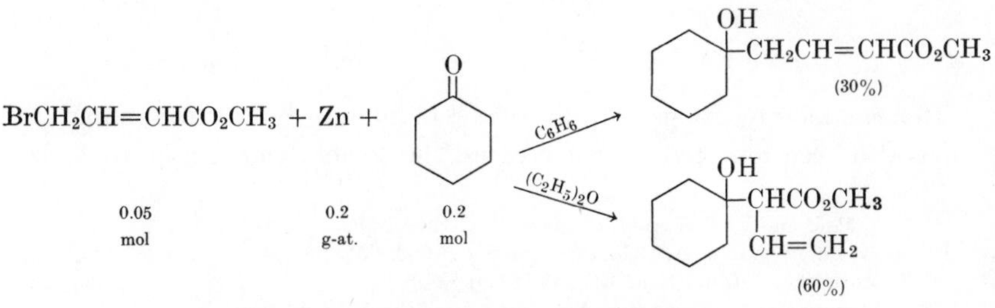

$$\text{BrCH}_2\text{CH}=\text{CHCO}_2\text{CH}_3 + \text{Zn} + $$

---

[34] J. Colonge and J. Varagnat, *Bull. Soc. Chim. Fr.*, **1961**, 234.
[35] J. Colonge and S. P. Cayrel, *Bull. Soc. Chim. Fr.*, **1965**, 3596.
[36] A. S. Dreiding and R. J. Pratt, *J. Amer. Chem. Soc.*, **75**, 3717 (1953).

other ketones, such as 2-methylcyclohexanone, the normal product is obtained in either solvent.

Reformatsky reactions with unsaturated haloesters have been widely used in the synthesis of natural products.[37-44]

Acceptors

The Reformatsky reaction with α,β-unsaturated carbonyl compounds may furnish either 1,2- or 1,4-addition products. Reaction of a variety of methyl vinyl ketones with bromoesters and zinc metal in refluxing ether is reported to give only 1,2-addition products.[45] On the other hand, reaction

$$RBrCHCO_2C_2H_5 + Zn + CH_3CR_1=CR_2COCH_3 \xrightarrow{(C_2H_5)_2O}$$
1.0 mol   1.0 g-at.   1.05 mol

$$\overset{OH}{\underset{CH_3CR_1=CR_2\overset{|}{C}CHRCO_2C_2H_5}{\underset{\overset{|}{C}H_3}{}}}$$
(59–70%)

of unsaturated ketones with ethyl bromoisobutyrate and zinc in refluxing tetrahydrofuran gave, after saponification, products of 1,4 addition.[46]

$$(CH_3)_2CBrCO_2C_2H_5 + Zn + C_6H_5CH=CHCOR \xrightarrow[2.\ OH^-]{1.\ THF^*}$$
0.133 mol   0.133 g-at.   0.1 mol

$$\underset{\overset{|}{C}_6H_5}{RCOCH_2\overset{|}{C}HC(CH_3)_2CO_2H}$$
(R = CH₃, 57%; R = C₂H₅, 60%)

It is likely that hindered haloesters such as ethyl α-bromoisobutyrate will generally give increased amounts of 1,4 addition; however, it is possible that a mixture of both 1,2 and 1,4 products are formed in many reactions of unsaturated carbonyl compounds with only the major product being isolated.[47]

---

* THF is tetrahydrofuran.

[37] I. M. Heilbron, E. R. Jones, and D. G. O'Sullivan, *J. Chem. Soc.*, **1946**, 866.

[38] J. W. Cook and R. Philip, *J. Chem. Soc.*, **1948**, 162.

[39] R. C. Fuson and P. L. Southwick, *J. Amer. Chem. Soc.*, **66**, 679 (1944).

[40] S. H. Harper and J. F. Oughton, *Chem. Ind.* (London), **1950**, 574.

[41] L. Canonica and M. Martinolli, *Gazz. Chim. Ital.*, **83**, 431 (1953).

[42] K. Tanabe, *Pharm. Bull.* (Japan), **3**, 25 (1955) [*C.A.*, **50**, 1677a (1956)].

[43] W. C. J. Ross, Brit. Pat. 626,712 [*C.A.*, **44**, 4039i (1950)].

[44] W. H. Linnell and C. C. Shen, *J. Pharm. Pharmacol.*, **1**, 971 (1949).

[45] J. Colonge and J. Varagnat, *Bull. Soc. Chim. Fr.*, **1961**, 237.

[46] J. C. Dubois, J. P. Guette, and H. B. Kagan, *Bull. Soc. Chim. Fr.*, **1966**, 3008.

[47] P. deTribolet, G. Gamboni, and H. Schinz, *Helv. Chim. Acta*, **41**, 1587 (1958).

Various acyl derivatives, including esters,[48] nitriles,[49-53] and acid chlorides,[54] may react with Reformatsky reagents to produce β-ketoesters. The highest yields appear to be obtained with nitriles.[50]

$$(CH_3)_2CBrCO_2C_2H_5 + Zn + C_6H_5CN \xrightarrow[\text{2. } H_2O]{\text{1. } C_6H_6} C_6H_5COC(CH_3)_2CO_2C_2H_5$$
$$\text{1.5 mol} \quad\quad \text{1.5 g-at.} \quad \text{1.0 mol} \quad\quad\quad\quad\quad\quad (57\%)$$

Reaction of Reformatsky reagents with nitrile esters,[55] diesters,[55-58] imines,[59-61] carbon dioxide,[62] epoxides,[63-64] and chlorosilanes[65] have been described.

Sequences similar to that of the Reformatsky reaction have been reported for a variety of halogen-substituted compounds including acetylenes,[66] amides,[67-68] ketones,[69] diesters,[70] fluoroesters,[71] and nitriles.[72]

Dehydration of β-Hydroxyesters

Because a major synthetic use of the Reformatsky reaction is the preparation of α,β-unsaturated esters, a discussion of the dehydration process is in order.

The usual acid-catalyzed dehydration of β-hydroxyesters normally leads to a mixture of α,β- and β,γ-unsaturated esters.[5] The nonconjugated isomer may be formed in significant amounts under either kinetic or thermodynamic conditions. In favorable cases it is possible to equilibrate the mixture of isomers produced in a kinetically controlled dehydration

[48] M. S. Bloom and C. R. Hauser, *J. Amer. Chem. Soc.*, **66**, 152 (1944).
[49] H. B. Kagan and Y. Suen, *Bull. Soc. Chim. Fr.*, **1966**, 1819.
[50] J. Cason, K. L. Rinehart, Jr., and S. D. Thornton, *J. Org. Chem.*, **18**, 1594 (1953).
[51] K. L. Rinehart, Jr., *Org. Syntheses Coll. Vol.*, **4**, 120 (1963).
[52] A. Horeau, J. Jacques, H. Kagan, and Y. Suen, *Bull. Soc. Chim. Fr.*, **1966**, 1823.
[53] L. Arsenijevic and V. Arsenijevic, *Bull. Soc. Chim. Fr.*, **1968**, 3403.
[54] P. L. Bayless and C. R. Hauser, *J. Amer. Chem. Soc.*, **76**, 2306 (1954).
[55] L. Arsenijevic and V. Arsenijevic, *Bull. Soc. Chim. Fr.*, **1968**, 4943.
[56] H. Lapin and A. Horeau, *Compt. Rend.*, **253**, 477 (1961).
[57] H. Lapin and A. Horeau, *Chimia* **15**, 551 (1961).
[58] H. Lapin and A. Horeau, *Gazz. Chim. Ital.*, **93**, 451 (1963).
[59] H. Gilman and M. Speeter, *J. Amer. Chem. Soc.*, **65**, 2255 (1943).
[60] H. B. Kagan, J. J. Bassalier, and J. L. Luche, *Tetrahedron Lett.*, **1964**, 941.
[61] E. Cuingnet, D. Paulain, and M. Tarterat-Adalberon, *Bull. Soc. Chim. Fr.*, **1969**, 514.
[62] G. Battaccio and G. P. Chiusoli, *Chem. Ind.* (London), **1966**, 1457.
[63] D. S. Deorha and P. Gupta, *Chem. Ber.*, **98**, 1722 (1965).
[64] S. Julia, C. Neuville, and R. Kevorkian, *C. R. Acad. Sci., Ser. C.*, **258**, 5900 (1964).
[65] R. J. Fessenden and J. S. Fessenden, *J. Org. Chem.*, **32**, 3535 (1967).
[66] H. B. Henbest, E. R. H. Jones, and I. M. S. Walls, *J. Chem. Soc.*, **1950**, 3646.
[67] N. L. Drake, C. M. Eaker, and W. Shenk, *J. Amer. Chem. Soc.*, **70**, 677 (1948).
[68] J. Cure and M. Gaudemar, *C. R. Acad. Sci., Ser. C*, **264**, 97 (1967).
[69] R. Kuhn and H. A. Staab, *Angew. Chem.*, **65**, 371 (1953).
[70] R. Gelin and S. Gelin, *C. R. Acad. Sci., Ser. C*, **255**, 1400 (1962).
[71] E. T. McBee, O. R. Pierce, and D. L. Christman, *J. Amer. Chem. Soc.*, **77**, 1581 (1955).
[72] L. K. Vinograd and N. S. Vul'fson, *Zh. Obshch. Khim.*, **29**, 2690 (1959) [*C.A.*, **54**, 1094e (1960)].

and obtain a higher yield of the conjugated isomer.[73] β,γ-Unsaturated acids or esters may be selectively converted to the corresponding γ-lactones on refluxing with acidic ethylene glycol.[73, 74] The conjugated isomers are inert under these conditions and separation becomes possible.

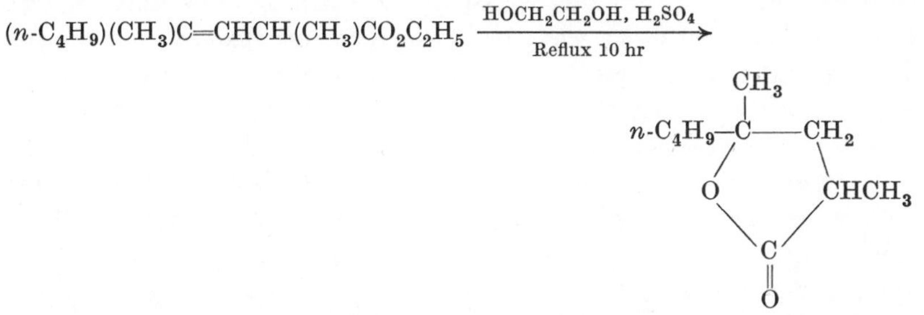

$$(n\text{-}C_4H_9)(CH_3)C\!\!=\!\!CHCH(CH_3)CO_2C_2H_5 \xrightarrow[\text{Reflux 10 hr}]{\text{HOCH}_2\text{CH}_2\text{OH, H}_2\text{SO}_4}$$

Base-catalyzed elimination of acetate derivatives produces conjugated esters in high purity.[75] The acetates may be prepared directly by addition of the Reformatsky reaction mixture to acetyl chloride. In the example illustrated, acid-catalyzed dehydration of the corresponding hydroxyester produces the conjugated ester of at best 70 % purity.[5]

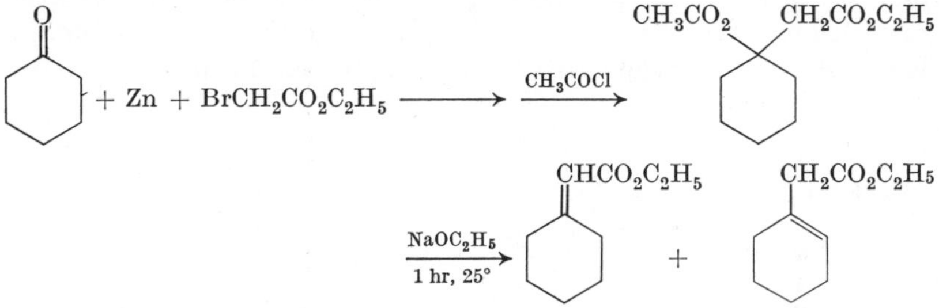

(99:1)

EXPERIMENTAL PROCEDURES

Chapter 1 in Volume 1 of this series contains a detailed description of the more commonly used Reformatsky procedures.[5] This section describes new methods for the preparation of zinc, promoters for the reaction, solvents, and two-step procedures. Experimental details for some of the new Reformatsky procedures are also given.

Preparation of Zinc

Several methods are used to prepare zinc metal for use in a Reformatsky reaction. In a widely used procedure 20-mesh zinc metal is treated with

[73] J. Cason and K. L. Rinehart, Jr., *J. Org. Chem.*, **20**, 1591 (1955).
[74] K. L. Rinehart, Jr., *Org. Syn. Coll. Vol.*, **4**, 444 (1963).
[75] K. H. Fung, K. J. Schmalzl, and R. N. Mirrington, *Tetrahedron Lett.*, **1969**, 5017.

dilute hydrochloric acid and then washed successively with water, acetone, and ether.[76] The treated metal is dried in a vacuum desiccator. Similar procedures have been utilized with zinc dust.[16, 77] The necessity of obtaining neutral and dry zinc metal for the reaction of haloesters with aliphatic aldehydes has been stressed.[16] Addition of white paraffin wax (5 g/l.) in the final ether wash has been recommended to preserve the activity of the zinc.[78] Freshly sand-papered zinc foil has also been used in Reformatsky reactions.[79]

In a careful preparation of the discrete Reformatsky reagent of ethyl α-bromoisobutyrate, Vaughan and his associates heated 20-mesh zinc with a few drops of concentrated nitric acid in concentrated sulfuric acid for 15 minutes at 100°.[9] The cooled zinc was washed free of acid with three portions of distilled water, then three portions of acetone and, finally, three portions of ether. The zinc was dried in an oven at 110° overnight before use.

Promoters

Many attempts have been made to increase the yields of Reformatsky reactions by addition of a variety of materials. Many of these promoters probably activate the zinc and produce a faster reaction with the haloester. Iodine is probably the most frequently used promoter. Addition of a few crystals of iodine suppresses enolization and leads to increased yields.[14]

$$CH_3CHBrCO_2C_2H_5 + Zn +$$

15–20% excess   Excess

(38% with iodine activation)

(23% without iodine) (plus 42% recovered ketone)

Addition of 10 to 20% of copper powder to the zinc has been reported to give increased yields in some reactions.[80, 81]

[76] M. S. Newman and F. J. Arens, Jr., J. Amer. Chem. Soc., 77, 946 (1955).

[77] C. R. Hauser, Org. Syn., Coll. Vol., 3, 408 (1955).

[78] F. S. Huber, Chemist-Analyst., 41, 62 (1952).

[79] K. L. Rinehart, Org. Syn., Coll. Vol., 4, 120, 440 (1963).

[80] R. E. Miller and F. F. Nord, J. Org. Chem., 16, 728 (1951).

[81] Z. Horii, H. Kugita, and T. Takeuchi, J. Pharm. Soc. Jap. 73, 895 (1953) [C.A., 48, 11329g (1954)].

$$\text{ClCH}_2\text{CO}_2\text{CH}_3 + \text{Zn} + \text{C}_6\text{H}_5\text{COCH}_3 \xrightarrow[\text{C}_6\text{H}_5\text{CH}_3]{\text{C}_6\text{H}_6}\longrightarrow \text{C}_6\text{H}_5\text{CH=CHCO}_2\text{CH}_3$$

Large excess   Large excess   (50% with copper; 10% without copper)

The intramolecular Reformatsky reaction of 2-bromoacetoxybenzaldehyde did not occur with zinc alone but proceeded readily with added zinc bromide to give modest yields of coumarin aud *trans*-coumaric acid.[82]

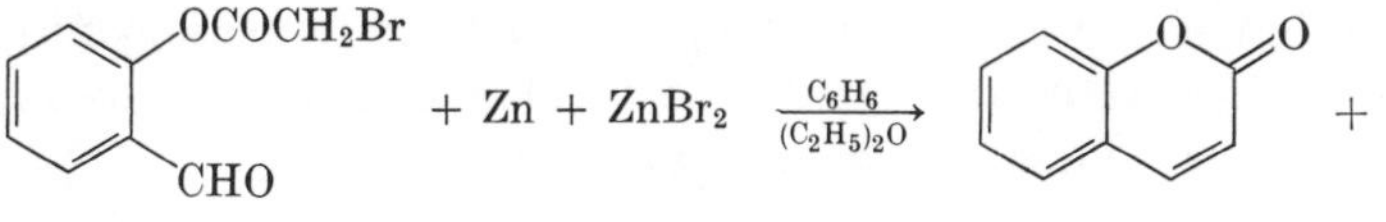

0.04 mol   0.06 g-at.   0.04 mol   (26%)

$$\text{o-HOC}_6\text{H}_4\text{CH=CHCO}_2\text{H}$$

(9%)

Mercuric chloride was found to be one of the more effective promoters in a study of a variety of additives.[80]

Solvents

The classical solvent for the Reformatsky reaction is benzene. In some cases, particularly with less reactive ketones, better yields are obtained in a mixture of benzene and ether.[33, 76]

$$\text{CH}_3\text{CHBrCO}_2\text{C}_2\text{H}_5 + \text{Zn} + (i\text{-C}_4\text{H}_9)_2\text{CO} \longrightarrow$$

0.2 mol   0.2 g-at.   0.04 mol

$$(i\text{-C}_4\text{H}_9)_2\text{CHOHCH}(\text{CH}_3)\text{CO}_2\text{C}_2\text{H}_5$$

(68% in benzene-ether; 25% in benzene alone)

t-Butyl bromoesters are inert to zinc in the usual solvents; however, deaction occurs smoothly in tetrahydrofuran.[83, 84] *t*-Butyl esters possess a number of advantages when the β-hydroxy acid is desired.[83] For example *t*-butyl esters can be hydrolyzed under relatively mild acidic conditions or, in some cases, the β-hydroxy acid can be obtained directly as shown in the accompanying reaction.

$$\text{BrCH}_2\text{CO}_2\text{C}(\text{CH}_3)_3 + \text{Zn} + \text{C}_6\text{H}_5\text{CHO} \xrightarrow[]{\text{THF}} \xrightarrow[\text{C}_6\text{H}_6]{\text{Heat}} \text{C}_6\text{H}_5\text{CHOHCH}_2\text{CO}_2\text{H}$$

0.055 mol   0.065 g-at.   0.036 mol   (91%)

A mixed solvent of tetrahydrofuran and trimethyl borate provides increased yields of β-hydroxyesters from carbonyl compounds susceptible

[82] R. C. Fuson and N. Thomas, *J. Org. Chem.*, **18**, 1762 (1953).
[83] D. A. Cornforth, A. E. Opara, and G. Read, *J. Chem. Soc., C*, **1969**, 2799.
[84] A. E. Opara and G. Read, *J. Chem. Soc., D*, **1969**, 679.

to self-condensation.[85a] Apparently the trimethyl borate functions by neutralizing the zinc alkoxide formed in the reaction.

$$BrCH_2CO_2C_2H_5 + Zn + CH_3CHO \xrightarrow{25°} \longrightarrow CH_3CHOHCH_2CO_2C_2H_5$$

0.1 mol   0.1 g-at.   0.1 mol   (95% in tetrahydrofuran-trimethyl borate 35% in tetrahydrofuran alone)

A similar result is reported for the Reformatsky reaction of a keto acid using a mixed solvent of benzene and dimethoxyethane.[85b] Superior yields of the desired lactone ester (10) are obtained and formation of the unsaturated lactone 11 is suppressed. The result is attributed to strong solvation of zinc bromide by the dimethoxyethane.

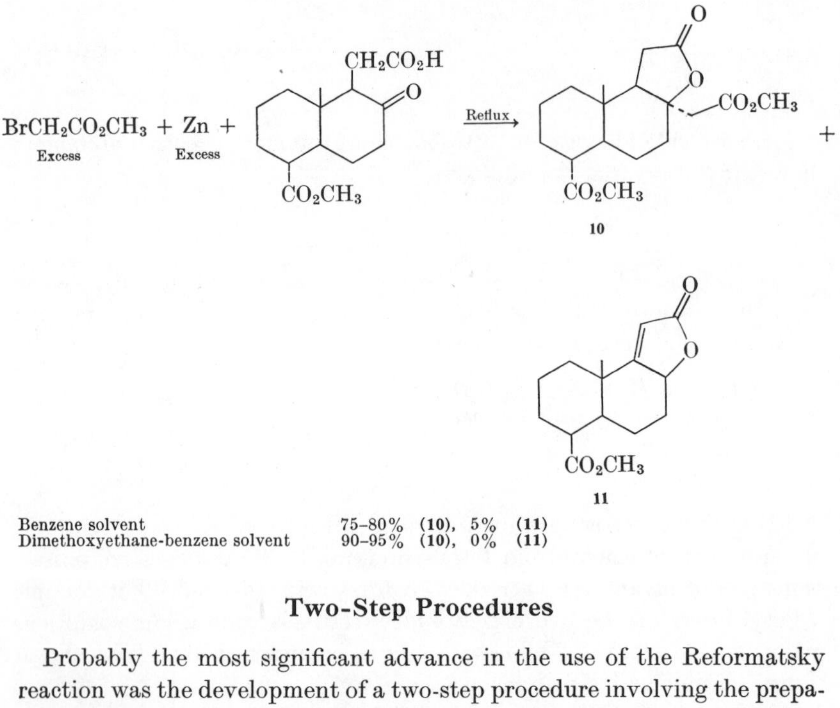

10

11

Benzene solvent   75–80% (10), 5% (11)
Dimethoxyethane-benzene solvent   90–95% (10), 0% (11)

Two-Step Procedures

Probably the most significant advance in the use of the Reformatsky reaction was the development of a two-step procedure involving the preparation of the zinc-haloester reagent in an initial step.[7, 8] The procedure is especially useful with carbonyl substrates that are subject to reduction by zinc metal.[7] A technique involving the titration of the Reformatsky reagent with fluorenone should be useful in applications of the two-step procedure.[9]

[85a] M. W. Rathke and A. Lindert, *J. Org. Chem.*, **35**, 3966 (1970).
[85b] R. A. Bell, M. B. Gravestock, and V. Y. Taguchi, *Can. J. Chem.*, **50**, 3749 (1972).

$$\text{BrCH}_2\text{CO}_2\text{C}_2\text{H}_5 + \text{Zn} \xrightarrow{(\text{C}_2\text{H}_5)_2\text{O}} \text{BrZnCH}_2\text{CO}_2\text{C}_2\text{H}_5$$

(26%)

Dimethoxymethane appears to be an excellent solvent for the two-step Reformatsky reaction.[86] Zinc reacts with ethyl bromoacetate in this solvent to provide an almost quantitative yield of the Reformatsky reagent. Subsequent reactions with aldehydes and ketones give high yields of the corresponding β-hydroxyesters.[86, 87]

Reactions

Two procedures for Reformatsky reactions are described in *Organic Syntheses*. Ethyl 3-phenyl-3-hydroxypropanoate was prepared from ethyl bromoacetate and benzaldehyde using zinc powder and a benzene-ether solvent; the total yield of hydroxy ester was 61 to 64%. 4-Ethyl-2-methyl-2-octenoic acid was prepared from ethyl 2-bromopropanoate and 2-ethyl-hexanal using zinc foil and benzene solvent; the overall yield of the pure acid was 30 to 35%.

The following representative examples illustrate procedures for the preparation of 3-hydroxy-3-isopropyl-4-methylpentanoic acid using a Reformatsky reaction with an α-halo t-butyl ester in tetrahydrofuran solution; for the preparation of ethyl 3-phenyl-3-hydroxypropanoate using a Reformatsky reaction in tetrahydrofuran-trimethyl borate solution; and for the preparation of ethyl 3-hydroxy-3-[2-furyl]propanoate using a two-step Reformatsky reaction in dimethoxymethane.

3-Hydroxy-3-isopropyl-4-methylpentanoic Acid.[83] A mixture of activated zinc[9] (4.2 g, 0.065 g-atom), a small crystal of iodine, and 50 ml of tetrahydrofuran was stirred and heated to reflux. A solution of 4.1 g (0.036 mol) of diisopropyl ketone and 10.7 g (0.055mol) of t-butyl bromo-acetate in 50 ml of tetrahydrofuran was added over a period of 30 minutes. The mixture was refluxed for an additional hour and the tetrahydrofuran then removed by distillation. Dry benzene (100 ml) was added and the mixture refluxed for 2 hours. Removal of the benzene left a residue of the acid salt which was converted to the pure acid by standard procedures. The yield of 3-hydroxy-3-isopropyl-4-methylpentanoic acid was 5.2 g (85%), mp 87–90°.

[86] M. Gaudemar and J. Cure, *C. R. Acad. Sci.*, *Ser. C*, **262**, 213 (1966).

[87] J. Cure and M. Gaudemar, *Bull. Soc. Chim. Fr.*, **1969**, 2471.

Ethyl 3-Phenyl-3-hydroxypropanoate.[85]   Granulated zinc (6.54 g, 0.1 g-atom) and a solution of 10.6 g (100 mmol) of benzaldehyde in 25 ml of tetrahydrofuran and 25 ml of trimethyl borate were placed in a flask immersed in a 25° water bath. The mixture was stirred with a magnetic bar and 11.1 ml (100 mmol) of ethyl bromoacetate injected all at once. The reaction mixture was stirred for 12 hours and then hydrolyzed by addition of 25 ml of concentrated ammonium hydroxide and 25 ml of glycerol. Extraction with ether and distillation gave 18.5 g (95%) of ethyl 3-phenyl-3-hydroxypropanoate, bp 105°/0.2 mm.

Ethyl 3-Hydroxy-3-[2-furyl]propanoate.[87]   Zinc metal (26 g, 0.4 g-atom) was covered with a small amount of dimethoxymethane. The solvent was heated to reflux and a solution of 33.4 g (0.2 mol) of ethyl bromoacetate in 150 ml of dimethoxymethane was added over a period of 20 minutes. After an additional 30 minutes of reflux the mixture was cooled to 0° and 19.2 g (0.2 mol) of furfural was added dropwise. Hydrolysis and distillation gave 16.8 g (50%) of ethyl 3-hydroxy-3-[2-furyl]propanoate, bp 85–87°/0.1 mm, n^{22}D 1.476.

COMPARISON WITH OTHER METHODS

Preparation of β-Hydroxyesters

At present a number of methods for converting aldehydes and ketones to β-hydroxyesters are available. An attempt has been made to compare some of the more generally applicable of these with the Reformatsky procedures. The methods are briefly described below and a comparison of published yields obtained with similar carbonyl compounds is given in Table I.

Method A. A single-step Reformatsky reaction conducted in benzene at room temperature using 20-mesh granulated zinc.[85a,b]

$$BrCH_2CO_2C_2H_5 + Zn + R_2CO \xrightarrow[25°]{C_6H_6} \xrightarrow[H_2O]{NH_4OH} R_2C(OH)CH_2CO_2C_2H_5$$
1.0 mol   1.0 g-at.   1.0 mol

Method B. A two-step Reformatsky reaction conducted in dimethoxy-methane.[87]

$$BrCH_2CO_2C_2H_5 + Zn \xrightarrow{CH_2(OCH_3)_2} BrZnCH_2CO_2C_2H_5$$

$$BrZnCH_2CO_2C_2H_5 + R_2CO \xrightarrow{0°} \xrightarrow{H_3O^+} R_2C(OH)CH_2CO_2C_2H_5$$

Method C. A single-step condensation of t-butyl bromoesters with carbonyl compounds and magnesium metal.[29]

$$BrCH_2CO_2C_4H_9\text{-}t + Mg + R_2CO \xrightarrow{(C_2H_5)_2O} \xrightarrow{H_2O, H_2SO_4}$$
1.1 mol   1.6 g-at.   1.0 mol

$$R_2C(OH)CH_2CO_2C_4H_9\text{-}t$$

Method D. A two-step condensation of carbonyl compounds with lithium ester enolates generated from esters and 2 equivalents of lithium amide in liquid ammonia.[88]

$$CH_3CO_2C_2H_5 + LiNH_2 \xrightarrow[\text{Reflux}]{NH_3} LiCH_2CO_2C_2H_5 + NH_3$$
1.0 mol   2.1 mol

$$LiCH_2CO_2C_2H_5 + R_2CO \xrightarrow[\text{Reflux}]{NH_3} \xrightarrow{NH_4Cl} R_2C(OH)CH_2CO_2C_2H_5$$
1.0 mol

Method E. A two-step condensation of carbonyl compounds with lithio ethyl acetate generated by reaction of ethyl acetate with 1 equivalent of lithium bis[trimethylsilyl]amide in tetrahydrofuran.[89]

$$CH_3CO_2C_2H_5 + LiN[Si(CH_3)_3]_2 \xrightarrow[-78°]{THF} LiCH_2CO_2C_2H_5 + HN[Si(CH_3)_3]_2$$
1.0 mol   1.0 mol

$$LiCH_2CO_2C_2H_5 + R_2CO \xrightarrow[-78°]{THF} \xrightarrow[H_2O]{HCl} R_2C(OH)CH_2CO_2C_2H_5$$
1.0 mol

The generation of ester enolates by means of lithium amide bases (Methods D and E) presents an attractive alternative to the Reformatsky

TABLE I. SYNTHESIS OF β-HYDROXYESTERS

Carbonyl Compound	Product	Method	Yield (%)	Ref.
CH_3COCH_3	$(CH_3)_2C(OH)CH_2CO_2R$	A	90	85
		B	66	87
		C	69	29
		D	—	
		E	90	89
$C_2H_5COC_2H_5$	$(C_2H_5)_2C(OH)CH_2CO_2R$	D	65	88
		A	50	85
		B	65	87
		C	—	
		D	31	88
		E	93	89
		A	80	85
		B	60	87
		C	80	29
		D	69	88
		E	91	89
C_6H_5CHO	$C_6H_5CHOHCH_2CO_2R$	A	84	85
		B	72	87
		C	—	
		D	37	88
		E	80	89

[88] W. R. Dunnavant and C. R. Hauser, *J. Org. Chem.*, **25**, 503 (1960).
[89] M. W. Rathke, *J. Amer. Chem. Soc.*, **92**, 3222 (1970).

procedure. In general, these reactions can be conducted in less time than the Reformatsky reaction and they do not require the preparation of the halogen derivative of the ester. The generation of lithio ethyl acetate by means of lithium bis[trimethylsilyl]amide (Method E) in particular gives good yields of β-hydroxyesters. Unfortunately, this method fails with other than simple acetate esters. However, lithium isopropylcyclohexylamide permits the quantitative conversion of a wide variety of esters to the corresponding lithio ester enolate.[90] These ester enolates react with carbonyl compounds in a fashion analogous to that of Method E and give good yields of β-hydroxyesters.[91]

$$R^1R^2CHCOC_2H_5 + LiN \underset{-78°}{\overset{THF}{\longrightarrow}} LiR^1R^2CCO_2C_2H_5 + HN$$

$$LiR^1R^2CCO_2C_2H_5 + R_2CO \underset{-78°}{\overset{THF}{\longrightarrow}} \overset{H_3O^+}{\longrightarrow} R_2\overset{OH}{\underset{|}{C}}R^1R^2CCO_2C_2H_5$$

Preparation of α,β-Unsaturated Esters

A major disadvantage of the Reformatsky method for preparing α,β-unsaturated esters is that dehydration of the intermediate β-hydroxyesters normally produces a mixture of conjugated and nonconjugated isomers.[5] The phosphonate modification of the Wittig reaction appears to overcome this difficulty and produce only conjugated unsaturated esters.[92] However, in at least one case, the nonconjugated isomer is reported as an impurity with the phosphonate method.[75] Another method, the boron trifluoride-

$$(C_2H_5O)_2\overset{O}{\overset{\uparrow}{P}}\overset{-}{CHCO_2C_2H_5} + R_2CO \longrightarrow R_2C{=}CHCO_2C_2H_5$$

catalyzed reaction of ethoxyacetylene with aldehydes or ketones, produces α,β-unsaturated esters, although the yields generally appear to be lower than those obtained by the phosphonate method.[93]

$$HC{\equiv}COC_2H_5 + R_2CO \xrightarrow[2.\ H_3O^+]{1.\ BF_3,\ (C_2H_5)_2O} R_2C{=}CHCO_2C_2H_5$$

The lithium derivative of ethoxyacetylene reacts with aldehydes or ketones to furnish acetylenic ethers which may be converted with dilute

[90] M. W. Rathke and A. Lindert, J. Amer. Chem. Soc., 93, 2318 (1971).

[91] M. W. Rathke and D. F. Sullivan, unpublished results.

[92] W. S. Wadsworth, Jr., and W. D. Emmons, J. Amer. Chem. Soc., 83, 1733 (1961).

[93] H. Vieregge, H. M. Schmidt, J. Renema, H. J. T. Bos, and J. F. Arens, Rec. Trav. Chim. Pay-Bas, 85, 929 (1966).

sulfuric acid to α,β-unsaturated esters.[94] This method is considered to be

$$Li C{\equiv}COC_2H_5 + R_2CO \xrightarrow{(C_2H_5)_2O} R_2\overset{\displaystyle OLi}{\underset{\displaystyle |}{C}}C{\equiv}COC_2H_5 \xrightarrow{H_2SO_4} R_2C{=}CHCO_2C_2H_5$$

less convenient than the boron trifluoride-catalyzed reaction with ethoxy-acetylene.[95]

Reaction of α-metalated carboxylate salts[96] with aldehydes or ketones furnishes β-hydroxy carboxylic acids.[97] The reaction with formaldehyde

$$RCH_2CO_2H + 2 LiNR'_2 \xrightarrow{THF} 2 HNR'_2 + RCHCO_2Li_2$$

$$RCHCO_2Li_2 + H_2CO \longrightarrow LiOCH_2CHRCO_2Li \xrightarrow{H_3PO_4} CH_2{=}CRCO_2H$$
$$8$$

followed by acid dehydration gives high yields of α-alkylacrylic acids.[9] The yields of β-hydroxy acid appear to decrease greatly with increasing steric hindrance of the aldehyde or ketone.[97]

A disadvantage of these alternative methods is that they do not appear to be so widely applicable as the Reformatsky reaction. The few examples reported for the base-catalyzed elimination of acetates of β-hydroxyesters[75] indicate that this route coupled with the Reformatsky or other method for preparing β-hydroxyesters may be the preferred procedure for preparing α,β-unsaturated esters.

TABULAR SURVEY

Tables II through VIII list examples of the Reformatsky reaction reported from 1942 (the year of the original chapter in this series) through December 1973. The tables are intended to be representative rather than exhaustive.

In each table the substrates (acceptors) are listed in order of increasing number of carbon atoms. The tables are short enough so that it was not necessary to set up rigorous rules for the ordering of substrates containing the same number of carbon atoms. However, acyclic substrates precede cyclic.

In the Product column of the tables the following terms require definition.

"Unsaturated ester" indicates the α,β-unsaturated ester.

[94] J. C. W. Postma and J. F. Arens, *Rec. Trav. Chim. Pay-Bas*, **75**, 1408 (1956).

[95] J. F. Arens, *Advan. Org. Chem.*, **2**, 161 (1960).

[96] P. L. Creger, *J. Amer. Chem. Soc.*, **92**, 1396 (1970).

[97] G. W. Moersch and A. R. Burkett, *J. Org. Chem.*, **36**, 1149 (1971).

[98] P. E. Pfeffer, E. Kinsel, and L. S. Silbert, *J. Org. Chem.*, **37**, 1256 (1972).

"Unsaturated esters" indicates a mixture of α,β- and β,γ-unsaturated esters.

"Normal ester" indicates that bond formation has taken place at the carbon atom bearing the halogen.

"Abnormal ester" indicates that bond formation has taken place at a carbon atom vinylogous to the halogen-bearing carbon atom.

TABLE II. Reformatsky Reactions with Aldehydes

No. of carbon atoms	Aldehyde	α-Haloester	Product(s) and Yield(s) (%)		Refs.
1	HCHO	n-$C_{14}H_{19}CHBrCO_2CH_3$	Hydroxyester	(—)	99
2	CF_3CHO	$BrCH_2CO_2C_2H_5$	Hydroxyester	(87)	100
4	n-C_3F_7CHO	$BrCH_2CO_2C_2H_5$	Hydroxyester	(94)	100
	n-C_3H_7CHO	$BrCH_2CO_2C_2H_5$	Hydroxyester	(58)	101
	⟨S⟩—CHO	$BrCH_2CO_2C_2H_5$	Hydroxyester	(62)	102
	n-$C_6H_{13}CHO$	$CH_3CHBrCO_2C_2H_5$	Hydroxyester (64–73)		33
	n-$C_4H_9CH(CH_3)CHO$	$CH_3CHBrCO_2C_2H_5$	Hydroxyester	(87)	103
	p-$O_2NC_6H_4CHO$	$BrCH_2CO_2C_2H_5$	Hydroxyester	(50)	104
	C_6H_5CHO	$CH_3CHBrCO_2C_2H_5$	Hydroxyester	(62)	105, 106
		$(+)CH_3CHBrCO_2CH_3$	Hydroxyesters (erythro: threo, 59:41)		103a
	n-$C_4H_9CH(C_3H_7$-$i)CHO$	$CH_3CHBrCO_2C_2H_5$	Hydroxyester	(61)	107
	$C_6H_5CH(CH_3)CHO$	$(CH_3)_2CBrCO_2C_2H_5$	Hydroxyester	(42)	108
10	n-$C_4H_9CH(C_4H_9$-$t)CHO$	$CH_3CHBrCO_2C_2H_5$	Unsaturated esters (62)		107
11	1-$C_{10}H_7CHO$	$CH_3CHBrCO_2C_2H_5$	Hydroxyester	(81)	109
	⟨⟩—Fe—⟨⟩CHO	$BrCH_2CO_2C_2H_5$	Unsaturated ester (75)		110
13	n-$C_{10}H_{21}CH(CH_3)CHO$	$CH_3CHBrCO_2C_2H_5$	Unsaturated esters (36), hydroxy acid (28)		103
	p-$C_6H_5C_6H_4CHO$	$BrCH_2CO_2C_2H_5$	Hydroxyester	(62)	111
16	n-$C_{15}H_{31}CHO$	$BrCH_2CO_2C_2H_5$	Hydroxyester (100)		112
	$(n$-$C_7H_{15})_2CHCHO$	$CH_3CHBrCO_2C_2H_5$	Unsaturated esters (90)		103
	p-$C_6H_5C_6H_4CH(C_2H_5)CHO$	$BrCH_2CO_2C_2H_5$	Hydroxyester	(68)	111

Note: References 99–237 are on pp. 458–460.

TABLE III. REFORMATSKY REACTIONS WITH KETONES

No. of Carbon Atoms	Ketone	α-Haloester	Product(s) and Yield(s) (%)	Ref.
C_3	FCH_2COCH_2F	$BrCH_2CO_2C_2H_5$	Hydroxyester (37)	113
	$ClCH_2COCH_3$	$BrCH_2CO_2CH_3$	Hydroxyester (38)	114
		$BrCH_2CO_2C_2H_5$	Hydroxyester (47)	115
	FCH_2COCH_3	$BrCH_2CO_2C_2H_5$	Hydroxyester (40)	113
C_5	$CH_3COC_3H_7\text{-}i$	$BrCH_2CO_2C_2H_5$	Hydroxyester (56)	116
		$CH_3CHBrCO_2CH_2C_3H_7\text{-}i$	Hydroxyester (36)	117
	Cyclopentanone	$BrCH_2CO_2C_2H_5$	Unsaturated ester (17)	118
		$C_2H_5CHBrCO_2C_2H_5$	Hydroxyester (44)	119
C_6	$CH_3COC_4H_9\text{-}i$	$BrCH_2CO_2C_2H_5$	Hydroxyester (74)	120
	$CH_3COC_4H_9\text{-}t$	$BrCH_2CO_2CH_3$	Hydroxyester (66)	121
		$BrCH_2CO_2C_2H_5$	Hydroxyester (73)	122
	Cyclohexanone	$n\text{-}C_4H_9CHBrCO_2C_2H_5$	Hydroxyester (57)	119
		$(CH_3)_2CHBrCO_2C_2H_5$	Hydroxyester (50)	119
	3-Methylcyclopentanone	$CH_3CHBrCO_2C_2H_5$	Hydroxyester (84)	123
C_7	$n\text{-}C_3H_7COC_3H_7\text{-}n$	$CH_3CHBrCO_2C_2H_5$	Hydroxyester (66–80)	133
	$C_2H_5OCH_2CH_2CH_2COCH_3$	$BrCH_2CO_2C_2H_5$	Hydroxyester (60)	124
	Cycloheptanone	$(CH_3)_2CBrCO_2C_2H_5$	Hydroxyester (83)	123
	3-Methylcyclohexanone	$C_2H_5CHBrCO_2C_2H_5$	Hydroxyester (59)	119
C_8	$n\text{-}C_6H_{13}COCH_3$	$CH_3CHBrCO_2C_2H_5$	Hydroxyester (79)	33
		$(CH_3)_2CBrCO_2C_2H_5$	Hydroxyester (78)	33
	$i\text{-}C_6H_{13}COCH_3$	$BrCH_2CO_2C_2H_5$	Hydroxyester (61)	122
	$C_6H_5COCH_2Cl$	$BrCH_2CO_2C_2H_5$	Hydroxyester (67)	125
	[bicyclic thiophene-fused cyclohexanone structure]	$BrCH_2CO_2C_2H_5$	Unsaturated esters (73)	126
C_9	$i\text{-}C_4H_9COC_4H_9\text{-}i$	$CH_3CHBrCO_2C_2H_5$	Hydroxyester (24–68)	33
	[2,2,6-trimethylcyclohexanone structure]	$BrCH_2CO_2C_2H_5$	[CH_3, HO, $CH_2CO_2C_2H_5$, CH_3, CH_3 substituted cyclohexane structure] (62)	126a
	$C_6H_5COCH_2CH_3$	$BrCH_2CO_2C_2H_5$	Hydroxyester (73)	127
	$C_6H_5COCH_2OCH_3$	$BrCH_2CO_2C_2H_5$	Hydroxyester (68)	128
	Indan-1-one	$BrCH_2CO_2C_2H_5$	Unsaturated acid (25)	129
	Indan-2-one	$BrCH_2CO_2C_2H_5$	Hydroxyester (33)	129

Note: References 99–237 are on pp. 458–460.

TABLE III. REFORMATSKY REACTIONS WITH KETONES (*Continued*)

of on ns Ketone	α-Haloester	Product(s) and Yield(s) (%)	Refs.
α-Tetralone	$BrCH_2CO_2C_2H_5$	Hydroxyester (90)	129
β-Tetralone	$BrCH_2CO_2C_2H_5$	Hydroxyester (50)	130, 129
4-Methylindan-1-one	$BrCH_2CO_2C_2H_5$	Unsaturated acid (45)	131
	$BrCH_2CO_2C_2H_5$	Hydroxyester (86)	132
	$BrCH_2CO_2C_2H_5$	Unsaturated ester (85)	133
7-Methyl-1-tetralone	$BrCH_2CO_2CH_5$	Unsaturated acids (57–72)	134
7-Methoxy-1-tetralone	$BrCH_2CO_2CH_3$	Unsaturated acids (97)	135
7-Ethyl-1-tetralone	$BrCH_2CO_2CH_3$	Unsaturated acids (57–72)	134
3,5-Dimethyl-1-tetralone	$BrCH_2CO_2C_2H_5$	Unsaturated ester (46)	136
5,8-Dimethyl-1-tetralone	$CH_3CHBrCO_2C_2H_5$	Unsaturated acids (—)	137
1,1-Dimethyl-2-tetralone	$BrCH_2CO_2C_2H_5$	Hydroxyester (50)	129
Acetylferrocene	$BrCH_2CO_2C_2H_5$	Unsaturated ester (—)	110
$C_6H_5COC_6H_5$	$BrCH_2CO_2C_2H_5$	Hydroxyester (57)	122
	$BrCH_2CO_2C_2H_5$	Hydroxyester (93)	138, 127
	$CH_3CHBrCO_2C_2H_5$	Hydroxyester (86)	139
2-Phenylcycloheptanone	$BrCH_2CO_2C_2H_5$	Unsaturated ester (75)	140
7-Isopropyl-1-tetralone	$BrCH_2CO_2CH_3$	Unsaturated acids (57–72)	134
	$BrCH_2CO_2CH_3$	Unsaturated acids (84)	141
	$BrCH_2CO_2C_2H_5$	Unsaturated ester (77), hydroxy- ester (10)	142

TABLE III. REFORMATSKY REACTIONS WITH KETONES (*Continued*)

No. of Carbon Atoms	Ketone	α-Haloester	Product(s) and Yield(s) (%)	Refs
C_{14} (*contd.*)		$BrCH_2CO_2CH_3$	Hydroxyester (—)	143,
		$CH_3CHBrCO_2CH_3$	Unsaturated acid (70)	145
	7-*t*-Butyl-1-tetralone	$BrCH_2CO_2CH_3$	Unsaturated ester (84)	146
C_{15}		$BrCH_2CO_2C_2H_5$	Unsaturated ester (85)	147
	2-Phenyl-1-indanone	$BrCH_2CO_2C_2H_5$	Unsaturated acid (85)	148
		$BrCH_2CO_2CH_3$	Hydroxyester (84)	149
C_{16}	$C_6H_5CCOC_6H_5$	$BrCH_2CO_2C_2H_5$	Hydroxyester (47)	150
		$BrCH_2CO_2CH_3$	Unsaturated acid (—)	151,
C_{17}	*n*-$C_{15}H_{31}COCH_3$	$CH_3CHBrCO_2C_2H_5$	Unsaturated esters (52)	33
	Benzoylferrocene	$CH_3CHBrCO_2C_2H_5$	Hydroxyester (—)	153

Note: References 99–237 are on pp. 458–460.

TABLE III. Reformatsky Reactions with Ketones (*Continued*)

of bon ms	Ketone	α-Haloester	Product(s) and Yield(s) (%)	Refs.
td.)		$BrCH_2CO_2C_2H_5$	Hydroxyester (90)	154
	$C_6H_4CH_3\text{-}p$	$BrCH_2CO_2C_2H_5$	Unsaturated acid (—)	155
	$p\text{-}(t\text{-}C_4H_9)C_6H_4COC_6H_5$	$BrCH_2CO_2C_2H_5$	Hydroxyester (29)	156
		$CH_3CHBrCO_2C_2H_5$	Unsaturated acid (33)	157
	C_6H_5	$BrCH_2CO_2C_2H_5$	Unsaturated ester (72)	158
		$BrCH_2CO_2C_2H_5$	Unsaturated acid (30)	159
	$C_{10}H_7\text{-}2$	$BrCH_2CO_2CH_3$	Unsaturated acid (60)	160
	$CH_3\quad CH_3$	$BrCH_2CO_2C_2H_5$	Hydroxyester (—)	161
	CH_3O	$BrCH_2CO_2C_2H_5$	Hydroxy acid, β-OH (—)	162

Note: References 99–237 are on pp. 458–460.

TABLE III. REFORMATSKY REACTIONS WITH KETONES *(Continued)*

No. of Carbon Atoms	Ketone	α-Haloester	Product(s) and Yield(s) (%)	Refs.
C_{21}	$n\text{-}C_{10}H_{21}COC_{10}H_{21}\text{-}n$	$BrCH_2CO_2C_2H_5$	Unsaturated esters (80)	163
	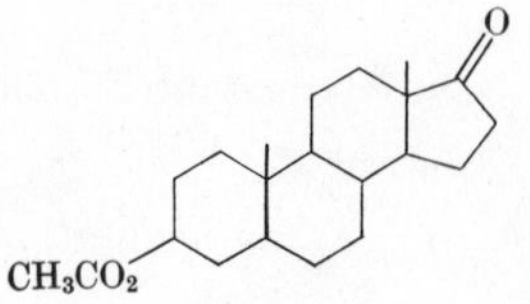	$CH_3CHICO_2C_2H_5$	Hydroxy acid, β-OH (32)	164
		$CH_3CHBrCO_2CH_3$	Hydroxyesters, β-OH (42), α-OH (28)	165
C_{22}	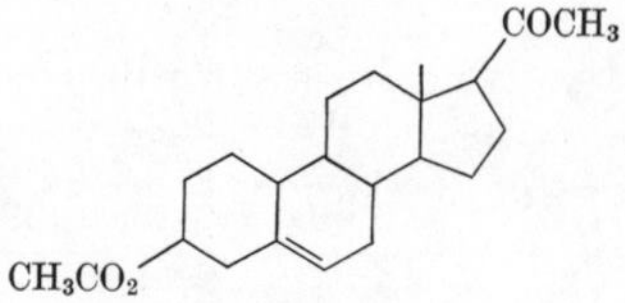	$BrCH_2CO_2C_2H_5$	Dihydroxy acid (—)	166
C_{23}	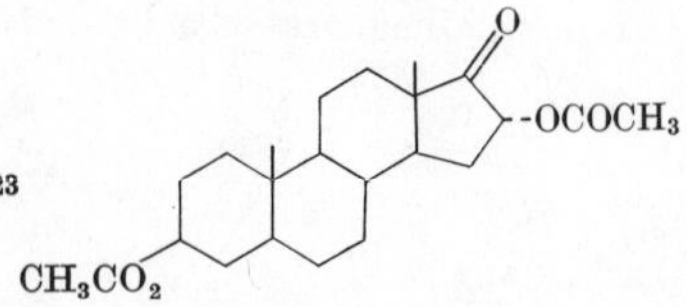	$CH_3CHBrCO_2C_2H_5$	Hydroxyester, β-OH (65)	167

Note: References 99–237 are on pp. 458–460.

No. of Carbon Atoms	Carbonyl Substrate	α-Haloester	Product(s) and Yield(s) (%)	Refs.
C_3	$Cl_2C{=}CHCHO$	$BrCH_2CO_2C_2H_5$	Hydroxyester (39)	168
C_4	$CH_2{=}CHCOCH_3$	$BrCH_2CO_2C_2H_5$	Hydroxyester (30)	169
C_5	$CH_3CH{=}CHCOCH_3$	$BrCH_2CO_2C_2H_5$	Hydroxyester (68)	45
		$CH_3CHBrCO_2C_2H_5$	Hydroxyester (59)	45
	$(CH_3)_2C{=}CHCHO$	$CH_3CHBrCO_2C_2H_5$	Unsaturated ester (—)	170
C_6	$CH_3CH{=}C(CH_3)COCH_3$	$BrCH_2CO_2C_2H_5$	Hydroxyester (70)	45
		$CH_3CHBrCO_2C_2H_5$	Hydroxyester (65)	45
		$C_2H_5CHBrCO_2C_2H_5$	Hydroxyester (63)	45
	$(CH_3)_2C{=}CHCOCH_3$	$BrCH_2CO_2C_2H_5$	Hydroxyester (73)	45
		$CH_3CHBrCO_2C_2H_5$	Hydroxyester (69)	45
		$C_2H_5CHBrCO_2C_2H_5$	Hydroxyester (60)	45
		$(CH_3)_2CBrCO_2C_2H_5$	Hydroxyester (—)	45
C_8	$n\text{-}C_5H_{11}C{\equiv}CCHO$	$BrCH_2CO_2C_2H_5$	Hydroxyester (47)	171
C_9	$C_6H_5C{\equiv}CCHO$	$BrCH_2CO_2C_2H_5$	Hydroxyester (55)	172
	[structure] —Fe(CO)₃	$BrCH_2CO_2CH_3$	Hydroxyester (74)	172a
C_{10}	$(CH_3)_2C{=}CHCH_2\text{-}$ $CH_2C(CH_3){=}CHCHO$	$BrCH_2CO_2C_2H_5$	Unsaturated ester (60)	173, 174
	$C_6H_5CH{=}CHCOCH_3$	$BrCH_2CO_2C_2H_5$	Unsaturated acids (—)	175
		$CH_3CHBrCO_2C_2H_5$	Hydroxyester (86)	176
C_{11}	$C_6H_5CH{=}CHCOC_2H_5$	$(CH_3)_2CBrCO_2C_2H_5$	Keto acid, 1,4 addition (60)	46
	$C_6H_5CH{=}C(CH_3)COCH_3$	$BrCH_2CO_2CH_3$	Hydroxyester (87)	177
C_{13}	[structure with CH₃, CH₃, CH=CHCOCH₃, CH₃]	$BrCH_2CO_2C_2H_5$	Unsaturated ester (50)	178
	[structure with CH₃, CH₃, CH=CHCOCH₃, CH₃]	$BrCH_2CO_2C_2H_5$	Unsaturated ester (—)	179
C_{14}	[structure with COC₂H₅, CH₃O]	$(CH_3)_2CBrCO_2C_2H_5$	Lactone, 1,4 addition (—)	180
C_{23}	[steroid structure with COCH₃, CH₃CO₂]	$(CH_3)_2CBrCO_2C_2H_5$	Lactone, 1,4 addition (75)	181
		$C_2H_5CHBrCO_2C_2H_5$	Hydroxyester (—)	181

Note: References 99–237 are on pp. 458–460.

TABLE V. REFORMATSKY REACTIONS WITH DIKETONES AND/OR THEIR MONOKETALS AND ALDEHYDIC OR KETONIC ESTERS

No. of Carbon Atoms	Substrate	α-Haloester	Product(s) and Yield(s) (%)	Refs.
C5	$CH_3COCH_2COCH_3$	$BrCH_2CO_2C_2H_5$	Dihydroxydiester (48)	182
C6	$CHOCH_2CH_2CO_2C_2H_5$	$BrCH_2CO_2C_2H_5$	Hydroxydiacid (—)	183
	$CH_3COCH_2CO_2C_2H_5$	$BrCH_2CO_2C_2H_5$	Hydroxydiester (13.7)	184
	$C_2H_5COCH_2OCOCH_3$	$BrCH_2CO_2C_2H_5$	Hydroxyester (60)	185
	$CH_3COCH_2CH_2OCOCH_3$	$BrCH_2CO_2C_2H_5$	Hydroxyester (51)	186
C7	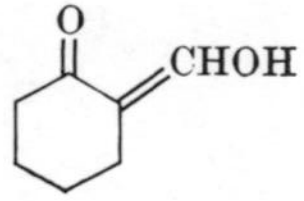 (cyclohexanone with =CHOH substituent)	$BrCH_2CO_2CH_3$	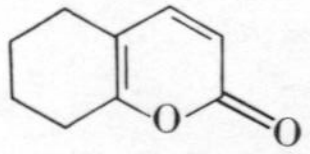 (bicyclic lactone structure) (25), unsaturated diesters (30)	187
C8	$CH_3CO(CH_2)_4COCH_3$	$BrCH_2CO_2C_2H_5$	Dihydroxydiester (64)	182
	$C_2H_5OCOCOCHFCO_2C_2H_5$	$BrCH_2CO_2C_2H_5$	Hydroxydiester (12)	188
	$CH_3COC(CH_3)_2CH_2CO_2C_2H_5$	$BrCH_2CO_2C_2H_5$	Ester lactone (50)	189
C9	$t\text{-}C_4H_9COCOC_4H_9\text{-}t$	$BrCH_2CO_2C_2H_5$	Hydroxyester (39)	190
C10	$C_6H_5COCO_2C_2H_5$	$BrCH_2CO_2C_2H_5$	Dihydroxydiester (62)	191
	$C_6H_5COCH_2OCOCH_3$	$BrCH_2CO_2C_2H_5$	Lactone (44.5)	192
	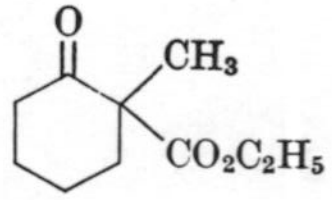 (cyclohexanone with CH_3 and $CO_2C_2H_5$ substituents)	$BrCH_2CO_2C_2H_5$	Hydroxydiester (88)	193
	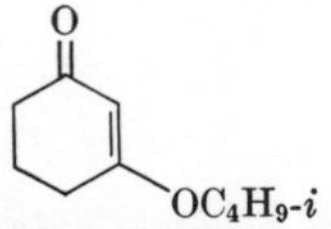 (cyclohexenone with $OC_4H_9\text{-}i$ substituent)	$BrCH_2CO_2C_2H_5$	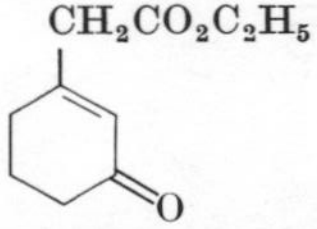 (cyclohexenone with $CH_2CO_2C_2H_5$ substituent) (—)	194

450

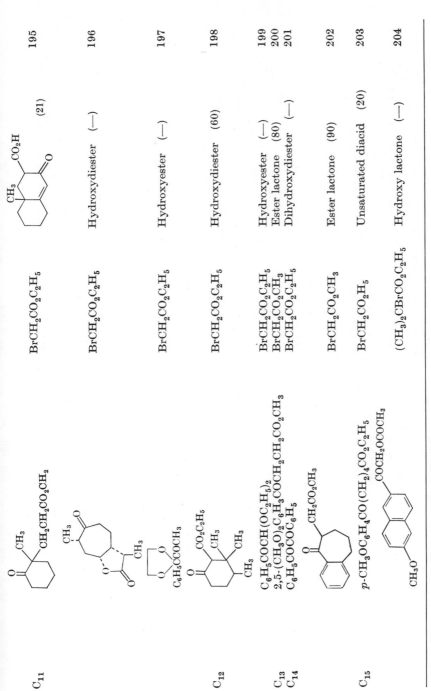

C_{11}		$BrCH_2CO_2C_2H_5$	(21)	195
		$BrCH_2CO_2C_2H_5$	Hydroxydiester (—)	196
		$BrCH_2CO_2C_2H_5$	Hydroxyester (—)	197
		$BrCH_2CO_2C_2H_5$	Hydroxydiester (60)	198
C_{12}		$BrCH_2CO_2C_2H_5$	Hydroxyester (—)	199
		$BrCH_2CO_2CH_3$	Ester lactone (80)	200
		$BrCH_2CO_2C_2H_5$	Dihydroxydiester (—)	201
C_{13}		$BrCH_2CO_2CH_3$	Ester lactone (90)	202
C_{14}		$BrCH_2CO_2C_2H_5$	Unsaturated diacid (20)	203
C_{15}		$(CH_3)_2CBrCO_2C_2H_5$	Hydroxy lactone (—)	204

Note: References 99–237 are on pp. 458–460.

451

TABLE V. REFORMATSKY REACTIONS WITH DIKETONES AND/OR THEIR MONOKETALS AND ALDEHYDES OR KETONIC ESTERS (Continued)

No. of Carbon Atoms	Substrate	α-Haloester	Product(s) and Yield(s) (%)	Refs.
C_{16}	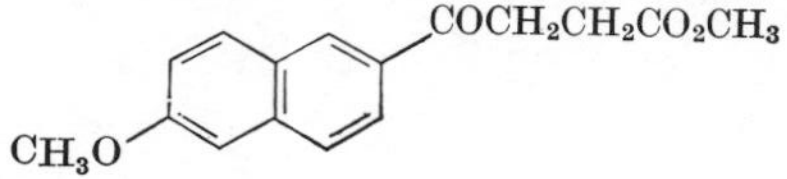 (6-methoxynaphthyl)–$COCH_2CH_2CO_2CH_3$	$BrCH_2CO_2CH_3$	Ester lactone (84)	205
C_{18}	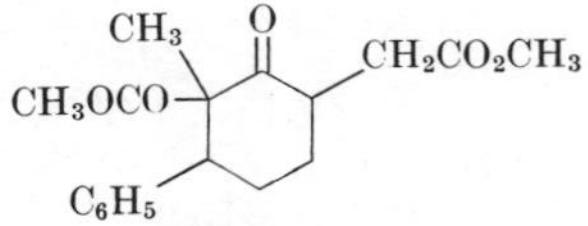 cyclohexanone with CH_3, CH_3OCO, C_6H_5, $CH_2CO_2CH_3$ substituents	$BrCH_2CO_2C_2H_5$	Diester lactone (—)	206
	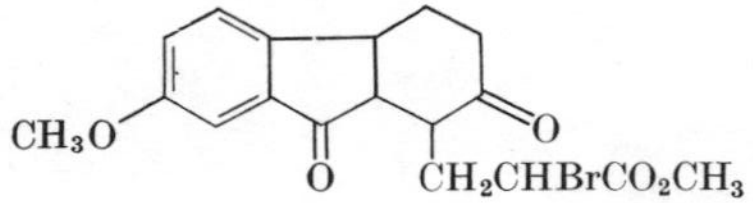 tricyclic structure, $CH_2CHBrCO_2CH_3$, CH_3O	—	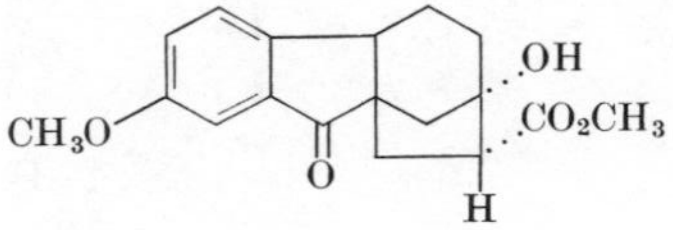 (tricyclic OH, CO_2CH_3, H, O, CH_3O structure) (44)	206a
C_{19}	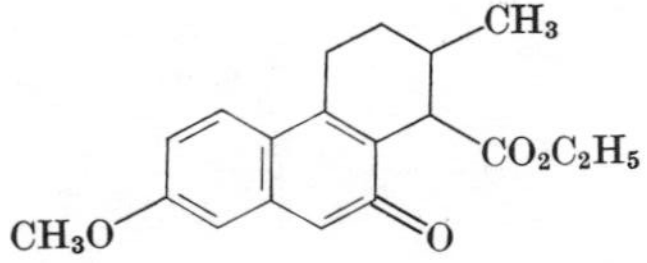 phenanthrene-type structure, CH_3, $CO_2C_2H_5$, O, CH_3O	$BrCH_2CO_2C_2H_5$	Hydroxydiester (—)	207
C_{20}	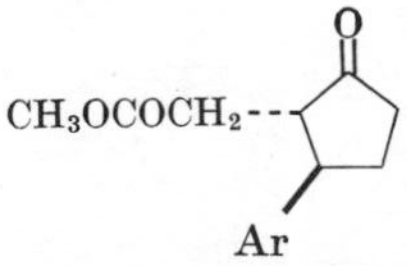 $CH_3OCOCH_2\cdots$ cyclopentanone with Ar; Ar = 6-methoxy-2-naphthyl	$BrCH_2CO_2CH_3$	Ester lactone (46), hydroxy-diester (18)	208
	p-$CH_3OC_6H_4CH(CH_2CO_2C_2H_5)$-$COC_6H_4OCH_3$-$p$	$BrCH_2CO_2C_2H_5$	Ester lactone (—)	209

Note: References 99–237 are on pp. 458–460.

TABLE VI. REFORMATSKY REACTIONS OF UNSATURATED HALOESTERS

No. of Carbon Atoms	Substrate	Haloester	Product(s) and Yield(s) (%)	Refs.
C_5	$(CH_3)_2C=CHCHO$	$BrCH_2C(CH_3)=CHCO_2CH_3$	Abnormal unsaturated esters[a] (—)	210
	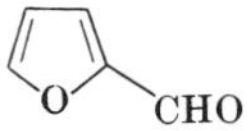	$BrCH_2CH=CHCO_2C_2H_5$	Normal hydroxyester (35)	211
	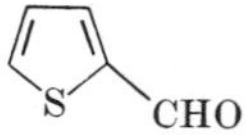	$BrCH_2CH=CHCO_2C_2H_5$	Normal unsaturated ester (33)	212
C_6	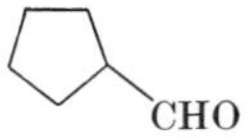	$BrCH_2CH=CHCO_2C_2H_5$	Normal hydroxyester (19), abnormal hydroxyester (49)	213, 211
		$BrCH_2CH=CHCO_2C_2H_5$	Normal hydroxyester (26), abnormal hydroxyester (7)	214
C_7	C_6H_5CHO	$BrCH_2CH=CHCO_2CH_3$	Normal hydroxyester (25), abnormal hydroxyester (2)	215
	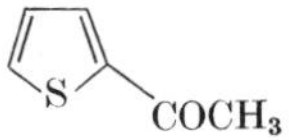	$ICH_2C(CH_3)=CCO_2C_2H_5$	Normal unsaturated ester (—)	216
	2-Methylcyclohexanone	$BrCH_2C=CCO_2CH_3$	Normal hydroxyester (40)	67
		$BrCH_2CH=CHCO_2CH_3$	Normal hydroxyester (—)	38

[a] Normal and abnormal are defined on page 442.

Note: References 99–237 are on pp. 458–460.

TABLE VI. REFORMATSKY REACTIONS OF UNSATURATED HALOESTERS (*Continued*)

No. of Carbon Atoms	Substrate	α-Haloester	Product(s) and Yield(s) (%)	Refs.
C₇ (*contd.*)	(2-thienyl)CH=CHCHO	$BrCH_2CH=CHCO_2C_2H_5$	Normal unsaturated acid (40)	212
		$BrCH_2CH=CHCH=CHCO_2CH_3$	None	212
C₉	(cyclohexanone with CO_2CH_3, CH_3 substituents)	$BrCH_2CH=CHCO_2CH_3$	Normal hydroxydiester (84)	217
	2,3-Dimethoxybenzaldehyde	$BrCH_2CH=CHCO_2CH_3$	Normal unsaturated ester (20)	218
C₁₀	1-Decalone	$BrCH_2CH=CHCO_2CH_3$	Normal hydroxyester (—)	219
	(chromanone, CH_3O)	$BrCH_2CH=CHCO_2CH_3$	Normal unsaturated ester (55)	219a
C₁₁	6-Methoxy-1-tetralone	$BrCH_2CH=CHCO_2CH_3$	Normal unsaturated ester (48)	220
	(benzosuberone structure)	$BrCH_2CH=CHCO_2CH_3$	Normal unsaturated ester (50)	221
C₁₃	$C_6H_5COC_6H_5$	$BrCH_2CH=CHCO_2CH_3$	Normal unsaturated ester (32–38)	214
	(cyclohexene with CH_3, $CH=CHCOCH_3$ substituents)	$BrCH_2CH=CHCO_2CH_3$	Normal unsaturated acid (15)	37
C₁₅	$p\text{-}CH_3OC_6H_4COC_6H_4OCH_3\text{-}p$	$BrCH_2CH=CHCO_2CH_3$	None	214

TABLE VII. REFORMATSKY REACTIONS WITH MISCELLANEOUS HALOGEN COMPOUNDS

No. of Carbon Atoms	Substrate	Halogen Compounds	Product(s) and Yield(s) (%)	Ref.
C_2	CH_3CHO	$BrCH(CO_2C_2H_5)_2$	Acetate of hydroxydiester (60)	222
		$Br_2C(CO_2C_2H_5)_2$	Acetate of hydroxy-bromodiester (71)	223
C_3	CH_3COCH_3	$Cl_2CHCO_2C_2H_5$	$(CH_3)_2C=CHCO_2C_2H_5$ (31)	80
		$Cl_3CCO_2C_2H_5$	Hydroxy-dichloroester (81)	224
C_4	$n\text{-}C_3H_7CHO$	$Br_2CH(CO_2C_2H_5)_2$	Hydroxy-bromodiester (—)	225
	$CH_3CH=CHCHO$	$(CH_3)_2CBrCN$	Hydroxynitrile (83)	226
C_6	$CH_3COCH_2CH_2OCOCH_3$	$BrFC(CO_2C_2H_5)_2$	$CH_3C=CHCH_2OCOCH_3$ (37) $\quad\vert$ $CF(CO_2C_2H_5)_2$	227
	Cyclohexanone	$C_2H_5OCOCH_2CHBrCO_2C_2H_5$	Ester lactone (51)	228
		$C_2H_5OCO(CH_2)_3CHBrCO_2C_2H_5$	Hydroxydiester (37)	70
		$BrCH_2CON(C_2H_5)_2$	Hydroxyamide (60)	67
C_7	C_6H_5CHO	$BrFCHCO_2C_2H_5$	Hydroxyfluoroester (68)	71
		$(C_2H_5O)_2CCBrCO_2C_2H_5$	$C_2H_5OCH=CHCO_2C_2H_5$ (—)	229
		$C_2H_5CHBrCO_2ZnBr$	Hydroxyacid (92)	230
		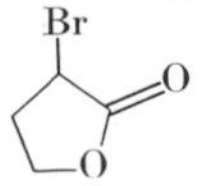	Hydroxy lactone (—)	231
C_8	$C_6H_5COCH_3$	$BrCH_2CON(C_2H_5)_2$	Hydroxyamide (64)	68

Note: References 99–237 are on pp. 458–460.

455

TABLE VIII. REFORMATSKY REACTIONS WITH MISCELLANEOUS ACCEPTORS

No. of Carbon Atoms	Acceptor	α-Haloester	Product(s) and Yield(s) (%)	Ref.
C_1	CO_2	$(CH_3)_2CBrCO_2CH_3$	$(CH_3)_2C(CO_2H)CO_2CH_3$ (50)	62
C_2	$CH_2{=}C{=}O$	$BrCH_2CO_2C_2H_5$	$CH_3COCH_2CO_2C_2H_5$ (23)	232
	CH_3CN	$CH_3CHBrCO_2C_2H_5$	Keto ester (16)	50
	$(ClCH_2)_2O$	$(CH_3)_2CBrCO_2C_2H_5$	$O[CH_2C(CH_3)_2CO_2C_2H_5]_2$ (66)	232a
C_3	$O{=}C{=}C{=}O$	$BrCH_2CO_2C_2H_5$	$HO_2CCH_2COCH_2CO_2C_2H_5$ (—)	233
	$(CH_3)_3SiCl$	$BrCH_2CO_2C_2H_5$	$(CH_3)_3SiCH_2CO_2C_2H_5$ (72)	65
C_6	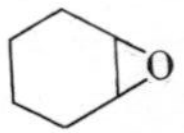	$BrCH_2CO_2C_2H_5$	 $CHOHCH_2CO_2C_2H_5$ (–)	63
C_7	C_6H_5COCl	$(CH_3)_2CBrCO_2C_2H_5$	$C_6H_5COC(CH_3)_2CO_2C_2H_5$ (57)	54
	$(CH_3)_2CBrCO_2CH_2CH_2CH{=}CH_2$	—	$CH_2{=}CHCH_2C(CH_3)_2CO_2ZnBr$ (100)	233a
C_8	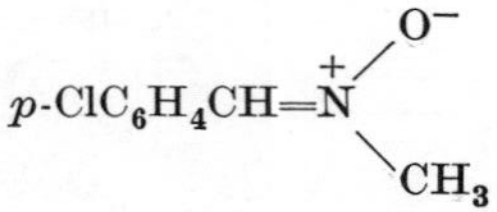	$BrCH_2CO_2C_2H_5$	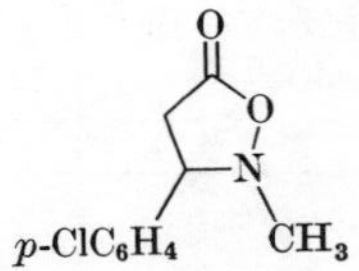 (34)	234

Phthalimide	$BrCH_2CO_2C_2H_5$	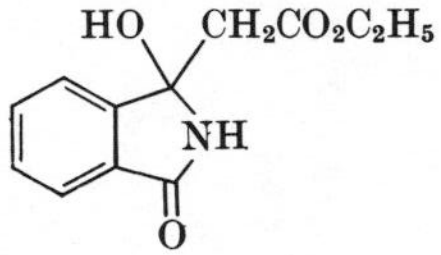 (structure with HO, $CH_2CO_2C_2H_5$, NH, O)	(−)	235
C_{10} $NCH_2OC_4H_9\text{-}n$ (piperidine)	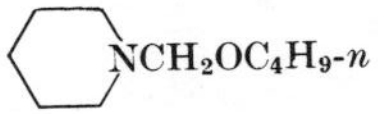 $(CH_3)_2CBrCO_2CH_3$	$N-CH_2C(CH_3)_2CO_2CH_3$	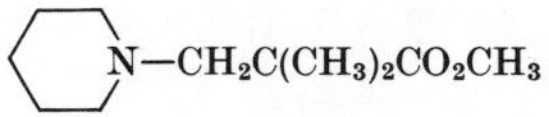 (56)	236
C_{11} (benzene with CH_2CN, $CO_2C_2H_5$)	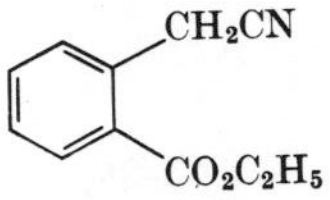 $(CH_3)_2CBrCO_2C_2H_5$	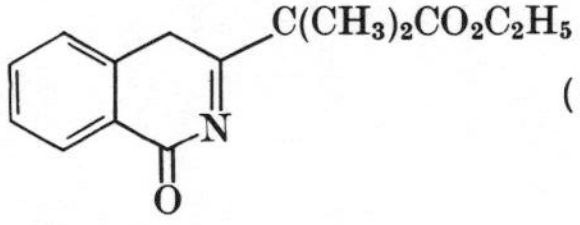 ($C(CH_3)_2CO_2C_2H_5$, N, O)	(−)	53
C_{13} $C_6H_5CH{=}NC_6H_5$	$BrCH_2CO_2C_2H_5$	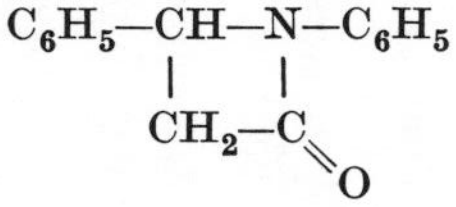 $C_6H_5-CH-N-C_6H_5$ / $CH_2-C{=}O$	(56)	59
	$i\text{-}C_3H_7CBrCO_2CH_3$	$C_6H_5CHNHC_6H_5$ / $i\text{-}C_3H_7CHCO_2CH_3$	(100)	237

Note: References 99–237 are on pp. 458–460.

457

References to Tables II–VIII

99 C. Malani, D. E. Minnikin, and N. Polgar, *J. Chem. Soc.*, **1965**, 5562.

100 E. T. McBee, O. R. Pierce, and D. D. Smith, *J. Amer. Chem. Soc.*, **76**, 3722 (1954).

101 H. B. Henbest and B. Nicholls, *J. Chem. Soc.*, **1959**, 227.

102 R. D. Scheutz and W. H. Houff, *J. Amer. Chem. Soc.*, **77**, 1836 (1955).

103 J. Cason and K. L. Rinehart, *J. Org. Chem.*, **20**, 1591 (1955).

103a T. Matsumoto, I. Tanaka, and K. Fukui, *Bull. Chem. Soc., Japan*, **44**, 3378 (1971).

104 L. K. Vinograd and N. S. Vul'fson, *Dokl. Akad. Nauk. SSSR*, **123**, 97 (1958) [*C. A.*, **53**, 5179c (1959)].

105 G. Bucchi, C. G. Inman, and E. S. Lipinsky, *J. Amer. Chem. Soc.*, **76**, 4327 (1954).

106 S. G. Cohen and S. Y. Weinstein, *J. Amer. Chem. Soc.*, **86**, 725 (1964).

107 K. L. Rinehart and L. J. Dolby, *J. Org. Chem.*, **22**, 13 (1957).

108 C. G. Overberger and P. V. Bonsignore, *J. Amer. Chem. Soc.*, **80**, 5427 (1958).

109 A. G. Georgiev and L. D. Zhelyazkov, *Dokl. Akad. Nauk SSSR*, **164**, 132 (1964) [*C. A.*, **60**, 9213c (1964)].

110 P. DaRe and E. Sianesi, *Experientia*, **21**, 648 (1965).

111 G. Cavallini, E. Massarini, D. Nardi, and R. D'Ambrosio, *J. Amer. Chem. Soc.*, **79**, 3514 (1957).

112 E. F. Jenny and C. A. Grob, *Helv. Chim. Acta*, **36**, 1936 (1953).

113 E. D. Bergmann, S. Cohen, E. Hoffman, and Z. Rand-Meir, *J. Chem. Soc.*, **1961**, 3453.

114 S. Akiyoshi, K. Okuno, and S. Nagahama, *J. Amer. Chem. Soc.*, **76**, 902 (1954).

115 R. Rambaud and D. Besserre, *Bull. Soc. Chim. Fr.*, **1962**, 13.

116 L. I. Smith, W. L. Kohlhase, and R. J. Brotherton, *J. Amer. Chem. Soc.*, **78**, 2532 (1956).

117 M. A. Perry, F. C. Canter, R. E. Debusk, and A. G. Robinson, *J. Amer. Chem. Soc.* **80**, 3618 (1958).

118 H. Pines, H. G. Rodenberg, and V. N. Ipatieff, *J. Amer. Chem. Soc.*, **76**, 771 (1954).

119 J. Maillard, B. Bernard, and R. Morin, *Bull. Soc. Chim. Fr.*, **1958**, 244.

120 M. Julia and J. M. Surzur, *Bull. Soc. Chim. Fr.*, **1956**, 1620.

121 M. S. Newman and R. Roshev, *J. Org. Chem.* **9**, 221 (1944).

122 R. Heilmann and R. Glenat, *Bull. Soc. Chim. Fr.*, **1955**, 1586.

123 F. Korte, J. Falbe, and A. Zschocke, *Tetrahedron*, **6**, 201 (1959).

124 J. Crowley, D. J. Millin, and N. Polgar, *J. Chem. Soc.*, **1957**, 2931.

125 W. W. Epstein and A. C. Sontag, *J. Org. Chem.*, **32**, 3390 (1967).

126 M. C. Kloetzel, J. E. Little, and D. M. Frisch, *J. Org. Chem.*, **18**, 1511 (1953).

126a T. Matsumoto, G. Sakata, Y. Tackibana, and K. Fukui, *Bull. Chem. Soc., Japan*, **45**, 1147 (1972).

127 J. A. Barltrop, R. M. Acheson, P. G. Philpott, K. E. McPhee, and J. S. Hunt, *J. Chem. Soc.*, **1956**, 2928.

128 M. Rubin, W. Paist, and R. C. Elderfield, *J. Org. Chem.*, **6**, 260 (1941).

129 A. Hafez and N. Campbell, *J. Chem. Soc.*, **1960**, 4115.

130 W. G. Dauben and R. Teranishi, *J. Org. Chem.* **16**, 550 (1951).

131 D. A. H. Taylor, *J. Chem. Soc.*, **1960**, 2805.

132 R. M. Acheson, K. E. McPhee, P. G. Philpott, and J. A. Barltrop, *J. Chem. Soc.*, **1956**, 698.

133 A. G. Anderson and S. Y. Wang, *J. Org. Chem.*, **19**, 277 (1954).

134 A. J. M. Wenham and J. S. Whitehurst, *J. Chem. Soc.*, **1956**, 3857.

135 W. S. Johnson and H. J. Gleen, *J. Amer. Chem. Soc.*, **71**, 1087 (1949).

136 E. E. Royals and J. L. Greene, *J. Org. Chem.*, **23**, 1437 (1958).

137 M. S. Newman and A. S. Hussey, *J. Amer. Chem. Soc.*, **69**, 3023 (1947).

138 C. D. Gutsche, N. N. Saha, and H. E. Johnson, *J. Amer. Chem. Soc.*, **79**, 4441 (1957).

139 M. Harferist and E. Magnien, *J. Amer. Chem. Soc.*, **78**, 1060 (1956).

140 C. D. Gutsche, *J. Amer. Chem. Soc.*, **73**, 786 (1951).

141 W. E. Bachmann and R. E. Holman, *J. Amer. Chem. Soc.*, **73**, 3660 (1951).

142 M. Rosenblum and F. W. Abbate, *J. Amer. Chem. Soc.*, **88**, 4178 (1966).

143 P. P. Gardner and W. J. Horton, *J. Amer. Chem. Soc.*, **74**, 657 (1952).

[144] P. D. Gardner, C. E. Wulfam, and C. L. Osborn, *J. Amer. Chem. Soc.*, **80**, 143 (1958).
[145] M. S. Newman, *J. Org. Chem.*, **9**, 518 (1944).
[146] J. M. Wenham and J. S. Whitehurst, *J. Chem. Soc.*, **1957**, 4037.
[147] E. D. Bergmann and R. Ikan, *J. Amer. Chem. Soc.*, **80**, 5803 (1958).
[148] N. Campbell and E. Ciganek, *J. Chem. Soc.*, **1956**, 3834.
[149] A. C. Cope and R. D. Smith, *J. Amer. Chem. Soc.*, **78**, 1012 (1956).
[150] J. G. Bennett and S. C. Bruce, *J. Org. Chem.*, **25**, 73 (1960).
[151] E. Bograchov, *J. Amer. Chem. Soc.*, **66**, 1613 (1944).
[152] G. R. Clemo and N. D. Ghatge, *J. Chem. Soc.*, **1956**, 1068.
[153] J. W. Huffman and R. L. Asbury, *J. Org. Chem.*, **30**, 3941 (1965).
[154] N. Campbell and J. R. Gorrie, *J. Chem. Soc.*, *C.*, **1968**, 1887.
[155] M. F. Ansell, G. T. Brooks, and B. A. Knights, *J. Chem. Soc.*, **1961**, 212.
[156] J. W. Wilt and J. L. Finnerty, *J. Org. Chem.*, **26**, 2173 (1961).
[157] J. L. Comp and G. H. Daub, *J. Amer. Chem. Soc.*, **80**, 6049 (1958).
[158] E. Buchta and D. Kiessling, *Ann.*, **709**, 219 (1967).
[159] D. D. Phillips and D. N. Chatterjee, *J. Amer. Chem. Soc.*, **80**, 4360 (1958).
[160] M. S. Newman, *J. Org. Chem.*, **9**, 518 (1944).
[161] M. S. Newman and R. M. Wise, *J. Amer. Chem. Soc.*, **78**, 450 (1956).
[162] M. N. Huffman and J. W. Salder, *J. Org. Chem.*, **18**, 919 (1953).
[163] N. Polgar and R. Robinson, *J. Chem. Soc.*, **1943**, 615.
[164] D. H. Hey, J. Honeyman, and W. J. Peal, *J. Chem. Soc.*, **1954**, 185.
[165] M. Tanabe and R. H. Peters, *J. Org. Chem.*, **36**, 2403 (1971).
[166] L. Ruzicka, P. A. Plattner, and J. Pataki, *Helv. Chim. Acta*, **25**, 425 (1942).
[167] N. Danieli, Y. Mazur, and F. Sondheimer, *Chem. Ind.* (London), **1958**, 1724.
[168] M. Julia and J. Bullot, *Bull. Soc. Chim. Fr.*, **1960**, 23.
[169] I. Heilbron, E. R. H. Jones, M. Julia, and B. C. L. Weedon, *J. Chem. Soc.*, **1949**, 1823.
[170] L. Crombie, S. H. Harper, and K. C. Sleep, *J. Chem. Soc.*, **1957**, 2743.
[171] L. Crombie, *J. Chem. Soc.*, **1955**, 1007.
[172] R. H. Wiley and C. E. Staples, *J. Org. Chem.*, **28**, 3408 (1963).
[172a] J. Lewis, R. Cowles, B. Johnson, and A. Parkins, *J. Chem. Soc.*, *Dalton Trans.*, **1972**, 1768.
[173] E. E. Royals, *J. Amer. Chem. Soc.*, **69**, 841 (1947).
[174] M. Mousseron and M. Mousseron-Canet, *C. R. Acad. Sci. Ser. C*, **247**, 1937 (1958).
[175] R. H. Wiley, *J. Chem. Soc.*, **1958**, 3831.
[176] R. H. Wiley, C. E. Staples, and T. H. Crawford, *J. Org. Chem.*, **29**, 2986 (1964).
[177] R. H. Wiley and C. E. Staples, *J. Org. Chem.*, **28**, 3408 (1963).
[178] W. G. Young, L. J. Andrews, and S. J. Cristol, *J. Amer. Chem. Soc.*, **66**, 520 (1944).
[179] G. Lowe, F. G. Torto, and B. C. L. Weedon, *J. Chem. Soc.*, **1958**, 1855.
[180] J. P. Guette, A. Horeau, and J. Jacques, *Bull. Soc. Chim. Fr.*, **1962**, 2030.
[181] C. Gandolfi, G. Doria, M. Amendola, and E. Dradi, *Tetrahedron Lett.*, **1970**, 3923.
[182] E. Buchta and K. Burger, *Ann.*, **580**, 125 (1953).
[183] A. Siegel, F. Wessely, P. Stockhammer, F. Antony, and P. Klezl, *Tetrahedron*, **4**, 49 (1958).
[184] R. Adams and B. L. VanDuuren, *J. Amer. Chem. Soc.*, **75**, 2377 (1953).
[185] P. S. Steyn, W. J. Conradie, C. F. Garbers, and M. J. DeVries, *J. Chem. Soc.*, **1965**, 3075.
[186] C. H. Hoffman, A. F. Wagner, A. N. Wilson, E. Walton, C. H. Shunk, D. E. Wolf, F. W. Holly, and K. Folkers, *J. Amer. Chem. Soc.*, **79**, 2316 (1957).
[187] A. S. Dreiding and A. J. Tomasewski, *J. Amer. Chem. Soc.*, **76**, 6388 (1954).
[188] D. E. A. Rivett, *J. Chem. Soc.*, **1953**, 3710.
[189] H. E. Baumgarten and D. C. Gleason, *J. Org. Chem.*, **16**, 1658 (1951).
[190] M. S. Newman and F. R. Kahle, *J. Org. Chem.*, **23**, 666 (1958).
[191] C. S. Rondestvedt and A. H. Filbey, *J. Org. Chem.*, **19**, 119 (1954).
[192] R. G. Linville and R. C. Elderfield, *J. Org. Chem.*, **6**, 270 (1941).
[193] W. E. Bachmann and S. Kushner, *J. Amer. Chem. Soc.*, **65**, 1963 (1943).
[194] J. J. Panouse and C. Sannie, *Bull. Soc. Chim. Fr.*, **1956**, 1272.

[195] A. S. Dreiding and A. J. Tomasewski, *J. Org. Chem.*, **19**, 241 (1954).
[196] O. S. Bhanot, J. R. Mahajan, and P. C. Dutta, *J. Chem. Soc. C.* **1968**, 1128.
[197] G. N. Walker, *J. Org. Chem.*, **23**, 34 (1958).
[198] H. Favre and H. Schinz, *Helv. Chim. Acta*, **41**, 1368 (1958).
[199] J. V. P. Torrey, J. A. Kuck, and R. Elderfield, *J. Org. Chem.*, **6**, 289 (1941).
[200] A. S. Dreiding and A. J. Tomasewski, *J. Amer. Chem. Soc.*, **76**, 540 (1954).
[201] H. W. Bost and P. S. Bailey, *J. Org. Chem.*, **21**, 803 (1956).
[202] W. J. Horton, H. W. Johnson, and J. L. Zollinger, *J. Amer. Chem. Soc.*, **76**, 4587 (1954).
[203] C. D. Gutshe, H. F. Strohmayer, and J. M. Chang, *J. Org. Chem.*, **23**, 1 (1958).
[204] A. Horeau and R. E. Miliozzi, *Bull. Soc. Chim. Fr.*, **1957**, 381.
[205] W. E. Bachmann and R. D. Morin, *J. Amer. Chem. Soc.*, **66**, 553 (1944).
[206] D. L. Turner, *J. Amer. Chem. Soc.*, **79**, 2271 (1957).
[206a] F. Ziegler and M. Condon, *J. Org. Chem.*, **36**, 3707 (1971).
[207] K. Miescher, *Helv. Chim. Acta*, **27**, 1727 (1944).
[208] L. J. Chinn, E. A. Brown, R. A. Mikulec, and R. B. Garland, *J. Org. Chem.*, **27**, 1733 (1962).
[209] L. Goldberg and R. Robinson, *J. Chem. Soc.*, **1941**, 575.
[210] R. H. Wiley, E. Imoto, R. P. Houghton, and P. Veeragu, *J. Amer. Chem. Soc.*, **82**, 1413 (1960).
[211] J. English and J. D. Gregory, *J. Amer. Chem. Soc.*, **69**, 2123 (1947).
[212] R. E. Miller and F. F. Nord, *J. Org. Chem.*, **16**, 1720 (1951).
[213] J. English, J. D. Gregory, and J. R. Trowbridge, *J. Amer. Chem. Soc.*, **73**, 615 (1951).
[214] R. D. Schuetz and W. H. Houff, *J. Amer. Chem. Soc.*, **77**, 1839 (1955).
[215] E. R. H. Jones, D. G. O'Sullivan, and M. C. Whiting, *J. Chem. Soc.*, **1949**, 1415.
[216] C. D. Hurd and H. E. Winberg, *J. Amer. Chem. Soc.*, **73**, 917 (1951).
[217] W. E. Bachman and A. S. Dreiding, *J. Org. Chem.*, **13**, 317 (1948).
[218] G. N. Walker, *J. Amer. Chem. Soc.*, **78**, 3201 (1956).
[219] W. E. Bachman and N. L. Wendler, *J. Amer. Chem. Soc.*, **68**, 2580 (1946).
[219a] S. Ramdas and J. Radhakrisknan, *Indian J. Chem.*, **10**, 351 (1972).
[220] G. Stork, *J. Amer. Chem. Soc.*, **69**, 2936 (1947).
[221] T. A. Crabb and K. Schofield, *J. Chem. Soc.*, **1958**, 4276.
[222] F. Gaudemar-Bardone and M. Gaudemar, *Bull. Soc. Chim. Fr.*, **1969**, 2878.
[223] F. Gaudemar-Bardone and M. Gaudemar, *Bull. Soc. Chim. Fr.*, **1971**, 3316.
[224] B. Castro, J. Villieras, and N. Ferracutti, *Bull. Soc. Chim. Fr.*, **1969**, 3521.
[225] M. Gaudemar, *Bull. Soc. Chim. Fr.*, **1966**, 3113.
[226] N. Goasdoue and M. Gaudemar, *J. Organometal Chem.*, **28**, C9 (1971).
[227] E. D. Bergmann and S. Cohen, *J. Chem. Soc.*, **1961**, 3457.
[228] J. Blanc and B. Gastambide, *Bull. Soc. Chim. Fr.*, **1962**, 2055.
[229] W. Oroshnik and P. E. Spoerri, *J. Amer. Chem. Soc.*, **67**, 721 (1945).
[230] M. Bellassoued, R. Couffignal, and M. Gaudemar, *J. Organometal. Chem.*, **36**, C33 (1972).
[231] H. Torabi, R. L. Evans, and H. E. Stavely, *J. Org. Chem.*, **34**, 3792 (1969).
[232] B. N. Dashkevich and V. V. Tsmur, *Zh. Obshch. Khim.*, **25**, 932 (1955) [*C. A.*, **50**, 4023c (1956)].
[232a] P. Johnson and J. Zitsman, *J. Org. Chem.*, **38**, 2346 (1973).
[233] L. B. Dashkevich and V. M. Siraya, *Trans. Leningr. Khim.-Farmatsevt. Inst.*, **16**, 51 (1962) [*C. A.*, **61**, 1751f (1964)].
[233a] J. Baldwin and J. Walker, *J. Chem., Soc., Chem. Commun.*, **1973**, 117.
[234] H. Stamm and J. Hoenicke, *Ann.*, **749**, 146 (1971).
[235] L. Arsenijevic and D. Stefanovic, *Glas. Hem. Drus., Beograd*, **35**, 209 (1970) [*C. A.*, **74**, 141205k (1971)].
[236] J. Canceill and J. Jacques, *Bull. Soc. Chim. Fr.*, **1965**, 903.
[237] J. L. Liche and H. B. Kagan, *Bull. Soc. Chim. Fr.*, **1971**, 2260.

AUTHOR INDEX, VOLUMES 1–22

Adams, Joe T., 8
Adkins, Homer, 8
Albertson, Noel F., 12
Allen, George R., Jr., 20
Angyal, S. J., 8
Archer, S., 14

Bachmann, W. E., 1, 2
Baer, Donald, R., 11
Behr, Lyell C., 6
Bergmann, Ernst, D., 10
Berliner, Ernst, 5
Blatchly, J. M., 19
Blatt, A. H., 1
Blicke, F. F., 1
Bloomfield, Jordan J., 15
Boswell, E. A., Jr., 21
Brand, William W., 18
Brewster, James H., 7
Brown, Herbert C., 13
Brown, Weldon G., 6
Bruson, Herman Alexander, 5
Bublitz, Donald E., 17
Buck, Johannes S., 4
Butz, Lewis W., 5

Cairns, Theodore L., 20
Carmack, Marvin, 3
Carter, H. E., 3
Cason, James, 4
Cope, Arthur C., 9, 11
Corey, Elias J., 9
Cota, Donald J., 17
Crounse, Nathan N., 5

Daub, Guido H., 6
Dave, Vinod, 18
Denny, R. W., 20
DeTar, DeLos F., 9
Djerassi, Carl, 6
Donaruma, L. Guy, 11
Drake, Nathan L., 1
DuBois, Adrien S., 5

Eliel, Ernst L., 7
Emerson, William S., 4
England, D. C., 6

Fieser, Louis F., 1
Folkers, Karl, 6
Fuson, Reynold C., 1

Geissman, T. A., 2
Gensler, Walter J., 6
Gilman, Henry, 6, 8
Ginsburg, David, 10
Govindichari, Tuticorin R., 6
Gutsche, C. David, 8

Hageman, Howard A., 7
Hamilton, Cliff S., 2
Hamlin, K. E., 9
Hanford, W. E., 3
Harris, Constance M., 17
Harris, J. F., Jr., 13
Harris, Thomas M., 17
Hartung, Walter H., 7
Hassall, C. H., 9
Hauser, Charles R., 1, 8
Heldt, Walter Z., 11
Henne, Albert L., 2
Hoffman, Roger A., 2
Hoiness, Connie M., 20
Holmes, H. L., 4, 9
Houlihan, William, J., 16
House, Herbert O., 9
Hudson, Boyd E., Jr., 1
Huyser, Earl S., 13

Ide, Walter S., 4
Ingersoll, A. W., 2

Jackson, Ernest L., 2
Jacobs, Thomas L., 5
Johnson, John R., 1
Johnson, William S., 2, 6
Jones, G., 15

461

CHAPTER AND TOPIC INDEX, VOLUMES 1–22

Many chapters contain brief discussions of reactions and comparisons of alternative synthetic methods which are related to the reaction that is the subject of the chapter. These related reactions and alternative methods are not usually listed in this index. In this index the volume number is in BOLDFACE, the chapter number in ordinary type.

SUBJECT INDEX, VOLUME 22

Since the table of contents provides a quite complete index, only those items not readily found from the contents pages are listed here. Numbers in BOLDFACE type refer to experimental procedures.